INTEGRATION TECHNOLOGIES FOR INDUSTRIAL AUTOMATED SYSTEMS

INDUSTRIAL INFORMATION TECHNOLOGY SERIES

Series Editor
RICHARD ZURAWSKI

Published Books

Industrial Communication Technology Handbook
Edited by Richard Zurawski

Embedded Systems Handbook
Edited by Richard Zurawski

Electronic Design Automation for Integrated Circuits Handbook
Edited by Luciano Lavagno, Grant Martin, and Lou Scheffer

Integration Technologies for Industrial Automated Systems
Edited by Richard Zurawski

Forthcoming Books

Automotive Embedded Systems Handbook
Edited by Nicolas Navet, and Françoise Simonot-Lion

INTEGRATION TECHNOLOGIES FOR INDUSTRIAL AUTOMATED SYSTEMS

Edited by

Richard Zurawski

ISA Corporation, Alameda, California

CRC Press
Taylor & Francis Group
Boca Raton London New York

CRC Press is an imprint of the
Taylor & Francis Group, an **informa** business

CRC Press
Taylor & Francis Group
6000 Broken Sound Parkway NW, Suite 300
Boca Raton, FL 33487-2742

First issued in paperback 2019

© 2007 by Taylor & Francis Group, LLC
CRC Press is an imprint of Taylor & Francis Group, an Informa business

No claim to original U.S. Government works

ISBN-13: 978-0-8493-9262-7 (hbk)
ISBN-13: 978-0-367-39052-5 (pbk)

Library of Congress Cataloging-in-Publication Data

Zurawski, Richard.
    Integration technologies for industrial automated systems / Richard Zurawski.
        p. cm. -- (Industrial information technology series ; 3)
    Includes bibliographical references and index.
    ISBN 0-8493-9262-4
    1. Computer integrated manufacturing systems. 2. Manufacturing processes--Automation. I. Title.
II. Series.

TS155.63.Z87 2006
670.42'7--dc22   2006043861

Visit the Taylor & Francis Web site at
http://www.taylorandfrancis.com

and the CRC Press Web site at
http://www.crcpress.com

Dedication

To my sons Richard and Martin

Preface

Introduction to the Book

The book is designed to cover a very wide range of topics describing technologies and solutions involved in the integration of industrial automated systems and enterprises. The emphasis is on advanced material to cover recent significant research results and technology evolution and developments. The book is primarily aimed at experienced professionals from industry and academia, but will be also useful to novices with some university background.

The book extensively covers e-technologies, software and IT technologies, network-based integration technologies, agent-based technologies, and security topics. The book contains 23 chapters, written by leading experts from industry and academia directly involved in the creation and evolution of the ideas and technologies treated in the book. Most contributions are from industry and industrial research establishments at the forefront of the developments shaping the field of industrial automation, such as ABB Corporate Research Center, Germany; ABB Corporate Research Center, Switzerland; Austrian Academy of Sciences, Austria; FIDIA, Italy; Carmeq, Germany; Centre Suisse d'Electronique et de Microtechnique, Switzerland; Institut für Automation und Kommunikation eV – IFAK, Germany; National Institute of Standards and Technology, United States; Rockwell Automation, Germany; SCC, Germany; Siemens AG, Germany; Singapore Institute of Manufacturing Technology, Singapore; Softing AG, Germany; Yokogawa America, United States.

The presented material is in the form of tutorials, surveys, and technology overviews. The contributions are grouped into sections for cohesive and comprehensive presentation of the treated areas. The reports on recent technology developments, deployments, and trends frequently cover material released to the profession for the first time. The book can be used as a reference (or prescribed text) text for university (post)graduate courses with a focus on integration technologies in industrial automated systems.

The material covered in this book will be of interest to a wide spectrum of professionals and researchers from industry and academia, as well as graduate students in the fields of electrical and computer engineering, manufacturing, and software engineering, as well as mechatronic engineering.

This book is an indispensable companion for those who seek to learn more on technologies, solutions, and trends in the integration of distributed decentralized industrial systems, and those who want to remain up-to-date with recent technical developments in the field.

Organization of the Book

The aim of the Organization section is to provide highlights of the contents of the individual chapters to assist readers in identifying material of interest, and to put topics discussed in a broader context. Where appropriate, a brief explanation of the topic under treatment is provided, particularly for chapters describing novel trends, and keeping novices in mind.

The book is organized into six parts: (1) Introduction; (2) E-Technologies in Enterprise Integration; (3) Software and IT Technologies in Integration of Industrial Automated Systems; (4) Network-Based Integration Technologies in Industrial Automated Systems; (5) Agent-Based Technologies in Industrial Automation; and (6) Security in Industrial Automation.

Part 1: Introduction

Chapter 1, "Integration Technologies for Industrial Automated Systems: Challenges and Trends," with a focus on selected integration issues, technologies, and solutions, offers a framework for the material presented in subsequent chapters.

Part 2: E-Technologies in Enterprise Integration

An introduction to e-manufacturing is presented in Chapter 2, entitled "Introduction to e-Manufacturing." This material provides an overview of the e-manufacturing strategies, fundamental elements, and requirements to meet the changing needs of the manufacturing industry in transition to an e-business environment. It covers e-manufacturing, e-maintenance, e-factory, and e-business.

Part 3: Software and IT Technologies in Integration of Industrial Automated Systems

This section contains eight contributions discussing the use of XML and Web services, component-based and Java technologies, MMS in factory floor integration, standards for the design of automated systems, and other IT-based solutions for enterprise integration. The first contribution, "Enterprise-Manufacturing Data Exchange Using XML" (Chapter 3), introduces the World Batch Forum's (WBF) Business to Manufacturing Markup Language (B2MML), which is a set of XML schemas based on the ISA-95 Enterprise-Control System Integration Standards. The material demonstrates how B2MML can be used to exchange data between the business/enterprise and manufacturing systems. The contribution also gives a roundup of the ISA-95 standard.

Web services technology provides means for the implementation of open and platform-independent integrated automation systems. The challenges for using Web services in automated systems, solutions, and future trends are discussed in Chapter 4, "Web Services for Integrated Automation Systems – Challenges, Solutions, and Future," by experts from ABB Corporate Research Center.

OLE for Process Control (OPC), the standard interface for access to Microsoft's Windows-based applications in automation, is one of the most popular industrial standards among users and developers of human–machine interfaces (HMIs), supervisory control and data acquisition (SCADA), and distributed control systems (DCS) for PC-based automation, as well as Soft PLCs. OPC is discussed in detail in the subsection *Component Technologies in Industrial Automation and Enterprise Integration* in Chapter 5.

The subsection *MMS in Factory Floor Integration* focuses on the highly successful international standard MMS (manufacturing message specification), which is an open systems interconnection (OSI) application layer messaging protocol designed for the remote control and monitoring of devices such as remote terminal units (RTUs), programmable logic controllers (PLCs), numerical controllers (NCs), robot controllers (RCs), etc. This subsection features Chapter 6, "The Standard Message Specification for Industrial Automation Systems: ISO 9506 (MMS)," which gives a fairly comprehensive introduction to the standard and illustrates its use.

An overview of Java technology, real-time extensions, and prospects for applications in controls and industrial automation is given in Chapter 7, "Java Technology in Industrial Applications." This contribution provides a roundup of two different real-time extensions for the Java language: Real-Time Specification for Java (RTSJ), developed by the Real-Time for Java Expert Group under the auspices of Sun Microsystems and reference implemented by TimeSys Corp.; and Real-Time Core Extensions, developed by the Real-Time Java Working Group operating within J-Consortium. This chapter also introduces the Real-Time Data Access (RTDA) specification, developed by the Real-Time Data Access Working Group (RTAWG), and operating within J-Consortium. The RTDA specification focuses on API for accessing I/O data in typical industrial and embedded applications.

Chapter 8, "Achieving Reconfigurability of Automation Systems Using the New International Standard IEC 61499: A Developer's View," introduces the IEC 61499 standard, which defines a reference architecture for open and distributed control systems to provide the means for compatibility among the automation systems of different vendors.

The final contribution in this section, Chapter 9, presents IT-based connectivity solutions for interfacing production and business systems. The presented concepts and architectures are a result of an extensive study and prototyping efforts conducted by ABB in the search for cost-effective approaches leveraging existing mainstream technologies such as enterprise application integration (EAI), Web Services and XML, and emerging industry standards such as ISA 95 and CIM.

Part 4: Network-Based Integration Technologies in Industrial Automated Systems

The aspects of network-based integration technologies are presented in five subsections: *Field Devices — Technologies and Standards*; *Fieldbus Technology*; *Real-Time Ethernet*; *Wireless Technology*; and *SEMI*.

Fieldbus technology is overviewed in four chapters. This subsection begins with Chapter 13, "Fieldbus Systems: History and Evolution," presenting an extensive introduction to fieldbus technology, a comparison and critical evaluation of the existing technologies, as well as the evolution and emerging trends. This chapter is a must for anyone with an interest in the origins of the current fieldbus technology landscape. It is also compulsory reading for novices to understand the concepts behind fieldbuses.

The next two chapters present an overview of some of the most widely used fieldbus technologies. Chapter 14, "PROFIBUS: Open Solutions for the World of Automation," presents a description of PROFIBUS (PROFIBUS DP). This is a comprehensive overview of PROFIBUS DP, one of the leading players in the fieldbus application area. It includes material on HART on PROFIBUS DP, applications and master and system profiles, and integration technologies such as GSD (general station description), EDD (electronic device description), and DTM (device type manager).

Chapter 15, "The CIP Family of Fieldbus Protocols," introduces the following CIP (Common Industrial Protocol-based networks) family of fieldbus protocols: DeviceNet, a CIP implementation employing a CAN data-link layer; ControlNet, implementing the same basic protocol on new data-link layers that allow for much higher speed (5 Mbps), strict determinism, and repeatability while extending the range of the bus (several kilometers with repeaters); and EtherNet/IP, in which CIP runs over TCP/IP. The chapter also introduces CIP Sync, which is a CIP-based communication principle that enables synchronous low jitter system reactions without the need for low-jitter data transmission. This is important in applications that require much tighter control of a number of real-time parameters characterizing hard real-time control systems. The chapter also overviews CIP Safety, a safety protocol that adds additional services to transport data with high integrity.

The issues involved in the configuration (setting up a fieldbus system) and management (diagnosing, monitoring, and adding new devices to the network, to mention some activities) of fieldbus systems are presented in Chapter 16, "Configuration and Management of Fieldbus Systems," which concludes the subsection on the fieldbus technology.

Ethernet, the backbone technology for office networks, is increasingly being adopted for communication in factories and plants at the fieldbus level. The random and native CSMA/CD arbitration mechanism is being replaced by other solutions allowing for deterministic behavior required in real-time communication to support soft and hard real-time deadlines, for instance, time synchronization of activities required to control drives, and for the exchange of small data records characteristic of monitoring and control actions. The direct support for Internet technologies allows for vertical integration of various levels of industrial enterprise hierarchy, to include seamless integration between automation and business logistic levels to exchange jobs and production (process) data; transparent data interfaces for all stages of the plant life cycle; the Internet- and Web-enabled remote diagnostics and maintenance; as well as electronic orders and transactions.

This subsection begins with Chapter 17, "The Quest for Real-Time Behavior in Ethernet," which discusses various approaches to ensure real-time communication capabilities, including those that support probabilistic as well as deterministic analysis of the network access delay. This chapter also presents a brief description of the Ethernet protocol.

The next chapter, "Principles and Features of PROFInet' (Chapter 18), presents a new automation concept, and the technology behind it, that has emerged as a result of trends in automation technology toward modular, reusable machines, and plants with distributed intelligence. PROFInet is an open standard for industrial automation based on the industrial Ethernet. The material is presented by researchers from the Automation and Drives Division of Siemens AG, the leading provider of automation solutions within Siemens AG.

Although the use of wireline-based field area networks is dominant, wireless technology offers a range of incentives in a number of application areas. In industrial automation, for example, wireless device (sensor/actuator) networks can provide the support for mobile operation required in the case of mobile robots, monitoring and control of equipment in hazardous and difficult-to-access environments, etc. The use of wireless technologies in industrial automation is covered in two chapters (Chapters 19 and 20).

Chapter 19, "Wireless Local and Wireless Personal Area Network Technologies for Industrial Deployment," presents a comprehensive overview of the commercial-of-the-shelf wireless technologies including IEEE 802.15.1/Bluetooth, IEEE 802.15.4/ZigBee, and IEEE 802.11 variants. The suitability of these technologies for industrial deployment is evaluated, including aspects such as application scenarios and environments, coexistence of wireless technologies, and implementation of wireless fieldbus services.

The means for interconnecting wire fieldbuses to wireless ones in the industrial environment, various design alternatives, and their evaluation are presented in Chapter 20, "Interconnection of Wireline and Wireless Fieldbuses." This is one of the most comprehensive and authoritative discussions of this topic, as presented by one of the leading authorities on fieldbus technology.

The final subsection is on SEMI and features Chapter 21, "SEMI Interface and Communication Standards: An Overview and Case Study." This is an excellent introduction to SEMI, providing an overview of the fundamentals of the SEMI Equipment Communication Standard, commonly referred to as SECS, its interpretation, the available software tools, and case study applications. The material was written by experts from the Singapore Institute of Manufacturing Technology, who were involved in a number of SEMI technology developments and deployments.

Part 5: Agent-Based Technologies in Industrial Automation

The high degree of complexity of manufacturing systems, coupled with the market-dictated requirements for agility, led to the development of new manufacturing architectures and solutions called agents, based on distributed, autonomous, and cooperating units, integrated by the plug-and-play approach. This section comprises a single chapter, "From Holonic Control to Virtual Enterprises: The Multi-Agent Approach" (Chapter 22), which offers a comprehensive treatment of the topic by presenting an introduction to the concept of agents and technology, cooperation and coordination models, interoperability, and applications to manufacturing systems.

Part 6: Security in Industrial Automation

With the growing trend for networking of industrial automated systems and their internetworking with LAN, WAN, and the Internet (for example, there is a growing demand for remote access to process data at the factory floor — assisted by embedded Web servers), many of those systems may become exposed to potential security attacks, which can compromise their integrity and cause damage as a result. The topic of IT security in automation systems is thoroughly explored in Chapter 23, "IT Security for Automation Systems." This chapter gives an overview of the IT security technologies, discusses best practices for industrial communication system security, and introduces some standardization activities in the area. It discusses security objectives, types of attacks, and the available countermeasures for general IT systems. The presented concepts and elements of IT security for industrial and utility communication systems are illustrated with case studies.

Locating Topics

To assist readers in locating material, a complete table of contents is presented at the front of the book. Each chapter begins with its own table of contents. Two indexes are provided at the end of the book. The Contributor Index lists contributors to this book, together with the titles of their contributions; there is also a detailed subject index.

Acknowledgments

I would like to express gratitude to my publisher Nora Konopka, and other CRC Press staff involved in this book's production — in particular, Jessica Vakili, Elizabeth Spangenberger, Melanie Sweeney, and Glenon Butler.

Richard Zurawski

Editor

Dr. Richard Zurawski, President of ISA Group (San Francisco and Santa Clara, California) is involved in providing solutions to Fortune 100 companies. Prior to that, he held various executive positions with San Francisco Bay area based companies. He was also a full-time R&D advisor with Kawasaki Electric (Tokyo) and held a regular professorial appointment at the Institute of Industrial Sciences, University of Tokyo. During the 1990s he participated in a number of Japanese Intelligent Manufacturing Systems programs, as well as in IMS.

He is editor of three major handbooks: *The Industrial Information Technology Handbook* (CRC Press, Boca Raton, Florida; 2004); *The Industrial Communication Technology Handbook* (CRC Press, Boca Raton, Florida; 2005); and *Embedded Systems Handbook* (CRC Press, Boca Raton, Florida; 2005).

Dr. Zurawski served as Associate Editor for *Real-Time Systems; The International Journal of Time-Critical Computing Systems* (Kluwer Academic Publishers), and *The International Journal of Intelligent Control and Systems* (World Scientific Publishing Company). He was a guest editor of four special sections in *IEEE Transactions on Industrial Electronics,* and a guest editor of a special issue on Industrial Communication Systems in the *Proceedings of the IEEE* (June 2005). In 1998, he was invited by *IEEE Spectrum* to contribute an article on Java technology to *"Technology 1999: Analysis and Forecast Issue."*

Dr. Zurawski is editor for the Industrial Information Technology book series, CRC Press, Boca Raton, Florida. He has served as editor at large for *IEEE Transactions on Industrial Informatics*, and Associate Editor for *IEEE Transactions on Industrial Electronics.*

Dr. Zurawski has served as a vice president of the IEEE Industrial Electronics Society (IES) and as the chairman of the IEEE IES Technical Committee on Factory Automation. He was on a steering committee of the ASME/IEEE *Journal of Microelectromechanical Systems.* In 1996, he received the Anthony J. Hornfeck Service Award from the IEEE Industrial Electronics Society.

Dr. Zurawski has established two major technical events: the Workshop on Factory Communication Systems, the only IEEE event dedicated to industrial communication networks; and the International Conference on Emerging Technologies and Factory Automation, the largest IEEE conference dedicated to factory and industrial automation. He served as a general, program, and track chair for a number of IEEE, IFAC, and other technical societies' conferences and workshops, including a conference organized for Sun Microsystems.

His research interests include formal methods, embedded and real-time systems, microelectromechanical systems (MEMS), hybrid systems and control, control of large-scale systems, human-oriented mechatronics and systems, bioelectronics, and electromagnetic fields (EMF). Dr. Richard Zurawski received a M.Sc. in electronics from the University of Mining & Metallurgy in Krakow, Poland; and a Ph.D. in computer science from La Trobe University in Melbourne, Australia.

Contributors

Luis Almeida
University of Aveiro
Aveiro, Portugal

Pulak Bandyopadhyay
GM R&D Center
Warren, Michigan

Ralph Büsgen
Siemens AG
Furth, Germany

Jean-Dominique Decotignie
Centre Suisse d'Electronique et de
 Microtechnique
Neuchatel, Switzerland

Christian Diedrich
Institut für Automation und
 Kommunikation eV – IFAK
Magdeburg, Germany

Wilfried Elmenreich
Vienna University of Technology
Vienna, Austria

David Emerson
Yokogawa America
Denison, Texas

Joachim Feld
Siemens AG
Nuremberg, Germany

A.M. Fong
Singapore Instiute of
 Manufacturing Technology
Singapore

Alberto Fonseca
University of Aveiro
Aveiro, Portugal

K.M. Goh
Singapore Institute of
 Manufacturing Technology
Singapore

Hans-Michael Hanish
University of Halle-Wittenberg
Halle, Germany

Zaijun Hu
ABB Corporate Research
 Center
Mannheim, Germany

Frank Iwanitz
Softing AG
Munchen, Germany

Ulrich Jecht
UJ Process Analytics
Baden-Baden, Germany

Muammer Koç
University of Michigan
Ann Arbor, Michigan

Eckart Kruse
ABB Corporate Research
 Center
Ladenburg, Germany

Juergen Lange
Softing AG
Munchen, Germany

Jay Lee
University of Cincinnati
Cincinnati, Ohio

Kang Lee
National Institute of
 Manufacturing Technology
Gaithersburg, Maryland

Y.G. Lim
Singapore Institute of
 Manufacturing Technology
Singapore

Arnd Luder
University of Magdeburg
Magdeburg, Germany

Vladimir Marik
Czech Technical University of
 Prague
Prague, Czech Republic

Kirsten Matheus
Carmeq GmbH
Berlin, Germany

Fabrizio Meo
FIDIA
San Mauro Torinese, Italy

Martin Naedele
ABB Research Center
Baden-Daettwil Switzerland

Jun Ni
University of Michigan
Ann Arbor, Michigan

P. Pedreiras
University of Aveiro
Aveiro, Portugal

Jorn Peschke
University of Magdeburg
Magdeburg, Germany

Stefan Pitzek
Vienna University of
 Technology
Vienna, Austria

Manfred Popp
Siemens AG
Furth, Germany

Thilo Sauter
Austrian Academy of Sciences
Wiener Naustadt, Austria

Victor Schiffer
Rockwell Automation
Hann, Germany

Karlheinz Schwarz
Schwarz Consulting Company
Karlsruhe, Germany

Wolfgang Stripf
Siemens AG
Karlsruhe, Germany

O. Tin
Singapore Institute of
  Manufacturing Technology
Singapore

Peter Wenzel
PROFIBUS International
Karlssruhe, Germany

Claus Vetter
ABB Corporate Research
  Center
Baden, Switzerland

Valeriy Viatkin
University of Auckland
Auckland, New Zealand

Pavel Vrba
Rockwell Automation
Prague, Czech Republic

Thomas Werner
ABB Corporate Research
  Center
Baden, Switzerland

K. Yi
Singapore Institute of
  Manufacturing Technology
Singapore

Richard Zurawski
ISA Group
Alameda, California

Contents

Part 1

Introduction

1

Integration Technologies for Industrial Automated Systems: Challenges and Trends

Richard Zurawski
ISA Group, U.S.A.

1.1 Introduction

One of the fundamental tenets of the integration of industrial automated enterprises is unrestricted and timely flow of data between applications at different levels of the enterprise hierarchy — for example, between shop-floor and enterprise level — as well as between different applications at the same level. This data exchange takes place among various IT infrastructure elements functionality and performance requirements of which are determined by their level in the hierarchy and the application they support. They may be controllers and operator workstations at the manufacturing/process level; workstations supporting the Manufacturing Execution System application; gateway servers in between control networks and the plant network; workplaces at the enterprise or business level supporting, for example, the Manufacturing Resource Planning application; etc. The primary conduit of data exchange in modern automated systems is a specialized communication infrastructure that takes on a hierarchical arrangement, with individual networks reflecting to a large extent the needs of applications at different levels — in terms of functionality and performance (data size, throughput, delay, availability, etc.). The life cycle of a plant spans typically many decades of operation, resulting in heterogeneity of the manufacturing/process equipment installed, supporting IT infrastructure, and applications to operate and maintain the plant. This translates into a diversity of field devices and supporting industrial networks, software platforms supporting applications, and languages used to develop those applications. Integration of the communication infrastructure of a plant and applications (largely implemented in software) is needed to achieve the required seamless and timely data flow throughout the entire enterprise. This is the focus of this chapter and a large portion of the book.

Section 1.2 gives an overview of selected integration issues, followed by a section (1.3)that provides an overview of the fieldbus networks and real-time Ethernet with a focus on standards. Subsequently, wireless local and personal area networks, and wireless sensors and wireless networks in factory automation are presented, followed by selected security issues in automation networks. Because the chapter aims at providing a framework for the book, ample references are provided to cover individual topics.

1.2 Integration Issues

Advances in the design of integrated circuits and embedded systems, tools availability, and falling fabrication costs of semiconductor devices and systems (system-on-chip, SoC) have allowed for an infusion of intelligence, such as sensors and actuators into field devices. The controllers used with these devices typically provide on-chip signal conversion, data and signal processing, and communication functions. The increased functionality and processing capabilities of controllers have been largely instrumental in the emergence of a widespread trend for the networking of field devices around specialized networks, frequently referred to as field area networks [1].

One of the main reasons for the emergence of field area networks in the first place was an evolutionary need to replace point-to-point wiring connections with a single bus, thus paving the road for the emergence of distributed systems and, subsequently, networked embedded systems with the infusion of intelligence into the field devices. A detailed description of the co-evolution of field area networks and plant automation concepts is provided in Chapter 13. A typical network architecture in industrial plant automation is shown in Figure 1.1.

The network — or a system of networks — may consist of a number of different types of networks to meet the functional and performance requirements of the enterprise hierarchy to be deployed. For example, a variety of field area networks, and sensor networks, are used at the manufacturing/process level. They are designed to support the exchange of small data records characteristic of monitoring and control actions, and are connected to process controllers. The traffic, which exhibits low data rates, is frequently subject to determinism of data transfer. To ensure the determinism, if mandated, the networks can be segmented to distribute the load. The control network(s) are used to exchange real-time data among controllers and operator workstations used for process control and supervision. There is a growing tendency for this level of networks to be based on the Ethernet and TCP/IP protocol suite. The major role play here field area networks that incorporate Ethernet for the lower two layers in the OSI model, such as PROFInet or EtherNet/IP; these are discussed in more detail in the following sections. Enterprise-level networks are typically used for manufacturing/process execution and various enterprise management applications. The traffic is characterized by high data rates and large packets; determinism of data transfer is largely not an issue. These networks are predominantly based on the Ethernet and TPC/IP protocol suite.

The use of propriety field devices (sensors/actuators), machining tool controllers, and manufacturing/process machinery typically leads to the deployment of dedicated field area and control networks, developed to link specific devices and systems. This creates "islands of automation" integrated locally around specific and frequently incompatible network technologies and data representations. The integration solutions involve both communication infrastructure, and applications interfaces and data representation. The integration, in the context of communication aspects, involving different plant automation units or even separate automation sections within a unit, is frequently referred to as *horizontal integration*. The term *vertical integration* refers to the integration among different levels of the plant or enterprise hierarchy, from field devices via manufacturing execution systems to business applications. In general, the integration of the communication infrastructure can be achieved using, for example, generic concepts of gateways and protocol tunneling [2]; the ANSI/EIA-852 standard is discussed in Reference [3]. The use of "industrial Ethernet," or Real-Time Ethernet (RTE), which supports real-time communication at the factory floor, is the emerging trend in both horizontal and vertical integration.

In RTE, the random and native CSMA/CD arbitration mechanism is being replaced by other solutions, allowing for deterministic behavior required in real-time communication to support soft and hard

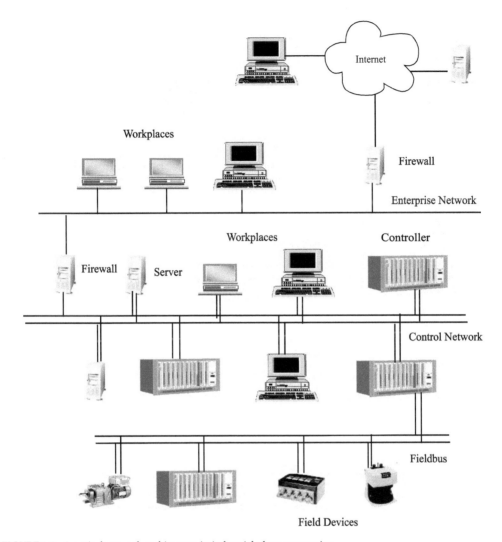

FIGURE 1.1 A typical network architecture in industrial plant automation.

real-time deadlines, for example, time synchronization of activities required to control drives, and for exchange of small data records characteristic of monitoring and control actions. The direct support for the Internet technologies allows for vertical integration of various levels of the industrial enterprise hierarchy to include seamless integration between automation and business logistic levels to exchange jobs and production (process) data, transparent data interfaces for all stages of the plant life cycle, Internet- and web-enabled remote diagnostics and maintenance, and electronic orders and transactions. In addition, the use of standard components such as protocol stacks, Ethernet controllers, bridges, etc., allows for mitigating the ownership and maintenance cost.

The two most widely used industry standards intended to provide interfaces to hide the details of device-dependent communication protocols are the Manufacturing Messaging Specification (MMS) [4, 5] and OLE for Process Control (OPC) of the Open Control Foundation [6].

MMS is an application layer messaging protocol for communication to and from field devices such as remote terminal units, programmable logic controllers, numerical controllers, robot controllers, etc. MMS adopts the client/server model to describe the behavior of the communicating devices. The central element of this model is the concept of the Virtual Manufacturing Device (VMD), which embeds (abstract) objects representing physical devices such as sensors and actuators, for example. MMS defines a wide range of services to allow access to the VMD and manipulation of its objects, to mention some of the functions. Separate companion standards are required for the definition of application-specific objects. Most of the recent MMS implementations are built on top of TCP. A comprehensive overview of the MMS standards is presented in Chapter 6. Kim and Haas [7] reported on the use of MMS on top of TCP/IP in the implementation of a virtual factory communication system.

OPC is an application layer specification for communication, or data exchange, between software applications in automation systems. Being originally built on Microsoft's COM model, the use of OPC implementations has been in practice until recently restricted to platforms supporting COM. The move by OPC to base future specifications on XML and Web services should remove this impediment and make OPC implementations platform independent. The OPC DA (Data Access) specification defines standardized read operations to transfer real-time data from process and control devices, together with time stamp and status information, to higher-level applications such as process supervision and manufacturing execution systems. In addition, it also allows for the location of OPC servers and browsing in the namespaces of the OPC servers. The OPC DX (Data Exchange) specification allows for server-to-server noncritical data exchange over Ethernet networks, for example, between controllers of different manufacturers. Other OPC specifications define alarms and event notifications, access to historical data, etc. [8]. OPC standards, implementation issues, and applications are discussed in Chapter 5.

CORBA, also based on the component model, has been primarily used in applications such as Enterprise Resource Planning or Supply Chain Management. Reports on applications of CORBA at the automation level are scarce. The evaluation of real-time implementations of CORBA for use with NC controllers is presented in Chapter 12. CORBA in manufacturing is overviewed in Reference [9].

Another approach to achieve seamless data exchange among applications is based on Web Services, which offers platform independence and programming language neutrality. The use of Web Services in industrial automation and arising challenges are comprehensively overviewed in Chapter 4. References [10, 11] offer an authoritative introduction to Web Services and programming.

The use of the Simple Network Management Protocol (SNMP), Lightweight Directory Access Protocol (LDAP), and Web-based approaches to exchange data between gateways and the Ethernet and TCP/IP protocol suite based control or plant networks is discussed in Reference [2].

1.3   Industrial Communication Systems: An Overview

1.3.1   Field Area Networks

Field area networks, or fieldbuses [12] (a fieldbus is, in general, a digital, two-way, multi-drop communication link) as they are commonly referred to, are, in general, networks connecting field devices such as sensors and actuators with field controllers (for example, programmable logic controllers [PLCs] in industrial automation), as well as man–machine interfaces. The field area networks are used in a variety of application domains: industrial and process automation, building automation, automotive and railway applications, aircraft control, control of electrical substations, etc. The benefits are numerous, including increased flexibility; improved system performance; and ease of system installation, upgrade, and maintenance. Unlike LANs, due to the nature of the communication requirements imposed by applications, fieldbus area networks, by contrast, have low data rates, small data packet size, and typically require real-time capabilities that mandate determinism of data transfer. However, data rates greater than 10 Mbit/s, typical of LANs, have become commonplace in field area networks. The field area networks employ, either directly or in combination, three basic communication paradigms: (1) client-server, (2) producer-consumer, and (3) publisher-subscriber models. The use of these models reflects intimately the requirements and constraints of an application domain or a specific application.

Although for the origins of field area networks, one can look back as far as the late 1960s in the nuclear instrumentation domain, CAMAC network [13], and the early 1970s in avionics and aerospace applications, MIL-STD-1553 bus [14], it was the industrial automation area that brought the main thrust of development. The need for integration of heterogeneous systems, difficult at that time due to the lack of the standards, resulted in two major initiatives that have had a lasting impact on the integration concepts and architecture of the protocol stack of field area networks. These initiatives were the TOP (Technical and Office Protocol) [15] and MAP (Manufacturing Automation Protocol) [16] projects. These two projects exposed some of the pitfalls of full seven-layer stack implementations (ISO/OSI model [17]) in the context of applications in industrial automation. As a result, typically, only layers 1 (physical layer); 2 (data link layer, including implicitly the medium access control layer); and 7 (application layer, which also covers the user layer) are used in field area networks [18], also prescribed in the international fieldbus standard, IEC 61158 [19]. In IEC 61158, the functions of layers 3 and 4 are recommended to be placed either in layer 2 or layer 7; the functions of layers 5 and 6 are always covered in layer 7.

The evolution of fieldbus technology, which begun well over two decades ago, has resulted in a multitude of solutions reflecting the competing commercial interests of their developers and standardization bodies, both national and international: IEC [20], ISO [21], ISA [22], CENELEC [23], and CEN [24]. This is also reflected in IEC 61158 (adopted in 2000), which accommodates all national standards and user organization championed fieldbus systems. Subsequently, implementation guidelines were compiled into Communication Profiles, IEC 61784-1 [25]. Those Communication Profiles identify seven main systems (or Communication Profile Families) known by the brand names: Foundation Fieldbus (H1, HSE, H2), used in process and factory automation; ControlNet and EtherNet/IP, both used in factory automation, and PROFIBUS (DP, PA), used in factory and process automation, respectively; PROFInet, used in factory automation; P-Net (RS 485, RS 232), used in factory automation and shipbuilding; WorldFIP, used in factory automation; INTERBUS, INTERBUS TCP/IP, and INTERBUS Subset, used in factory automation; and Swiftnet *transport*, Swiftnet *full stack*, used by aircraft manufacturers. The listed application areas are the dominant ones.

1.3.2   Real-Time Ethernet (RTE)

In the RTE, the random and native CSMA/CD arbitration mechanism is being replaced by other solutions, allowing for the deterministic behavior required in real-time communication. A variety of solutions have been proposed to achieve this goal. Some can coexist with regular Ethernet nodes; some reuse the same hardware but are incompatible; some are compatible but cannot offer guarantees in the presence of nodes that do not implement the same modifications — as classified in Decotignie [26].

The RTE, under standardization by the IEC/SC65C committee, is a fieldbus technology that incorporates Ethernet for the lower two layers in the OSI model. There are already a number of implementations that use one of the three different approaches to meet real-time requirements. The first approach is based on retaining the TCP/UDP/IP protocol suite unchanged (subject to nondeterministic delays); all real-time modifications are enforced in the top layer. Implementations in this category include Modbus/TPC [27] (defined by Schneider Electric and supported by Modbus-IDA [28]), EtherNet/IP [29] (defined by Rockwell and supported by the Open DeviceNet Vendor Association [ODVA] [30] and ControlNet International [31]), P-Net (on IP) [32] (proposed by the Danish P-Net national committee), and Vnet/IP [33] (developed by Yokogawa, Japan). In the second approach, the TCP/UDP/IP protocols suite is bypassed, and the Ethernet functionality is accessed directly — in this case, RTE protocols use their own protocol stack in addition to the standard IP protocol stack. The implementations in this category include Ethernet Powerlink (EPL) [34] (defined by Bernecker and Rainer [B&R], and now supported by the Ethernet Powerlink Standardization Group [35]); TCnet (a Time-Critical Control Network) [36] (a proposal from Toshiba); EPA (Ethernet for Plant Automation) [37] (a Chinese proposal); and PROFIBUS CBA (Component-Based Automation) [38] (defined by several manufacturers, including Siemens, and supported by PROFIBUS International [39]). Finally, in the third approach, the Ethernet mechanism and infrastructure are modified. The implementations include SERCOS III [40] (under development by SERCOS), EtherCAT [41] (defined by Beckhoff and supported by the EtherCat Technology Group [42]), PROFInet IO [43] (defined by several manufacturers, including Siemens, and supported by PROFIBUS International).

1.3.3   Wireless Technologies and Networks

The use of wireless links with field devices, such as sensors and actuators, allows for flexible installation and maintenance, allows for mobile operation required in the case of mobile robots, and alleviates the problems associated with cabling. For a wireless communication system to operate effectively in an industrial/factory floor environment, it must guarantee high reliability, low and predictable delay of data transfer (typically, less than 10 ms for real-time applications), support for a high number of sensor/actuators, and low power consumption, to mention a few. In industrial environments, the wireless channel characteristic degradation artifacts can be compounded by the presence of electric motors or a variety of equipment causing the electric discharge, which contribute to even greater levels of bit error and packet losses. Improving channel quality and designing robust and loss-tolerant applications, both the subject of extensive research and development, seem to have the potential to alleviate these problems to some extent [44].

In addition to peer-to-peer interaction, the sensor/actuator stations may communicate with the base station(s), which may have its transceiver attached to the cable of a fieldbus, thus resulting in a hybrid wireless-wireline fieldbus system [45]. To leverage low cost, small size, and low power consumption, Bluetooth 2.4 GHz radio transceivers can be used as the sensor/actuator communication hardware. To meet the requirements for high reliability, low and predictable delay of data transfer, and support for a high number of sensor/actuators, custom optimized communication protocols may be required for the operation of the base station, as the commercially available solutions such as IEEE 802.15.1/ Bluetooth [46, 47], IEEE 802.15.4/ZigBee [48], and IEEE 802.11 [49–51] variants may not fulfill all the requirements.

A representative example of this kind of system is a wireless sensor/actuator network developed by ABB and deployed in a manufacturing environment [52]. The system, known as WISA (wireless sensor/actuator), has been implemented in a manufacturing cell to network proximity switches, which are some of the most widely used position sensors in automated factories to control positions of a variety of equipment, including robotic arms. The sensor/actuator communication hardware is based on a standard Bluetooth 2.4 GHz radio transceiver and low power electronics that handle the wireless communication link. The sensors communicate with a wireless base station via antennas mounted in the cell. For the base station, a specialized RF front end was developed to provide collision-free air access by allocating a fixed Time Division Multiple Access (TDMA) time slot to each sensor/actuator. Frequency hopping (FH) was employed to counter both frequency-selective fading and interference effects, and operates in combination with automatic retransmission requests (ARQs). The parameters of this TDMA/FH scheme were chosen to satisfy the requirements of up to 120 sensor/actuators per base station. Each wireless node has a response or cycle time of 2 ms, to make full use of the available radio band of 80 MHz width. The frequency hopping sequences are cell specific and were chosen to have low cross-correlations to permit parallel operation of many cells on the same factory floor with low self-interference. The base station can handle up to 120 wireless sensor/actuators and is connected to the control system via a (wireline) field bus. To increase capacity, a number of base stations can operate in the same area. WISA provides wireless power supply to the sensors, based on magnetic coupling [53].

In the future, different wireless technologies will be used in the same environment. This may pose some problems with coexistence if networks are operated in the same frequency band. A good overview of this issue is presented in Reference [44].

1.3.4   Security in Industrial Networks

The growing trend for horizontal and vertical integration of industrial automated enterprises, largely achieved through internetworking of the plant communication infrastructure, coupled with a growing demand for remote access to process data at the factory floor level, exposes automation systems to potential electronic security attacks that might compromise the integrity of these systems and endanger plant safety. Safety, or the absence of catastrophic consequences for humans and environment, is, most likely, the most important operational requirement for automation and process control systems. Another important requirement is system/plant availability; the automation system and plant must be operationally safe over extended periods of time, even if they continue to operate in a degraded mode in the presence of a fault. With this requirement, security software updates in the running field devices may be difficult or too risky. As pointed out in Dzung et al. [54], "security is a process, not a product." This motto embeds the practical wisdom that solutions depend on specific application areas, systems, and devices.

The limited computing, memory, and communication bandwidth resources of controllers embedded in the field devices pose considerable challenge for the implementation of effective security policies, which, in general, are resource demanding. This limits the applicability of the mainstream cryptographic protocols, even vendor-tailored versions. The operating systems running on small footprint controllers tend to implement essential services only, and do not provide authentication or access control to protect mission- and safety-critical field devices. In applications restricted to the Hypertext Transfer Protocol (HTTP), such as embedded Web servers, Digest Access Authentication (DAA) [55], a security extension to HTTP, may offer an alternative and viable solution.

Fieldbuses, in general, do not have any security features. Because they are frequently located at the premises requiring access permit, eavesdropping or message tampering would require physical access to the medium. Potential solutions to provide a certain level of security were explored in Palensky and Sauter [56] and Schwaiger and Treytl [57], where the focus was on the fieldbus-to-Internet gateway.

The emerging Ethernet-based fieldbuses are more vulnerable to attack owing to the use of the Ethernet and TCP/IP protocols and services. Here, the general communication security tools for TCP/IP apply [54].

Local area wireless sensor/actuator networks are particularly vulnerable to DoS (denial-of-service) attacks by radio jamming and even eavesdropping. The details on protection solutions for this class of networks are extensively discussed in Dzung et al. [54] and Schaefer [58].

The security issues as applied to middleware applications are discussed in some detail in Dzung et al. [54].

References

1. Zurawski, R. ed., The Industrial Communication Technology Handbook CRC Press, Boca Raton, FL, 2005.
2. Sauter, T., Linking Factory Floor and the Internet, in *The Industrial Communication Technology Handbook*, Ed. R. Zurawski, CRC Press, Boca Raton, FL, 2005, pp. 24-1 to 24-19.
3. Loy, D. and S., Soucek, Extending EIA-709 Control Networks across IP Channels, in *The Industrial Communication Technology Handbook*, Ed. R. Zurawski, CRC Press, Boca Raton, FL, 2005, pp. 25-1 to 25-17.
4. ISO 9506-1, Manufacturing Message Specification (MMS): Part 1: Service Definition, 2003.
5. ISO 9506-1, Manufacturing Message Specification (MMS): Part 2: Protocol Definition, 2003.
6. Online]: www.opcfoundation.com.
7. Kim, D.-S. and J. Haas, Virtual Factory Communication System Using ISO 9506 and Its Application to Networked Factory Machine, in *The Industrial Communication Technology Handbook*, Ed. R. Zurawski, CRC Press, Boca Raton, FL, 2005, pp. 37-1 to 37-10.
8. Iwanitz, F. and J. Lange, OLE for Process Control, Huthig, Heidelberg, Germany, 2001.
9. Barretto, M.R.P., P.M.P. Blanco, and M.A. Poli, CORBA in Manufacturing — Technology Overview, in *The Industrial Information Technology Handbook*, Ed. R. Zurawski, CRC Press, Boca Raton, FL, 2004, pp. 6-1 to 6-23.
10. Eckert, K.-P., The Fundamentals of Web Services, in *The Industrial Information Technology Handbook*, Ed. R. Zurawski, CRC Press, Boca Raton, FL, 2004, pp. 10-1 to 10-13.
11. Eckert, K.-P., Programming Web Services with .Net and Java, in *The Industrial Information Technology Handbook*, Ed. R. Zurawski, CRC Press, Boca Raton, FL, 2004, pp. 11-1 to 11-17.
12. Thomesse, J.-P., Fieldbus Technology in Industrial Automation, *Proceedings of the IEEE*, Vol. 93, No. 6, June 2005, pp. 1073–1101.

13. Costrell, R., CAMAC Instrumentation System — Introduction and General Description". *IEEE-Transactions-on-Nuclear-Science.* April 1971, NS-18(2), pp. 3–8.

14. Gifford, C.A., A Military Standard for Multiplex Data Bus, Proceedings of the IEEE–1974, National Aerospace and Electronics Conference, May 13–15, 1974, Dayton, OH, 1974, pp. 85–88.

15. Dillon, S.R., Manufacturing Automation Protocol and Technical and Office Protocols – Success through the OSI Model, in *Proceedings COMPCON Spring'87*, 1987, pp. 80–81.

16. Schutz, H.A., The Role of MAP in Factory Integration, *IEEE Transactions on Industrial Electronics*, 35(1): 6–12, 1988.

17. Zimmermann, H., OSI Reference Model: The ISO model of architecture for open system interconnection, *IEEE Transactions on Communications*, 28(4): 425–432, 1980.

18. Pleinevaux, P. and J.-D. Decotignie, Time Critical Communication Networks: Field Buses, *IEEE Network*, 2: 55–63, 1988.

19. International Electrotechnical Commission, IEC 61158-1, Digital Data Communications for Measurement and Control — Fieldbus for Use in Industrial Control Systems, Part 1: Introduction, 2003.

20. International Electrotechnical Commission [IEC]. [Online] www.iec.ch.

21. International Organization for Standardization [ISO]. [Online] www.iso.org

22. Instrumentation Society of America [ISA]. [Online] www.isa.org

23. Comité Européen de Normalisation Electrotechnique [CENELEC]; [Online] www.cenelec.org

24. European Committee for Standardization [CEN]; [Online] www.cenorm.be

25. International Electrotechnical Commission, IEC 61784-1, Digital Data Communications for Measurement and Control — Part 1: Profile Sets for Continuous and Discrete Manufacturing Relative to Fieldbus Use in Industrial Control Systems, 2003.

26. Decotignie, J.-D., Ethernet-based real-time and industrial communications, *Proceedings of the IEEE*, 93(6): 1103–1117, June 2005.

27. IEC: Real-Time Ethernet Modbus-RTPS, Proposal for a Publicly Available Specification for Real-Time Ethernet, Document IEC 65C/341/NP, date of circulation: 2004-06-04.

28. [Online] www.modbus-ida.org

29. EC: Real Time Ethernet: EtherNet/IP with Time Synchronization, Proposal for a Publicly Available Specification for Real-Time Ethernet, Document IEC, 65C/361/NP, date of circulation: 2004-12-17.

30. [Online] www.odva.org

31. [Online] www.controlnet.org

32. IEC: Real-Time Ethernet: P-NET on IP, Proposal for a Publicly Available Specification for Real-Time Ethernet, Document IEC, 65C/360/NP, date of circulation: 2004-12-17.

33. IEC: Real-Time Ethernet Vnet/IP, Proposal for a Publicly Available Specification for Real-Time Ethernet, Document IEC, 65C/352/NP, Date of circulation: 2004-11-19.

34. IEC: Real-Time Ethernet EPL (ETHERNET Powerlink), Proposal for a Publicly Available Specification for Real-Time Ethernet, Document IEC, 65C/356a/NP, date of circulation: 2004-12-03.

35. [Online] www.ethernet-powerlink.org

36. IEC: Real-Time Ethernet TCnet (Time-Critical Control Network), Proposal for a Publicly Available Specification for Real-Time Ethernet, Document IEC, 65C/353/NP, date of circulation: 2004-11-19.

37. IEC: Real-Time Ethernet EPA (Ethernet for Plant Automation), Proposal for a Publicly Available Specification for Real-Time Ethernet, Document IEC 65C/357/NP, date of circulation: 2004-11-26.

38. Feld, J., PROFINET — Scalable Factory Communication for all Applications, *2004 IEEE International Workshop on Factory Communication Systems*, September 22–24, 2004, Vienna, Austria, pp. 33–38.

39. [Online] www.profibus.org

40. IEC: Real-Time Ethernet SERCOS III, Proposal for a Publicly Available Specification for Real-Time Ethernet, Document IEC, 65C/358/NP, date of circulation: 2004-12-03.

41. IEC: Real-Time Ethernet Control Automation Technology (ETHERCAT), Proposal for a Publicly Available Specification for Real-Time Ethernet, Document IEC, 65C/355/NP, date of circulation: 2004-11-19.

42. [Online] www.ethercat.org

43. IEC: Real-Time Ethernet PROFINET IO, Proposal for a Publicly Available Specification for Real-Time Ethernet, Document IEC, 65C/359/NP, date of circulation: 2004-12-03.

44. Willig, A., K. Matheus, and A. Wolisz, Wireless Technology in Industrial Networks, *Proceedings of the IEEE*, 93(6): 1130–1151, June 2005.

45. Decotignie, J.-D., Interconnection of Wireline and Wireless Fieldbuses, in *The Industrial Communication Technology Handbook*, Ed. R. Zurawski, CRC Press, Boca Raton, FL, 2005, pp. 26-1 to 26-13.

46. Bluetooth Consortium, *Specification of the Bluetooth System*, 1999; [Online] www.bluetooth.org.

47. Bluetooth Special Interest Group, Specification of the Bluetooth System, Version 1.1, December 1999.

48. LAN/MAN Standards Committee of the IEEE Computer Society, IEEE Standard for Information Technology — Telecommunications and Information Exchange between Systems — Local and Metropolitan Area Networks — Specific Requirements — Part 15.4: Wireless Medium Access Control (MAC) and Physical Layer (PHY) Specifications for Low Rate Wireless Personal Area Networks (LR-WPANs), October 2003.

49. LAN/MAN Standards Committee of the IEEE Computer Society, IEEE Standard for Information Technology — Telecommunications and Information Exchange between Systems — Local and Metropolitan Networks — Specific Requirements — Part 11: Wireless LAN Medium Access Control (MAC) and Physical Layer (PHY) Specifications: Higher Speed Physical Layer (PHY) Extension in the 2.4 GHz band, 1999.

50. LAN/MAN Standards Committee of the IEEE Computer Society, Information Technology — Telecommunications and Information Exchange between Systems — Local and Metropolitan Area Networks — Specific Requirements — Part 11: Wireless LAN Medium Access Control (Mac) and Physical Layer (Phy) Specifications, 1999.

51. Institute of Electrical and Electronic Engineering, Part 11: Wireless LAN Medium Access Control (MAC) and Physical Layer (PHY) Specifications, Amendment 4.

52. Further Higher Data Rate Extension in the 2.4 GHz Band, June 2003, a NSI/IEEE Std 802.11.

53. Apneseth Christoffer, Dacfey Dzung, Snorre Kjesbu, Guntram Scheible, and Wolfgang Zimmermann, Introducing Wireless Proximity Switches, ABB Review, (2), 42–49, 2002, www.abb.com/review.

54. Dzung Dacfey, Christoffer Apneseth, Jan Endresen, and Jan-Erik Frey, Design and Implementation of a Real-Time Wireless Sensor/Actuator Communication System, in Proceedings of the IEEE ETFA 2005, Catania, Italy, September 19–23, 2005.

55. Dzung, D., M. Naedele, T.P. von Hoff, and M. Cervatin, Security for Industrial Communication Systems, *Proceedings of the IEEE*, 93(6), 1152–1177, June 2005.

56. Cervatin, M. and T.P. von Hoff, HTTP Digest Authentication for Embedded Web Servers, in *Embedded Systems Handbook*, Ed. R. Zurawski, CRC-Taylor & Francis, Boca Raton, FL, 2005, pp. 45-1 to 45-14.

57. Palensky, P. and T. Sauter, Security Considerations for FAN–Internet Connections, *IEEE Intern. Workshop on Factory Communication Systems*, Porto, 2000, pp. 27–35.

58. Schwaiger, C. and A. Treytl, Security Topics and Solutions for Automated Networks, in *The Industrial Communication Technology Handbook*, Ed. R. Zurawski, CRC-Taylor & Francis, Boca Raton, FL, 2005, pp. 27-1 to 27-16.

59. Schaefer, G., Sensor Network Security, in *Embedded Systems Handbook*, Ed. R. Zurawski, CRC-Taylor & Francis, Boca Raton, FL, 2005, pp. 39-1 to 39-23.

Part 2

E-Technologies in Enterprise Integration

2

Introduction to e-Manufacturing

Muammer Koç
University of Michigan – Ann Arbor

Jun Ni
University of Michigan – Ann Arbor

Jay Lee
University of Cincinnati

Pulak Bandyopadhyay
GM R&D Center

2.1   Introduction

For the past decade, the impact of web-based technologies has added "velocity" to the design, manufacturing, and aftermarket service of a product. Today's competition in manufacturing industry depends not just on lean manufacturing but also on the ability to provide customers with total solutions and life-cycle costs for sustainable value. Manufacturers are now under tremendous pressure to improve their responsiveness and efficiency in terms of product development, operations, and resource utilization with a transparent visibility of production and quality control. Lead times must be cut short to their extreme extent to meet the changing demands of customers in different regions of the world. Products are required to be made-to-order with no or minimum inventory, requiring (a) an efficient information flow between customers, manufacturing, and product development (i.e., plant floor, suppliers, and designers); (b) a tight control between customers and manufacturing; and (c) near-zero downtime of the plant floor assets. Figure 2.1 summarizes the trends in manufacturing and function of predictive intelligence as an enabling tool to meet the needs [1–4].

With emerging applications of Internet and tether-free communication technologies, the impact of e-intelligence is forcing companies to shift their manufacturing operations from the traditional factory integration philosophy to an e-factory and e-supply chain philosophy. It transforms companies from a local factory automation to a global enterprise and business automation. The technological advances for achieving this highly collaborative design and manufacturing environment are based on multimedia-type information-based engineering tools and a highly reliable communication system for enabling distributed procedures in concurrent engineering design, remote operation of manufacturing processes, and operation of distributed production systems. As shown in Figure 2.2, e-manufacturing fills the gaps existing in the traditional manufacturing systems. The gaps between product development and supply chain consist of lack of life-cycle information and lack of information about supplier capabilities. Hence, designers, unless with years of experience, work in a vacuum; design the product according to the specification given; and wait for the next step. Most of the time, the design made according to specifications is realized to be infeasible for manufacturing with suppliers' machinery. As a result, lead times

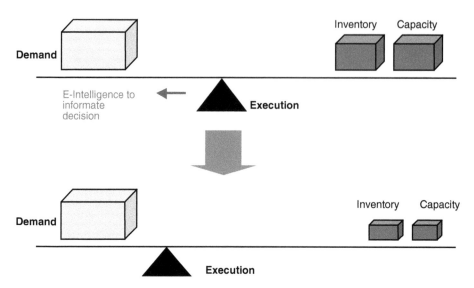

FIGURE 2.1   The transformation of e-Manufacturing for unmet needs.

become longer. Similarly, for instance, because of the lack of information and synchronization between suppliers and assembly plants, just-in-time manufacturing and on-time shipment become possible only with a substantial amount of inventory whereas with e-manufacturing, real-time information regarding reliability and status of supplier's equipment will also be available as a part of the product quality information. With these information and synchronization capabilities, less and less inventory will be necessary, contributing to the profitability of the enterprise.

2.2   e-Manufacturing: Rationale and Definitions

e-Manufacturing is a transformation system that enables the manufacturing operations to achieve predictive near-zero-downtime performance as well as to synchronize with the business systems through the use of web-enabled and tether-free (i.e., wireless, web, etc.) infotronics technologies. It integrated information and decision-making among data flow (of machine/process level), information flow (of factory and supply system level), and cash flow (of business system level) [5–7]. e-Manufacturing is a business strategy as well as a core competency for companies to compete in today's e-business environment. It is aimed to complete integration of all the elements of a business including suppliers, customer service network, manufacturing enterprise, and plant floor assets with connectivity and intelligence brought by the web-enabled and tether-free technologies and intelligent computing to meet the demands of e-business/e-commerce practices that gained great acceptance and momentum over the last decade. e-Manufacturing is a transformation system that enables e-Business systems to meet the increasing demands through tightly coupled supply chain management (SCM), enterprise resource planning (ERP), and customer relation management (CRM) systems as well as environmental and labor regulations and awareness, (Figure 2.3) [4–7].

e-Manufacturing includes the ability to monitor the plant floor assets, predict the variation of product quality and performance loss of any equipment for dynamic rescheduling of production and maintenance operations, and synchronize with related business services to achieve a seamless integration between manufacturing and higher level enterprise systems. Dynamically updated information and knowledge about the capabilities, limits, and variation of manufacturing assets for various suppliers guarantee the best decisions for outsourcing at the early stages of design. In addition, it enables customer orders autonomously across the supply chain, bringing unprecedented levels of speed, flexibility, and visibility to the production process reducing inventory, excess capacity, and uncertainties.

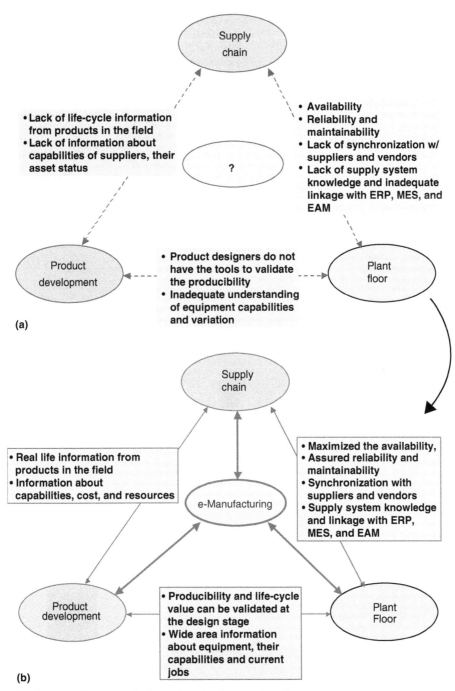

FIGURE 2.2 The transformation of e-Manufacturing for unmet needs.

The intrinsic value of an e-Manufacturing system is to enable real-time decision making among product designers, process capabilities, and suppliers as illustrated in Figure 2.4. It provides tools to access life-cycle information of a product or equipment for continuous design improvement. Traditionally, product design or changes take weeks or months to be validated with suppliers. With the e-Manufacturing system platform, designers can validate product attributes within hours using the actual process characteristics and machine capabilities. It also provides efficient configurable information exchanges and synchronization with various e-business systems.

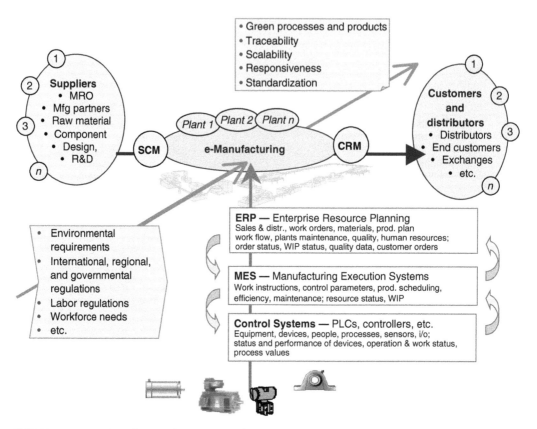

FIGURE 2.3 Integration of e-Manufacturing into e-business systems to meet the increasing demands through tightly coupled SCM, ERP, and CRM systems as well as environmental and labor regulations and awareness.

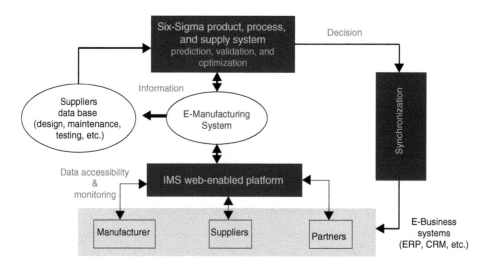

FIGURE 2.4 Using e-Manufacturing for product design validation.

2.3   e-Manufacturing: Architecture

Currently, manufacturing execution systems (MES) enable the data flow among design, process, and manufacturing systems. The ERP systems serve as an engine for driving the operations and the supply chain systems. However, the existing structure of the ERP and MES cannot informate (i.e., communicate the information in real-time) the decision across the supply chain systems. The major functions and objectives of e-Manufacturing are to:

1. enable an only handle information once (OHIO) environment;
2. predict and optimize total asset utilization in the plant floor;
3. synchronize asset information with supply chain network; and
4. automate business and customer service processes.

The proposed e-manufacturing architecture in this position paper addresses the above needs.

To address these needs, an e-Manufacturing system should offer comprehensive solutions by addressing the following requirements:

1. development of intelligent agents for continuous, real time, remote, and distributed monitoring of devices, machinery, and systems to predict machine's performance status (health condition) and to enable capabilities of producing quality parts;
2. development of infotronics platform that is scalable and reconfigurable for data transformation, prognostics, performance optimization, and synchronization; and
3. development of virtual design platform for collaborative design and manufacturing among suppliers, design, and process engineers as well as customers for fast validation and decision making.

Figure 2.5 illustrates the proposed e-Manufacturing architecture and its elements [5–7].

Data gathering and transformation: This has already been done at various levels. However, massive raw data are not useful unless it is reduced and transformed into useful information format (i.e., XML) for responsive actions. Hence, data reconfiguration and mining tools for data reduction, representation for plant floor data need to be developed. An infotronics platform, namely, Device-to-Business (D2BÔ) has been developed by the Intelligent Maintenance Systems (IMS) Center. To make pervasive impacts to different industrial applications, existing industrial standards should be used (i.e., IEEE 802.xx standard committees, MIMOSA, etc.)

Prediction and optimization: Advanced prediction methods and tools need to be developed in order to measure degradation, performance loss, or implications of failure, etc. For prediction of degradation on components/machinery, computational and statistical tools should be developed to measure and predict the degradation using intelligent computational tools.

Synchronization: Tools and agent technologies are needed to enable autonomous business automation among factory floor, suppliers, and business systems. Embedded intelligent machine infotronics agent that links between the devices/machinery and business systems and enables products, machinery, and systems to (1) learn about their status and environment, (2) predict degradation of performance, (3) reconfigure itself to sustain functional performance, and (4) informate business decisions directly from the device itself [1–7].

Under this architecture, many web-enabled applications can be performed. For example, we can perform remote machine calibration and experts from machine tool manufacturers can assist users to analyze machine calibration data and perform prognostics for preventive maintenance. Users from different factories or locations can also share this information through these web tools. This will enable users to exchange high-quality communications since they are all sharing the same set of data formats without any language barriers.

Moreover, by knowing the degradation of machines in the production floor, the operation supervisor can estimate their impacts to the materials flow and volume and synchronize it with the ERP systems. The revised inventory needs and materials delivery can also be synchronized with other business tools such as CRM system. When cutting tools wear out on a machining center, the information can be directly

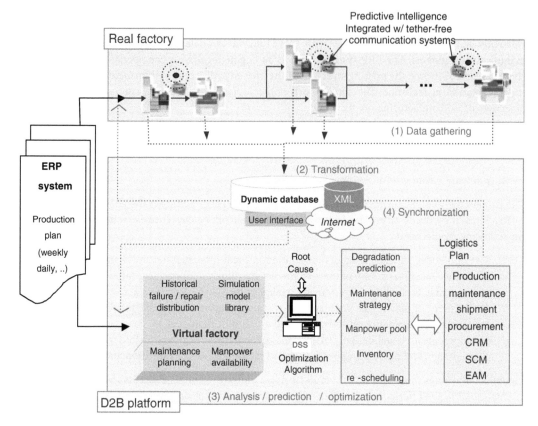

FIGURE 2.5 An e-Manufacturing architecture that comprises (1) and (2) data gathering and transformation, (3) prediction and optimization, and (4) synchronization [5].

channeled to the tool providers and update the tool needs for tool performance management. In this case, the cutting tool company is no longer selling cutting tools, but instead, selling cutting time. In addition, when the machine degrades, the system can initiate a service call through the service center for prognostics. This will change the practices from MTTR to MTBD (mean time between degradation) [10–13]. Figure 2.6 shows an integrated e-Manufacturing system with its elements.

2.4   Intelligent Maintenance Systems and e-Maintenance Architecture

Predictive maintenance of plant floor assets is a critical component of the e-Manufacturing concept. Predictive maintenance systems, also referred to as e-Maintenance in this document, provides manufacturing and operating systems with near-zero downtime performance through use and integration of (a) real-time and smart monitoring, (b) performance assessment methods, and (c) tether-free technologies. These systems can compare a product's performance through globally networked monitoring systems to shift to the degradation prediction and prognostics rather than fault detection and diagnostics. To achieve maximum performance from plant floor assets, e-Maintenance systems can be used to monitor, analyze, compare, reconfigure, and sustain the system via a web-enabled and infotronics platform. In addition, these intelligent decisions can be harnessed through web-enabled agents and connect them to e-business tools (such as customer relation management systems, ERP systems, and e-commerce systems) to achieve smart and effective service solutions. Remote and real-time assessment of machine's performance requires

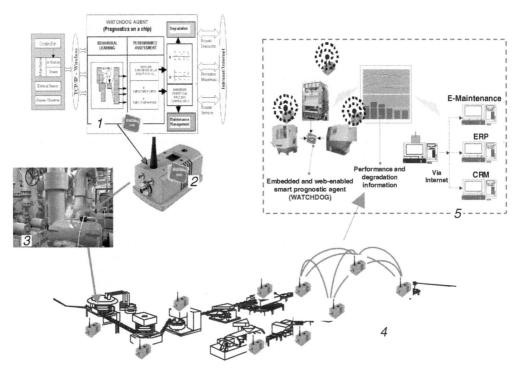

FIGURE 2.6 Various elements of an e-Manufacturing system: (1) data gathering and predictive intelligence- D2B™ platform and Watchdog Agent™, (2)–(4) tether-free communication technologies, and (5) optimization and synchronization tools for business automation.

an integration of many different technologies including sensory devices, reasoning agents, wireless communication, virtual integration, and interface platforms [14–17].

Figure 2.7 shows an intelligent maintenance system with its key elements. The core-enabling element of an intelligent maintenance system is the smart computational agent that can predict the degradation or performance loss (Watchdog Agent™), not the traditional diagnostics of failure or faults. A complete understanding and interpretation of states of degradation is necessary to accurately predict and prevent failure of a component or a machine once it has been identified as a critical element to the overall production system. The degradation is assessed through the performance assessment methods explained in the previous sections. A product's performance degradation behavior is often associated with multi-symptom-domain information cluster, which consists of degradation behavior of functional components in a chain of actions. The acquisition of specific sensory information may contain multiple behavior information such as nonlinear vibration, thermal or materials surface degradation, and misalignment. All of the information should be correlated for product behavior assessment and prognostics.

2.5 Conclusions and Future Work

This chapter introduced an e-Manufacturing architecture, and outlined its fundamental requirements and elements as well as expected impact to achieve high-velocity and high-impact manufacturing performance. Web-enabled and infotronics technologies play indispensable roles in supporting and enabling the complex practices of design and manufacturing by providing the mechanisms to facilitate and manage the integrated system discipline with the higher system levels such as SCM and ERP. e-Maintenance is a major pillar that supports the success of the integration of e-Manufacturing and e-business. Figure 2.8 shows the integration among e-Maintenance, e-Manufacturing, and e-business systems. If implemented

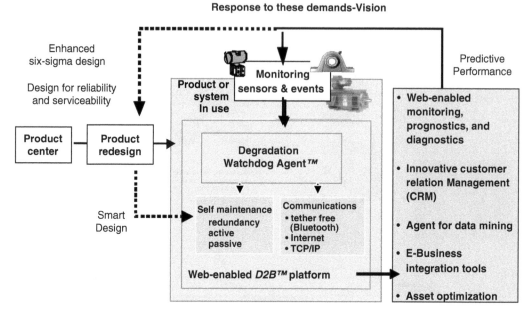

FIGURE 2.7   An intelligent e-Maintenance system.

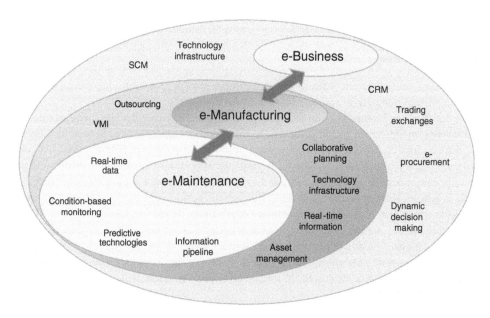

FIGURE 2.8   e-Manufacturing and its integrations with e-Maintenance and e-business.

properly, manufacturers and users will benefit from the increased equipment and process reliability with optimal asset performance and seamless integration with suppliers and customers.

In order to further advance the development and deployment of the e-Manufacturing system, research needs can be summarized as follows:

1. Predictive intelligence (algorithms, software, and agents) with a focus on degradation detection on various machinery and products.
2. Mapping of relationship between product quality variation and machine and process degradation.

3. Data mining, reduction, and data-to-information-to-knowledge conversion tools.
4. Reliable, scalable, and common informatics platform between devices and business, including implementation of wireless, Internet, and Ethernet networks in the manufacturing environment to achieve flexible and low-cost installations and commissioning.
5. Data/information security and vulnerability issues at the machine/product level.
6. Distributed and web-based computing and optimization and synchronization systems for dynamic decision making.
7. Education and training of technicians, engineers, and leaders to make them capable of pacing with the speed of information flow and understanding the overall structure.
8. Develop a new enterprise culture that resonates the spirit of e-manufacturing.

References

1. Zipkin, P., Seminar on the Limits of Mass Customization, Center for Innovative Manufacturing and Operations Management (CIMOM), Apr. 22, 2002.
2. Waurzyniak, P., Moving towards e-factory, *SME Manufacturing Magazine,* 127(5), November 2001, http: //www.sme.org/gmn/mag/2001/01nom042/01nom042.pdf.
3. Waurzyniak, P., Web tools catch on, *SME Manufacturing Magazine,* 127(4), Oct. 2001, www.sme.org.
4. Rockwell Automation e-Manufacturing Industry Road Map, http://www.rockwellautomation.com.
5. Koç, M. and J. Lee, e-Manufacturing and e-Maintenance — Applications and Benefits, *International Conference on Responsive Manufacturing (ICRM) 2002,* Gaziantep, Turkey, June 26–29, 2002.
6. Koç, M. and J. Lee, A System Framework for Next-Generation e-Maintenance System, EcoDesign 2001, *Second International Symposium on Environmentally Conscious Design and Inverse Manufacturing,* Tokyo Big Sight, Tokyo, Japan, Dec. 11–15, 2001.
7. Lee, J., Ahad Ali, and M. Koç, e-Manufacturing — Its Elements and Impact, *Proceedings of the Annual Institute of Industrial Engineering (IIE) Conference, Advances in Production Session,* Dallas, TX, May 21–23, 2001.
8. Albus, J.S., A new approach to manipulator control: the CMAC, *Journal of Dynamic Systems and Control, Transactions of ASME,* Series, G., 97, 220, 1975.
9. Lee, J., Measurement of machine performance degradation using a neural network model, *Journal of Computers in Industry,* 30, 193, 1996.
10. Wong, Y. and S. Athanasios, Learning convergence in the CMAC, *IEEE Transactions on Neural Networks,* 3, 115, 1992.
11. Lee, J. and B. Wang, *Computer-Aided Maintenance: Methodologies and Practices,* Kluwer Academic Publishing, Dordrecht, 1999.
12. Lee, J., Machine Performance Assessment Methodology and Advanced Service Technologies, *Report of Fourth Annual Symposium on Frontiers of Engineering,* National Academy Press, Washington, D.C., 1999, pp. 75–83.
13. Lee, J. and B.M. Kramer, Analysis of machine degradation using a neural networks based pattern discrimination model, *Journal of Manufacturing Systems,* 12, 379–387, 1992.
14. Maintenance is not as mundane as it sounds, *Manufacturing News,* 8(21), Nov. 30, 2001. http://www.manufacturingnews.com/news/01/1130/art1.html.
15. Society of Manufacturing Engineers (SME), Less factory downtime with 'predictive intelligence,' *Manufacturing Engineering Journal,* Feb. 2002, www.sme.org.
16. Lee, J., e-Intelligence heads quality transformation, *Quality in Manufacturing Magazine,* March/April 2001, www.manufacturingcenter.com.
17. How the machine will fix itself in tomorrow's world, *Tooling and Productions Magazine,* Nov. 2000, www.manufacturingcenter.com.

Part 3

Software and IT Technologies in Integration of Industrial Automated Systems

Section 3.1

XML in Enterprise Integration

3

Enterprise–Manufacturing Data Exchange using XML

David Emerson
Yokogawa America

3.1 Introduction

The integration of enterprise-level business systems with manufacturing systems is increasingly an important factor driving productivity increases and making businesses more responsive to supply chain demands. As a result, more and more businesses are making integration a priority and are searching for standards and tools to make integration projects easier. The World Batch Forum's (WBF) Business To Manufacturing Markup Language (B2MML) is an Extensible Markup Language (XML) vocabulary, based upon the ANSI-ISA 95 (ISA-95) standards and the international equivalent IEC/ISO 62264-1 standard.

3.2 Integration Challenges

While there are many software tools that provide varying levels of assistance for integrating systems, integration projects typically require extensive labor in order to overcome differences between terminology, data formats, interfaces, and communications options in the systems to be integrated.

A significant difference in the software tools commonly used within the enterprise and manufacturing domains exists. For example, many enterprises use middleware products that provide robust communications between systems from the same and different vendors. This type of middleware is not commonly seen in manufacturing systems, primarily due to the cost of the software and the technical expertise required by the middleware. Higher level manufacturing systems, such *as plant information management systems* that collect and aggregate data from manufacturing systems, are sometimes used with enterprise middleware systems, although often the interfaces with enterprise systems are a custom-developed or single vendor solution.

At this point in time, manufacturing systems are predominantly based upon Microsoft Windows. This is in large part a result of the constant drive to reduce manufacturing costs, the mature nature of the manufacturing system marketplace, and the reluctance to replace manufacturing systems with newer versions or operating systems. These factors often preclude the use of enterprise middleware solutions and have fostered a *de facto* industry standard for communication called OPC.

OPC is a set of communication protocols developed by the OPC Foundation for the exchange of manufacturing data using Microsoft's Distributed Common Object Model (DCOM) technology. The OPC Foundation is a nonprofit organization, primarily funded by manufacturing system vendors, that develops and maintains the OPC protocols. While limited to Microsoft platforms, the OPC Foundation is starting to develop web service implementations of their protocols to enable cross-platform connectivity. While OPC is a common tool for interoperability in the manufacturing domain, it is infrequently seen in the enterprise domain and is not a match for high-end middleware products.

The use of the World Wide Consortium's (W3C) XML is a common trait in many recent integration projects. As a mainstream technology, XML offers universal support by software vendors, a large number of tools for developing and using it, and the promise of a common language that will work with disparate systems. However, while XML provides interoperability on the protocol level, there remains an application-level integration issue of what structure the XML being exchanged should have, the elements/attributes to use, and the organization of the data.

When integration projects settle on communication formats and protocols, there is still the need to identify the data to be exchanged followed by a data mapping exercise. When cross-functional project teams assemble, there is usually a learning curve as team members from different parts of the organization learn the type of data available and needed in other parts of the organization.

3.3   Solutions

While the communication issues involving the physical and transport layers should be resolved as appropriate for each project, taking into account corporate and local requirements and infrastructure, it is assumed that the resulting architecture will utilize XML for the protocol layer. XML provides the following benefits as the protocol for integration projects:

- It is a mainstream technology supported by all major operating system and application software vendors.
- Numerous tools are available for manipulating XML making the task of data mapping/conversion simpler.
- As a mainstream technology, it has a better chance of providing a longer lived technology than proprietary and older technologies. This is an important consideration in determining the total cost of ownership of a solution.

With XML as the common protocol for an integration project, the issue of standardizing the XML vocabulary for the project becomes critical. B2MML provides a solution to this issue. Also, B2MML is based on the ISA-95 and IEC/ISO 62264-1 standards. Coupled together, B2MML and ISA-95 permit designers to define the data mapping using a standardized, common terminology and models that can be carried over to the B2MML XML vocabulary.

If custom interface development is required to integrate a computer system, there is a long-term benefit to using ISA-95 and B2MML. By interfacing individual systems to B2MML, a single format is used for all data received by a system, the number of interfaces is reduced, programmers may more easily move between interfaces, and the same terminology used for designing the data mapping is used in the interfaces. These factors will reduce software maintenance costs, make the integrated system easier to upgrade, and integrate new systems. Figure 3.1 shows a comparison of the number of interfaces required

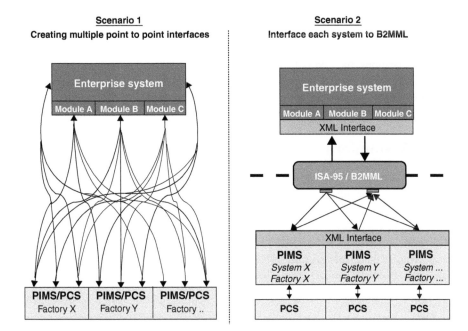

FIGURE 3.1 Point-to-point vs. common interfaces.

when point-to-point interfaces are used as in scenario 1 versus fewer interfaces required when a common format such as B2MML is used as in scenario 2.

3.4 B2MML

The B2MML is a set of XML schemas that are based on the ISA-95 Enterprise-Control System Integration Standards. The XML schemas comply with the W3C's XML schema format and define a vocabulary using the terminology and models in the ISA-95 standard. XML documents based on B2MML may be used to exchange data between the business/enterprise and manufacturing systems.

B2MML was developed by a group of volunteers working for the World Batch Forum (WBF), a nonprofit educational professional organization. While the WBF is the owner of B2MML, the licensing terms make the schemas available royalty free for any use. B2MML was created with the intention of fostering the use of the ISA-95 standards by providing XML schemas that could be used, and modified as necessary, for integration projects.

The existence of a core set of ISA-95 XML schemas is critical since without it each company, or even work group, would have to develop their own definitions of XML elements and types based on the ISA-95 standards. This would inevitably lead to numerous variations with enough structural and nomenclature differences to make the exchange of data using XML more difficult than expected. The creation of B2MML will not make XML-based data exchange easy, but it should make it easier. Even when the B2MML schemas are used to derive proprietary schemas that extend and constrain the originals if the B2MML element names and type definitions are used, there will be a common footing that can be used to establish a data mapping between applications.

B2MML has advantages over proprietary interfaces in that it is independent of any one vendor; is based on an international standard; and representatives from the manufacturing domain, both vendors and end users, have been very active in its development.

As a vendor-independent and standards-based XML vocabulary, B2MML can be used to implement ISA-95-based designs using most XML-enabled middleware and application interfaces. This provides the ability for project teams to use a vendor-independent framework during analysis and design and the ability to carry it directly to the implementation phase.

While other organizations, such as the Open Applications Group (OAG), provide standard interfaces for enterprise applications, they do not provide the level of detail and completeness required for full functioned interfaces with manufacturing systems that B2MML provides. Where OAG's OAGIS XML schemas provide interfaces primarily for within the enterprise domain, B2MML's interfaces are totally focused on the exchange of data between the enterprise and manufacturing domains.

3.5   ISA-95 Standard

In order to understand B2MML, one must have a basic understanding of the ISA-95 standards. A complete explanation of the ISA-95 standards is beyond the scope of this chapter; however, a brief overview of the standards is provided.

ISA is a nonprofit educational organization that serves instrumentation, systems, and automation professionals in manufacturing industries. ISA is an accredited standards body under agreement with the American National Standards Institute (ANSI). ISA develops standards relating to the manufacturing industry, primarily process manufacturing. The ISA-95 standards that B2MML is based upon are:

- ANSI/ISA-95.00.01-2000 — Enterprise-Control System Integration Part 1: Models and Terminology and
- ANSI/ISA-95.00.02-2001 — Enterprise-Control System Integration Part 2: Object Model Attributes.

The Part 2 standard provides attributes for the object models defined in Part 1. Since B2MML uses the models and terminology defined in Part 1 and the attributes defined in Part 2, it is said to be based upon the ISA-95 standards.

After ISA-95 was accepted as a U.S. standard by ANSI, it was submitted to the IEC and ISO for acceptance as international standards. While slight modifications were made to the standards, the international versions are substantially the same as the ISA version. IEC and ISO agree to release the international standard as a dual logo standard; therefore, it is available from either organization. The international version of Part 1 is called IEC/ISO 62264-1. At the time of writing, the Part 2 version of the international standard was progressing through the joint IEC/ISO working group; hence, it is not yet a released international standard.

ISA-95 builds upon existing work; its models are based upon "The Purdue Reference Model for CIM" developed in the 1990s by a group of chemical company representatives under the leadership of Dr. Theodore Williams at Purdue University (Purdue Model). The MESA International Functional Model as defined in "MES Functionality and MRP to MES Data Flow Possibilities — White Paper Number 2 (1994); and IEC 61512-1 batch control — Part 1: models and terminology (ANSI/ISA-88)." The value of ISA-95 is in providing a more comprehensive and detailed definition of the data exchange between the enterprise and manufacturing domains than the previous works.

The terms "enterprise" and "manufacturing domains" are defined in ISA-95 in order to put terms to the reality of the different business issues, needs, and drivers at the different levels of a business. ISA-95 uses the levels defined in the Purdue Model as shown in Figure 3.2.

Levels 0–2 represent process control and supervisory functions and are not addressed in the standard. Level 3, manufacturing operations and control, is considered the manufacturing domain and represents the highest level of manufacturing functions. Level 4, business planning and logistics, encompasses all enterprise- or business-level functions that interact with manufacturing and is referred to as the enterprise domain. The focus of the standards is on the interfaces between Levels 3 and 4.

Is it important to note that the ISA-95 enterprise and manufacturing domains refer to functions, not organizations, individuals, or computer systems. Any one organization, person, or computer system may perform functions in both domains.

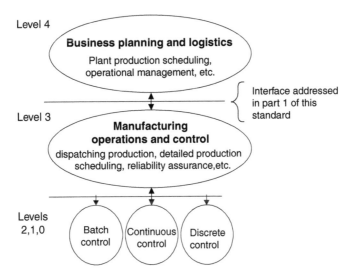

FIGURE 3.2 Levels in a manufacturing enterprise.

In practice, there is no single boundary between domains that applies to all industries, companies, divisions, and manufacturing plants. The standards draw an arbitrary line based upon commonly accepted practices. In recognition of the flexible boundaries between domains, the ISA is currently working on further parts of the standards that will define the functions and data flows inside level 3 so that when different boundaries exist, the standard may be used to identify the data flows that cross the specific Level 3–4 boundary in use.

Figure 3.3 illustrates the concept of the flexible boundary between the enterprise and manufacturing domains. When boundary #1 between the enterprise and manufacturing domains is used, functions 1 and 3 in the enterprise domain must interface with functions 4 and 5 in the manufacturing domain. Functions 2 and 6 do not interface with functions in the other domains and therefore would not be the focus of an integration project. However, when boundary #2 is used, all the functions except for 1 and 2 would be the focus of an integration project since they interface with functions in the other domain.

Drawing heavily upon the Purdue Model, ISA-95 defines data flows that may cross between the enterprise and manufacturing domains. These data flows are grouped into categories of information, which are the foundations of the ISA-95 models and the B2MML schemas. The key information categories are listed in Figure 3.4.

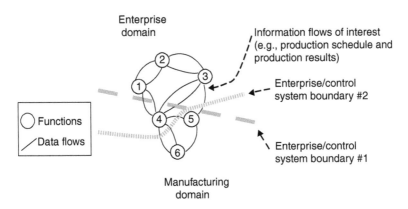

FIGURE 3.3 Flexible domain boundary.

Production capability	Information describing the manufacturing capability for a period of time. Total capability is the sum of committed, available and unattainable capabilities. This information is used to inform enterprise systems of a manufacturing area's ability to produce which is required to develop accurate plans and schedules.
Product definition	Information describing how a product is produced. When product definitions are maintained at the enterprise level this information must be sent to the manufacturing domain when product modifications are made or new products introduced.
Production information	Information instructing the manufacturing domain what to make and when in the form of a schedule and the report by the manufacturing domain up to the enterprise domain of actual production accomplishments including material usage, units of labor and equipment used for a product.

FIGURE 3.4 Key information categories between the enterprise and manufacturing domains.

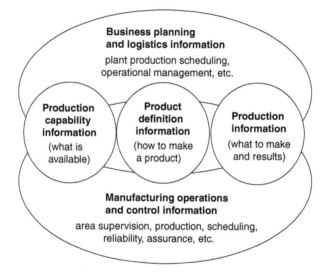

FIGURE 3.5 Categories of information.

Figure 3.5 illustrates the overlap of information in the enterprise and manufacturing domains and how the three key information categories provide a conduit for the flow of information between domains.

In addition to the three categories of information, three types of resources used by each category are identified in the standard as personnel, equipment, and material.

Each category of information may include information about some or all of the resource types and may include information about multiple instances of each resource type (Figure 3.6).

3.6   ISA-95 Models

ISA-95 defines nine object models defining the structure of the categories of information and resources. Each object model defines the data associated with the category of information or resource. The models are listed in Figure 3.7.

Communicating actual production results from the manufacturing domain to the enterprise domain is one of the most common and important goals of integration projects. The production performance model, shown in Figure 3.8, addresses this function.

This model is typical of the category of information models in that it defines a hierarchy built upon resources.

Personnel	Individuals, or classes of people, with certain qualifications may be identified as a capability, required as part of a product definition, scheduled or reported as units of labor for production performance.
Equipment	Pieces of equipment, or classes of equipment, with certain characteristics may be identified as a capability, required as part of a product definition, scheduled or reported as utilized as part of production performance.
Material	Material sublots, lots, material definitions or material classes with certain properties may be identified as a capability, required as part of a product definition, scheduled or reported as consumed or produced as part of production performance.
	Note: The standards considers energy to be a material.

FIGURE 3.6 Resources used in the three categories of information.

Production capability model

Process segment capability model

Process segment model

Product definition model

Production schedule model

Production performance model

Personnel model

Equipment model

Material model

FIGURE 3.7 List of ISA-95 object models.

The hierarchy starts with the production performance object, which is made up of one or more production responses. This permits manufacturing requests from the enterprise domain to be split into multiple elements, for example, if the request was for more than is manufactured at one time. In this case, each production response would report the results of an element of the manufacturing request, with the sum of all the production responses making up the production performance associated with the manufacturing request.

Moving down the hierarchy each production response is made up of one or more segment responses. A segment response is the "production response for a specific segment of production." The production capability and process segment models are used to define segments for each application and would map into production performance at this level. The objects that each segment response consists of are listed in Figure 3.9.

Taken together, the production performance object defines the actual performance of the process that is reported by the manufacturing domain to the enterprise domain, most likely in response to a production request.

Each of the category of information models is constructed in a similar manner. The resource models for personnel, equipment, and material are themselves similar. The material model, shown in Figure 3.10, is a good example of this.

Reading the material model from left to right shows a hierarchy of material information.

Material classes (e.g., oils) define a grouping of material definitions (e.g., peanut oil) that are used to define material lots that may be made up of material sublots. Sublots may themselves by made up of multiple sublots. Material classes, definitions, and lots are further defined by lists of properties. Sublots

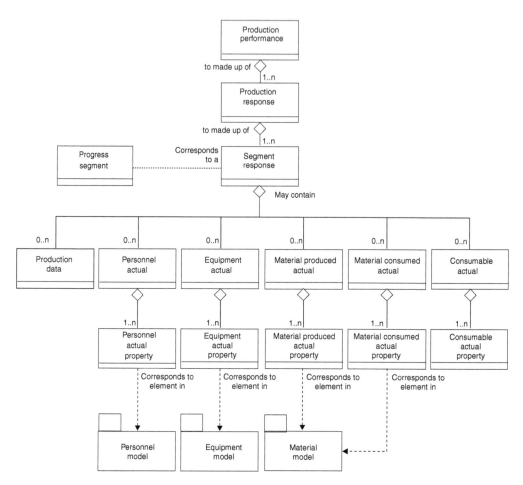

FIGURE 3.8 ISA-95 production performance model.

Production data	Data associated with the products being produced, the process segment or waste material but not directly identified as a resource.
Personnel actual	Units of labor for the personnel classes or persons related with the process segment.
Equipment actual	Equipment or classes of equipment used by the segment.
Material produced actual	Material produced by the segment. This may include one or multiple products or intermediate materials as well as byproducts and waste products. Material may be identified by sublot, lot, material definition or material class.
Material consumed actual	Material consumed by the segment. Material may be identified by sublot, lot, material definition or material class.
Consumable actual	Material not tracked by lots, not included in bills of material, or not individually tracked that have been consumed by the segment.

FIGURE 3.9 Objects that make up the segment response object.

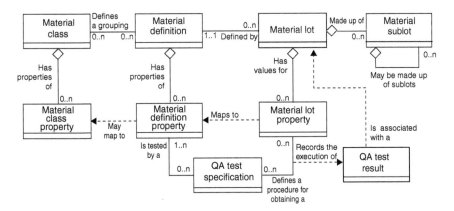

FIGURE 3.10 ISA-95 material model.

do not have properties since each sublot must have the same properties of the parent lot. The QA objects provide a means to document test specifications and results for each property.

When used with the category of information models, any of the four levels of material may be referenced as appropriate. For example, production performance typically references specific lots and sublots used in production. Production schedule may reference a material definition, or for tracking purposes, a material lot or sublot. Production capability and product definition would probably reference material classes and definitions since they deal with more abstract information.

While each of the models may be used by itself when used together, they are able to provide an integrated set of data exchanges. The interrelationships of the nine models are shown in Figure 3.11. Below each model title is a summary of the model's purpose. The horizontal dashed lines indicate how from the right side each model builds upon the model to the left. Note that the process segment capability model and the process segment model have been combined under Process Capability. In the standard, these two models were shown separately in order to make them clearer, but they both define the capabilities of the manufacturing process.

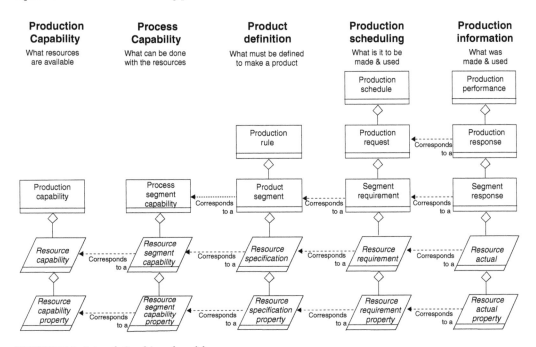

FIGURE 3.11 Interrelationships of models.

3.7   B2MML Architecture

B2MML is a collection of XML schemas organized to align with the ISA-95 standard's object models. The basis for each of the schemas, including the mapping to the standard's data models, is listed in Figure 3.12.

There is a separate schema for each model, with the exception of the equipment model, which has two schemas, one for the equipment objects, the other for the maintenance objects. This was done to provide the flexibility of using equipment and maintenance objects separately.

The separate schemas permit applications to only reference the schemas required, thereby eliminating unused elements from populating application namespaces.

The common schema, B2MML-V02-Common.xsd, does not directly relate to an ISA-95 model; rather it is used to contain type definitions, which are referenced by more than 1 schema.

The internal structure of the model-related schemas follows the ISA-95 standard's object model structures. The root element in a schema is named after each data model's root element, and each object in an object model is generally represented as an XML element.

The standard's technique of using application-specific properties is implemented in the schemas using property types, which may be used to list any number of application-specific properties in an XML document.

All elements in the schemas are declared using simple and complex types. The common schema is included by each of the other schemas that use the types as needed. Any type that is used in only one schema is defined in that schema.

In B2MML, only a few elements are declared globally, meaning they may be used in other schemas or XML documents. Generally, the objects in the ISA-95 standard that represent data to be exchanged between systems are implemented as global elements with the addition of a few container elements for the equipment, personnel, and material models. The other objects, which are generally part of the exchanged objects, are defined as local elements. The global elements are listed in Figure 3.13.

Most of the elements in the schemas are optional. This enables XML documents based upon them to only contain the elements applicable to the application, resulting in more concise XML documents.

The B2MML schemas permit most XML types to be expanded with additional elements. This is accomplished by placing an element called "Any" as the last element in a type's definition. The "Any"

B2MML schema	Schema basis
B2MML-V02-Common.xsd	All elements and types used in more than 1 other schema are defined here
B2MML- V02-Personnel.xsd	ISA-95 Personnel Model
B2MML- V02-Equipment.xsd	ISA-95 Equipment Model (except for the maintenance objects)
B2MML- V02-Maintenance.xsd	ISA-95 Equipment Model (maintenance objects only)
B2MML- V02-Material.xsd	ISA-95 Material Model
B2MML- V02-ProcessCapability.xsd	ISA-95 Production Capability Model
	ISA-95 Process Segment Capability Model
B2MML- V02-ProcessSegment.xsd	ISA-95 Process Segment Model
B2MML- V02-ProductDefinition.xsd	ISA-95 Product Definition Model
B2MML- V02-ProductionSchedule.xsd	ISA-95 Production Schedule Model
B2MML- V02-ProductionPerformance.xsd	ISA-95 Production Performance Model

FIGURE 3.12 B2MML schemas corresponding to ISA-95 models.

Equipment	PersonnelClass
EquipmentCapabilityTestSpecification	PersonnelInformation
EquipmentClass	ProcessSegment
EquipmentInformation	ProcessSegmentInformation
MaintenanceInformation	ProductInformation
MaintenanceRequest	ProductionCapability
MaintenanceResponse	ProductionPerformance
MaintenanceWorkOrder	ProductionRequest
MaterialClass	ProductionResponse
MaterialDefinition	ProductionSchedule
MaterialInformation	ProductDefinition
MaterialLot	QAMaterialTestSpecification
MaterialSubLot	QualificationTestSpecification
Person	

FIGURE 3.13 B2MML global elements.

type is defined using the "AnyType," which is based upon the XML schema wildcard component ##any. The XML schema wildcard component permits any element to be included inside the "Any" element that is at the end of the type's list of elements.

The use of this wildcard is a compromise between maintaining the ability to rigorously validate XML documents against the schema and the pragmatic recognition that diverse integration projects have unique requirements that can best be served by permitting application-specific elements to be used to extend B2MML types.

While the application-specific addition of elements can hurt interoperability, this can be limited by having XML processors expect to find either nothing or some unknown (from the B2MML viewpoint) element after the last standard B2MML element in each type. This technique will make XML processors more robust and ensure that the standard B2MML data can be processed.

3.8   Using the B2MML Schemas in XML Documents

The root element in a B2MML XML document must be a globally defined element. For example, a production performance document may use either the ProductionPerformance or ProductionResponse element while a material document may use one of the MaterialClass, MaterialDefinition, Material-Information, MaterialLot, or MaterialSubLot elements.

Individual XML documents may reference one or more of the model-based, resource, or common schemas as required. This is done by placing namespace references attributes in the root element as shown in Figure 3.14.

In Figure 3.14, the xmlns attribute declares the namespace for the document. While the string http://www.wbf.org/xml/b2mml-v02 has the form of a URL, it is merely a unique string used to identify the version of B2MML used by the document.

A simple B2MML document is shown in Figure 3.15.

```
<MaterialInformation
  xmlns               = "http://www.wbf.org/xml/b2mml-v02"
  xmlns:xsi           = "http://www.w3.org/2001/XMLSchema-instance"
  xsi:schemaLocation  = "http://www.wbf.org/xml/b2mml-v02
                        B2MML-V02-Material.xsd">
```

FIGURE 3.14 Sample XML root element with namespace references.

```
<?xml version="1.0" encoding="UTF-8"?>
<MaterialInformation
  xmlns="http://www.wbf.org/xml/b2mml-v02"
  xmlns:xsi="http://www.w3.org/2001/XMLSchema-instance"
  xsi:schemaLocation="http://www.wbf.org/xml/b2mml-v02
                      B2MML-V02-Material.xsd">

  <MaterialLot>
    <ID>M-1215</ID>
    <Description>Sample Lot</Description>
    <MaterialDefinitionID>M42</MaterialDefinitionID>
    <Status>Available</Status>
    <StorageLocation>T-942</StorageLocation>
  </MaterialLot>
</MaterialInformation>
```

FIGURE 3.15 Simple B2MML material information XML document.

This document uses MaterialInformation as the root element, and references the B2MML namespace and schema location, as well as the standard XML W3C namespace. The data contents of the file provide information about the material lot with an ID of M-1215. Note that many optional elements in MaterialLot are not in this file; this is an example of how unneeded optional elements can be omitted.

Many elements are based upon types whose content has been restricted to be an enumerated list. This means that the value of the element must be one of the values listed in the schema. For example, EquipmentElementLevel is based upon EquipmentElementLevelType, which in turn is based upon EquipmentElementLevel1Type. These two types are shown in Figure 3.16.

Whenever there is an enumerated list in B2MML, a simple type's content is restricted to the values in the enumerated list and a companion complex type is declared, which extends the simple type by adding an attribute named OtherValue. This is required in order to provide XML document authors the ability to extend the list. The enumerated list may be extended by an XML document author by giving the EquipmentElementLevel element a content of "Other" and an attribute of "OtherValue" whose content is the extended value. Figure 3.17 contains a sample B2MML document that demonstrates the use of the enumeration list extension method.

In Figure 3.17, the bold text **other** is one of the permitted enumerated values for Equipment Element Level. When this value is used, the XML processor must look for the attribute OtherValue, which in this case has the value of Work Center and then uses the attributes value as the value of the element. This technique may be used on any of the enumerated lists.

The B2MML schemas have been designed to permit most XML types to be expanded with additional elements. The element "Any" that appears as the last element in most complex types serves as a container for any other elements the XML document author wants to insert into an element. The Any element is based upon the AnyType complex type that is defined in the B2MML common schema. The Any element and AnyType ComplexType declarations are shown in Figure 3.18.

The string "##any" seen in Figure 3.18 is a W3C XML schema wildcard component that permits any other element to be added to the end of the type's list of elements. The use of this wildcard is a compromise between maintaining the ability to rigorously validate XML documents against the schema and the pragmatic recognition that diverse integration projects have unique requirements that can best be served by permitting application-specific elements to be used to extend B2MML types.

When elements are added to an existing B2MML element, they must be added within the Any element; otherwise, the XML documents will not be valid. While it is good practice for all added elements to use

```
<xsd:simpleType name = "EquipmentElementLevel1Type">
  <xsd:restriction base = "xsd:string">
    <xsd:enumeration value = "Enterprise" />
    <xsd:enumeration value = "Site" />
    <xsd:enumeration value = "Area" />
    <xsd:enumeration value = "ProcessCell" />
    <xsd:enumeration value = "Unit" />
    <xsd:enumeration value = "ProductionLine" />
    <xsd:enumeration value = "WorkCell" />
    <xsd:enumeration value = "ProductionUnit" />
    <xsd:enumeration value = "Other" />
  </xsd:restriction>
</xsd:simpleType>

<xsd:complexType name = "EquipmentElementLevelType">
  <xsd:simpleContent>
    <xsd:extension base = "EquipmentElementLevel1Type">
      <xsd:attribute name = "OtherValue" type = "xsd:string"/>
    </xsd:extension>
  </xsd:simpleContent>
</xsd:complexType>
```

FIGURE 3.16 Use of enumerated lists in type declarations.

```
<?xml version="1.0" encoding="UTF-8"?>
<MaterialInformation
  xmlns="http://www.wbf.org/xml/b2mml-v02"
  xmlns:xsi="http://www.w3.org/2001/XMLSchema-instance"
  xsi:schemaLocation="http://www.wbf.org/xml/b2mml-v02
                      B2MML-V02-Material.xsd">

  <MaterialLot>
    <ID>M-1215</ID>
    <Description>Sample Lot</Description>
    <MaterialDefinitionID>M42</MaterialDefinitionID>
    <Status>Available</Status>
    <Location>
      <EquipmentID>T-942</EquipmentID>
      <EquipmentElementLevel OtherValue="WorkCenter">
        Other
      </EquipmentElementLevel>
    </Location>
    <StorageLocation>T-942</StorageLocation>
  </MaterialLot>
</MaterialInformation>
```

FIGURE 3.17 Example of extending an enumerated list in an XML document.

```
Any element declaration:

<xsd:element name="Any" type="AnyType"
                        minOccurs="0"
                        maxOccurs="unbounded" />




Any type complex type declaration:

<xsd:complexType name="AnyType" >
    <xsd:sequence>
        <xsd:any namespace="##any" processContents="skip"
                                   minOccurs = "0"
                                   maxOccurs="unbounded"/>
    </xsd:sequence>
</xsd:complexType>
```

FIGURE 3.18 Any element and AnyType ComplexType declarations.

a prefix that identifies the XML schema, they are defined in this reference as optional and not required by B2MML.

While the application-specific addition of elements can hurt interoperability, this has been limited by having XML processors expect to find either nothing or some unknown (from the B2MML viewpoint) element after the last standard B2MML element in each type. This technique will make XML processors more robust and ensure that the standard B2MML data can be processed.

Figure 3.19 contains an example of adding elements not in B2MML to a B2MML element.

In this case, three elements not defined in B2MML are included as part of the MaterialLot element by placing them inside the Any element. The extended elements have a prefix of "ext:" which is defined in the namespace declarations at the top of the document. There the "ext:" prefix is defined to point to the ExtensionExample.xsd XML schema. The figure contains this schema, which has been used to declare the simple types used in the XML document. Of note is the fact that while XML processors will check for well-formed XML, they will not validate the content within the Any element since the AnyType has been defined with the attribute processContents= "skip."

3.9   Usage Scenario

The following scenario provides an example of using B2MML's Production Performance schema to report production results from a manufacturing system to an enterprise system. Figure 3.20 lists the manufacturing data to be reported.

Figure 3.21 contains a production performance XML document containing these results. The document has been broken into parts for clarity and for reference in the description below. If the XML in each box were concatenated, it would create one production performance document.

Header: The header information in Figure 3.21 includes an XML declaration, the start of the document's root element, ProductionPerformance, and attributes declaring XML namespaces, identification of the XML schema the document is based upon, and a suggestion as to the schema's location. To further understand the XML syntax, refer to the W3C's XML and XML schema recommendations.

Production performance and response information: The production performance and production response elements provide information to the receiving system regarding where this information fits into the overall production performance data. There may be one or many production performance XML documents per lot of product. Therefore, sufficient information must be included in the document to permit the receiving system to know where to store or send each piece of data.

XML Document with extensions:

```
<?xml version="1.0" encoding="UTF-8"?>
<MaterialInformation
    xmlns="http://www.wbf.org/xml/b2mml-v02"
    xmlns:xsi="http://www.w3.org/2001/XMLSchema-instance"
    xmlns:ext="ExtensionExample"
    xsi:schemaLocation="http://www.wbf.org/xml/b2mml-v02 B2MML-V02-Material.xsd">

    <MaterialLot>
        <ID>M-1215</ID>
        <Description>Sample Lot</Description>
        <MaterialDefinitionID>M42</MaterialDefinitionID>
        <Status>Available</Status>
        <MaterialLotProperty>
            <ID>Purity</ID>
            <Description>Measurement of purity</Description>
            <Value>
                <ValueString>99.4</ValueString>
                <DataType>float</DataType>
                <UnitOfMeasure>Percent</UnitOfMeasure>
            </Value>
        </MaterialLotProperty>
        <Location>
            <EquipmentID>T-942</EquipmentID>
            <EquipmentElementLevel>Unit</EquipmentElementLevel>
        </Location>
        <StorageLocation>T-942</StorageLocation>
        <Quantity>
            <QuantityString>200</QuantityString>
            <DataType>float</DataType>
            <UnitOfMeasure>Kg</UnitOfMeasure>
        </Quantity>
        <Any>
            <ext:ExtendedElement1>sample content</ext:ExtendedElement1>
            <ext:ExtendedElement1>sample content</ext:ExtendedElement1>
            <ext:ExtendedElement2>472.5</ext:ExtendedElement2>
        </Any>
    </MaterialLot>
</MaterialInformation>
```

Custom schema (extensionexample.xsd) containing definitions of extended elements

```
<?xml version="1.0"?>
<xsd:schema targetNamespace="http://Extensions"
            xmlns="http://Extensions"
            xmlns:xsd=http://www.w3.org/2001/XMLSchema
            elementFormDefault="qualified"
            attributeFormDefault="unqualified">

    <xsd:simpleType name="ExtendedElement1">
        <xsd:restriction base="xsd:string"/>
    </xsd:simpleType>

    <xsd:simpleType name="ExtendedElement2">
        <xsd:restriction base="xsd:float"/>
    </xsd:simpleType>

</xsd:schema>
```

FIGURE 3.19 B2MML document with extended elements.

Production data:

Date	Time	Event	Temperature 1 (Deg C)	Temperature 2 (Deg C)
2003-08-05	14:34:03	Start Charging Milk	25.0	26.4
2003-08-05	14:39:29	End Charging Milk	19.6	21.3

Material used

Date	Time	Material	Target (Kg)	Actual Quantity (Kg)
2003-08-05	14:34:03	Milk	400	402.4
2003-08-05	14:59:43	Flour	750	750.3

FIGURE 3.20 Production data to be reported to an enterprise system.

In this case, the overall production performance ID of MT593 is a batch ID and the production response, MT593-1, is a subdivision of the batch operating on one unit.

Segment response information: The segment response information identifies the product or process segment within the production response. In this case, the segment maps to a product segment since that element is used and the process segment element is not. The actual start and end times provide potentially important information that can be used by the enterprise system for costing or utilization purposes.

Production data: This section contains four production data elements from the Production Data table above. Each measurement has been placed in its own ProductionData element with a unique ID and containing its value, data type, and units of measure.

Material consumed — milk: The material consumed — milk section is used to transmit the amount of milk actually added to the process, the target (i.e., amount of milk that was supposed to be added), and the time the milk was added.

The MaterialConsumedActual element contains identifying information about the material, the location the material was added from, the amount added, and properties of the material consumed. The properties have been used to convey the time the milk was consumed and the target amount. In any integration project, the sending and receiving systems must be programmed to use the same property IDs as part of the data-mapping exercise.

This is an example of how an element's properties can be used to provide extended information without using the Any element. This type of extension should be easier for receiving systems since properties will be expected.

Material consumed — flour: This section is similar to the milk material consumed section, except it refers to the addition of flour. This is an example of how each material consumed may be documented.

End of elements: These three lines indicate the end of each of the elements opened in the earlier sections. </ProductionPerformance> indicates the end of the XML document.

Many optional elements have been omitted from this example, as will often be the case in actual implementations. When empty elements are shown above, it is because they are required by the B2MML schemas.

While elements such as MaterialProducedActual, PersonnelActual, and EquipmentActual have not been shown, their usage closely follows the above example.

3.10 Schema Customization

While the ISA-95 standards provide a firm basis for many integration projects, they cannot satisfy every requirement. If the addition of elements using the "Any" type is insufficient, the schemas may be used

Header

```
<?xml version="1.0" encoding="UTF-8"?>
<ProductionPerformance
    xmlns="http://www.wbf.org/xml/b2mml-v02"
    xmlns:xsi="http://www.w3.org/2001/XMLSchema-instance"
    xsi:schemaLocation="http://www.wbf.org/xml/b2mml-v02 B2MML-V02-ProductionPerformance.xsd">
```

Production performance and response information

```
<ID>B-1</ID>
<PublishedDate>2003-08-05T15:12:34-05:00</PublishedDate>
<ProductionScheduleID>MT593</ProductionScheduleID>

<ProductionResponse>
    <ID>UR1</ID>
    <ProductionRequestID>MT593-1</ProductionRequestID>
```

Segment response information

```
<SegmentResponse>
    <ID>SR1</ID>
    <ProductSegmentID>UR1-Charge Milk</ProductSegmentID>
    <ActualStartTime>2003-08-05T14:34:03-05:00</ActualStartTime>
    <ActualEndTime>2003-08-05T14:39:29-05:00</ActualEndTime>
```

Production data

```
<ProductionData>
    <ID>Charge Start Temp 1</ID>
    <Value>
        <ValueString>25.0</ValueString>
        <DataType>float</DataType>
        <UnitOfMeasure>Deg C</UnitOfMeasure>
    </Value>
</ProductionData>

<ProductionData>
    <ID>Charge Start Temp 2</ID>
    <Value>
        <ValueString>26.4</ValueString>
        <DataType>float</DataType>
        <UnitOfMeasure>Deg C</UnitOfMeasure>
    </Value>
</ProductionData>

<ProductionData>
    <ID>Charge End Temp 1</ID>
    <Value>
        <ValueString>19.6</ValueString>
        <DataType>float</DataType>
        <UnitOfMeasure>Deg C</UnitOfMeasure>
    </Value>
</ProductionData>

<ProductionData>
    <ID>Charge End Temp 2</ID>
    <Value>
        <ValueString>21.3</ValueString>
        <DataType>float</DataType>
        <UnitOfMeasure>Deg C</UnitOfMeasure>
    </Value>
</ProductionData>
```

Material consumed - milk

```
<MaterialConsumedActual>
    <MaterialClassID>Milk</MaterialClassID>
    <MaterialDefinitionID>Milk Low-Fat</MaterialDefinitionID>
    <MaterialLotID>MLF-3948</MaterialLotID>
    <Location>
        <EquipmentID>T-19</EquipmentID>
        <EquipmentElementLevel OtherValue="EquipmentModule">
            Other
        </EquipmentElementLevel>
    </Location>
```

FIGURE 3.21 B2MML production performance document.

```
                    <Quantity>
                        <QuantityString>402.4</QuantityString>
                        <DataType>float</DataType>
                        <UnitOfMeasure>Kg</UnitOfMeasure>
                    </Quantity>
                    <MaterialConsumedActualProperty>
                        <ID>Time Consumed</ID>
                        <Value>
                            <ValueString>2003-08-05T14:39:29-05:00</ValueString>
                            <DataType>time</DataType>
                            <UnitOfMeasure></UnitOfMeasure>
                        </Value>
                    </MaterialConsumedActualProperty>

                    <MaterialConsumedActualProperty>
                        <ID>Target</ID>
                        <Value>
                            <ValueString>400</ValueString>
                            <DataType>float</DataType>
                            <UnitOfMeasure>Kg</UnitOfMeasure>
                        </Value>
                    </MaterialConsumedActualProperty>
                </MaterialConsumedActual>
```

Material consumed - flour

```
                <MaterialConsumedActual>
                    <MaterialClassID>Flour</MaterialClassID>
                    <MaterialDefinitionID>Enriched Flour</MaterialDefinitionID>
                    <MaterialLotID>EF-382</MaterialLotID>
                    <Location>
                        <EquipmentID>R-43</EquipmentID>
                        <EquipmentElementLevel>Unit</EquipmentElementLevel>
                    </Location>
                    <Quantity>
                        <QuantityString>750.3</QuantityString>
                        <DataType>float</DataType>
                        <UnitOfMeasure>Kg</UnitOfMeasure>
                    </Quantity>
                    <MaterialConsumedActualProperty>
                        <ID>Time Consumed</ID>
                        <Value>
                            <ValueString>2003-08-05T14:59:43-05:00</ValueString>
                            <DataType>time</DataType>
                            <UnitOfMeasure></UnitOfMeasure>
                        </Value>
                    </MaterialConsumedActualProperty>
                    <MaterialConsumedActualProperty>
                        <ID>Target</ID>
                        <Value>
                            <ValueString>750</ValueString>
                            <DataType>float</DataType>
                            <UnitOfMeasure>Kg</UnitOfMeasure>
                        </Value>
                    </MaterialConsumedActualProperty>
                </MaterialConsumedActual>
```

End of elements

```
            </SegmentResponse>
        </ProductionResponse>
    </ProductionPerformance>
```

FIGURE 3.21 (Continued)

to derive custom corporate or application-specific schemas. While the derivation of new schemas may seem contradictory to the use of a standard, it is a pragmatic recognition that companies have requirements beyond the core functionality of the standards and B2MML.

B2MML types and elements may be referenced or included in other schemas. This may be done to build new types that are extensions or restrictions of B2MML types or to include B2MML elements inside corporate or project-specific schemas.

Since the B2MML schemas are freely distributed with no restrictions placed on their use, each user is free to change their contents or include them in other work. It is strongly recommended that if modifications are made to the B2MML types, they be done as part of another schema using a different namespace and filename. If a B2MML schema file has its contents changed without the namespace and filename being changed, there is an increased risk of errors in the future from incompatible versions of the same file being mixed up.

3.11 Conclusion

B2MML, the Business To Manufacturing Markup Language, is an XML-based implementation of the ISA-95 standard. This industry markup language will enable the use of mainstream information technology with a standards-based approach to integrating enterprise and manufacturing systems.

References

ISA, www.isa.org

OPC Foundation, www.opcfoundation.org

Open Applications Group, www.oag.org

Using XML with S88.02, by David Emerson, presented at the *World Batch Forum 2000 European Conference*, Brussels, Belgium, Oct., 2000.

World Batch Forum, www.wbf.org

XML Schema Part 0: Primer; W3C Recommendation, 2 May 2001, http://www.w3.org/TR/2001/REC-xmlschema-0-20010502/

XML Schema Part 1: Structures; W3C Recommendation 2 May 2001, http://www.w3.org/TR/2001/REC-xmlschema-1-20010502/

XML Schema Part 2: Datatypes, W3C Recommendation 02 May 2001, http://www.w3.org/TR/2001/REC-xmlschema-2-20010502/

Section 3.2

Web Services in Enterprise Integration

4

Web Services for Integrated Automation Systems — Challenges, Solutions, and Future

Zaijun Hu
ABB Corporate Research Center

Eckhard Kruse
ABB Corporate Research Center

4.1 Introduction

Integrated automation systems are gaining more and more momentum in the automation industry. They address not only the vertical integration that covers layers from devices via manufacturing execution systems to business applications but also the horizontal integration ranging from design, engineering, and operation to maintenance and support. The emerging Web Services technology with growing acceptance in industry is a good way to create an open, flexible, and platform-neutral integrated system. In this chapter, we attempt to analyze and describe the main challenges in using Web Services for integrated automation systems. We believe that performance, client compatibility, client addressability, object designation, and ability to deal with multiple structures are important and essential issues for deploying Web Services in automation systems. We present some solution concepts including architecture, mechanisms and methods such as the structure cursor, Web Services bundling, the event service, the object designator, and so forth. Finally, we discuss the future of using Web Services, where Ontology will play an important role for efficient system engineering and assembling.

4.2   Background

Integration is a strong trend in the current development of the automation technology. Integrated control systems, integrated factories, or integrated manufacturing are some examples. A large integrated automation system covers not only the whole production life cycle including purchase, design, engineering, operation, and maintenance but it also involves the different control levels ranging from the field device layer to Enterprise Resource Planning (ERP) layer [1–3]. The creation of such systems thus poses the challenge of addressing various requirements from different areas at the same time, of assembling heterogeneous applications, integrating data models, and binding the applications to the data models. Typically, the diverse applications developed for handling issues of different business areas such as purchase, design, and engineering are distributed via network and unstructured. It is difficult for an engineer to find a suitable application for his specific purpose. There is no common and structured way of organizing or describing the applications in the automation area. Another challenge for large integrated automation system is the heterogeneity of platforms on which applications are developed. On the one hand, Microsoft's COM technology is widely used to create applications for traditional automation systems such as human–machine interface (HMI) or Supervisory Control and Data Acquisition (SCADA). OLE for Process Control (OPC), originally based on the COM technology, provides a standard specification for data access. It greatly facilitates interoperability for access to control instruments and devices. On the other hand, many applications in other areas such as ERP or Supply Chain Management (SCM) are based on CORBA or EJB. Interoperability between heterogeneous platforms is always a headache for integration — a uniform base would extremely reduce development costs. Appropriate data models are another challenge when building large integrated automation system. A unified description method, easy transformation and mapping, and efficient navigation mechanisms are natural requirements on data modeling. Last but not the least, the binding of applications to data models of a large automation system is crucial. The engineering costs for finding appropriate applications for specific data is quite high. An efficient way to reduce costs will greatly influence the development direction of automation technology. Web Services will play an ever-more important role in addressing the challenges in integrated automation systems due to their open, flexible, standard, and service-oriented architecture.

4.3   ABB Industrial IT Platform

To address the integration challenges in automation systems, ABB has created an integration platform for integrated automation systems called Aspect Integrator Platform (AIP), which follows the paradigm of decoupling the data model from its computational model. It is subjected to IEC 61346 [13]. The basic elements in the model are Aspect Object, Aspect, and Structure. An *Aspect Object* in AIP is a container that holds different parts of an object in an automation system. Such an object might be, for example, a reactor, a pump, or a node (computer). The Aspect Object covers data modeling, including data type, relationship among data, and its structure. The Aspect represents operations that are associated with an object. It can contain its own data. Examples of aspects are signal flow diagram, CAD drawing, analysis program, simulation, trend display, and so on. The Aspect focuses on the operational aspect. Figure 4.1 shows an AIP example.

To create an automation system based on the AIP platform, a data model should usually be built at first. Then, the engineer chooses suitable applications in the form of aspects and binds them to the data model. For an integrated automation system, which covers the whole production life cycle and all control levels, a large number of Aspects and Aspect Objects result. Thus, binding suitable Aspects to a certain Aspect Object requires significant engineering effort. Web Services could help to simplify this process. By creating an additional layer to cover Aspects of AIP, they could be searched and accessed in a unified way, using the standard Web Service discovery and description mechanisms. Besides, it should be noted that an Aspect itself can also be a Web Service.

A *Structure* — another element defined in AIP architecture and conforming to IEC 61346 — represents the semantic relationship of a data model. In IEC 61346 it is separated from the objects and expressed

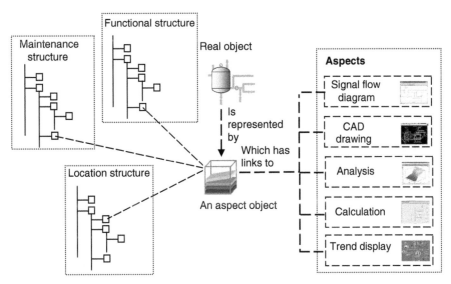

FIGURE 4.1   An example showing AIP architecture concept.

through an additional aspect such that an object can be organized in different structures at the same time. IEC 61346 presents three examples of information structures that are important for design, engineering, operation, and maintenance: function-oriented, location-oriented, and product-oriented structures. A structure is determined through a defined hierarchy, which describes the semantic relationships between Aspect Objects from a certain point of view. For example, the function-oriented structure organizes objects based on their purpose or function in the system, while the location-oriented structure results from the spatial -constitution relationship, for example, ground area, building, floor, room, and so on. IEC 61346 provides the structure concept to address the semantics of a data model, but it does not define mechanisms to describe the semantics in different structures. Figure 4.2 shows three structures regarding function, location, and maintenance. The maintenance structure presented in the figure is useful for a maintenance engineer.

4.4   Web Services

4.4.1   Definition

Web Services can be defined in different ways [5]. From a business point of view, Web Services present a common service-oriented architecture for companies and organizations to provide their key businesses in the form of services. From an application point of view, Web Services create a platform-independent and programming language-neutral middleware for interoperable interaction among applications. In this chapter, we concentrate on the technical aspect of Web Services and use the definition from the W3C [4]. "A Web service is a software system identified by a URI [RFC 2396], whose public interfaces and bindings are defined and described using XML. Its definition can be discovered by other software systems. These systems may then interact with the Web Service in a manner prescribed by its definition, using XML based messages conveyed by Internet protocols." Web Services have the following key features:

They can be described according to their nonoperational service information and operational information. The nonoperational information includes service category, service description, and expiration date, as well as business information about the service provider (e.g., company name, address, and contact information). The typical description language used for the nonoperational information is Universal Description, Discovery, and Integration (UDDI). The operational information describes the behaviors of Web Services. It covers dynamical aspects such as service interface, implementation binding, interaction protocol, and the invoking endpoint (URL). Web Service Description Language (WSDL) is usually used to describe the operational information.

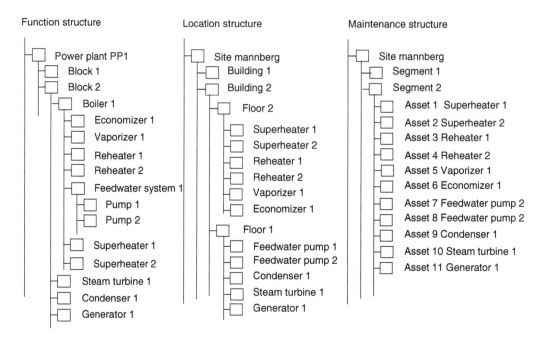

FIGURE 4.2 Example for functional, location, and maintenance structure.

Web Services have repositories for storing the nonoperational and operational information of Web Services. By means of the repositories, Web Services can be published, located, or discovered anywhere and anytime. They can also be invoked over a network such as the World Wide Web. SOAP is used to describe messages for Web Services. HTTP, TCP/IP, etc., can be used as communication protocols.

Web Services are standard-based, platform, and programming language-independent. They use standards for the description of services.

In comparison with the traditional middleware and component-based technologies, the differentiating features of Web Services are description and discovery mechanisms based on standards that enable the platform- and programming language-neutrality. Web Services provide a way for integrating applications developed on different platforms; it is thus a natural choice to use them within integrated automation systems.

4.4.2 Architecture

4.4.2.1 Basic Components

The Web Service architecture consists of a set of building blocks, which represent different roles. The key components are service provider, service requester, and service broker. Their relationship is illustrated in Figure 4.3.

The *service provider* deploys and publishes services by registering them with the server broker. It provides an environment for running Web Services so that consumers can use them. The *service broker* has a repository to register and manage the service description including nonoperational and operational information. It can also provide some mechanism for efficiently organizing and structuring Web Services. The *service requester* finds required services using the service broker, binds them to the service provider, and then uses them.

4.4.2.2 Technology Stacks

The Web Service concept comprises different aspects such as description, discovery, composition, management, interaction, and communication. These are addressed through different layered and interrelated technologies. Figure 4.4 gives an overview of their relationship.

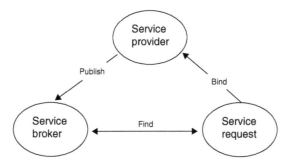

FIGURE 4.3 Basic components of Web Services.

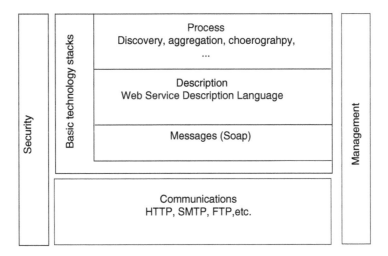

FIGURE 4.4 Technology stacks.

Figure 4.4 shows how Web Services include the basic technology stacks and communications. The security and management of Web Services are also important for the development of a Web Service system. The process stack in the basic technology stacks is responsible for the discovery, aggregation, and choreography of Web Services, while the description stack defines how to describe a Web Service. The messages stack is related to the method of exchanging information between Web Services.

4.4.2.3   Web Service Style

There are two Web Service styles: remote-procedure-call (RPC)-style and message-style.

RPC-style: A remote procedure call (RPC)-style Web Service is like a remote object for a client application. When the client application invokes a Web Service, it sends parameter values to the Web Service, which executes the required methods and then sends back the return values. Because of this back and forth conversation between the client and the Web Service, RPC-style Web Services are tightly coupled and resemble traditional distributed object paradigms, such as RMI or DCOM. RPC-style Web Services are synchronous, meaning that when a client sends a request, it waits for a response before doing anything else.

Message-style: Message-style Web Services are loosely coupled and document-driven rather than being associated with a service-specific interface. When a client invokes a message-style Web Service, the client typically sends an entire document, for example, a purchase order, rather than a discrete set of parameters. The Web Service accepts the entire document, processes it, and may or may not return a result message. Because there is no tightly coupled request–response between the client and the Web Service, message-style Web Services provide a looser coupling between the client and the server. Message-style Web Services are usually asynchronous, meaning that a client that invokes a Web Service does not wait for a response before it does something else. The response from the Web Service, if any, can appear hours or days later, making interactions more efficient. Asynchronous Web Services may be a requirement for enterprise-class Web Services.

4.5 Challenges of Using Web Services for Integrated Automation Systems

Web Services provide many attractive features for integrated automation systems, but there are also several challenges.

4.5.1 Multiple Structures

For automation systems, it is crucial to have close integration of the information, which is carefully selected for specific purposes. Multiple structures of industrial information systems characterize the current trend in information modeling in the automation industry. They vertically cover different business layers from the process, sensor/actuator, field bus, HMI to manufacturing execution system (MES), and ERP. Horizontally, they address different life-cycle phases ranging from ordering, design, engineering, to operation, optimization, and maintenance.

The information required for each business layer or life-cycle phase is different. It has to be structured accordingly, taking into account the specific properties of the layers and phases, and it has to be provided in a consistent way for the integrated solution. In this context, a multiple structural representation is inevitable. Even for the same business layer or the same life-cycle phase, multi-structure or multiple views of information are sometimes desired to provide an insight into the system from different points of view. The challenges in handling multiple structures can be characterized as follows:

- Information structures and hierarchies are closely related and interactive, that is, they are subject to the global goal of the information system.
- Connection points are clearly defined. The connection points determine how the multiple structures are associated and interrelated. For example, for a plant-centric automation architecture, the connection points are plant objects such as valves, pumps, etc.
- It is possible to navigate between structures.
- Each structure has a clearly defined semantic and serves one purpose. For example, for the engineering process, the product information structure is used to organize the product information.
- Designation is required to uniquely identify an information entity or an object of an information system.
- A multiple hierarchies-based information model should enable the integration of the computation model that provides the operation model for information manipulation and utilization. Additionally, it supports the integration of external computation applications that can use or process the information.

Web Services cover the dynamic aspect of an application, namely functions in the form of interfaces described by WSDL. They promote the separation of operations from the data model. For data presented in the form of multiple structures, it is necessary to provide the corresponding services for navigation, identification of data entities, and moving from one structure to another.

4.5.2 Client Compatibility

One benefit of Web Services is that they allow to integrate heterogeneous applications and to "webalize" legacy applications. A legacy automation system usually has client–server and peer-to-peer architecture using a defined protocol for the communication such as a socket or COM/DCOM. Web Services do not destroy this client–server architecture style. They only change the way of message and interface description and make it independent of the platform and the programming language. One possible scenario is the migration of a legacy automation system to Web service-based architecture. Web Services are usually used to provide mediator-like interfaces to the clients of the legacy automation system. One requirement could be that the clients who are using the legacy systems should not be required to do any adaptation, at least in the earlier phase of the migration. For example, process graphics displaying process data from the process server machines should not be changed if the data server of an automation system is just wrapped in a web service. This client compatibility guarantees low development cost and incremental evolution of an automation system. Interfaces between clients and servers do not have to be changed, including data types, data models, and invocation methods. For example, if a COM-based application provides an automation model to its clients, the client compatibility requires that the clients can use Web Services in the same way as if nothing has been changed.

4.5.3   Performance

An automation system is a real-time system with a large amount of process data, which changes over time. Thus, data transfer capacity and speed are two essential quality attributes.

4.5.3.1   Selective Data Access and Presentation

Web Services usually use Simple Object Access Protocol (SOAP) to describe messages exchanged between service requestors and service providers. Different communication protocols (Figure 4.4) can be used. When Web Services are invoked via the Internet or Intranet, the time for communication may be considerably longer than the time for data access, processing, and presentation. For monitoring and controlling an automation system, for example, a SCADA system, data access and presentation are typical functions. Here, it is not necessary to constantly obtain all data from the data server that is connected to instrumentation and devices. Efficient data access and presentation are required. However, in the Web Service environment due to SOAP, much data overhead is introduced. Time-costly roundtrips may occur frequently if no optimization is applied.

Additionally, Web Services communicate with the external world by sending XML messages, which have the advantage of being a platform-independent textual representation of information. Consequently, for the communication between the service provider and the service requestor, it is necessary to package the message, to transfer it to the service provider, and to unpack or parse it. Again, this might take a considerable amount of time and is opposed to the high-performance requirements of a real-time automation system.

An intelligent caching mechanism can help to tackle these problems by creating efficient data access and presentation, and by reducing the data overhead and round-trip time.

4.5.3.2   "Chatty" Interfaces

Multiple sequential calls between an interface and a business logic layer are acceptable in a stand-alone application for an automation system, but they cause a large performance loss when it comes to Web Services. In an automation system such as SCADA or an HMI system, the amount of process data exchanged between clients and servers is large. Transferring these data via Internet or Intranet through sequential calls will lead to large performance loss. This is a challenge for Web Services applications. Avoiding repetitive data transmission and reducing the number of interactions between the service requestor and the service provider are important issues to be solved.

4.5.4   Object Designation

Web Services are usually stateless, meaning that after the invocation of a Web Service, all state-related data created during the Web Service call are deleted and thus not available anymore. One way to address this problem is to use session management. Each Web Service requestor is allocated with a session on the server side that manages all client-specific state-related data such as intermediate variables, global variables for the client, and so on. But creation of a sessions on the server side for a client is always a burden for the server, impairs the scalability of the system, and thus should be avoided whenever possible. An integrated automation system usually has a structured data repository containing asset- and process-related data, which are organized in different structures for satisfying a variety of requirements. The ABB Industrial IT platform is such an example. The repository is a kind of data pool that is connected to processes over OPC or other communication channels. Careful design of methods for designating a data entity on the server is essential for efficient data browsing, navigation, and access. Uniqueness and multiple-structure characteristics of data should be taken into account. Well-designed designation methods are a condition for the use of stateless Web Services.

4.5.5   Client Addressability

In a client–server application, the client usually initiates the communication. It sends requests to the server, and the server responds and gives the requested data back to the client. In an automation system, sometimes it is required that the server triggers the interaction between the client and the server. An alarm and event server is such an example, which informs a client that a process parameter such as pressure or temperature has exceeded an upper or lower limit value. For this purpose, it supplies condition-related events.

There are also simple and tracking-related events. For example, a message about the failure of a unit can be represented by a simple event. The information about intervention in a process (corrective action on site) can be represented by a tracking-related event.

Events are organized in the event space. There are a variety of methods by which the client can influence the behavior of the server. Condition-related events, for example, can be enabled, disabled, and acknowledged. Web Services uses SOAP for description of messages and usually HTTP as a communication protocol. But HTTP is not good at delivering event notifications to clients or supporting long-lived message exchanges.

4.5.6   Security

Security is a very important aspect, especially in automation systems. Exposing a Web Service entails that the location and execution mechanism of the code changes, and this change requires a revision of the security policies mechanism. All data sent and received by a Web Service are formatted using SOAP on top of an XML specification. SOAP messages are easily readable; thus, it is necessary to encrypt certain data such as passwords. In this chapter, security is not the focus and thus is not discussed in detail.

4.6   Solution Concepts

We have listed some challenges for using Web Services with integrated automation systems. In this section, we are proposing concepts and solutions to address these challenges.

4.6.1   Overall Architecture of Web Services for an Automation System

Web Services are usually implemented based on the client-server architecture. They require client-side proxies and server-side implementations of the Web Service interfaces. For an integrated automation system, some special Web Services such as the structure navigation service, the service request unpacking service, the event service, and so on are needed on the service provider side. For each Web Service, there exists a service proxy on the service requestor side. Figure 4.5 illustrates the overall architecture of such a system.

Figure 4.5 also contains typical components in a traditional automation system, HMI client, communication channels over control bus (FDT, Profibus) and OPC, and devices. Usually in a client–server architecture, the connection between the HMI client and the automation system server is established through certain programming interfaces, as shown in Figure 4.5. For an automation system with Web

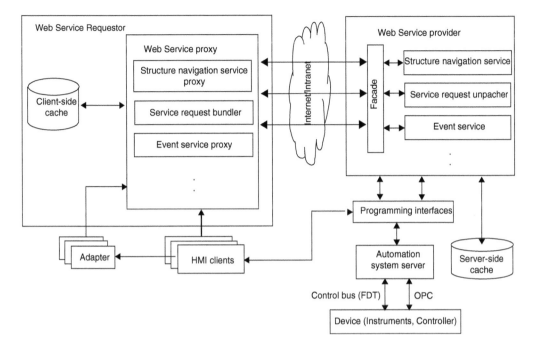

FIGURE 4.5 The overall architecture of an automation system with Web Service Implementation.

Service support, the connections between the client and the server can be realized through Web Services. The façade pattern [8] on the service provider side controls the communication and request handling to simplify the implementation. The client- and server-side caches improve the performance of Web Services. The adapter pattern on the service requestor side solves the problem of client compatibility.

The main purpose of this architecture is to migrate an integrated automation system to the Web Service platform to improve interoperability of systems. The service broker (one of the basic components of Web Services) and the process stack including service discovery (one of basic technology stacks of Web Services) are not exploited here. Both must not necessarily be deployed for interoperability improvement.

4.6.2   Structure Cursor

The structure cursor is a concept for implementing the structure navigation service. As mentioned before, multiple structures are a characteristic of an integrated automation system. The structure cursor is used to navigate in a certain structure or to move from one structure to another. It enables access to all information in multiple structures such as product information, real-time information, asset information, location information, and so on. The structure cursor has two parts: the structure navigation service on the service provider side and the structure navigation service proxy on the service requestor side. There could be different structure cursors for different purposes [9] so that the structure-specific semantics can be taken into account in navigation.

4.6.3   Client Compatibility

An adapter [8] is an effective way to address client compatibility. Figure 4.5 shows the role of adapters. Their original purpose is to "convert the interface of a class into another interface clients expect." An adapter lets classes work together that otherwise could not because of incompatible interfaces. For the migration of legacy automation systems to the Web Service platform, the implementation usually begins with the server side of automation systems, which provides the processing functionalities. To avoid the forced change on the client-side, the adapter pattern can be used. It addresses incompatibility in interface, type, access logic, and processing logic. For example, if an automation system provides a COM automation object model for accessing information in the automation system, the adapter pattern can be used to solve incompatibility problems when the COM interfaces are converted to the Web Service interfaces. The same also applies for type. The adapter pattern does not add any new functionality; it just plays a role of conversion or transformation.

Another kind of incompatibility in exposing an automation system to its clients by means of Web Services is the possibly different access logic. We could consider COM automation object model as an example again. A COM automation object model is a way to expose the functionalities of an application to its environment so that clients can exploit the functionalities or that the application can be controlled from outside. A COM automation object model usually contains a set of classes, which implement a set of COM interfaces. It is object-oriented, that is, it can be used to navigate the whole object tree to access the information of a concrete object. Identification of objects is accomplished by object names. In contrast, Web Services use URLs to identify themselves, and do not automatically provide mechanisms for identifying an object. Therefore, two kinds of potential incompatibility may occur: identification of objects and navigation in the object tree. One way to work around the potential problems is to use one-to-one mapping and introduce an object designator for each function defined in Web Services. The one-to-one mapping means that each interface implemented in the automation object model is exposed as a Web Service identified through a URL.

4.6.4   Design for Performance

To address the performance challenge when using Web Services for an automation system, it is essential to design a proper mechanism for handling roundtrips and the amount of data transferred between the service requestor and the service provider. The following recommendations should be considered when designing a Web Service.

4.6.4.1   Caching

Caching is an effective mechanism for increasing performance. Web Services performance in an integrated automation system can be maximized by carefully studying the data characteristics and correctly using data caching. There are three major choices to use caches: near the service client or consumer (client-oriented), near the service provider (provider-oriented), or at strategic points in the network [7]. In this chapter, only client- and server-oriented caches are considered (Figure 4.5).

A client-oriented cache intercepts the requests from a client, and if it finds the requested objects in the cache it returns them to the client. The content to be cached fully depends on the client's needs. Data requested by the client can also be prefetched and stored in the cache if necessary. Typical client-oriented caching is the proxy caching, transparent caching, and so forth [7].

The provider-oriented cache is located on the server side. It is content-dependent, meaning that if many clients require the same data (or the same data are required repeatedly), these data can be put into the cache for sharing. Examples for the provider-oriented cache are the reverse proxy cache and the push cache [7]. The provider-oriented cache is useful if it is very time-consuming to make data available on the provider-side for transfer to the client. An example of such a case is large simulation programs, which require intensive computation and thus a long processing time. In this case, the provider-oriented cache can avoid unnecessary redundant computations and thus reduce the waiting time.

To properly cache data, it is necessary to take the following issues into account:

- *What kind of data can be cached?* You should consider using caching in a Web Service when the service's requested information is primarily read only. An integrated automation system contains not only the real-time process data such as temperature, pressure, and flow rate but also other static data that includes information of equipment or components (name, size, location, etc.), data on the producer of the equipment, data on the features of the equipment, price information, and so on. For a client, it is not necessary to update the static data constantly. Such a kind of data can be cached on the client side. For the provider-oriented cache, it is necessary to identify which data can be shared by many clients, or which are required repeatedly. The key criterion is how long it will take to make data available for the client.

- *Data marking*: Data marking is a mechanism for identifying data entities to be cached. Therefore, object designation plays an important role. This task becomes difficult if the data are organized in multiple structures. Caching can be used for a single property of a data entity, for the whole data entity, or for a structure, such as the functional structure, location structure, or maintenance structure, which contains a group of data entities for a certain purpose.
- *Time window*: It should be possible to define a time window for caching. The time window defines a range in which data should not be obtained from the Web Services server. Only after the defined time window the system refreshes the data. The time window is similar to an aging mechanism. It is necessary to have an update or cleanup mechanism to force the refreshing of the cache. The time window should also allow to differently deal with slowly changing and quickly changing data, as typically both types coexist in automation systems. For example, the temperature of a boiler changes at a comparably slower rate than the pressure in response to the disturbance. Process data changing at a slower rate can have a relatively longer time window.
- *Data model for cache*: As mentioned above, Web Services are usually stateless. Web Services represent a set of functions that can be invoked by the service requestor. Caches deal only with data. To associate functions represented by Web Services and the data to be accessed, the object identification is needed, meaning that each function should contain object identification to specify which objects are treated in the function. Different data models for caching can be used, for example, hierarchical data structures (trees) or hash tables. Today, many libraries are available for the implementation of such data models.
- *Granularity of data*: To avoid unnecessary roundtrips in the client–server communication, it is important to find an optimal granularity of the data handled by the Web Services and transferred between the service requestor and the service provider. While fine-grained data entities lead to smaller sets of data, coarse-grained entities create relatively large data chunks. For Web Services using SOAP as a protocol, each invocation of a Web service needs to parse the XML document request and construct the XML response. Fine grained granularity may cause more roundtrips and more efforts for parsing and constructing the XML data. A tradeoff between the fine- and coarse-grained strategy will help to increase performance. For the coarse-grained strategy, the service provider may provide more information than the client needs for a particular request. However, if the client issues similar requests, caching the data may improve response time. This is especially true for clients making synchronous requests, since they must consider the time to construct the response in addition to the time to transfer the data.
- For an automation system, the proper granularity can be found based on an analysis of data, regarding which data are logically related and typically used together. These data entities can be put together and transferred as a chunk. For example, if for monitoring and controlling some process variables such as temperature and pressure shall be displayed together, they can be grouped into a data chunk. For manufacturing execution systems, all information on the work order can build in a single data set.

4.6.4.2   Bundling Web Services

Another way to improve performance and to reduce the number of roundtrips is to bundle Web Service calls. As already mentioned, each web service invocation requires dealing with XML data, including constructing the XML request and response as well as XML document parsing. This may become critical if a great number of sequential Web Service calls are involved to fulfill a task. A potential solution to improve performance is to bundle the sequential Web Service calls to create one single call. The Service Request Bundler on the client side and the Service Request Unpacker on the server side (Figure 4.5) can be used to implement this mechanism.

4.6.4.3 Serialization

Complex objects and data structures must be serialized to be transmitted, causing an overhead for serialization and deserialization and the volume of serialized data. There are two kinds of serialization:

1. *XML Serialization*: This is the default serialization model. When a Web Service returns a complex data structure, it is serialized to XML, producing a significant overhead in the size of the data being transmitted. XML, or SOAP serialization, is platform-independent. This kind of serialization can be used for static data such as asset information, plant structure, and so on.
2. *Binary serialization*: Objects are serialized into a sequence of bytes, and transmitted inside a SOAP envelope. This reduces the overhead introduced by XML serialization, but platform independence is lost. Also, it is necessary to introduce some code to manage the serialization and deserialization processes. This kind of serialization is especially suited for real-time data.

4.6.5 Object Designator

The object designator [9] identifies objects and their properties, which need to be processed by Web Services, both on the service requestor and the service provider side. As mentioned earlier, all functions in a Web Service should have an object designator as one of their parameters. There are two methods to identify an object in an information system: direct and indirect.

The *direct identification* method uses a globally unique ID (GUID) to reference an object. The prerequisite is that all information objects or entities are assigned such GUID when they are created. The information system also has to provide the structure to access objects by using GUIDs. Direct identification methods are a very easy way to identify objects, because a client can obtain an object by just supplying a GUID without any complex or complicated navigation. Another advantage is that the server running the information system can be switched to a backup system without affecting the current clients if the same GUIDs are used in both systems. A drawback, however, is the consumption of additional memory and hard disk capacity to manage the potentially large number of GUIDs.

The indirect identification method uses relationships such as aggregation, composition, etc., among objects to identify an object. The indirect identification method usually needs less memory and hard disk capacity, but the reference may be much more complicated.

The object designator can also be used to identify the position of an object in the structures. Obviously, the designator depends on the structure it is addressing, that is, it is structure-specific.

4.6.6 Client Addressability

In an automation system, client addressability regards

- how the service provider finds the suitable service requestors and
- how the service provider informs its service requestors of what has happened on the server side.

An alarm and event server is an example where client addressability is important. Two basic mechanisms are necessary to have client addressability in an automation system with Web Service support. The first mechanism allows for subscribing and unsubscribing events so that the service requestor can be notified about messages coming from the service provider. The second one regards cyclically querying (polling) the service provider to check if any events or messages have occurred. The basic components are the event service proxy on the client-side and the event service on the server-side (Figure 4.5). The event service proxy on the client-side deals with registering, polling, and managing event handlers for the service requestor. The event service on the provider-side is responsible for event queue management and functionalities like registering and managing event handlers. The Event Service Proxy and the Event Service in Figure 4.6 are the basis for the implementation of event-handling mechanisms.

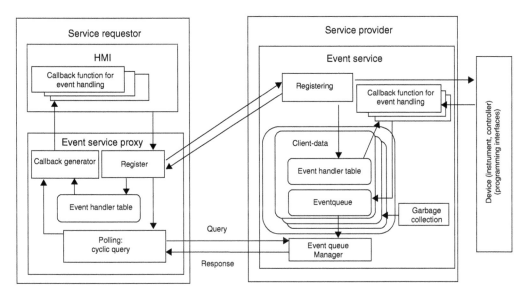

FIGURE 4.6 Event Service architecture.

4.7 Future

We have discussed various challenges and solution concepts to address interoperability — a key issue in integrated automation systems. Platform and programming language neutrality is the important feature of Web Services to improve interoperability among various applications for automation tasks such as simulation, data processing, presentation, and management. With increasing complexity and size of integrated automation systems, especially when more and more applications from business management, MES, and different phases of product lifecycle are involved, efficient system engineering and assembling may emerge as a new challenge. Data modeling, computation modeling, association of the computation elements (software applications, components, process modules) to data models [10], efficient composition of systems with existing applications and modules that are implemented in the form of Web Services are just a few examples for the new challenges.

The further development of Web Services technology — automatic Web Service discovery, automatic Web Service execution, and automatic Web Service composition and interoperation [11] — will help to address these issues.

Ontology as an explicit specification of conceptualization [12] will likely play a more important role in the development of Web Service and integration technology. The traditional software (applications, modules, components), especially component-based software, uses the Interface Description Language (IDL) to describe functionalities. It provides a way to describe the semantics of applications or components on a very low level. The description is platform- and programming language-dependent and can be only used for certain platforms such as COM or CORBA. Web Services use WSDL that is based on XML and thus independent from platform and programming language. From that point of view, it is better than IDL but still cannot address the description of semantics on a high level such as relationships among Web Services, domain knowledge, concepts, and so forth. Web Ontology [13] is a natural next step in technology development to address this problem. It uses controlled vocabularies or terms to encode classes and subclasses of concepts and relations. It can be used as an additional semantic layer that may sit on top of the data model and computation model including software applications, modules, or components, as illustrated in Figure 4.7. In this way, data model and computation model may share the same ontologies, or the ontologies used for data and computation model can be mapped or transformed in a simple way. The essential issue for successful use of ontologies is efficient ontology engineering. It

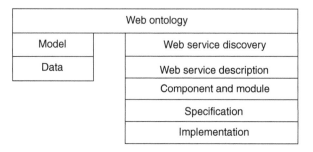

FIGURE 4.7   Data and computation model with ontology layer.

includes creation of unified and standards-based ontologies, ontology management, ontology mapping and transformation, ontology matching, and so forth. Obviously, ontology engineering, which is aimed at creation of unified and widely accepted ontologies, is not an easy work. It is a long-term process and needs cooperation from different related stakeholders. Creation of ontologies based on existing standards could be an effective way. For the integration in automation systems, different ontologies such as domain ontologies and computation ontologies are needed.

For engineering, the full use of other Web Services features — namely the service broker and service discovery, aggregation, and choreography — will facilitate searching for suitable components and aggregating applications to build an integrated automation system.

4.8   Conclusion

In this chapter, we have discussed the use of Web Services for implementing integrated automation systems. We believe that performance, client compatibility, client addressability, object designation, structure navigation, and security are very important for integrated automation systems and pose major challenges. The solution concepts based on Web Services have been presented, such as the structure cursor, the client- and server-oriented cache, Web Services bundling, the object designator, and the event service. More efficient system engineering and assembling will definitely benefit from the further development of Web Services technology such as automatic discovery, execution, composition, and interoperation.

References

1. Ragaller, K., An Inside Look at Industrial IT Commitment, ABB Technology Day, 14 November, 2001.
2. Krantz, L., Industrial IT — The next way of thinking, ABB Review, 1, pp. 4–10, 2000.
3. Bratthall, L.G., R. van der Geest, H. Hofmann, E. Jellum, Z. Korendo, R. Martinez, M. Orkisz, C. Zeidler, and J.S. Andersson, Integrating Hundreds of Products through One Architecture — The Industrial IT Architecture, ICSE 2002.
4. Web Services Architecture Requirements, http://www.w3.org/TR/wsa-reqs#id2604831
5. Thompson, M., Defining Web Services, TECH/CPS 1004, Butler Direct Limited by Addax Media Limited, December 2001.
6. Booth, D., H. Haas, F. McCabe, E. Newcomer, M. Champion, C. Ferris, and D. Orchard, Web Services Architecture, http://www.w3.org/TR/2003/WD-ws-arch-20030808/
7. Barish, G. and K. Obraczka, World Wide Web Caching: Trends and Techniques, IEEE Communications Magazine, Internet Technology Series, May 2000.
8. Gamma, E., R. Helm, R. Johnson, and J. Vlissides, Design Patterns Elements of Reusable Object-Oriented Software, Addison-Wesley Publishing Company, Reading, MA, 1995.

9. Hu, Z., A Web Service Model for the Industrial Information System with Multi-Structures, The International Association of Science and Technology for Development, Tokyo, Japan, September 25–27, 2002.
10. Hu, Z., E. Kruse, and L. Draws, Intelligent binding in the engineering of automation systems using ontology and web services, IEEE SMC Transactions Part C, 33, pp. 403–412, August 2003.
11. Mcilraith, S.A., T.C. Son, and H. Zeng, Semantic Web Services, IEEE Intelligent Systems, Vol. 16(2), 46–53, March/April 2001.
12. Gruber, T.R., A translation approach to portable ontology specifications, Knowledge Acquisition, 5, 199–220, 1993.
13. International Electrical Commission (IEC), IEC 1346-1, "Industrial Systems, Installations and Equipment and Industrial Products — Structuring Principles and Reference Designations," 1st ed., 1996.

Section 3.3

Component Technologies in Industrial Automation and Enterprise Integration

5

OPC — Openness, Productivity, and Connectivity

Frank Iwanitz
Softing AG

Jürgen Lange
Softing AG

5.1   Introduction

This chapter provides an introduction in the Openness, Productivity, and Connectivity (OPC) technology. After explaining the history of OPC, the structure of the OPC Foundation, use, cases, and advantages of OPC, the introduction of the specifications follows. The chapter closes with an outlook into the future.

5.2   Open Standards — Automation Technology in Flux

The pace of change in industrial control and automation technology is accelerating. Demands on machines and systems concerning flexible retrofit, production speed, and fail safety are increasing, as are cost pressures. Software is becoming more and more the essential factor with products, systems, and complete plants.

At the same time, changes in the field of automation brought about by the use of the PC as an automation component, by the Internet, and by the tendency to more open standards, can be clearly seen — to the benefit of the user and the manufacturer.

The PC is more and more used for visualization, data acquisition, process control, and the solution of further tasks in automation. It complements or replaces the traditional PLC and the operator terminal. The reasons for this are the continual decrease in the price of the mass-produced PC, the permanent multiplication of the computing capacity of the CPU, the availability of even more efficient and comfortable software components, and the ease of integration with Office products.

Efficiency and cost savings are achieved through the reuse of software components and the flexible compilation of such components into distributed automation solutions.

Horizontal integration of the automation solutions through communication between the distributed components also plays an important role. Immense additional savings are achieved by means of vertical integration by optimizing the process of product planning, development, manufacturing, and sales. This optimization is realized through a consistent data flow, permanent data consistency, and the availability of data on the field level, control level, and office level.

The use of standardized interfaces by several manufacturers is a prerequisite for the flexible compilation and integration of software components.

OPC is now generally accepted as one of the most popular industrial standards among users and also among developers. Most of the Human Machine Interface (HMI), Supervisory Control and Data Acquisition (SCADA), and Distributed Control System (DCS) manufacturers in the field of PC-based automation technology, as well as the manufacturers of Soft PLCs, are offering OPC client and/or OPC server interfaces with their products. The same is true for suppliers of devices and interface cards. In the last few years, OPC servers widely replaced Dynamic Data Exchange (DDE) servers and product-specific drivers in this field.

Today, OPC is the standard interface for access to Windows-based applications in automation technology. Most of the OPC specifications are based on the Distributed Component Object Model (DCOM), Microsoft's technology for the implementation of distributed systems. Besides the DCOM-based communication in future, in the context of new OPC concepts, more and more data will be exchanged via Web Services.

OPC specifications define an interface between clients and servers as well as servers and servers for different fields of application — access to real-time data, monitoring of events, access to historical data, and others. Just as any modern PC can send a print task to any printer, thanks to the integration of printer drivers, software applications can have access to devices of different manufacturers without having to deal with the distinct device specifications. OPC clients and servers can be combined and linked like building blocks using OPC technology.

At present, OPC clients and servers are mainly available on PC systems with Windows 9X/Me/NT/2000/XP and x86 processors. Due to the availability of Web Services for multiple operating systems, the use of OPC components in different environments and in embedded systems will become more important in the future.

Why is OPC so successful? The approach of the OPC Foundation has always been to avoid unnecessarily detailed discussions and political disputes and to create practical facts within a very short time.

The development of the Data eXchange specification can serve as an example. More than 30 companies joined the effort and delivered the specification 18 months after starting.

OPC has succeeded in defining a uniform standard worldwide, which has been adopted by manufacturers, system integrators, and users.

5.3 History of OPC

Since reusable software components made their entry into automation technology and replaced mono-lithic, customized software applications, the question of standardized interfaces between components has increased in significance. If such interfaces are missing, every integration is connected with cost-intensive and time-consuming programming supporting the respective interface. If a system consists of several software components, these adaptations have to be carried out several times.

Following the immense distribution of Windows operating systems and their coherent Win32-API in the PC area, different technologies were created to enable communication between software modules by means of standardized interfaces. A first milestone was DDE, which was complemented later on by the more efficient technology Object Linking and Embedding (OLE). With the introduction of the first HMI and SCADA programs based on PC technology between 1989 and 1991, DDE was used for the first time as an interface for software drivers to access the process periphery.

During the development of Windows NT, the DCOM was developed as a continuation of the OLE technology. Windows NT was rapidly accepted by industry. In particular, the highly expanding HMI, SCADA, and DCS systems were made available for NT.

With the increased distribution of their products and the growing number of communication protocols and bus systems, software manufacturers faced more and more pressure to develop and maintain hundreds of drivers. A large part of the resources of these enterprises had to be set aside for the development and maintenance of communication drivers.

In 1995, the companies Fisher-Rosemount, Intellution, Intuitive Technology, Opto22, Rockwell, and Siemens AG decided to work out a solution for this growing problem, and they formed the OPC Task Force. Members of Microsoft staff were also involved and supplied technical assistance.

The OPC Task Force assigned itself the task to work out a standard for accessing real-time data under Windows operating systems, based on Microsoft's (OLE/)DCOM technology OLE for Process Control, or OPC. The members of the OPC Task Force worked intensively, so that already in August 1996 the OPC Specification Version 1.0 [3] was available. In September 1996, during the ISA Show in Chicago, the OPC Foundation was established; it has been coordinating all specification and marketing work since then.

An important task of the OPC Foundation is to respond to the requirements of the industry and to consider adding them as functional extensions of existing or newly created OPC specifications. The strategy is to extend existing specifications, to define basic additions in new specifications, and to carry out modifications with the aim of maximum possible compatibility with existing versions. In September 1997, a first update of the OPC Specification was published in the form of version 1.0A [4]. This specification was no longer named "OPC Specification" but, more precisely, "Data Access Specification." It defined the fundamental mechanisms and functionality of reading and writing process data. This version also served as the basis for the first OPC products, which were displayed at the ISA Show 1997. Consideration of further developments in Microsoft DCOM and industry requirements led to the creation of the Data Access Specification version 2.0 [5] in October 1998.

Already, rather early, after the release of version 1.0A, it could be seen that there was a need for the specification of an interface for monitoring and processing events and alarms. A working group formed to solve this problem worked out the "Alarms and Events Specification," which was published in January 1999 as version 1.01. Version 1.10 of the "Alarms and Events Specification" has been available since October 2002 [9].

In addition to the acquisition of real-time data and the monitoring of events, the use of historical data offers another large field of application in automation. The work on the Historical Data Access Specification already began in 1997 and was completed in September 2000 [11].

Defining and implementing security policies for use with OPC components is also of great importance. A corresponding specification has been available since September 2000 and is titled "OPC Security Specification" [15].

In particular, from the field of industrial batch processing, additional requirements have been forwarded to the OPC Foundation, which have led to the OPC Batch Specification [13].

During work on version 2.0 of the Data Access Specification and the other specifications, it emerged that there are elements common to all specifications. These elements have been combined in two specifications. The document OPC Overview [1] contains explanatory aspects only, while the OPC Common Definitions and Interfaces Specification [2] contains normative definitions.

With the increasing implementation of the OPC specifications in products, and their application in multiple environments, further requirements arose. New working groups have been created by the OPC Foundation. The Data Access 3.0 working group extended the existing Data Access Specification with further functionality. The specification [7] has been available since March 2003. The OPC and XML group defined a way to read and write data by using Web Services, and enables the use of OPC components via the Internet and on operating system platforms without DCOM. The specification Version 1.0 [16] has been available since July 2003. The OPC DX working group defined a specification for server-to-server communication without using a client. It has been available since March 2003 [8].

The rapid growth in the number of OPC products, from only a few in 1997 to some thousands in 2003, shows the enormous acceptance of this technology. OPC has succeeded in developing from a concept to an industrial standard within only three years.

5.4   OPC — An Overview

OPC is the technological basis for the convenient and efficient link of automation components with control hardware and field devices. Furthermore, it provides the condition for integration of Office products and information systems on the company level, such as Enterprise Resource Planning (ERP) and Manufacturing Execution Systems (MES).

Today, OPC, on the one hand, is based on Microsoft's DCOM; on the other hand, OPC XML-DA uses the concept of Web Services. The DCOM describes an object model for the implementation of distributed applications as per the client–server paradigm. A client can use several servers at the same time and a server can provide its functionality to several clients at the same time.

At the core of DCOM is the term "interface." DCOM objects provide their services through interfaces. An interface describes a group of related methods (functions).

The most diverse OPC components of different manufacturers can work together. No additional programming for adaptation of the interfaces between the components is necessary. Complex correlation, for example, dependencies of the software component on hardware components, remains concealed behind this abstract interface. Complete components (hardware and software) can be exchanged, provided the interface described in the specifications is supported.

The OPC standards are freely accessible technical specifications that define sets of standard interfaces for different fields of application in automation technology. These interfaces allow a highly efficient data exchange between software components of different manufacturers.

Figure 5.1 shows the currently available specifications or those under work and their relations to each other. These specifications concern different fields of application and are thus largely independent of one another. However, it is possible to combine them in one application.

5.4.1   Areas of OPC Use

What is the industrial environment for using OPC products?

Today, and especially in the future, information drives more and more production. This information exists in different forms in various devices at different levels of production. And this information must be available in different forms at different places. OPC provides a way to access and deliver data at the different levels. Figure 5.2 shows an example for the use of OPC in real applications.

OPC technology is well established in a number of industries (energy, building automation, chemical engineering, etc.).

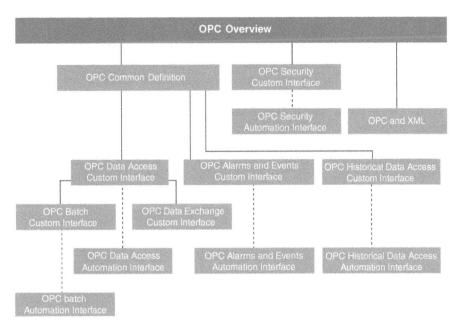

FIGURE 5.1 Available and in-progress specifications.

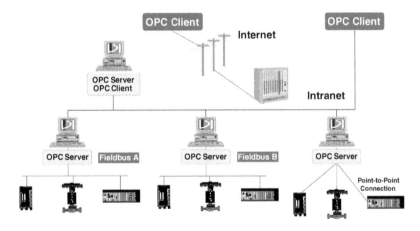

FIGURE 5.2 Examples for the use of OPC in real applications.

OPC technology can be used to immediately monitor and influence the production process. Process information is visualized, and control information is sent to the devices.

In other areas (administration, planning), there is more interest in aggregated information (machine use per hour, etc.).

There are a number of OPC specifications and products that can be used in various areas (production, administration, planning).

5.5 OPC: Advantages for Manufacturers and Users

For hardware manufacturers, for example, manufacturers of devices (PLCs, barcode readers, measurement devices, embedded devices, etc.) or PC interface boards (fieldbus interfaces, data acquisition systems, etc.), usage of OPC technology provides a number of advantages:

1. The product can be used by all OPC-compatible systems in the market and is not limited to an individual system for which a corresponding solution (i.e., specific drivers) must be developed. Due to the existence of standardized interfaces and the interoperability related to them, there is no need to become familiar with the specific requirements of other systems.
2. The time-to-market for new device generations is significantly reduced as only one OPC server has to be updated, not a large number of drivers.
3. The effort needed for support is also reduced as less products have to be supported.
4. Manufacturers of software applications for data acquisition, visualization, or control benefit like hardware manufacturers by a clear encapsulation of the software interface from the specific features of the accessed hardware. The product can be used with all devices and communication protocols on the market that make an OPC interface available. The manufacturer no longer has to develop corresponding solutions (specific drivers). Due to the existence of standardized interfaces and the interoperability related to them, there is no need to become familiar with the specifications of other devices and communication protocols.
5. The time needed for support is considerably reduced as many products to be supported (product-specific drivers up to now) no longer exist.
6. Using the OPC technology brings much benefit to system integrators. Their flexibility in the choice of products for their project is considerably increased. Consequently, the number of projects that can be processed increases considerably.
7. The time needed for integration and training is considerably reduced, as OPC provides a standardized interface that remains the same for all products.
8. Last but not the least, many advantages are the result for the end user by the usage and the huge distribution of OPC: OPC provides additional flexibility (distribution of components, use of new technologies, choice between products, etc.) during the design of the overall system as products of various manufacturers can be combined.

5.6   Structure and Tasks of the OPC Foundation

An important prerequisite for the success of a standardization initiative is an authority coordinating the interests of the members involved. The task of this authority is to protect the common objective from the political interests of individuals. Specification work has to be initiated and marked with clear mission statements in order to avoid a proliferation of variants and derivatives. Furthermore, public relations have to supply the market with information and support the common standard.

The OPC Foundation was founded in 1996 as an independent nonprofit organization with the aim of further developing and supporting the new OPC standard.

Besides specification development, other tasks are distributed to different offices and persons within the OPC Foundation.

The Board of Directors is the Foundation's decision-making body. It is elected once a year at the General Assembly by the members entitled to vote.

Once a year, the Board of Directors appoints the persons who safeguard the interests of the Foundation between the general meetings.

The Technical Steering Committee (TSC) establishes working groups for several target projects. It consists of representatives of the same companies as the Board of Directors plus the chairmen of the working groups of the OPC Foundation.

There are two kinds of OPC Foundation members: the OPC technology users and the OPC technology providers. The technology provider companies are further categorized as profit and nonprofit organizations. The latter do not have voting rights.

The annual membership fees for technology users and nonprofit organizations are independent of their size. The annual membership fees for the technology providers depend on their annual turnover.

In March 2003, the OPC Foundation had over 300 members worldwide, from North America, Europe, and the Far East.

FIGURE 5.3 Logo to be used by OPC Foundation Members.

The OPC Foundation has an Internet Web site. At www.opcfoundation.org, visitors can find information regarding the organization, its members, the working groups, and current events. Furthermore, visitors can download released specifications and technical reports in the form of "white papers." An electronic product catalog allows visitors to search for OPC subjects under several headings, such as manufacturer, client, server, development tool, and training. In addition, it is possible to exchange questions and opinions in discussion forums.

From the central Web site of the OPC Foundation, the links www.opceurope.org, www.opcjapan.org, and www.opcchina.org lead to the sites of the European, Japanese, and Chinese subcommittees.

The OPC Foundation provides a membership application form on its Web site. Members of the OPC Foundation may use the logo shown in Figure 5.3 for public relations activities.

Specification work counts among the tasks of the OPC Foundation with the greatest importance. The specification process has to be clearly defined, progress has to be monitored, and the results have to be released.

First of all, the Board of Directors defines the specification issue in the form of a mission statement and appoints the chairman of the working group. The chairman addresses the member companies of the OPC Foundation and asks them to cooperate. Interested companies then appoint members. The working group meets several times and prepares the specification. Creation of a sample code, which proves the principal implementation possibility and use of the new specification, is effected in parallel. The specification and the sample code are then passed to the Technical Steering Committee for approval. If approved, the specification and code are submitted to the Board of Directors for release. Otherwise, they are returned to the working group for further processing.

5.7   Technological Basis of OPC

As it has already been mentioned, today's (and future) OPC specifications are based on two technologies — DCOM and Web Services. That is why a short introduction into these technologies is provided, before explaining the specifications.

5.7.1   DCOM

The DCOM describes Microsoft's solution for the implementation of distributed, object-oriented applications in heterogeneous environments. A component object is the basic component of such applications. It has one or more interfaces providing methods that permit access to data and functionality of the object (reading and writing of data, accessing properties, adding and deleting of objects).

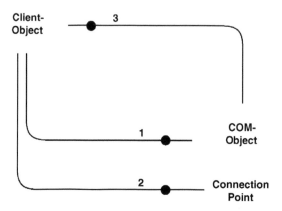

FIGURE 5.4 Creation and use of a callback connection between server and client.

One or more component objects belong to one server component providing a large number of services. To make use of these services, a client accesses methods at interfaces of the server's component objects. The individual services to be provided are described in specifications (e.g., OPC specifications). The structure of the component objects, of their interfaces, methods, and parameters is defined in an IDL file, which describes the contract between a client and server. DCOM ensures binary interoperability at runtime. A client can query at runtime whether the server supports a certain interface.

A number of identifiers are used to designate server and interfaces uniquely, for example, CLSID (ClassID) and IID (InterfaceID). The life cycle of an individual component object and of the server is managed via reference counters.

DCOM permits the implementation of interactions between client and server as well as between server and client (callback). This possibility is provided by using connection points. Figure 5.4 shows the relations. The client queries the server whether it supports a specific connection point (1). In this case, the client transmits a reference to an interface on its side to the server (2). The server can later call methods at this interface (3).

5.8   XML, SOAP, and Web Services

There are a large number of products that implement DCOM-based OPC Specifications. But there are also some restrictions that have to be considered during the development and use of this kind of products.

- DCOM does not pass firewalls, that is, direct addressing of computers through the firewall is not possible. However, this is precisely what DCOM needs to perform an internal check.
- There are some devices and applications that provide or require data and that do not run on Microsoft systems. They include, for example, applications in the ERP or MDS areas as data consumers or Embedded Devices as data sources.

These restrictions have been the reason why OPC foundation has started the OPC XML-DA specification effort. This specification is no longer based on DCOM, but on a technology independent of a specific operating system. Since this specification is explained later, the relevant components of this technology will be introduced below.

The eXtensible Markup Language (XML) is a flexible data description language, which is easy to comprehend and learn. Information is exchanged by means of readable XML documents. An XML document is called *well formed* if it corresponds to the XML syntax; it is called *valid* if, in addition, it corresponds to a default schema. The creation of XML documents and schemas as well as validation and processing of the files are supported by a variety of tools.

Today, the support of XML is guaranteed by practically all systems. Thus, even heterogeneous systems can easily interact by exchanging XML documents.

The Simple Object Access Protocol (SOAP) is an interaction protocol that links two technologies: XML and HTTP. HTTP is used as the transport protocol. The parameters of the interactions are described with XML. SOAP is thus predestined specifically for the Internet. SOAP is a protocol independent of object architectures (DCOM, CORBA). A SOAP telegram consists of a part describing the structure of the HTTP call (request/response, host, content type, and content length). This part is included in all the HTTP telegrams. A UniversalResourceIdentifier (URI) was added, which defines the end point and the method to be called. The method parameters are transferred as XML. The programmer is responsible for mapping the SOAP protocol to a concrete implementation. In the meantime, SOAP has been submitted to the World Wide Web Consortium (W3C) for standardization. In this context, the name has changed to XML protocol. Version 1.2 has been available since December 2001.

Based on the technologies introduced above, it is already possible to implement distributed applications that interact via SOAP and are independent of the operating system and the hardware. However, something is still missing: a way of describing an application's interface and of generating program components from this description that are, on the one hand, compliant with the existing infrastructure (HTTP, etc.) and that, on the other, can be integrated in existing programs. This is where Web Services come into play.

The World Wide Web is used increasingly for application-to-application communication. The programmatic interfaces made available are referred to as Web Services. SOAP is used as the interaction protocol between components.

Web Services are described using XML. The language used is Web Services Description Language (WSDL), which is standardized in the W3C. An application interacting with the Web Service will deliver a valid XML message that is compliant with the schema. The function call is sent as an XML message. This also happens with the response and a possible error information. Components that support or use Web Services can be implemented on any platform supporting XML and HTTP.

The introduced technologies were not defined by individual companies or company groups, but by the W3C.

This fact is also of importance for future OPC specifications as shown in Table 5.1. In the past, the fact that OPC is only based on DCOM has been criticized. This point should disappear if specifications are based on XML and Web Services.

Table 5.2 shows the current status of the different specifications. "Recommendation" stands for agreed standard, and "Draft" stands for a standard that is still not agreed. "Note" stands for a rather detailed working paper.

TABLE 5.1   OPC Specifications — Contents and Release Status (Status July 2003)

Specification	Contents	Release Status
OPC Overview [1]	General description of the application fields of OPC specifications	Release 1.00
OPC Common Definitions and Interfaces [2]	Definition of issues concerning a number of specifications	Release 1.00
OPC Data Access Specification [7]	Definition of an interface for reading and writing real-time data	Release 3.0
OPC Alarms and Events Specification[9]	Definition of an interface for monitoring events	Release 1.1
OPC Historical Data Access Specification [11]	Definition of an interface for access to historical data	Release 1.1
OPC Batch Specification [13]	Definition of an interface for access to data required for batch processing. This specification is based on the OPC Data Access Specification and extends it	Release 2.0
OPC Security Specification [15]	Definition of an interface for setting and utilization of security policies	Release 1.0
OPC XML-DA Specification [16]	Integration of OPC and XML for the building of Web applications	Release 1.0
OPC Data eXchange (DX) Specification [8]	Communication between server and server in process	Release 1.0
OPC Complex Data [17]	Definition of possibilities to describe the structure of Complex Data and of ways to access this type of data.	Release 1.0

TABLE 5.2   XML Specifications — Release State (July 2003)

XML 1.0	W3C Recommendation
XML Schema Part 1 and 2 1.0	W3C Recommendation
SOAP/XMLP 1.2	W3C Recommendation
WSDL 1.1	W3C Draft

5.9   OPC Specifications

5.9.1   OPC Overview [1]

As it has already been mentioned, there are several OPC specifications for different applications in automation technology. All specifications describe software interfaces. The existing specifications and their relationships are shown in Figure 5.1.

The "OPC Overview" contains general, nonnormative guidelines for OPC. It contains, for example, facts about OPC applications and OPC basic technology.

5.9.2   OPC Common Definitions and Interfaces Specification [2]

Before the specifications for data accessing are explained, some remarks related to the content of the OPC Common Definitions and Interfaces Specification will be made.

In the preceding part of the chapter, some information about the history of OPC was provided. The specification discussed here came into being when the Data Access Specification was being prepared together with other specifications. During this process, the OPC Foundation members realized that there are some definitions relevant to all specifications. They are summarized in this specification and comprise:

- *Functionality to be provided by all servers.* This includes the possibility of adapting the server to the geographical area of application (setting the language for the textual messages from the server to the client).
- *The procedure of server recognition.* Entries in the registry database contain information necessary for starting the server. The entries are shown in Figure 5.5 and explained below. A client will search the database to obtain this information, which is simple in the local registry but difficult on remote computers. A component offering this functionality was specified and made available.
- *The procedure of installation.* There are several components (proxy/stub) used together by all servers and clients of one specification. These components must be available on the computer as long as OPC products are used.

An OPC Server is characterized by the following registry entries:

- *ProgId*: Every DCOM Server is characterized by a Program Identifier (ProgId). Rules exist regarding how to generate this Identifier. But these rules do not guarantee that this identifier is unique. Below the ProgId key, a specific key OPC exists. This entry is used to differentiate between OPC Servers and other DCOM servers.
- *CLSID*: This 128-bit-long numerical identifier uniquely describes a DCOM Server, that is, also an OPC Server. There is a way of generating this number that makes it unique. The implementers will use the generated number for exactly one server. The LocalServer32 key contains a reference to the location where the server executable can be found. The Implemented Categories keys contain information regarding which specification is implemented by a server. This is used by a client or the Server Enumerator.
- *AppId*: The Application Identifier contains more information about the server. This includes security settings. The AppId can be, but must not be, the same as the ClassId.

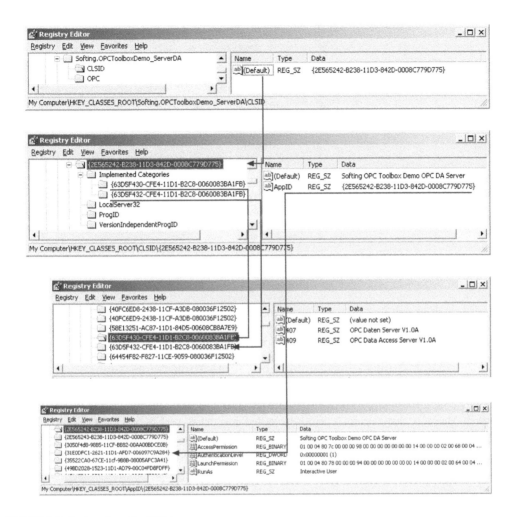

FIGURE 5.5 Registry keys for OPC Servers.

5.9.3 Data Access Specification [4, 5]

Data Access Servers permit transparent read and write access to any kind of values. These values can be made available by field devices and fetched via different communication systems. Also, servers permitting access to hardware and software (plug-in cards, other programs) in the PC are possible.

OPC specifications are supposed to support interoperability and plug'n'play. One prerequisite is that a client can obtain information on the available values in the server very conveniently. For this purpose, a namespace and functionality for browsing the namespace were defined. This functionality is implemented in the server and used by the client.

A client is not necessarily interested in all values. Different clients may want to register values under different aspects, for example, all temperature values. Therefore, the specification defines different COM objects organized in a hierarchy. By creating the specific objects, the client can adapt the server to its requirements.

Figure 5.6 shows the components that form part of a Data Access application — OPC Data Access Client and Data Access Server with namespace and object hierarchy.

A namespace can be hierarchical or flat. It can be a result of the commissioning process for the server. The specification does not define how the namespace has to be created, how many hierarchical levels the

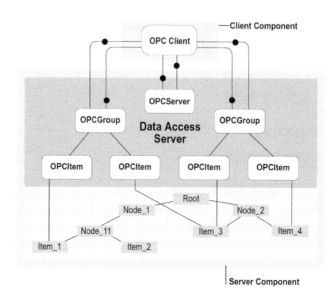

FIGURE 5.6 Components of a Data Access application.

namespace may have, or how the nodes and leaves are to be designated. Only methods at an interface are defined, which enable the client to read out the information.

The namespace is identical for all clients having access to the same server. The object hierarchy, in contrast, is specific to the client. After the server is started, the client has access to an interface of the OPCServer object. It can create OPCGroup objects and, in this way, determine how access to values is structured. The value that a specific client is interested in is represented by an OPCItem object. For the OPCGroup objects, the client can have OPCItem objects created by the server, which permit access to data. OPCItem objects have no interfaces. This is due to the requirement that several values have to be efficiently read and written at the same time. The necessary functionality is available at interfaces of the OPCGroup object.

A server can also call methods at interfaces of the client. It can, for instance, inform the client of process value changes.

Furthermore, methods with parameters have been specified for the various objects. Methods are grouped at interfaces.

At the OPCServer object methods for adding and removing of OPCGroup objects do exist. Other methods can be invoked by the client to store and load configuration information related to process communication (communication parameter). It is not intended to use this functionality to store the recent object hierarchy of the server. Methods for browsing the namespace are grouped at another interface of this object. The client can furthermore access state information of the server (vendor, version, and others).

The OPCGroup object provides access to methods that can be used to add and remove OPCItem objects. Different methods to read and write data are grouped at different interfaces. This also includes methods supporting the creation of callback connections. The state of the OPCGroup object can be obtained and influenced by various methods. Parameter values can be changed, which influence the data are acquired from the process.

One standardized format for exchanging data between server and client is necessary for interoperability. In automation technology, many different data types are used (IEC 61131, fieldbuses, visualization, etc.). The OPC specification creates a standardized representation by defining that DCOM data types are used when values are exchanged between server and client. The data type conversion between application data types and DCOM data types by the server and the client is application specific. The data format shown in Figure 5.7 contains in addition to the process value also a time stamp and quality information. The

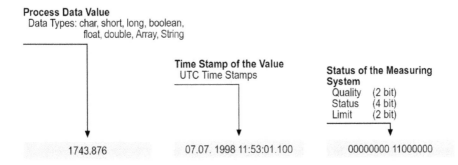

Process Data Value
Data Types: char, short, long, boolean,
float, double, Array, String

Time Stamp of the Value
UTC Time Stamps

**Status of the Measuring
System**
Quality   (2 bit)
Status   (4 bit)
Limit   (2 bit)

1743.876   07.07. 1998 11:53:01.100   00000000 11000000

FIGURE 5.7 Data format.

time stamp can be either generated in the server or, if it already exists, it can be transferred from the device. The quality information contains a description of the value validity (good/bad/uncertain). The quality is described in more detail by the status value (e.g., bad — not connected).

For the data exchange between client and server, there are a number of requirements that the specification takes into account in defining different types of data exchange.

A client is supposed to read or write values. For this purpose, it can use synchronous and asynchronous read and write requests. Reading can take place for the server cache or for the device (place where the process value is created, e.g., the device connected by a communications link). With synchronous calls, the request is processed completely before information is sent to the client. Under certain circumstances, this may not be efficient. Therefore, there is also a possibility of asynchronous reading and writing. Here, the client calls the specific method at the corresponding interface of the OPCGroup object and passes some information. The server then processes the request and provides the client with the remaining information at a later point in time.

Both types of request processing are also available for writing. However, a client can write only to the device and not to the cache.

For synchronous and asynchronous read requests, the client always has to address the corresponding OPCItem objects in the OPCGroup object. With a refresh, values of all OPCItem objects of an OPCGroup object can be requested either from the cache or from the device. Here, addressing is implicit.

In the procedures mentioned above, the client is always active — it polls for data. However, a type of data exchange where the server automatically transmits values to the client is required as well. The transmission is based on the evaluation of relevant default values. This procedure is also supported in the specification.

After this short overview about the Data Access Specification, some functionality will be explained in more detail in the next paragraphs.

As already mentioned, the namespace contains all data points provided by the server. From these points, the client selects those in which it is interested and requests the server to create OPCItem objects for them. As a criterion for the assignment of points in the namespace to OPCItem objects in the server, fully qualified ItemIds are used. They normally consist of sections of the namespace that uniquely identify a data point. In a hierarchical namespace, a fully qualified ItemId will therefore contain the identifier of the leaf representing the data point as well as one or more identifiers of nodes in the namespace. The different identifiers are separated by server-specific delimiters.

The client will first query the structure of the namespace and, in the next step, identifiers for leaves and nodes. By setting method parameters, the client can influence the return values. Different values for one parameter determine whether only node identifiers or leaf identifiers as well are to be passed by the server. Another parameter contains filters. They are used as a criterion: which leaf identifiers must be passed. A limitation in terms of data type, access rights, and character string is possible. The client can navigate in the namespace by indicating node identifiers and thus informing the server that this is the current view of the namespace. From a specific node, the client will finally query all identifiers for the items. In a real application, many different types of information may be relevant for clients. This concerns all values that change or are to be changed by a client. However, some rather static values are also of interest to clients (manufacturer, revision number of devices, description of measurement methods, telephone number of maintenance staff, etc.). If all this information were mapped to leaves in the namespace, the latter might become very large. Also, this information on devices and values (manufacturer, version) might occur repeatedly. For efficient access to this information, properties were introduced with version 2.0 of the Data Access Specification.

As already mentioned, the client has access to an interface of the OPCServer object right after the server component was started. After browsing the namespace, the client will start creating a corresponding object hierarchy in the server. To do this, it can generate one or more OPCGroup objects for structuring data access. Later, the client can assign all values to be read or written with a request to OPCGroup objects (i.e., create OPCItem objects). By defining three parameter values, the client determines how values are to be acquired automatically by the server. With an update rate (in msec), the client defines at which rate values are to be read and written into the cache. The value of PercentDeadband determines the conditions under which values are automatically sent to the client. An OPCGroup object can have the state "active" or "inactive." If the latter is the case, the values of the OPCItem objects are not obtained automatically.

In the last step, the client creates the OPCItem objects for the different OPCGroup objects, which can also be "active" or "inactive." This determines whether or not they are included in automatic data acquisition.

The object hierarchy and object properties can be changed at any time.

The client is able to read data from the server by invoking synchronous or asynchronous calls. A more efficient way to obtain data from the server will be explained in the remaining part of this section.

First, the client has to create the desired object hierarchy and to build the callback connection. This is used to pass values from the server to the client. Besides the hierarchy of the object, their behavior also influences the method of data exchange explained below. It works only if both OPCGroup and OPCItem objects have "Active" state.

Timeliness and sensitivity of the data access are influenced by the parameter values "UpdateRate" and "PercentDeadband" of the appropriate OPCGroup object. The "UpdateRate" (msec) value defines how often values of (active) OPCItem objects are automatically read. The "PercentDeadband" value (in %/100, e.g., 0.01) influences the sensitivity of the data exchange.

Besides changes of the value, state changes for the values are also reasons for passing data automatically from the server to the client.

A number of OPCGroup objects can be created, of course, at the same time, which support this way of data exchange.

The diagram in Figure 5.8 explains the automatic data exchange in detail.

The namespace was configured for the server. For different variables, the EngineeringUnit Type and EngineeringUnit information were defined. The following explanations will only be valid if "analog" is indicated as EU type and the value is of a simple data type.

The client has created an active OPCGroup object and assigned values for UpdateRate (1000 ms in the example) and PercentDeadband (0.1 in the example). Furthermore, the client has created an active OPCItem object for a temperature value. After all objects were generated and the callback connection exists, a value is immediately returned to the client via the callback connection. This value (44) is used as a reference value for the following calculations in the server. According to the given update rate, the temperature value is read from the process every 1000 ms. For each value that is read, a calculation takes place according to the following algorithm: first, the absolute difference between the values of the EU information is calculated (30 in the example). After creation of the OPCGroup object, this value is multiplied by the percent deadband (in the example, $0.1 \times 30 = 3$). For every scanning process, the absolute difference between the value last transmitted to the client and the current value is calculated and compared with the other calculated value. If the former is larger than the latter (after 6 sec, the result for the example is 4>3), the scanned value is sent to the client and used as a new reference value.

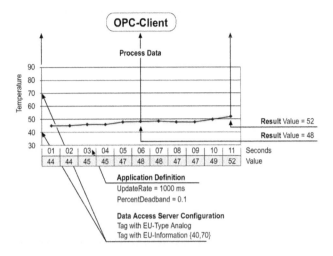

FIGURE 5.8 Automatic data exchange between server and client.

As has already been mentioned, the procedure in this form works only for values with simple data types and EU type "analog." For other EU types (e.g., "discrete") and structured variables, the value is transferred with each change that takes place. The sensitivity of the data transmission can be influenced by the EU information for every value or by the percent deadband for all OPCItem objects of an OPCGroup object. If the criterion applies to several OPCItem objects, all values are transferred with one call.

5.9.4 OPC Data Access 3.0 [7]

In Version 3.0 of the Data Access Specification, some new functionality and enhancements are defined. Products implementing Data Access 3.0 have to be able to interact with products implementing Data Access 2.0 or Data Access 1.0A, of course. The server is characterized by a CategoryId (refer to Section 5.9.2, "Common Definitions and Interfaces"). That is why the client already has a hint as to which functionality can be used.

The following specification issues are new for Data Access 3.0:

- Data can be read and written without the need to create OPCGroup and OPCItem objects. This concerns applications, where Data Access Server are used as an IO layer together with PC-based control systems.
- Deadband and SamplingRate can be set at OPCItem object level in addition to the settings for the OPCGroup object. This provides for a more precise setting as the way to set the value only at the OPCGroup level.
- Browser interface, with a functionality comparable to OPC XML-DA. It makes it easier to implement browsing at the client side.
- Connection monitoring functionality has been added to the specification.

5.9.5 OPC XML-DA [16]

The use of the Internet is a case of application for OPC XML-DA products. Therefore, the following conditions had to be taken into account when the specification was written:

- The interaction parameters are coded using XML, which leads to an overhead.
- Interactions take place by HTTP, which is a stateless protocol. Due to the desired scalability of the servers and the line costs, interactions in the Web mainly take place on a short-term basis. A client fetches information from the server; after this, the server "forgets" about the client. Some solutions implement state but this is a specific approach.

Therefore, the Data Access Specification model with an object hierarchy for each client and with callbacks cannot be applied to a Web Service.

An OPC XML-DA Service is stateless. There is no functionality for the creation of objects as defined in the Data Access Specification.

During browsing as well, no information about the position of the client in the namespace is stored in the OPC XML-DA Service, but all information about the namespace (or a defined part of it) is transferred to the client at the same time. The client can poll for values at the server, but it should also be possible to receive changed values automatically.

Nevertheless, the definition of subscriptions does provide the service with state information. The service must know which data the client is interested in, at what rate they have to be recorded (UpdateRate), and when they have to be transferred (deadband). Since, however, a server cannot call the client on its own initiative with HTTP (it does not know about the client), the specification defines a query of the subscription values, which is initiated by the client. If nothing has changed, the client will not receive any values.

Another important point is monitoring of the connection between client and server and, in case of a subscription, monitoring of the client's availability. Therefore, most function calls contain time values indicating the maximum time that the client will wait for a response from the server or the minimum number of times it will call the server (subscription).

Table 5.3 contains the defined methods.

How does the subscription work in OPC XML-DA?

The sequence of function calls is shown in Figure 5.9 and explained below.

TABLE 5.3 Methods of the OPC XML-DA Specification

GetStatus/GetStatusResponse	Client obtains server status
Browse/ BrowseResponse	Client obtains namespace information
GetProperties/GetPropertiesResponse	Client obtains property information
ReadRequest/ ReadResponse	Client reads data
WriteRequest/ WriteResponse	Client writes data
Subscribe (Client)/ SubscribeResponse (Server)	Client establishes subscription
SubscriptionCancel/Response	Client cancels subscription
SubscriptionPolledRefresh(Client)/ SubscriptionPolledRefreshResponse(Server)	Client initiates requests for the values that are provided in the subscription

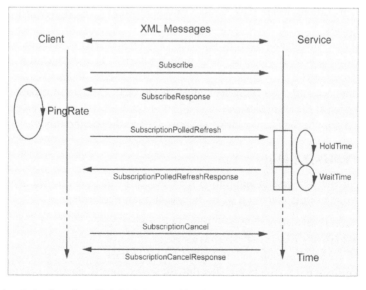

FIGURE 5.9 Subscription based on OPC XML-DA specification issues.

To set up a subscription, the client sends a request indicating the variables it is interested in, as well as the values for RequestedSamplingRate and Deadband. These two values can be defined both for a number of variables or single variables in one call. The SubscriptionPingRate defines with what frequency the service is supposed to check whether the client still exists.

The service responds with a subscribe response that contains the handle for the callback and the supported SubscriptionPingRate. Based on RequestedSamplingRate, the server acquires the values and decides based on Deadband regarding whether to send it to the client or not.

By sending a SubscriptionPolledRefreshRequest message to the server, the client requests for the data for a subscription. It receives the data with SubscriptionPolledRefreshResponse. In the request, the client can tell the service how long it will wait for the response and, in this way, determine how long the connection will remain open. The server will delay the sending of the response accordingly. The service will at least wait the Holdtime to send the SubscriptionPolledRefreshResponse to the client. It will continue to acquire data if no data have changed for another period — the WaitTime. If a value has changed, the service will immediately send the SubscriptionPolledRefreshResponse. If WaitTime also expires and nothing has changed, the service will send an "empty" SubscriptionPolledRefreshResponse, that is, a message without data.

The client will cancel a subscription by passing a SubscriptionCancelRequest message. The service responds with SubscriptionCancelResponse message.

5.9.6   OPC Data eXchange Specification [8]

Why do we need the OPC DX Specification?

As already mentioned, OPC has a huge installed base. There are cases when OPC Servers have to directly interchange data, for example, if information must be transferred from one area of production to another area. Currently, transferring data between servers is possible only using clients or proprietary bridges, which slows down the exchange and excludes overall solutions.

There are a number of fieldbus systems that use TCP/IP on Ethernet as their transport and network protocol. All of them use the same medium but cannot directly intercommunicate. Here also, OPC DX is supposed to offer a solution.

Figure 5.10 shows the structure of an OPC DX application. A DX Server can receive data from one (or more) DA Server(s) or from one (or more) DX Server(s). For this purpose, a DX Server has implemented DA Client functionality. By means of this client, the DX Server creates OPCGroup and OPCItem objects. The data point from which the data are recorded is called Sourceitem; the data point where they are written is called Targetitem. This linkage and their property are called "connection." All connections taken together are called "configuration." Connections are defined by a configuration client

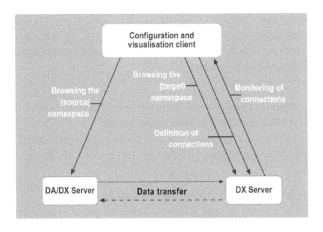

FIGURE 5.10   Structure of an OPC DX application.

and mapped in the namespace of the DX Server. A monitoring client monitors the connections. Data Access functionality is used for this purpose. Other clients (visualization) can access the existing items in a "normal" way.

The specification distinguishes between the configuration model and the run-time model. The configuration model defines the semantics of the connections as well as possibilities of creating, deleting, and changing connections and transferring events.

The run-time model defines the data transfer between DX and DA/DX Servers. Here, the manner in which the data are recorded as well as error and status monitoring are important.

The information on connections is stored in a defined branch of the namespace. The attributes of a connection and, consequently, the structure of this branch are specified. A Data Access Client is able to browse the namespace and create OPCItem objects for values of interest. A DX server therefore also implements Data Access Server functionality besides Data Access client functionality.

Connections are created, modified, and deleted by different commands issued by a configuration client.

The specification not only defines the configuration, but also the behavior of a DX Server at runtime, that is, during horizontal data exchange.

The DX Server can subscribe to values from a Data Access Server that supports the specification versions 2.04, 2.05, and 3.0. Another possibility is for the DX Server to support the OPC XML-DA Specification and to receive data from OPC XML-DA Servers.

The target item and the source item can have different data types. During creation of the OPCItem object, the DX Server can request the DA/DX Server to perform the conversion. If this is not supported, the DX Server must perform the conversion itself.

The specification also defines the runtime behavior of a DX Server if the connection to the source server is interrupted. In this case, the DX Server tries to reestablish the connection with the frequency of a ping rate, which can be set as desired. If this is not possible, this information is mapped to the corresponding attributes of the target item. Depending on the properties of the connection, the DX Server can set a substitute value instead of the actual value.

A monitoring client can sign up for the corresponding items in the configuration branch of the namespace and thus obtain information on the state of the connections.

The DX Server stores the entire current configuration at runtime. This means that all results of the calls for adding, modifying, and deleting connections are stored. The specification does not define how and where this is done. During starting of the DX Server, the current configuration is reloaded and available again.

5.9.7   Complex Data Specification [17]

The Data Access Server supports access to simple and complex data. A temperature value is an example of simple data, whereas records of diagnostic information serve as examples of complex data. Both data types (simple and complex) can be read or written in OPC Data Access applications. Yet, the existing OPC specifications do not define the means for a client and a server to exchange structural information about complex data. Therefore, the client can only pass complex data as octet strings to other applications (e.g., database and visualization applications).

The following behavior is requested:

- The client understands the structure of the data. In this case, the client may forward individual elements of the complex data item to other applications.
- The client understands the structure of the data and knows its semantics. In this case, the client is not only able to distinguish between elements, but also knows about their type and relations with each other.

The description of complex data type is a prerequisite for this behavior.

The new specification proposes two approaches for the description of complex data:

- Type descriptions defined within OPC specifications and
- Type descriptions defined outside OPC specifications (e.g., Fieldbus organizations)

TABLE 5.4   Properties Used to Describe Complex Data

PropertyId	Name	Meaning
109	Complex Data Type Description System	Identifies the type description system used (e.g., OPC, Fieldbus Consortium)
110	Complex Data Type Description ID	Identifies the type description of a complex data item (e.g., reference to an XML file, reference number related to a type description system of consortia)
111	Complex Data Type Description Version	Identifies the version of the type description
112	Complex Data Type Description	A BLOB that contains the information necessary for clients to interpret the value of the complex data item

OPC-type descriptions are concerned with the definition of the structure of complex data. This allows the client to know the elements of complex data items. The type description is effected using XML. This description system is not very flexible but it provides for easy implementation.

The type description systems of other organizations can also contain semantic definitions and allow a client to know the semantics. These descriptions can be more flexible but they increase the requirements for implementation, as client and server should be able to understand a number of type description systems.

The new specification defines that information on and about complex data items must be provided using defined properties. Table 5.4 shows the defined properties and their meanings.

Only items with complex data types must support these properties.

Besides the possibility of describing complex data types, the new specification also defines the behavior during writing of complex data.

Not all elements of the complex data item, once written, may be writable. In this case, the client supplies the entire buffer in a write request and the server ignores the values of nonwritable elements.

In a case where the value for one or more writable elements is invalid, or cannot be applied, the server rejects the entire write request.

5.9.8   OPC Alarms and Events [9]

By implementing Data Access Specification definitions, values can be automatically transmitted from the server to the client if these values have changed within a certain time period or if the values' states have changed. This is not sufficient or not efficient regarding the following requirements:

- It is not only important to be informed about changes of the value, but it is also important to be informed that a parameter has exceeded a certain limit.
- The value changes sometimes during the UpdateRate.
- Alarms have to be acknowledged.

The Alarms and Events specification has been written to fulfill these (and other) requirements.

The specification model differentiates between things that happen (simple events and tracking-related events) and things that exist (condition-related events, i.e., alarms).

The occurrence of an alarm can be acknowledged. An object hierarchy has been defined, where a client can adapt the server to its requirements.

The client normally will not be interested in all events that can be monitored by a server. It can select the interesting events by applying filter. The filter consists of parts from the event area and the filter space. The event area provides a topological structuring of events. The filter space provides event structuring based on event attributes (event type, event category, event severity, etc.). Both can be configured during adapting the server to the real environment.

Figure 5.11 shows the components that are part of an Alarms and Events application — OPC Alarms and Events Client and Server. Events can be structured in an event area that is always hierarchical. In the server itself, different DCOM objects with interfaces and methods must be implemented. After launching

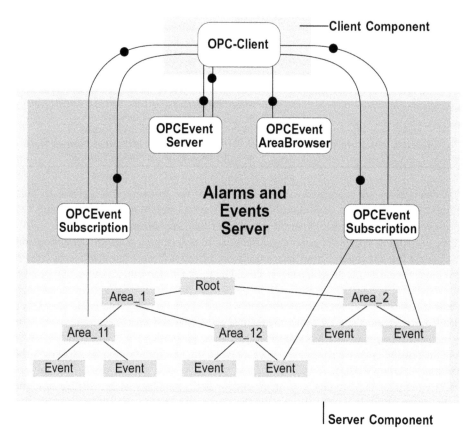

FIGURE 5.11 Components of an Alarms and Events application.

the server, the client has access to an interface of the OPCEventServer object. Then, the client can either create the OPCEventAreaBrowser object to obtain information on the content of the event area, or it can create OPCEventSubscription objects. These objects are used to monitor event sources and, in case of events, transmit relevant messages to the client. Events are assigned to the OPCEventSubscription objects via filters.

This specification also permits both directions of interaction (client → server and server → client).

Filters are used for assigning events to OPCEventSubscription objects. In this way, a client determines which events are to be monitored by the server. Values of filter parameters are derived from information that concerns the event area (fully qualified AreaId, fully qualified SourceId) or properties of events (type, category, etc.).

There are three types of events already defined in the specification. An example of a simple event would be a device failure, and that of a tracking-related event, the changing of a setpoint. An example of a condition-related event would be a positive or negative deviation from temperature limit values. For all these three types, there are different categories. The specification proposes only some settings for this value. More categories can be defined based on real application scenarios. For condition-related events, the different categories have conditions and subconditions. The specification already contains some proposals for categories, conditions, and subconditions. Events also have a priority. If necessary, application-specific priorities must be converted into OPC priorities.

A large variety of filters can be defined, for example, <type=condition-related events, category=level, priority=500–700, area=area_1, and source=source_A>.

In case of event occurrence, the server sends an event notification to the client, which has assigned the event to an OPCEventSubscription object via a filter. Depending on the type of event, the notification

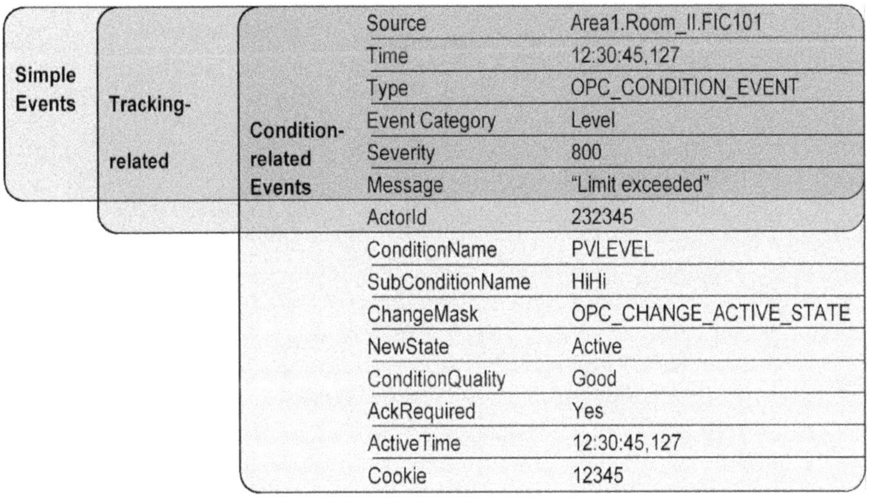

Simple Events	Tracking-related	Condition-related Events	Source	Area1.Room_II.FIC101
			Time	12:30:45,127
			Type	OPC_CONDITION_EVENT
			Event Category	Level
			Severity	800
			Message	"Limit exceeded"
			ActorId	232345
			ConditionName	PVLEVEL
			SubConditionName	HiHi
			ChangeMask	OPC_CHANGE_ACTIVE_STATE
			NewState	Active
			ConditionQuality	Good
			AckRequired	Yes
			ActiveTime	12:30:45,127
			Cookie	12345

FIGURE 5.12 Structure of an event notification.

can have different numbers of parameters. Figure 5.12 shows the mandatory parameters for the various event types. Examples in the figure are related to an event notification informing the client about the occurrence of a condition-related event.

During the notification about a simple event, values for the parameters Source (information from the event space), Time (when did the event occur?), Type, Event Category, Severity, and Message are transferred. The language of the message can be set by means of a method at the general interface mentioned in Part "Common Interfaces and Definitions."

For notification about a tracking-related event, there is the additional parameter ActorId. It contains a numerical identifier to indicate the cause of the event. The creation of the identifier and the meaning of the value are not described in the specification.

The largest number of parameters must be transferred for notification about a condition-related event. The following parameters are added to the ones already mentioned:

- *ConditionName:* Name of the condition from the event area.
- *SubConditionName:* Name of the subcondition, if any.
- *ChangeMask:* Indicates in what way the state of the condition has changed, for example, inactive → active.
- *State*: State of the condition, for example, active, acknowledged.
- *ConditionQuality:* The parameter can be compared with the quality of a value from the Data Access Specification. A notification is also sent if the state changes.
- AckRequired: The event must be acknowledged.
- *ActiveTime:* Indicates the time at which the state became active. This value is not identical to "Time" since the latter indicates the time of event occurrence. The receipt of an acknowledgment also leads to an event. After the receipt, the values of the parameters differ.
- *Cookie:* Used by the client in the acknowledge and by the server to relate the acknowledge to the event.

There may be other attributes in addition to these mandatory ones. They have to be supported for all events of a category, that is, they are defined for a category.

5.9.8 OPC Historical Data Access [11]

The Historical Data Access Specification defines an API to access historical data. Historical data has been collected over some time frame and is now available. The specification does not define how the data are

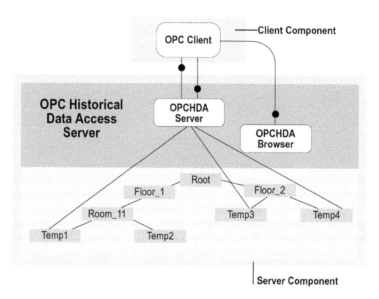

FIGURE 5.13 Components of an HDA application.

collected and stored. The recent specification only considers parameter values; access to stored event information is not specified.

Accessing data by using Data Access clients and storing it in a database could be one way to collect data later accessible for Historical Data Access Server. Other possibilities could include other ways of accessing data. Another use of case of interest would be access to data that has been stored over some time in a measuring device and is accessed only once a week, for example, environmental parameter. Figure 5.13 shows the object hierarchy and the relation to the namespace.

Different from Data Access Server, Historical Data Access Server will provide access to larger amounts of data. This is caused by the larger number of variables (all variables in a process) and by the amount of values per variable. Stored values do not change. But they can be deleted and new values or attributes can be added. Different from Data Access Server, the access to aggregated values plays an important role (average over a time frame, smallest value over a time frame, etc.).

Only two DCOM objects were specified for the server. Immediately after launching of the component, the client has access to the OPCHDAServer object where the entire functionality is available. The OPCH-DABrowser object is used for searching the namespace of an HDA Server. This namespace contains all data points for which values are available. Other than with the Data Access Server, there is no explicit object for the access to data points; this is not necessary since access takes place very rarely. For the client, methods for synchronous and asynchronous reading, recording, and changing entries in the database of historical data are available.

The tables in Figure 5.14 show a few aggregates and attributes for stored raw data. For both parameters, the specification already contains default values. The server manufacturer can also provide other possibilities.

An HDA Client can use four different ways to access historical data:

1. *Read*: Using this approach, the client can read raw values, processed values, values at a specific point in time (AtTime), modified values, or value attributes. The client can invoke synchronous or asynchronous calls.
2. *Update*: The client can insert, replace, or insert and replace values. This can be done for a time frame or for specific values. Synchronous and asynchronous calls are also available.
3. *Annotation*: The client can read and insert annotations. Again, this can be performed synchro-nously and asynchronously.

Raw Data

	1	2	3	4	5	6	7	8	9	10
Temp1	18	18	19	20	21	20	19	18	17	18
Temp2	17	17	18	19	20	17	17	17	18	18
Temp3	16	17	18	18	18	17	18	19	19	18
Temp4	17	18	18	19	20	20	20	20	19	19

Attributes

	Temp1	Temp2	Temp3	Temp4
Datatype	VT_I2	VT_I2	VT_I2	VT_I2
Description	Temperature in Room11	Temperature in Room12	Temperature in Room13	Temperature in Room14
NodeName	128.7.16.123	128.7.16.123	128.7.16.124	128.7.16.124

Aggregates

	Temp1	Temp2	Temp3	Temp4
Average	18,8	17,8	17,8	19
Delta	0	1	2	2
Minimum	17	17	16	17
Maximum	21	20	19	20
Range	4	3	3	3

FIGURE 5.14 Relations between raw data, attributes, and aggregated data.

4. *Playback*: The client can request the server to send values with a defined frequency for a defined time frame (every 15 sec values stored over 15 min). This can be applied both for raw and aggregated data.

5.9.9   OPC Batch [13]

The OPC Batch Specification defines an interface between clients and servers for a certain type of application called batch processing. With this procedure, recipes are processed by resources and reports are generated. Functionality that is made available by Data Access Specification is required. For batch processing, there is the international standard IEC 61512-1 defining a model for this kind of data exchange. Many products in this area have been implemented accordingly. Therefore, the idea was to combine the existing functionality of the Data Access Servers with the specifications in this standard. The consequence was an extension of the Data Access Specification, which defines a special namespace and some additional methods adapted optimally to the conditions during batch processing.

Figure 5.15 shows a part of the namespace that is specific to a server used in batch processing. The namespace is defined to be always hierarchical. Specific nodes exist on the highest hierarchy level (OPCB-PhysicalModel, OPCBBatchModel, and OPCBBatchList). These nodes represent models of IEC 61512-1. Batches contain parameters and results. For each batch, there are corresponding OPCBParameters and OPCBResults nodes. The parameters and results are represented by specific properties.

5.9.10   OPC Security [15]

OPC Clients and Server can run on different PCs. Security aspects have to be considered in such applications. Security settings can be applied in two different ways:

1. by using the utility program "dcomcnfg" and/or
2. by using the security API functionality of the operating system

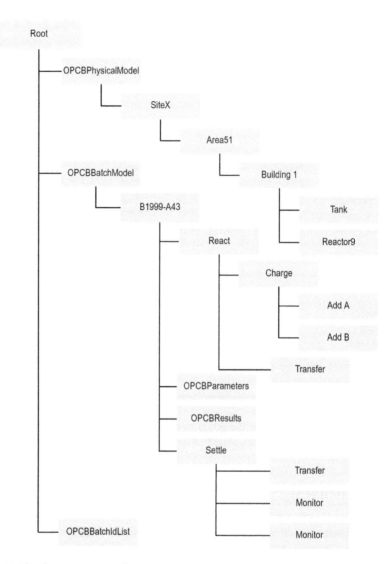

FIGURE 5.15　OPC batch server — part of a namespace.

By using "dcomcnfg," security settings are applied to the components, that is, the server or to the overall system. If smaller granularity is necessary (settings for objects, methods, data, ...), the security API must be used. It is the goal of the Security specification to foster interoperability between security-aware applications, that is, applications that use the security API.

The Security Specification defines a model for security, different levels of security, and possibilities of how the client and the server can exchange security information.

The specification does not define which objects are to be secured and how.

Figure 5.16 shows the Windows NT/2000 security model on which the specification is based.

The model differentiates between principals described by access certificates and security objects. All processes in Windows NT/2000 are principals. If a user logs in, a process is started. Depending on which group a user belongs to (administrators, guests, etc.) and on the corresponding access rights, an access certificate is assigned to the process. This is a kind of "ticket" describing the properties of the principal. On the other hand, there are security objects. These are objects to which access is monitored. If a principal tries to access a security object, the reference monitor decides, using the access control list, whether the principal may do that. If "Dcomcnfg" is used, the ACL is edited by means of this tool, and the reference

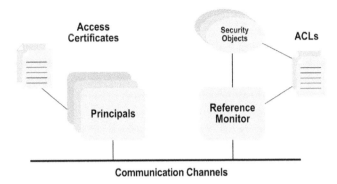

FIGURE 5.16 Windows security model.

monitor is part of the DCOM run-time environment. If COM Security API methods are used, the reference monitor and creation of the ACL must be implemented as parts of the server. In this case, security objects of any granularity can be implemented.

Furthermore, a differentiation is made in the specification regarding the way to characterize principals. Security decisions can be based on access certificates for NT user (NT Access Token) or specific OPC user. NT users are already known in the system. For each of these different certificates, one interface has been specified.

5.9.11 Compliance Test

The OPC Foundation defines compliance testing as a way to verify whether a server implementation conforms to the specification. There are no test tools to check clients.

Compliance testing is based on the following prerequisites:

- *Test cases*: A huge number of test cases have been defined, by generating a number of different parameter values for the method calls. These parameter values can be used to verify the server's behavior in response to valid, invalid, and impossible method calls. Default and extreme values have been defined for the various parameter.
- *Test system*: A Compliance Test Client has been developed by the OPC Foundation.
- *Test procedure*: During running the Compliance Test Client against a server, test result files are generated. These have to be submitted to the OPC Foundation. It will derive the result. The information is added to a page on the OPC Foundation Web site.

The Compliance Test Client supports the following test possibilities:

- *Stress tests*: Here, it can be determined that a selectable number of objects can be added and removed in the server component to be tested.
- *Logical tests*: Here, it is tested whether the Proxy/Stub-DLLs are installed correctly.
- *Interface tests*: Here, methods are tested at the interfaces of the objects. Valid, invalid, and meaningless parameter values are used for all the methods.

Configuration of the Compliance Test Client has to be carried out prior to starting the test run.

Now the test run can be started. The results of the proceeded test cases are displayed. If test cases have not been passed successfully, the implementation of the server has to be changed. At the end of the test, after passing all test cases successful, a binary result file will be generated by the test client. This file must be sent to the OPC Foundation. The result is then published on the OPC Foundation Web site. Figure 5.17 shows an example.

Table 5.5 shows the current availability of compliance tests for the different specifications.

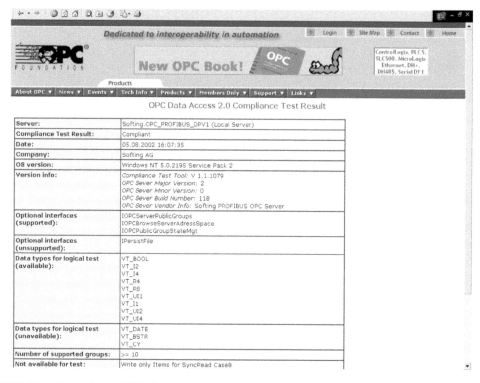

FIGURE 5.17 Test result as published on the OPC Foundation Web site.

TABLE 5.5   OPC Compliance Test — Release States

Specification	Compliance Test Availability
DA 2.05	V2.0.2.1105
DA 3.00	Beta
AE 1.1	V1.0.2.1105
HDA 1.1	Beta
DX	Test Specification Release Candidate
XML-DA	Test Specification Release Candidate

5.10   Implementation of OPC Products

With the release of the OPC XML-DA and OPC DX Specifications, not only the implementation of DCOM based Client and Servers have to be considered, it also becomes important to think about implementation based on Web Services.

There are some major differences between these two approaches:

- DCOM is available mainly on Microsoft systems. Portation exists for Unix systems, but only a few OPC products for this area are available.
- OPC XML-DA products can be easily made available everywhere where HTTP and XML are supported. Even if XML support does not exist, products can be implemented. Coding and decoding of XML messages can be implemented without a parser.

In the following paragraphs, first implementation of DCOM-based OPC products is considered and then Web Service-based solutions are considered.

5.10.1   OPC DCOM Server Implementation

OPC Servers can be implemented in three types:

1. *InProc Servers*: The server is implemented as a DLL and runs in the process space of the client. In this manner, high-speed data access is possible since no process or computer boundaries have to be overcome. There will only be restrictions if the server needs to access a protected kernel resource.
2. *OutProc Servers*: The server is implemented as a stand-alone executable. It can run on the same computer as the client or on a different one. The advantage is that several clients have access if protected resources are used. Since there are process and maybe even computer boundaries between client and server, data traffic is slower than with the InProc solution.
3. *Service*: This is a special kind of OutProc Server. Services can be configured in such a way that they are immediately started during booting without the necessity of user log-in. Thus, the initialization of communications links is shifted to the booting process of the computer. The advantage of this solution is that the server is available immediately after booting. The disadvantage is that services are only available for Windows NT/2000/XP. On other operating systems (Windows 98/ME), they run as a "normal" OutProc server.

5.10.2   OPC DCOM Client Implementation

An OPC Client must provide functionality in two phases of the life cycle of an automation system (during commissioning/configuration and online operation). It must be possible to:

- Find and start available OPC Servers during the configuration phase. The namespace/eventspace of the servers must be browsed to obtain fully qualified item identifier necessary to add OPCItem objects and to receive filter criteria. The same is true for properties. ValidateItem checks whether the passed parameters would result in a valid OPCItem object. To accomplish this, OPCGroup objects must also be able to be added. The client must query for supported filter criteria. The result of the configuration process can be stored somewhere. Access to other methods also needs to be implemented.
- At runtime, the OPC Client requests that all objects have to be created in the OPC Server. The OPC Client supplies the application with the requested data and monitors communication. During data exchange, it must also be possible to start additional Servers, add and manipulate OPCGroup/OPCEventSubscription objects, as well as add and manipulate OPCItem objects.

5.10.3   Creating OPC DCOM Components by Means of Tools

As far as DCOM knowledge is seen as core competence, the development of OPC clients and servers can be done by means of different developing environments. Detailed knowledge of DCOM and the OPC specifications is necessary for this.

A simpler way to create OPC components is to use toolkits. Toolkits encapsulate the DCOM and OPC functionality. They exchange information through an API with an application using the toolkits. Toolkits are available for client and server components. Most toolkits support Windows 9X/NT/2000/XP. Versions for Windows CE and Linux also exist.

The toolkits can be distinguished by their possibilities of optimization. There are compact solutions intended for a simple server type and more complex toolkits that permit extremely flexible solutions.

Characteristic of complex toolkits is that they can be combined with each other resp. are based on each other. With the OPC Toolbox shown in Figure 5.18, components can be developed that contain both Data Access and Alarms and Events functionality, which can be used on standard Windows operating systems but also on Windows CE.

There are toolkits for different use cases that are performed as a C++ class library, so-called C++ toolkits and those that provide a more or less complete OPC component in which only some few function calls must be integrated. ActiveX Controls is a special kind of OPC client implementation: the complete

FIGURE 5.18 OPC toolkit sample.

OPC Client functionality is provided as OPC Client Controls and can be used in Visual Basic, Excel, or other ActiveX container applications without any programming.

5.10.4 Implementation of OPC XML Servers and Clients

Three approaches can be distinguished for the implementation. They are described below.

1. *Starting from scratch — the most demanding approach.* This approach mainly has to be used for implementing OPC products on embedded devices. If HTTP and XML support is available, the infrastructure components for receiving and sending SOAP telegrams and coding XML messages must be implemented. These components must be combined with the remaining parts of the client and server application. If this is not available, the necessary parts must be ported. There are a number of solutions available in the public domain.
2. *Using toolkits supporting WSDL.* Toolkits exist in the marketplace that generate the necessary infrastructure components and templates to integrate with parts of client and server applications based on the WSDL file.
3. *Using OPC toolkits — the easiest approach.* These toolkits not only implement infrastructure components, but also the internal logic of client and server application.

5.11 Outlook into Future

The huge acceptance and fast penetration of the OPC technology in many areas of automation technology are unique. Many standards — for example, fieldbuses in industrial communication or in the area of PLC programming — have established in one decade or more or not at all. OPC has been much more successful. With the defined specifications, multiple areas of application from fast and equidistant data transfer (Data Access) via processing of large amounts of historical data (Historical Data Access) up to the collection and acknowledgment of volatile and critical events (Alarms and Events) can be covered.

The availability of Data eXchange and OPC XML-DA opens up further areas of application in vertical integration and peer-to-peer interaction. On the basis of the specifications, clients and servers can run in the same process on one PC (InProc Server) or in different processes (OutProc Server) on PCs that can be close together or far apart. One or several clients can access a server. A client, in turn, can communicate with one or more servers. With the availability of Web Services-based products, OPC can be used in embedded devices or any other kind of devices supporting this technology.

There are a variety of products that translate the OPC specifications into reality. The scope ranges from sample clients and servers to tools for creating OPC components and servers for a wide variety of communication systems and devices. A multitude of visualization systems, and systems for data acquisition and diagnosis, have been equipped with OPC Client interfaces. These products are used in a vast number of actual applications. A large variety of programming languages and development environments are available for the creation of OPC components. OPC products can be used on a wide range of hardware and software platforms.

Solutions for the compliance test are available. The use of successfully tested products raises the confidence of end users in product quality and interoperability.

A nonprofit organization, the OPC Foundation is in charge of marketing and developing the technology.

These points characterize the current actual state described in this chapter. One question still remains to be answered — what does OPC's future look like?

5.12 The Future of OPC

There are two issues that drive the future of OPC:

1. OPC XML-DA is available. The market will request to also provide other specifications on this technological basis, for example, OPC XML-AE.
2. The market will request for migration paths between technologies and solutions.

Microsoft pushes .NET. More and more potential OPC client providers will go this way. A request for an interoperable .NET interface for accessing the legacy DCOM OPC server arises. OPC XML-DA servers and clients on different platforms will be available in the future. Legacy DCOM clients and servers should be used in mixed applications. Wrappers between these components are necessary. Figure 5.19 shows the different constellations and the necessary wrapper. The OPC Foundation provides a skeleton for a .NET to DCOM wrapper. It implements the DCOM object view and makes it accessible from .NET languages. Wrappers between DCOM and Web Services and vice versa will be provided by different vendors.

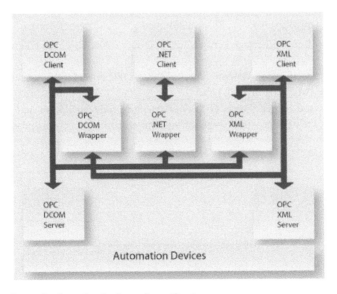

FIGURE 5.19 Migration paths for technologies and specifications.

It can be foreseen that many more OPC products will be developed in the future and the success story of OPC will continue.

References

1. OPC Overview, Version 1.0, October 27, 1998, OPC Foundation, www.opcfoundation.org
2. OPC Common Definitions and Interfaces Version 1.0, October 27, 1998, OPC Foundation, www.opcfoundation.org
3. OPC Specification Version, Version 1.0, September 1996, OPC Foundation, www.opcfoundation.org
4. OPC Data Access Specification Version 1.0A, September 1997, OPC Foundation, www.opcfoundation.org
5. OPC Data Access Custom Interface Standard Version 2.05a, December 2001, OPC Foundation, www.opcfoundation.org
6. OPC Data Access Automation Interface Standard Version 2.02, February 1999, OPC Foundation, www.opcfoundation.org
7. OPC Data Access Custom Interface Standard Version 3.00, March 2003, OPC Foundation, www.opcfoundation.org
8. OPC Data eXchange Specification Version 1.00, March 2003, OPC Foundation, www.opcfoundation.org
9. OPC Alarms and Events Custom Interface Standard Version 1.10, October 10 2002, OPC Foundation, www.opcfoundation.org
10. OPC Alarms and Events Automation Interface Standard Version 1.01, (Draft) December 15, 1999, OPC Foundation, www.opcfoundation.org
11. OPC Historical Data Access Custom Interface Standard Version 1.10, January 2001, OPC Foundation, www.opcfoundation.org
12. OPC Historical Data Access Automation Interface Standard Version 1.0, January 2001, OPC Foundation, www.opcfoundation.org
13. OPC Batch Custom Interface Specification, Version 2.00, July 2001, OPC Foundation, www.opcfoundation.org
14. OPC Batch Automation Interface Specification, Version 1.00, July 2001, OPC Foundation, www.opcfoundation.org
15. OPC Security Custom Interface Version 1.00, October 2000, OPC Foundation, www.opcfoundation.org
16. OPC XMLDA Specification Version 1.00, July 2003, OPC Foundation, www.opcfoundation.org
17. OPC Complex Data Version 1.00, Draft 02 August 2002, OPC Foundation.
18. Iwanitz, Frank and Jürgen Lange, *OPC Fundamentals, Implementation and Application*, 2nd ed., Hüthig Verlag, Heidelberg, 2002. www.softing.com/en/communications/products/opc/book.htm

Section 3.4

MMS in Factory Floor Integration

6

The Standard Message Specification for Industrial Automation Systems: ISO 9506 (MMS)

Karlheinz Schwarz
Schwarz Consulting Company
(SCC)

6.1   Introduction

The international standard Manufacturing Message Specification (MMS) [1, 2] is an Open Systems Interconnection (OSI) application layer messaging protocol designed for the remote control and monitoring of devices such as remote terminal units (RTUs), programmable logic controllers (PLCs), numerical controllers (NCs), or robot controllers (RCs). It provides a set of services allowing the remote manipulation of variables, programs, semaphores, events, journals, etc. MMS offers a wide range of services satisfying both simple and complex applications.

For years, the automation of technical processes has been marked by increasing requirements with regard to flexible functionalities for the transparent control and visualization of any kind of process [7, 8]. The mere cyclic data exchange will more and more be replaced by systems that join together independent yet coordinated systems — like communication, processing, open- and closed-loop control, quality protection, monitoring, configuring, and archiving systems. These individual systems are interconnected and work together. As a common component, they require a suitable real-time communication system with adequate functions.

The MMS standard defines common functions for distributed automation systems. The expression *manufacturing*, which stands for the first M in MMS, has been badly chosen. The MMS standard does not contain any manufacturing-specific definitions. The application of MMS is as general as the application of a personal computer. MMS offers a platform for a variety of applications.

The first version of MMS documents was published in 1990 by ISO TC 184 (Industrial Automation) as an outcome of the GM initiative *Manufacturing Application Protocols* (MAP). The current version was published in 2003:

- *Part 1: ISO 9506-1 Services*: Describes the services that are provided to remotely manipulate the MMS objects. For each service, a description is given of the parameters carried by the service primitives. The services are described in an abstract way that does not imply any particular implementation.
- *Part 2: ISO 9506-2 Protocol*: Specifies the MMS protocol in terms of messages. The messages are described with ASN.1, which gives the syntax.

Today, MMS is being implemented — unlike the practice 15 years ago and unlike the supposition still partly found today — on all common communication networks that support the safe transport of data. These can be networks like Transmission Control Protocol (TCP)/Internet Protocol (IP) or International Organization for Standardization (ISO)/OSI on Ethernet, a fieldbus, or simple point-to-point connections like high-level data link control (HDLC), RS 485, or RS 232. MMS is independent of a seven-layer stack. Since MMS was originally developed in the MAP environment, it was generally believed earlier that MMS could be used only in connection with MAP.

MMS is the basis of the international project *Utility Communication Architecture* (UCA™, IEEE TR 1550) [13], IEC 60870-6-TASE.2 (Inter-Control Center Communication) [9–12], IEC 61850 (Communication Networks and Systems in Substations) [16–20], and IEC 61400-25 (Communications for Monitoring and Control of Wind Power Plants) [15, 21].

This chapter introduces the basic concepts of MMS applied in the abovementioned standards.

6.2 MMS Client–Server Model

MMS describes the behavior of two communicating devices by the client–server model (Figure 6.1). The client can, for example, be an operating and monitoring system, a control center, or another intelligent device. The server represents one or several real devices or whole systems. MMS uses an object-oriented modeling method with object classes (named variable, domain, named variable list, journal, etc.), instances from the object classes, and methods (services like read, write, information report, download, read journal, etc.).

The standard is comprehensive. This does not at all mean that an MMS implementation must be complex or complicated. If only a simple subset is used, then the implementation can also be simple. Meanwhile, MMS implementations are available in the third generation. They allow the use of MMS on both PC platforms and embedded controllers.

The MMS server represents the objects that the MMS client can access. The virtual manufacturing device (VMD) object represents the outermost "container" in which all other objects are contained.

Real devices can play both roles (client and server) simultaneously. A server in a control center for its part can be a client with respect to a substation. MMS basically describes the behavior of the server. The server contains the MMS objects, and it also executes services. MMS can be regarded as server-centric.

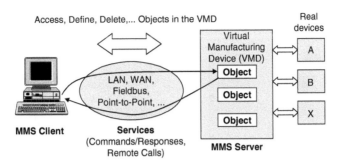

FIGURE 6.1 MMS client–server model.

In principle, in a system more devices are installed that function as server (for example, controllers and field devices) than devices that perform as clients (e.g., PC and workstation).

The calls that the client sends to the server are described in ISO 9506-1 (services). These calls are processed and answered by the server. The services can also be referred to as remote calls, commands, or methods. Using these services, the client can access the objects in the server. It can, for example, browse through the server, i.e., make visible all available objects and their definitions (configurations). The client can define, delete, change, or access objects via reading and writing.

An MMS server models real data (e.g., temperature measurement, counted measurand, or other data of a device). These real data and their implementation are concealed or hidden by the server. MMS does not define any implementation details of the servers. It only defines how the objects behave and represent themselves to the outside (from the point of view of the wire) and how a client can access them.

MMS provides the very common classes. The named variable, for example, allows the structuring of any information provided for access by an application. The content (the semantic of the exchanged information) of the named variables is outside the MMS standard. Several other standards define common and domain-specific information models.

IEC 61850 defines the semantics of many points in electric substations. For example, "Atlanta26/XCBR3.Pos.stVal" is the position of the third circuit breaker in substation Atlanta26. The names *XCBR*, *Pos*, and *stVal* are standardized names.

The coming standard IEC 61400-25 (Communication for Wind Power Plants) defines a comprehensive list of named points specific for wind turbines. For example, "Tower24/WROT.RotSpd.mag" is the (dead-banded) measured value of the rotor position of Tower24. "RotSpd.avgVal" is the average value (calculated based on a configuration attribute "avgPer"). These information models are based on common data classes like measured value, three-phase value (delta and Y), and single-point status.

6.3 Virtual Manufacturing Device

According to Figure 6.2, the real data and devices are represented — in the direction of a client — by the virtual manufacturing device. In this regard, the server represents a standard driver that maps the real world to a virtual one. The following definitions help to clarify the modeling in the form of a virtual device:

If it is there and you can see it	It is REAL
If it is there and you cannot see it	It is TRANSPARENT
If it is not there and you can see it	It is VIRTUAL
If it is not there and you cannot see it	It is GONE

— **Roy Wills**

The VMD can represent, for example, a variable "Measurement3" whose value may not permanently exist in reality; only when the variable is being read will measurement and transducer get started in

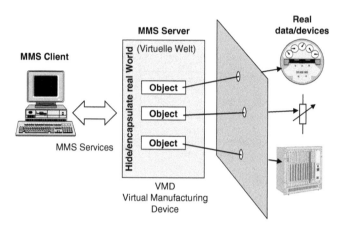

FIGURE 6.2   Hiding real devices in the VMD.

determining the value. All objects in a server can already be contained in a device before the delivery of a device. The objects are predefined in this case.

Independent of the implementation of a VMD, data and the access to data are always treated in the same way. This is completely independent of the operating system, the programming language, and memory management. Like printer drivers for a standard operating system hide the various real printers, so a VMD also hides real devices. The server can be understood as a communication driver that hides the specifics of real devices. From the point of view of the client, only the server with its objects and its behavior is visible. The real device is not visible directly.

MMS merely describes the server side of the communication (objects and services) and the messages that are exchanged between client and server.

The VMD represents the behavior of a real device as far as it is visible "over the wire." It contains, for example, an identification of manufacturer, device type, and version. The virtual device contains objects like variables, lists, programs, data areas, semaphores, events, journals, etc.

The client can read the attributes of the VMD (Figure 6.3); i.e., it can browse through a device. If the client does not have any information about the device, it can view all the objects of the VMD and their attributes by means of the different "get" services. With that, the client can perform a first plausibility check on a just-installed device by means of a "get(object-attribute)" service. It learns whether the installed device is the ordered device with the right model number (model name) and the expected issue number (revision). All other attributes can also be checked (for example, variable names and types).

The attributes of all objects represent a self-description of the device. Since they are stored in the device itself, a VMD always has the currently valid and thus consistent configuration information of the respective device. This information can be requested online directly from the device. In this way, the client always receives up-to-date information.

MMS defines some 80 functions:

- Browsing functions about the contents of the virtual device: Which objects are available?
- Functions for reading, reporting, and writing of arbitrarily structured variable values
- Functions for the transmission of data and programs, for the control of programs, and many other functions

The individual groups of the MMS services and objects are shown in Figure 6.4. MMS describes such aspects of the real device that shall be open, i.e., standardized. An open device must behave as described by the virtual device. How this behavior is achieved is not visible, nor is it relevant to the user that accesses the device externally. MMS does not define any local, specific interfaces in the real systems. The interfaces are independent of the functions that shall be used remotely. Interfaces in connection with MMS are always understood in the sense that MMS quasi-represents an interface between the devices

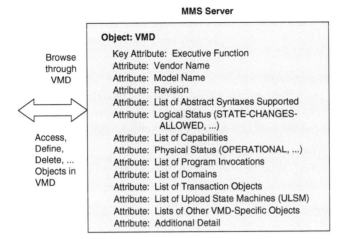

FIGURE 6.3 VMD attributes.

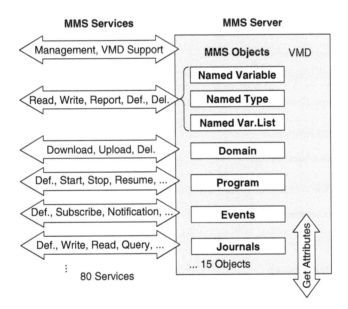

FIGURE 6.4 MMS objects and services.

and not within the devices. This interface could be described as an external interface. Of course, interfaces are also needed for implementations of MMS functions in the individual real devices. These shall not and cannot be defined by a single standard. They are basically dependent on the real systems — and these vary to a great extent.

6.3.1 MMS Models and Services

6.3.1.1 ISO 9506-1 (Part 1): Service Specification

6.3.1.1.1 Environment and General Management Services

Two applications that want to communicate with each other can set up, maintain, and close a logical connection (initiate, conclude, abort).

6.3.1.1.2 VMD Support

The client can thereby query the status of a VMD or the status is reported (unsolicited status); the client can query the different lists of the objects (get name list), the attributes of the VMD (identify), or change the names of objects (rename).

6.3.1.1.3 Domain Management

Using a simple flow control (download, upload, delete domains, etc.), programs and data of arbitrary length can be transmitted between client and server and also a third station (and vice versa). In the case of simple devices, the receiver of the data stream determines the speed of the transmission.

6.3.1.1.4 Program Invocation Management

The client can create, start, stop, and delete modularly structured programs (start, stop, resume, kill, delete, etc.).

6.3.1.1.5 Variable Access

This service allows the client to read and write variables that are defined in the server or a server is enabled to report the contents to a client without being requested (information report). The structures of these data are simple (octet string) to arbitrarily complex (structure of array of …). In addition, data types and arbitrary variables can be defined (read, write, information report, define variable, etc.). The variables constitute the core functionality of every MMS application; therefore, the variable access model will be explained in detail below.

6.3.1.1.6 Event Management

This allows an event-driven operation; i.e., a given service (e.g., read) is only carried out if a given event has occurred in the server. An alarm strategy is integrated. Alarms will be reported to one or more clients if certain events occur. These have the possibility to acknowledge the alarms later (define, alter event condition monitoring, get alarm summary, event notification, acknowledge event notification, etc.). This model is not explained further.

6.3.1.1.7 Semaphore Management

The synchronization of several clients and the coordinated access to the resources of real devices are carried out hereby (define semaphore, take/relinquish control, etc.). This model is not explained further.

6.3.1.1.8 Operator Communication

Simple services for communication with operating consoles integrated in the VMD (input and output). This model is not explained further.

6.3.1.1.9 Journal Management

Several clients can enter data into journals (archives, logbooks), which are defined in the server. Then these data can selectively be retrieved through filters (write journal, read journal, etc.). This model is not explained further.

6.3.1.2 ISO 9506-2 (Part 2): Protocol Specification

If a client invokes a service, then the server must be informed about the requested type of service. For a "read" service, e.g., the name of the variables must be sent to the server. This information, which the server needs for the execution, is exchanged in so-called protocol data units (PDUs). The set of all the PDUs that can be exchanged between client and server constitute the MMS protocol.

In other words, the protocol specification — using ISO 8824 (Abstract Syntax Notation One, ASN.1) and ISO 8825 (ASN.1 BER, the basic encoding rules for ASN.1) — describes the abstract and concrete syntax of the functions defined in Part 1. The syntax is explained below exemplarily.

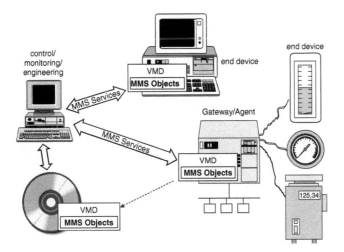

FIGURE 6.5 Location of VMDs.

6.4 Locality of the VMD

VMDs are virtual descriptions of real data and devices (e.g., protection devices, transducers, wind turbines, and any other automation device or system). Regarding the implementation of a VMD, there are three very different possibilities where a VMD can be located (Figure 6.5):

1. *In the end device:* One or several VMDs are in the real device, which is represented by the VMD. The implementations of the VMD have direct access to the data in the device. The modeling can be carried out in such a way that each application field in the device is assigned to its own VMD. The individual VMDs are independent of each other.
2. *In the gateway:* One or several VMDs are implemented in a separate computer (a so-called gateway or agent). In this case, all MMS objects that describe the access to real data in the devices are at a central location. While being accessed, the data of a VMD can be in the memory of the gateway — or it must be retrieved from the end device only after the request. The modeling can be carried out in such a way that for each device or application, a VMD of its own will be implemented. The VMDs are independent of each other.
3. *In a file:* One or several VMDs are implemented in a database on a computer, on a File Transfer Protocol (FTP) server, or on a CD-ROM (the possibilities under 1 and 2 are also valid here). Thus, all VMDs and all included objects with all their configuration information can be entered directly into engineering systems. Such a CD-ROM, which represents the device description, could also be used, for example, to provide a monitoring system with the configuration information: names, data types, object attributes, etc. Before devices are delivered, the engineering tools can already process the accompanying device configuration information (electronic data sheet). The configuration information can also be read later online from the respective VMDs via corresponding MMS requests.

The VMD is independent of the location. This also allows, for example — besides the support during configuration — that several VMDs can be installed for testing purposes on a computer other than the final system (Figure 6.6). Thus, the VMDs of several large robots can be tested in the laboratory or office. The VMDs will be installed on one or several computers (the computers emulate the real robots). Using a suitable communication (for example, intranet or a simple RS 232 connection — available on every PC), the original client (a control system that controls and supervises the robots) can now access and test the VMDs in the laboratory. This way, whole systems can be tested beforehand regarding the interaction of individual devices (for example, monitoring and control system).

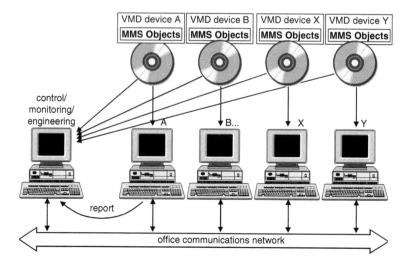

FIGURE 6.6 VMD testing using PC in an office environment.

If the Internet is used instead of the intranet, global access is possible to any VMD that is connected to the Internet. The author tested the access from Germany to a VMD that was implemented on a PLC in the U.S. Through standards like MMS and open transmission systems, it has become possible to set up global communication networks for the real-time process data exchange.

The previous statements about the VMD are also fully valid for all standards that are based on MMS.

6.5   Interfaces

The increasing distribution of automation applications and the exploding amount of information require more and more, and increasingly more complex interfaces for operation and monitoring. Complex interfaces turn into complicated interfaces very fast. Interfaces "cut" components in two pieces; through this, interactions between the emerged subcomponents — which were hidden in one component before — become visible. An interface discloses which functions are carried out in the individual subcomponents and how they act in combination.

Transmitters and receivers of information must likewise be able to understand these definitions. The request "Read T142" must be formulated understandably, transmitted correctly, and understood unambiguously (Figure 6.7). The semantics (named terms that represent something) of the services and the service parameters are defined in MMS. The content, e.g., of named variables is defined in domain-specific standards like IEC 61850.

FIGURE 6.7 Sender and receiver of information.

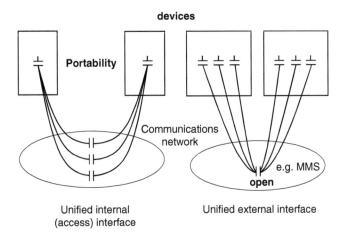

FIGURE 6.8 Internal and external interfaces.

Interfaces occur in two forms:

- Internal program — program interfaces or Application Programming Interfaces (APIs)
- External interfaces over a network (wide area network (WAN), local area network (LAN), fieldbus, etc.)

Both interfaces affect each other. MMS defines an external interface. The necessity of complex interfaces (complex because of the necessary functionality, not because of an end in itself) is generally known and accepted. To keep the number of complex interfaces as small as possible, they are defined in standards or industry standards — mostly as open interfaces. Open interfaces are in the meantime integral components of every modern automation. In mid-1997 it was explained in [22] that the trend in automation engineering obviously leads away from the proprietary solutions to open, standardized interfaces — i.e., to open systems.

The reason why open interfaces are complicated is not because they were standardized. Proprietary interfaces tend more toward being complicated or even very complicated. The major reasons for the latter observation are found in the permanent improvement of the interfaces, which expresses itself in the quick changes of version and in the permanent development of new — apparently better — interfaces. Automation systems of one manufacturer often offer — for identical functions — a variety of complicated interfaces that are incompatible with each other.

At first, interfaces can be divided up into two classes (Figure 6.8): internal interfaces (for example, in a computer) and external interfaces (over a communication network). The following consideration is strongly simplified because, in reality, both internally and externally several interfaces can lie one above the other. However, it nevertheless shows the differences in principle that must be paid attention to. MMS defines an external interface. Many understand MMS in such a way that it offers — or at least also offers — an internal interface. This notion results in completely false ideas. Therefore, the following consideration is very helpful.

The left-hand side of the figure shows the case with a uniform internal interface and varying external interfaces. This uniform internal interface allows many applications to access the same functions with the same parameters and perhaps the same programming language — independent of the external interface. Uniform internal interfaces basically allow the portability of the application programs over different external interfaces.

The right-hand side of the figure shows the case with the external interface being uniform. The internal interfaces are various (since the programming languages or the operating systems, for example, are various). The uniform external interface is independent of the internal interface. The consequence of this is that devices whose local interfaces differ and are implemented in diverse environments can communicate together. Differences can result, for example, from an interface being integrated into the application in a certain device, but being available explicitly in another device. The essential feature of this uniform external interface is the interoperability of different devices. The ISO/OSI Reference Model is aimed at exactly this feature.

The (internal) MMS interface, for example, in a client (perhaps $READ (Par. 1, Par. 2, ... Par. N)), depends on manufacturer, operating system, and programming language. MMS implementations are available for UNIX or Windows NT. On the one hand, this is a disadvantage because applications that want to access an MMS server would have to support, depending on environment, various real program interfaces. On the other hand, the MMS protocol is completely independent of the fast-changing operating system platforms. Standardized external interfaces like MMS offer a high degree of stability, because, in the first place, the communication can hardly be changed arbitrarily by a manufacturer and, in the second place, several design cycles of devices can survive.

Precisely, the stability of the communication, as it is defined in MMS, also offers a stable basis for the development of internal interfaces on the various platforms, such as under Windows 95, NT, or in UNIX environments.

Openness describes in the ISO/OSI world the interface on the wire. The protocol of this external interface executes according to defined standardized rules. For an interaction of two components, these rules have to be taken into account on the two sides; otherwise, the two will not understand each other.

6.6   Environment and General Management Services

MMS uses a connection-oriented mode of operation. That is to say, before a computer can read a value from a PLC for the first time, a connection must be set up between the two.

MMS connections have particular quality features, such as:

- Exclusive allocation of computer and memory resources to a connection. This is necessary to guarantee that all services (for example, five "reads," etc.) allowed to be carried out simultaneously find sufficient resources on both sides of the connection.
- Flow control in order to avoid blockages and vain transmissions if, e.g., the receive buffers are full.
- Segmentation of long messages.
- Routing of messages over different networks.
- Supervision of the connection if no communication takes place.
- Acknowledgment of the transmitted data.
- Authentication, access protection (password), and encoding of the messages.

Connections are generally established once and then remain established as long as a device is connected (at least during permanently necessary communication). If, for example, a device is only seldom accessed by a diagnostics system, a connection then does not need to be established permanently (waste of resources). It suffices to establish a connection and later to close it to release the needed resources again. The connection can remain established for rare but time-critical transmissions. The subordinate layers supervise the connection permanently. Through this, the interruption of a connection is quickly recognized.

The MMS services for the connection management are:

- *Initiate*: Connection setup
- *Conclude*: Orderly connection teardown — waiting requests are still being answered
- *Abort*: Abrupt connection teardown — waiting requests are deleted

Besides these services that are all mapped to the subordinate layers, there are two other services:

- *Cancel*
- *Reject*

After the MMS client has sent a read request to the MMS server, for example, it may happen that the server leaves the service in its request queue and, for whatever reason, does not process it. Using the service cancel, the client can now delete the request in the server. On the other hand, it may occur that the server shall carry out a service with forbidden parameters. Using reject, it rejects the faulty request and reports this back to the client.

Although MMS was originally developed for ISO/OSI networks, a number of implementations are available in the meantime that also use other networks, such as the known TCP/IP network. From the point of view of MMS, this is insignificant as long as the necessary quality of the connection is guaranteed.

6.7 VMD Support

The VMD object consists of 12 attributes. The key attribute identifies the executive function. The executive function corresponds directly with the entity of a VMD. A VMD is identified by a presentation address:

Object:	VMD
Key attribute:	Executive function
Attribute:	Vendor name
Attribute:	Model name
Attribute:	Revision
Attribute:	Logical status (STATE CHANGES ALLOWED, NO STATE CHANGES ALLOWED, LIMITED SERVICES SUPPORTED)
Attribute:	List of capabilities
Attribute:	Physical status (OPERATIONAL, PARTIALLY OPERATIONAL, INOPERABLE NEEDS COMMISSIONING)
Attribute:	List of program invocations
Attribute:	List of domains
Attribute:	List of transaction objects
Attribute:	List of upload state machines (ULSMs)
Attribute:	List of other VMD-specific objects

The attributes *vendor name, model name*, and *revision* provide information about the manufacturer and the device.

The *logical status* defines which services may be carried out. The status "limited services supported" allows that only such services may be executed that have read access to the VMD.

The *physical status* indicates whether the device works in principle.

Two services are used to get the status unsolicited (*unsolicited status*) or explicitly requested (*status*). Thus, a client can recognize whether a given server — from the point of view of the communication — works at all.

The list of capabilities offers clients and servers a possibility to define application-specific agreements in the form of features. The available memory of a device, for example, could be a capability. Through the "get capability list" service, the current value can be queried. The remaining attributes contain the lists of all the MMS objects available in a VMD. The VMD therefore contains an object dictionary in which all objects of a VMD are recorded.

The following three services complete the VMD:

- *Identify* supplies the VMD attributes vendor name, model name, and revision. With that, a plausibility check can be carried out from the side of the client.
- *Get name list* returns the names of all MMS objects. It can be selectively determined from which classes of objects (for example, named variable or event condition) the names of the stored objects shall be queried. Let us assume that a VMD was not known to the client until now (because it is, for example, a maintenance device); the client can then browse through the VMD and systematically query all names of the objects. Using the get services, which are defined for every object

class (e.g., get variable access attributes), the client can get detailed knowledge about a given object (for example, the named variable T142).

• *Rename* allows a client to rename the name of an object.

6.8   Domain Management

Domains are to be viewed as containers that represent memory areas. Domain contents can be interchanged between different devices. The object type domain with its 12 attributes and 12 direct operations, which create, manipulate, delete a domain, etc., are part of the model.

The abstract structure of the domain object consists of the following attributes:

```
Object:         Domain
Key attribute:  Domain name
Attribute:      List of capabilities
Attribute:      State (LOADING, COMPLETE, INCOMPLETE, READY, IN USE)
Constraint:     State (LOADING, COMPLETE, INCOMPLETE)
    Attribute:  Assigned application association
    Attribute:  MMS deletable
    Attribute:  Sharable (TRUE, FALSE)
    Attribute:  Domain content
    Attribute:  List of subordinate objects
Constraint:     State (IN USE)
    Attribute:  List of program invocation references
    Attribute:  Upload in progress
    Attribute:  Additional detail
```

The *domain name* is an identifier of a domain within a VMD.

Domain content is a dummy for the information that is within a domain. The contents of the data to be transmitted can be coded transparently or according to certain rules agreed upon before. Using the MMS version (2003), the data stream can be coded per default in such a way that a VMD can be transmitted completely, including all MMS object definitions that it contains. This means, on the one hand, that the contents of a VMD can be loaded from a configuration tool into a device (or saved from a device) and, on the other hand, that the contents can be stored on a disk per default.

Using a visible string, the list of capabilities describes which resources are to be provided — by the real device — for the domain of a VMD.

MMS deletable indicates whether this domain can be deleted by means of an MMS operation.

Sharable indicates whether a domain may be used by more than one program invocation.

List of program invocation lists those program invocation objects that use this domain.

List of subordinate objects lists those MMS objects (no domains or program invocations) that are defined within this domain: objects that were created (1) by the domain loading, (2) dynamically by a program invocation, (3) dynamically by MMS operations, or (4) locally.

State describes one of the ten states in which a domain can be.

Upload in progress indicates whether the content of this domain is being copied to the client at the moment. MMS defines loading in two directions:

• Data transmission from the client to the server (download)
• Data transmission from the server to the client (upload)

Three phases can be distinguished during loading:

• Open transmission
• Segmented transmission, controlled by the data sink
• Closed transmission

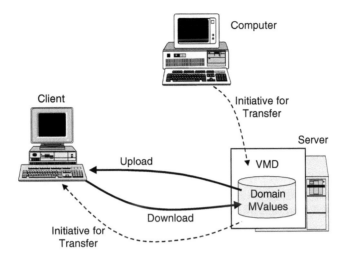

FIGURE 6.9 MMS domain transfer.

Transmission during download and upload is initiated by the client. If the server initiates transmission, then it has the possibility to initiate the transmission indirectly (Figure 6.9). For this purpose, the server informs the client that the client shall initiate the loading. Even a third station can initiate the transmission by informing the server, which then informs the client.

6.8.1 What Is the Domain Scope?

Further MMS objects can be defined within a domain: variable objects, event objects, and semaphore objects. A domain forms a scope (validity range) in which named MMS objects are reversibly unambiguous.

MMS objects can be defined in three different scopes, as shown in Figure 6.10. Objects with VMD-specific scope (for example, the variable Status_125) can be addressed directly through this name by all clients. If an object has a domain-specific scope such as the object Status_155, then it is identified by two identifiers: domain identifier Motor_2 and object identifier Status_155.

A third scope is defined by the application association. The object Status_277 is part of the corresponding connection. This object can only be accessed through this connection. When the connection is closed, all objects are deleted in this scope.

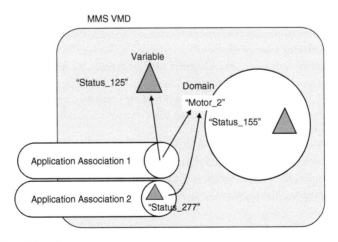

FIGURE 6.10 VMD and domain scope.

MMS objects can be organized using the different scopes. The object names (with or without domain scope) can be compounded from the following character set: the identifiers can contain 1 to 32 characters and they must not start with a number.

The object names can be structured by agreement in a further standard or other specification. Many standards that reference MMS make much use of this possibility. This way, all named variables with the prefix "RWE_" and similar prefixes, for example, could describe the membership of the data (in a trans-European information network) to a specific utility of an interconnected operation.

6.9   Program Invocation Management

A program invocation object is a dynamic element that corresponds with the program executions in multitasking environments. Program invocations are created by linking several domains. They are either predefined or created dynamically by MMS services or created locally.

A program invocation object is defined by its name, its status (idle, starting, running, stopping, stopped, resuming, unrunnable), the list of the domains to be used, and nine operations:

Object: Program invocation
Key attribute: Program invocation name
Attribute: State (IDLE, STARTING, RUNNING, STOPPING, STOPPED, RESUMING, RESET-
                   TING, UNRUNNABLE)
Attribute: List of domain references
Attribute: MMS deletable (TRUE, FALSE)
Attribute: Reusable (TRUE, FALSE)
Attribute: Monitor (TRUE, FALSE)
Constraint: Monitor = TRUE
     Attribute: Event condition reference
     Attribute: Event action reference
     Attribute: Event enrollment reference
     Attribute: Execution argument
     Attribute: Additional detail

Program invocations are structured flatly, though several program invocations can reference the same domains (shared domains). The contents of the individual domains are absolutely transparent both from the point of view of the domain and from the point of view of the program invocations. What is semantically connected with the program invocations is outside the scope of MMS. The user of the MMS objects must therefore define the contents; the semantics result from this context. If a program invocation connects two domains, then the domain contents must define what these domains will do together — MMS actually only provides a wrapper.

The *program invocation name* is a clear identifier of a program invocation within a VMD.

State describes the status in which a program invocation can be. Altogether, seven states are defined.

List of domains contains the names of the domains that are combined with a program invocation. This list also includes such domains that are created by the program invocation itself (this can be a domain into which some output is written).

MMS deletable indicates whether this program invocation can be deleted by means of an MMS operation.

Reusable indicates whether a program invocation can be started again after the program execution. If it cannot be started again, then the program invocation can only be deleted.

Monitor indicates whether the program invocation reports a transition to the client when exiting the running status.

Start argument contains an application-specific character string that was transferred to a program invocation during the last start operation; e.g., this string could indicate which function started the program last.

Additional detail allows the companion standards to make application-specific definitions [3–6].

6.9.1 Program Invocation Services

Create program invocations: This service arranges an operational program, which consists of the indicated domains, in the server. After installation, the program invocation is in the status idle, from where it can be started. The monitor and monitor type indicate whether and how the program invocation shall be monitored.

Delete program invocation: Deletable program invocations are deleted through this service. Primarily, the resources bound to a program invocation are released again.

Start: The start service causes the server to transfer the specified program invocation from the idle into the running state. Further information can be transferred to the VMD through a character string in the start argument. A further parameter (start detail) contains additional information that can be defined by companion standards.

Stop: The stop service changes a specified program invocation from the running to the stopped state.

Resume: The resume service changes a specified program invocation from the stopped to the running state.

Reset: The reset service changes a specified program invocation from the running or stopped to the idle state.

Kill: The kill service changes a specified program invocation from arbitrary states to the unrunnable state.

Get program invocation attribute: Through this service the client can read all attributes of a certain program invocation.

6.10 MMS Variable Model

MMS named variables are addressed using identifiers made up of the domain name and the named variable name within the domain. Components of an MMS named variable may also be individually addressed using a scheme called alternate access. The alternate access address of a component consists of the domain name and the named variable name, along with a sequence of enclosing component names of the path down to the target component.

The variable access services contain an extensive variable model, which offers the user a variety of advanced services for the description and offers access to arbitrary data of a distributed system.

A wide variety of process data is processed by automation systems. The data and their definitions and representation are usually oriented at the technological requirements and at the available automation equipment. The methods the components employ for the representation of their data and the access to them correspond to the way of thinking of their implementers. This has resulted in a wide variety of data representations and access procedures for one and the same technological datum in different components. If, for example a certain temperature measurement shall be accessed in different devices, then a huge quantity of internal details must generally be taken into account for every device (request, parameter, coding of the data, etc.).

As shown in Figure 6.11, the number of the protocols for the access of a client (on the left in the figure) to the data from n servers (S1–Sn) can be reduced to a single protocol (on the right in the figure). Through this, the data rate required for the communication primarily in central devices can be reduced drastically.

In programs, variables are declared; i.e., they get a name, a type, and a value. Described in a simplified way, both the name and the type are converted by the compiler into a memory location and into a reference that is only accessible to the compiled program. Without any further measures, the data of the variable are not identifiable outside the program. It is concealed from the user of the program how a compiler carries out the translation into the representation of a certain real machine.

The data are stored in different ways, depending on the processor; primarily, the data are stored in various memory locations. During the runtime of the program, only this representation is available.

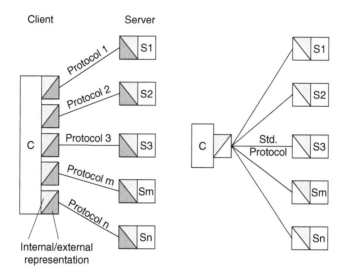

FIGURE 6.11 Unified protocols.

These data are not visible from the outside. They must first be made visible for access from the outside. To enable this, an entity must be provided in the implementation of the application. It is insignificant here whether this entity is separated from or integrated into the program. This entity is acting for all data that shall be accessible from the outside.

The following consideration is helpful in explaining the MMS variable model: What do protocols through which process data are accessed have in common? Figure 6.12 shows the characteristics in principle. On the right is the memory with the real process data, which shall be read. The client must be able to identify the data to be read (gray shade). For this purpose, pointer (start address) and length of the data must be known. By means of this information, the data can be identified in the memory.

Yet how can a client know the pointer of the data and what the length of the data is? It could have this information somehow and indicate it when reading. Yet if the data should move, then the pointer of the data is not correct anymore. There can also be the case where the data do not exist at all at the time of reading but must first be calculated. In this case, there is no pointer. To avoid this, references to

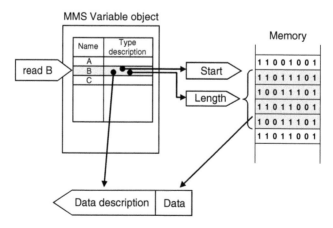

FIGURE 6.12 Data access principle.

the data, which are mapped to the actual pointers (table or algorithm), are used in most cases. In our case, the reference B is mapped by a table to the corresponding pointer and length.

The pointer and length are stored in the type description of the table. The pointer is a system-specific value that generally is not visible on the outside. The length is dependent on the internal representation of the memory and on the type. An individual bit can, for example, be stored as an individual bit in the memory or as a whole octet. However, this is not relevant from the point of view of the communication.

The data themselves and their descriptions are important for the message response of the read service. The question of the external representation (for example, an individual bit encoded as a bit or an octet) is — unlike the internal representation — of special importance here. The various receivers of the data must be able to interpret the data unambiguously. For this purpose, they need the representation that is a substantial component of the MMS variable model. The data description is therefore derived from the type description.

For a deeper understanding of the variable model, three aspects have to be explained more exactly:

- Objects and their functions
- Services (read, write, etc.) that access the MMS variable objects
- Data description for the transmission of the data

The object model of an MMS variable object is conceptually different from a variable according to a programming language. The MMS objects describe the access path to structured data. In this sense, they do not have any variable value.

6.10.1 Access Paths

The access path represents an essential feature of the MMS variable model. Starting from a more complex hierarchical structure, we will consider the concept. An abstract and extremely simplified example was deliberately chosen. Here, we are merely concerned about the principle.

Certain data of a machine shall be modeled using MMS methods. The machine has a tool magazine with n similar tools. A tool is represented, according to Figure 6.13, by three components (tool type, number of blades, and remaining use time).

The machine with its tool magazine M is outlined in the figure on the right. The magazine contains three tools: A, B, and C. The appropriate data structure of the magazine is shown on the left. The structure is treelike; the root M is drawn as the topmost small circle (node). M has three components (branches): A, B, and C, which are also represented as circles. These components in turn also have three components (branches). In this case, the branches end in a leaf (represented in the form of a square). Leaves represent

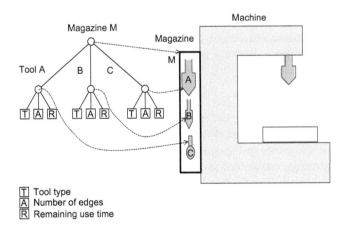

FIGURE 6.13 Data of a machine.

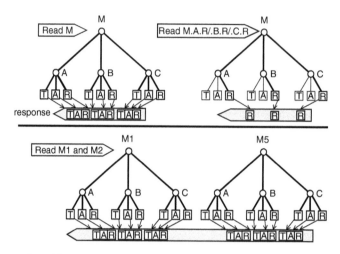

FIGURE 6.14 Access and partial access.

the endpoints of the branches. For each tool, three leaves are shown: T (tool type), A (number of blades), and R (remaining use time). The leaves represent the real data. The nodes ordered hierarchically are merely introduced for reasons of clustering. Leaves can occur at all nodes. A leaf with the information "magazine full/not full" could, for example, be attached at the topmost node.

With this structure, the MMS features are explained in more detail. The most essential aspect is the definition of the access path. The access to the data (and their use) can be carried out according to various task definitions:

1. Selecting all leaves for reading, writing, etc.
2. Selecting certain leaves for reading, writing, etc.
3. Selecting all leaves as components of a higher-level structure, for example, "machine" with the components magazine M and drilling machine.
4. Selecting certain leaves as components of a higher-level structure.

Examples of cases 1 and 2 are shown in Figure 6.14. The case that the complete structure is read is shown in the top left corner (the selected nodes, branches, and leaves are represented in bold lines or squares). All nine data are transmitted as response $(3 \times (T + A + R))$. At first, the representation during transmission is deliberately refrained from here and also in the following examples.

Only a part of the data is read in the top right corner of the figure: only the leaf R of all three components A, B, and C. The notation for the description of the subset M.A.R/.B.R/.C.R is chosen arbitrarily.

The subset M.A.R represents a (access) path that leads from a root to a leaf. It can also be said that one or several paths represent a part of a tree. The read message contains three paths that must be described in the request completely. A path, for example, can also end at A. All three components of A will be transmitted in this case.

Besides the possibility of describing every conceivable subset, MMS also supports the possibility of reading several objects M1 and M5 in a read request simultaneously (see lower half of figure). Of course, every object can only be partly read (not represented).

An example of case 4 is shown in Figure 6.15 (case 4 has to be understood as the generalization of case 3). Here a new structure was defined using two substructures. The object "machine" contains only the R component of all six tools of the two magazines M1 and M5. The object "machine" will not be mixed up with a list (named variable list). Read "machine" supplies the six individual R values.

The component names, such as A, B, or C, need to be unambiguous only below a node — and only at the next lower level. Thus, R can always stand for remaining lifetime. The position in the tree indicates the remaining lifetime of a particular tool. The new structure "machine" has all the features that were described in the previous examples for M1 and M5.

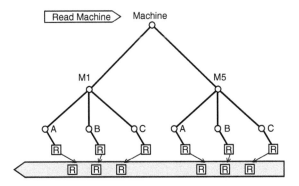

FIGURE 6.15 Partial trees.

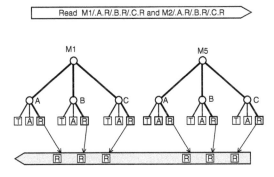

FIGURE 6.16 Partial trees used for read requests.

These features of the MMS variable objects can be applied to (1) the definition of new variable objects and (2) the access to existing variable objects. The second case is also interesting. As shown in Figure 6.16, the same result as in Figure 6.15 can be reached by enclosing the description (only the R components of the tools shall be read from the two objects M1 and M5) in the read request every time. The results (read answer) are absolutely identical in the two cases.

Every possibility to assemble hierarchies from other hierarchies or to read parts of a tree during the access has its useful application. The first case is important in order to avoid enclosing the complete path description every time when reading an extensive part of the tree. The second case offers the possibility of constructing complex structures based on standardized basic structures (for example, the structure of "tool data" consisting of the components T, A, and R) and of using them for the definition of new objects.

Summarizing, it can be stated that the access paths accomplish two tasks:

- Description of a subset of nodes, branches, and leaves of objects during reading, writing, etc.
- Description of a subset of nodes, branches, and leaves of objects during the definition of new objects

In conclusion, this may be expressed in the following way: path descriptions describe the way or ways to a single datum (leaf) or to several data (leaves). A client can read the structure description (complete tree) through the MMS service "get variable access attributes."

Another aspect is of special importance too. Until now, we have not considered the description of leaves. Every leaf has one of the following MMS basic data types:

- Boolean
- Bit
- Integer
- Unsigned

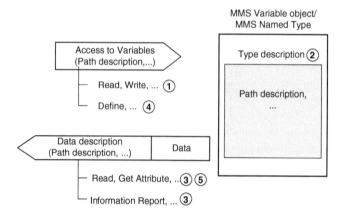

FIGURE 6.17 Application of path descriptions.

- Floating point
- Real
- Octet string
- Visible
- Generalized time
- Binary
- BCD

Every node in a tree is either an array or a structure. Arrays have n elements (0 to n − 1) of the same type (this can be a basic type, an array, or a structure). When describing a part of a tree, any number of array elements can be selected (e.g., one element, two adjacent elements, two arbitrary elements, etc.).

Structures consist of one or several components. Every component can be marked by its position and, if necessary, by an Identifier — the component name. This component name (e.g., A) is used for access to a component. Parts of trees can describe every subset of the structure.

The path description contains the following three elements:

- All of the possibilities of the description of the structures (individual and composite paths in the type description of the MMS variables) are defined in the form of an extensive abstract syntax.
- For every leaf of a structure (these are MMS basic data types), the value range (size) is also defined (besides the class). The value range of the class "integer" can contain one, two, or more octets. The value range four octets (often represented as Int32), e.g., indicates that the value cannot exceed these four octets. On the other hand, with ASN.1 BER coding (explained later) and the value range Int32, the decimal value 5 will be transmitted in only one octet (not in four). That is, only the length needed for the current value will be transmitted.
- The aspect of the representation of the data and their structuring during transmission on the line (communication in the original meaning) is dealt with below in the context of the encoding of messages.

The path description is used in five ways (see Figure 6.17):

- During access to and during the definition of variable objects
- In the type description of variable objects
- In the description of data during reading, for example
- During the definition of named type objects (an object name of its own is assigned to the type description — i.e., to one or several paths — of those objects)
- When reading the attributes of variables and named type objects

6.10.2   Objects of the MMS Variable Model

The five objects of the MMS variable model are:

Description of simple or complex values:
- Unnamed variable
- Named variable

List of several unnamed variables or named variables:
- Named variable list
- Scattered access (not explained here)

Description of the structure by means of a user-defined name:
- Named type

6.10.3   Unnamed Variable

The unnamed variable object describes the assignment of an individual MMS variable to a real variable that is located at a definite address in the device. An unnamed variable object can never be created or deleted. The unnamed variable has the following attributes:

Object:	Unnamed variable
Key attribute:	Address
Attribute:	MMS deletable (FALSE)
Attribute:	Access method (PUBLIC)
Attribute:	Type description

6.10.3.1   Address

Address is used to reference the object. There are three different kinds:

1. Numeric address (nonnegative integer values)
2. Symbolic address (character string)
3. Unconstrained address (implementation-specific format)

Even though in 2 the address is represented by character strings, this kind of addressing has absolutely to be distinguished from the object name of a named variable (see Section 6.8.1 and explanations below).

6.10.3.2   MMS Deletable

The attribute is always FALSE here.

6.10.3.3   Access Method

The attribute is always PUBLIC here.

The attribute points to the inherent abstract type of the subordinate real variable as it is seen by MMS. It specifies the class (bit string, integer, floating point, etc.), the range of the values, and the group formation of the real variable (arrays of structures). The attribute type description is completely independent of the addressing.

Figure 6.18 represents the unnamed variable roughly sketched. The unnamed variable with the address 62 (MMSString) has three components with the names value, quality, and time. These component names are only required if individual components (specifying the path, for example, 62/Value) shall be accessed.

6.10.4   MMS Address of the Unnamed Variable

The MMS address is a system-specific reference that is used by the system for the internal addressing — it is quasi-released for access via MMS. There the address can assume one of three forms (here ASN.1 notation is deliberately used for the first time):

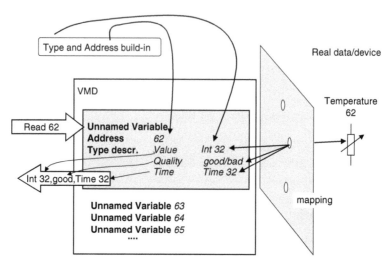

FIGURE 6.18 Unnamed variable object.

```
Address  ::= CHOICE {
numericAddress            [0]  IMPLICIT  Unsigned32
symbolicAddress           [1]  MMSString
unconstrainedAddress      [2]  IMPLICIT  OCTET  STRING
}
```

The definition above has to be read as follows: address defines as (::=) a selection (keyword CHOICE) of three possibilities. The possibilities are numbered here from [0] to [2] to be able to distinguish them. The keyword IMPLICIT is discussed later.

The numeric address is defined as Unsigned32 (four octets). Thus, the addresses can be defined as an index with a value range of up to 2**32. Since only the actual length (e.g., only one octet for the value 65) will be transmitted for an Unsigned32, the minimal length of the index, which can thus be used, is merely one octet. Already, 255 objects (of arbitrary complexity) can be addressed with one octet.

The symbolic address can transmit an arbitrarily long MMSString (for example, DB5_DW6).

The unconstrained address represents an arbitrarily long octet string (for example, 24FE23F2A1hex). The meaning and structure of these addresses are outside the scope of the standard.

These addresses can be used in MMS unnamed variable and named variable objects and in the corresponding services. MMS can neither define nor change these addresses. The address offers a possibility to reference objects by short indexes.

The addresses can be structured arbitrarily. Unnamed variables could, for example, contain measurements in the address range [1000 to 1999], status information in the address range [3000 to 3999], limit values in the address range [7000 to 7999], etc.

6.10.5   Services for the Unnamed Variable Object

6.10.5.1   Read

This service uses the "variable get" (V-Get) function to transmit the current value of the real variable, which is described by the unnamed variable object, from a server to a client. V-Get represents the internal, system-specific function through which an implementation gets the actual data and provides them for the communication.

6.10.5.2   Write

This service uses the V-Put function to replace the current value of the real variable, which is described by the unnamed variable object, by the enclosed value.

6.10.5.3 Information Report

Like "read," but without prior request by the client. Only the read.response is sent by the server to the client without being asked. The information report corresponds to a spontaneous message. The application itself determines when the transmission is to be activated.

6.10.5.4 Get Variable Access Attributes

Through this operation, a client can query the attributes of an unnamed variable object.

6.10.6 Explanation of the Type Description

Features of the structure description of MMS variable objects were explained in principle above. For those interested in the details, the formal definition of the MMS type specification is explained according to Figure 6.19.

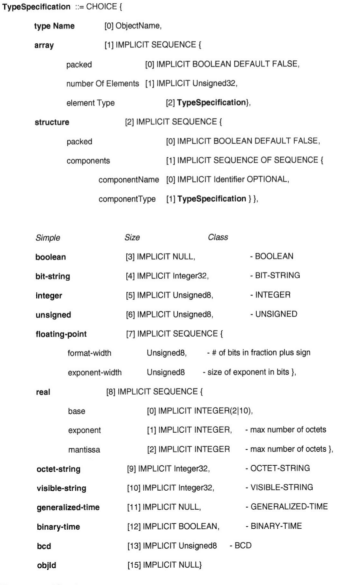

FIGURE 6.19 MMS type specification.

The description in ASN.1 was deliberately selected here too. The type specification is a CHOICE (selection) of 15 possibilities (tags [0] to [13] and [15]). Tags are qualifications of the selected possibility. The first possibility is the specification of an object name, a named type object. If we remember that one named type object describes one or several paths, then the use is obvious. The path description referenced by the name can be used to define a named variable object. Or, if during reading the path must be specified, it can be referenced by a named type object in the server.

Note that the ASN.1 definitions in MMS are comparable with Extensible Markup Language (XML) schema. ASN.1 BER provides very efficient message encoding compared to XML documents. The coming standard IEC 61400-25 applies ASN.1 as well as XML schema for the specification of messages.

The two next possibilities (array and structure) have a common feature. Both refer — through their element type or component type — back to the beginning of the complete definition (type specification). This recursive definition allows the definition of arbitrarily complex structures. Thus, an element of a structure can, in turn, be a structure or an array.

Arrays are defined by three features. *Packed* defines whether the data are stored optimized. *Number of elements* indicates the number of the elements of the array of equal type (*element type*).

The data of structures can also be saved as packed. Structures consist of a series of components (components [1] IMPLICIT SEQUENCE OF SEQUENCE). This series is marked by the keyword SEQUENCE OF, which describes a repetition of the following definition. Next in the list is SEQUENCE {component Name and component Type}, which describes the individual component. Since the SEQUENCE OF (repetition) can be arbitrarily long, the number of the components at a node is also arbitrary.

Then follow the simple data types. They start at tag [3]. The length of the types is typical for the simple data types. For example, integers of different lengths can be defined. The length (size) is defined as Unsigned8, which allows for an integer with the length of 255 octets.

It should be mentioned here that in the ASN.1 description of the MMS syntax, expressions like "integer" (written in small letters) show that they are replaced by another definition (in this case, by tag [5] with the IMPLICIT-Unsigned8 definition). Capital letters at the beginning indicate that the definition is terminated here; it is not replaced anymore. It is here a basic definition.

Figure 6.20 shows an example of an object defined in IEC 61850-7-4. The circuit breaker class is instantiated as XCBR1. The hierarchical components of the object are mapped to MMS (according to IEC 61850-8-1). The circuit breaker is defined as a comprehensive MMS named variable.

The components of the hierarchical model can be accessed by the description of the alternate access: XCBR1 component ST component Pos component stVal. Another possibility of mapping the hierarchy to a flat name is depicted in Figure 6.21. Each path is defined as a character string.

6.10.7   Named Variable

The named variable object describes the assignment of a named MMS variable to a real variable. Only one named variable object should be assigned to a real variable. The attributes of the object are as follows:

Object: Named variable
Key attribute: Variable name
Attribute: MMS deletable
Attribute: Type description
Attribute: Access method (PUBLIC, etc.)
     Constraint: Access method = PUBLIC
     Attribute: Address

6.10.7.1   Variable Name

The variable name unambiguously defines the named variable object in a given scope (VMD specific, domain specific, or application association specific). The variable name can be 32 characters long (plus 32 characters if the object has a domain scope).

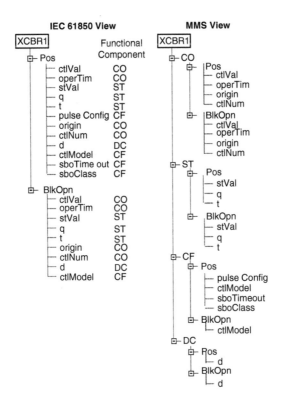

FIGURE 6.20   Example MMS named variable.

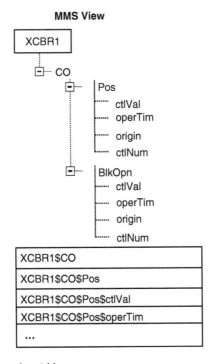

FIGURE 6.21   Example MMS named variable.

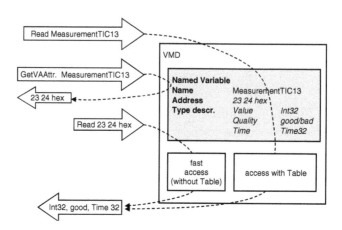

FIGURE 6.22 Address and variable name of named variable objects.

6.10.7.2 MMS Deletable

This attribute shows whether the object may be deleted using a service.

6.10.7.3 Type Description

This attribute describes the abstract type of the subordinate real variable as it represents itself to the external user. This attribute is not inherently in the system, unlike the unnamed variable object; i.e., this type description can be defined from the outside.

6.10.7.4 Access Method

This attribute contains the information that a device needs to identify the real variable. It contains values that are necessary and adequate to find the memory location. The contents lie outside MMS. A special method, the method PUBLIC, is standardized. The attribute address is also available in the case of PUBLIC. This is the address that identifies an unnamed variable object. Named variables can thus be addressed by the name and the ADDRESS (see Figure 6.22).

6.10.7.5 Address

See Section 6.10.3. Defining a named variable object does not allocate any memory because the real variable must already exist; it is assigned to the named variable object with the corresponding name.

Altogether, six operations are defined on the object:

Read: The service uses the V-Get function to retrieve the current value of the real data, which is described by the object.

Write: The service uses the V-Put function to replace the current value of the real data, which is described by the object, by the enclosed value.

Define named variable: This service creates a new named variable object, which is assigned to real data.

Get variable access attribute: Through this operation, a client can query the attributes of a named variable object.

Delete variable access: This service deletes a named variable object if attribute deletable is (=TRUE).

Figure 6.23 shows the possibilities to reference a named variable object by names and, if it is required, also by address (optimal access reference of a given system). For a given name, a client can query the address by means of the service "get variable access attribute."

This possibility allows access through technological names (MeasurementTIC13) or with the optimal (index-) address 23 24 hex.

As shown in Figure 6.23, an essential feature of the VMD is the possibility for the client application to define by request named variable objects in the server via the communication. This includes the definition of the name, the type, and the structure. The name by which the client would like to reference

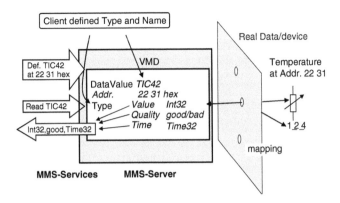

FIGURE 6.23 Client-defined named variable object.

the named variable later is TIC42 here. The first component, "value," is of the type Integer32, the second is "quality" with the values good or bad, and the third is "time" of the type Time32. The type of the data value object can be arbitrarily simple (flat) or complex (hierarchical). As a rule, the data value objects are implicitly created by the local configuring or programming of the server (they are predefined).

The internal assignment of the variable to the real temperature measurement is made by a system-specific, optimal reference. This reference, whose structure and contents are transparent, must be known when defining the named variable, though. The reference can, e.g., be a relative memory address (for example, DB5 DW15 of a PLC). So quick access to the data is allowed.

The named variable object describes how data for the communication are modeled, accessed, encoded, and transmitted. What is transmitted is described independently of the function. From the point of view of the communication, it is not relevant where the data in the server actually come from or where in the client they actually go to and how they are managed — this is deliberately concealed.

Figure 6.24 shows the concrete encoding of the information report message. The message is encoded according to ASN.1 BER.

The encoding using XML would be several times longer than using ASN.1 BER. These octets are packed into further messages that add lower-layer-specific control and address information (e.g., the TCP header, IP header, and Ethernet frames).

The receiving ID is able to interpret the report message according to the identifier, lengths, names, and other values. The interpretation of the message requires the same stack, i.e., knowledge of all layers involved, including the definitions of IEC 61850-7-4, IEC 61850-7-3, IEC 61850-7-2, and IEC 61850-7-1.

6.10.8 Access to Several Variables

6.10.8.1 Named Variable List

The named variable list allows the definition of a single name for a group of references to arbitrary MMS unnamed variables and named variables. Thus, the named variable list offers a grouping for the frequently repeated access to several variables (Figure 6.25). Although the simultaneous access to several MMS variables can also be carried out in a single service (read or write), the named variable list offers a substantial advantage.

When reading several variables in a read request, the individual names and the internal access parameters (pointers and lengths), corresponding to the names in the request, must be searched for in a server. This search can last for some time in the case of many names or a low processor performance. By using the named variable list object, the search is not required — except for the search of a single name (the name of the named variable list object) — if the references, for example, have been entered into the named variables on the list system specifically, and thus optimally. Once the name of the list has been found, the appropriate data can be provided quickly.

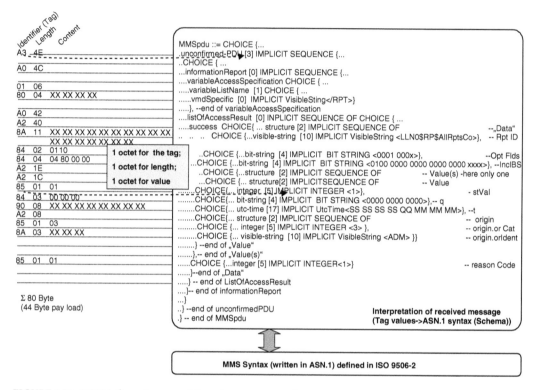

FIGURE 6.24 MMS information report (spontaneous message).

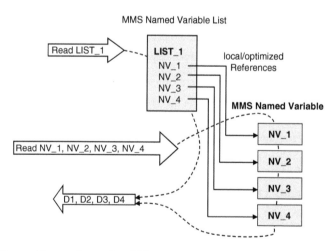

FIGURE 6.25 MMS named type and named variable.

Thus, the named variable list object provides optimal access features for the applications. This object class is used in the known applications of MMS very intensively.

The structure of the named variable list object is as follows:

Object:　　　　　Named variable list
Key attribute:　　Variable list name
Attribute:　　　　MMS deletable (TRUE, FALSE)
Attribute:　　　　List of variable

Attribute: Kind of reference (NAMED, UNNAMED, SCATTERED)
Attribute: Reference
Attribute: Access description

6.10.8.2 Variable List Name

The variable list name unambiguously identifies the named variable list object in a given scope (VMD specific, domain specific, or application association specific). See also MMS object names in Section 6.8.1.

6.10.8.3 MMS Deletable

This attribute shows whether the object may be deleted.

6.10.8.4 List of Variable

A list can contain an arbitrary number of objects (unnamed variable, named variable, or scattered access object).

6.10.8.5 Kind of Reference

Lists can refer to three object classes: named variables, unnamed variables, and scattered access. No named variable lists can be included.

6.10.8.6 Reference

An optimal internal reference to the actual data is assigned to every element of the list. If a referenced object is not (anymore) available, the entry into the list will indicate it. When accessing the list, for example, by "read," no data but an error indication will be transmitted to the client for this element.

6.10.8.7 Access Description

Each variable of the list may be the complete variable. The Access Description may reduce the referenced variable, i.e., only a part of the variable is made visible through the named variable list.

6.10.9 Services

6.10.9.1 Read

This service reads the data of all objects that are part of the list (unnamed variable, named variable, and scattered access object). For objects that are not defined, an error is reported in the corresponding place of the list of the returned values.

6.10.9.2 Write

This service writes the data from the write request into the objects that are part of the list (unnamed variable, named variable, and scattered access object). For objects that are not defined, an error is reported in the corresponding place of the list of the returned values.

6.10.9.3 Information Report

This is just like the "read" service, where the read data are sent by the server to the client without prior request (read request) by the client, i.e., as if only a "read response" would be transmitted.

6.10.9.4 Define Named Variable List

Using this service a client can create a named variable list object.

6.10.9.5 Get Named Variable List Attributes

This service queries the attributes of a named variable list object.

6.10.9.6 Delete Named Variable List

This service deletes the specified named variable list object.

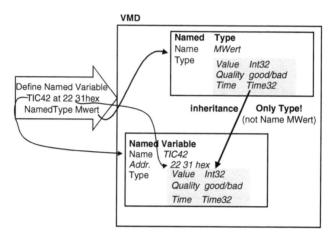

FIGURE 6.26 Inheritance of type of the MMS named type objects.

6.10.9.7 Named Type Object

The named type object merely describes structures. The object model is very simple:

Object:	Named type
Key attribute:	Type name
Attribute:	MMS deletable (TRUE, FALSE)
Attribute:	Type description

The essential attribute is the type description, which was already discussed before for named and unnamed variables. On the one hand, TASE.2 standard data structures can be specified by means of named types. This is the most frequent application of the named type objects. On the other hand, named types can be used for access to the server. The read request can refer to a named type object. Or the named type object will be used to define named variables.

Figure 6.26 describes the application of the named type objects for the definition of a named variable. A variable will be created by the request "define named variable."

It shall have the name TIC42, the address 22 31 hex, and the type that is defined in the named type object, MWert. The variable inherits the type from the named type object. The inheritance has the consequence that the variable will have only the type — not the name of the named type object. This inheritance was defined so strictly in order to avoid, through deleting the named type, the type of the variable becoming undefined or, by subsequent definition of a differently structured named type with the old type name, MWert, the type of the variable being changed (the named type object and the new type description would be referenced by the old name).

Perhaps it is objected now that this strict inheritance has the consequence that the type also would have to be saved for each variable (even though many variables have the same type description). Since these variables can internally be implemented in a system in whatever way the programmer likes, they can refer through an internal index to a single type description. He must only make sure that this type description is not deleted. If the accompanying named type object gets deleted, then the referenced type description must remain preserved for these many variables.

The disadvantage that the name of the "structure mother," i.e., the named type, is not known anymore as an attribute of the variables has been eliminated in the MMS revision.

6.10.9.8 Define Named Type

This service creates a named type object.

6.10.9.9 Get Named Type Attribute

This service delivers all attributes of a named type object.

Read, write, define named variable, define scattered access, define named variable list, and define named type use the type description of the named type object when carrying out their tasks.

6.11 Conclusion

MMS is a standard messaging specification (comparable to Web services), widely implemented by industrial device manufacturers like ABB, Alstom, General Electric, and Siemens. It solves problems of heterogeneity so often found in automation applications. MMS is the *lingua franca* of industrial devices.

MMS provides much more than TCP/IP, which essentially offers a transfer stream of bytes. MMS transfers commands with parameters between machines.

MMS allows a user to concentrate on the applications and the application data to be accessible — and not on communication problems, which are already solved. It provides a basis for the definition of common and domain-specific semantics. Examples are the standards IEC 60870-6 TASE.2, IEC 61850, and IEC 61400-25.

References

1. ISO 9506-1, Manufacturing Message Specification (MMS): Part 1: Service Definition, 2003.
2. ISO 9506-2, Manufacturing Message Specification (MMS): Part 2: Protocol Definition, 2003.
3. ISO/IEC 9506-3, Manufacturing Message Specification (MMS): Part 3: Companion Standard for Robotics, 1991.
4. ISO/IEC 9506-4, Manufacturing Message Specification (MMS): Part 4: Companion Standard for Numerical Control, 1992.
5. ISO/IEC CD 9506-5, Manufacturing Message Specification (MMS): Part 5: Companion Standard for Programmable Controllers, 1993.
6. ISO/IEC 9506-6, Manufacturing Message Specification (MMS): Part 6: Companion Standard for Process Control, 1993.
7. ESPRIT Consortium CCE-CNMA, Preston, U.K. (Editors), MMS: A Communication Language of Manufacturing, Berlin: Springer-Verlag, 1995.
8. ESPRIT Consortium CCE-CNMA, Preston, U.K. (Editors), CCE: An Integration Platform for Distributed Manufacturing Applications, Berlin: Springer-Verlag, 1995.
9. Inter-control center communication, *IEEE Transactions on Power Delivery*, 12, 607–615, 1997.
10. IEC 60870-6-503, Telecontrol Equipment and Systems: Part 6: Telecontrol Protocols Compatible with ISO Standards and ITU-T Recommendations: Section 503: Services and Protocol (ICCP Part 1), 1997.
11. IEC 60870-6-802, Telecontrol Equipment and Systems: Part 6: Telecontrol Protocols Compatible with ISO Standards and ITU-T Recommendations: Section 802: Object Models (ICCP Part 4), 1997.
12. März, W. and K. Schwarz, Powerful and open communication platforms for the operation of interconnected networks, in *Proceedings of ETG-Tage/IEEE PES*, Berlin, 1997.
13. IEEE Technical Report 1550, Utility Communications Architecture, UCA, http://www.nettedauto mation.com/standardization/IEEE_SCC36_UCA, 1999.
14. Becker, G., W. Gärtner, T., Kimpel, V. Link, W. März, W. Schmitz, and K. Schwarz, *Open Communication Platforms for Telecontrol Applications: Benefits from the New Standard IEC 60870-6 TASE.2 (ICCP), Report 32*, VDE-Verlag, Berlin, 1999.
15. Wind Power Communication: Verification Report and Recommendation, Elforsk rapport 02:14; Stockholm, April 2002, www.nettedautomation.com/download/02_14_rapport.pdf.
16. IEC 61850-7-1, Communication Networks and Systems in Substations: Part 7-1: Basic Communication Structure for Substation and Feeder Equipment: Principles and Models, 2003.

17. IEC 61850-7-2, Communication Networks and Systems in Substations: Part 7-2: Basic Communication Structure for Substation and Feeder Equipment: Abstract Communication Service Interface (ACSI), 2003.
18. IEC 61850-7-3, Communication Networks and Systems in Substations: Part 7-3: Basic Communication Structure for Substation and Feeder Equipment: Common Data Classes, 2003.
19. IEC 61850-7-4, Communication Networks and Systems in Substations: Part 7-4: Basic Communication Structure for Substation and Feeder Equipment: Compatible Logical Node Classes and Data Classes, 2003.
20. IEC 61850-8-1, Communication Networks and Systems in Substations: Part 8-1: Specific Communication Service Mapping (SCSM): Mappings to MMS (ISO/IEC 9506-1 and ISO/IEC 9506-2) and to ISO/IEC 8802-3, 2004.
21. IEC CD 61400-25, Wind Turbines: Part 25: Communications for Monitoring and Control of Wind Power Plants, 2004.

Further Resources

See the following Web pages for additional information:
http://www.livedata.com
http://www.sisconet.com
http://www.tamarack.com
http://www.scc-online.de
http://www.nettedautomation.com
http://www.nettedautomation.com/qanda/iec61850/information-service.html# (IEC 61850 circuit breaker model)

Section 3.5

Java Technology in Industrial Automation and Enterprise Integration

7

Java Technology and Industrial Applications

Jörn Peschke
Otto-von-Guericke University of Magdeburg, Germany

Arndt Lüder
Otto-von-Guericke University Magdeburg, Germany

7.1 Introduction New Programming Paradigms in Industrial Automation

Industrial automation is currently characterized by a number of trends induced by the current market situation. The main trends are the pursuit of high flexibility, good scalability, and high robustness of automation systems; the integration of new technologies; and the harmonization of used technologies in all fields and levels of automation. Of special interest is the integration of technologies, which were originally developed for the office world into the control area. This trend is characterized by the emergence of industrial PCs, operating systems for embedded devices like WindowsCE, embedded Linux and RTLinux, data presentation technologies like XML, and communication technologies such as Ethernet and Internet technologies.

In this context, object-oriented languages like Java have gained importance. Particularly Java, which has evolved with the Internet and related technologies, fits very well in different areas of industrial automation.

In the following sections, the boundary conditions, advantages, and problems of using Java in the area of industrial automation will be described in more detail. A special focus will be on the lower levels of the automation pyramid, where real-time requirements are of significance.

7.2 Requirements in Automation and Typical Application Areas of Java

Java can be characterized as a high-level, consequently object-oriented programming language with a strong type-system. One important feature is the idea of a Virtual Machine abstracting the concrete underlying hardware by translating Java programs to an intermediate language called Bytecode, which is executed on the Java Virtual Machine (JVM). Thereby, the concept "Write Once, Run Anywhere" (WORA) is realized, enabling a platform-independent application design. This means that Java applications are (of course, under consideration of different Java versions and editions) executable on every platform, which can run a JVM. Together with typical concepts of object-orientation, like encapsulation of functionalities in classes, inheritance, and polymorphism, Java opens many possibilities for a reusability of the code. Java also provides high stability of applications. This is realized by extensive checks (e.g., regarding type or array boundaries) during compile-, load-, and runtime. Due to the fact that error-prone concepts like direct pointer manipulations are beyond the scope of the language and memory allocation (and deallocation) is realized by an automatic memory management, the so-called Garbage Collection (GC), the efficiency of software development with Java is very high. Although it is not easy to specify this in figures, some sources state at least 20% improvement compared to C/C++ [9]. In contrast to the most used PLC programming languages specified in the IEC 61131, the efficiency increases by more than 50%. The extensive, easy-to-use networking abilities of Java can help to reduce the difficulty of programming distributed systems.

Besides general advantages, the typical potential application areas for Java in industrial automation and their specific requirements have to be considered. On the upper levels of the automation pyramid (ERP and SCADA/MES-Systems), Java is already used as one alternative. The requirements in this area are very similar to typical IT-applications and characterized by powerful hardware. Here, Java2SE (Standard Edition) and Java2EE (Enterprise Edition) with a wide variety of APIs supporting different technologies for communication, visualization, and database access are the proper Java platforms. Examples for Java applications can mainly be found for different interface realizations of ERPs and the implementation of advanced technologies on the MES level [11].

At the field device level, Java is still not a very common language as the requirements here are a problematic issue for standard Java versions. Many features of Java, normally responsible for typical advantages, cause problems at the field device level. The hardware on this level is very heterogeneous and most devices have only limited resources with respect to memory and computing power. For communication purposes, several different field bus protocols are used, although the increasing relevance of Ethernet-based protocols shows potentials for a common communication medium [15].

For indicator and control elements, Java provides powerful APIs for user-interface programming (ATW and Swing) and there are no special requirements regarding computing power. In contrast to this, for control devices like PLC, SoftPLC, or IPC, there are constraints that do not fit in every case to the features of standard Java. These devices are characterized by:

- real-time requirements for application parts that realize control functionalities (cyclic control program execution with a defined cycle time)
- direct hardware access to the I/O level
- today's usage of PLC typical programming concepts (e.g., IEC 61131)

A special case are smart I/O-devices where the computing resources are even more limited than in normal PLCs.

7.3 Problems of Using Java at the Field Level under Real-Time Conditions

It can be stated that for the use of standard Java in non-real-time conditions with limited resources, several problems have to be solved, even if the advantages mentioned above make Java an interesting programming language for control engineering.

Before describing the existing problems, the usual necessities for the design of applications in the field control area will be given. A (distributed) control application requires, of course, real-time behavior as well as the realization of communication with other applications, field I/O, and/or remote I/O. Normally, hardware access regarding memory allocation, access to local I/O, and similar things are also necessary. Figure 7.1 gives an overview of the strengths and problems of Java against the background of these requirements. In general, Java provides advantages for tasks like user interaction over an HMI or communication with other applications/remote I/Os over different protocols. The close relation of Java to the Internet world makes it easy to support communication via protocols like http, ftp, or SMTP.

Unfortunately, other features of Java make it difficult to use it in control engineering without modifications or enhancements. This concerns the resource consumption of Java as well as the real-time capabilities and direct access to the hardware [6]. In the following sections, the reasons for these problems shall be described more in detail.

7.3.1 Resource Consumption

With respect to resource consumption, for the use of Java, two requirements have to be taken into account: first, the necessity to run a JVM on the target system (depending on the system architecture) and second, the need for large standard libraries, if the full features of Java are to be used.

To ensure the applicability of Java on different systems in a common way, three Java2 Editions were introduced. One of them, Java2ME (Micro Edition), was specially developed with respect to limited and embedded devices. With Java2ME consisting of two different configurations and offering the possibility to add profiles and optional packages, it is possible to tailor a Java platform to the special needs of devices and application areas. While the Connected Device Configuration (CDC) is used for devices with a memory of 2 MB and more, the Connected Limited Device Configuration (CLDC) was designed for smaller devices. With enhancements like personal profile and RMI-optional packages, the CDC provides nearly the same functionality as J2SE. The CLDC-based K(kilo)-VM is much more limited but runs on devices with memory down to 128 KB.

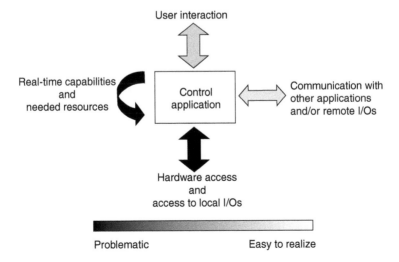

FIGURE 7.1 Requirements around a control application.

The more serious problems are the insufficient real-time capabilities of Java. Besides others, especially some features normally providing basic advantages of Java are problematic: the concept of JVM and GC.

7.3.2 Execution Speed and Predictability

Initially, the interpretation of Bytecode by the JVM was the rule. To improve the performance, later techniques such as just-in-time-compiler (JIT) were introduced. In a JIT, methods are translated into machine-code of the current platform during the execution. Hence, at their next invocation they can be executed much faster. For real-time systems, both variants are not sufficient. The interpretation may be too slow (depending on the speed requirements), and the JIT is highly nondeterministic as the execution time of a method can depend on how many times it was performed before! For a real-time system, the worst case is always important and here a JIT compiler is normally not better than an interpreter. Hence, the solution for such systems is to use ahead-of-time compiling (AOT). Here, the application is either translated completely *before* the execution or *during* the class-loading. Both methods enable a considerable optimization and especially the first one reduces the resource consumption on the target system. Unfortunately, with a complete compilation in advance, an important Java feature — the dynamic class loading — is no longer possible. The compilation during class loading retains this possibility but needs more resources on the target system (class loader).

7.3.3 Garbage Collection

Although the Java Language Specification (JLS) does not require a GC, most JVM implementations use one for automatic memory management. A GC scans the memory for objects with no references and frees the memory, mostly also followed by a defragmentation of memory.

The GC is normally not interruptible (except incremental GC, which is rarely used) and so it stops any execution of application threads at an unpredictable point of time for an unknown time span. This is an extremely nondeterministic behavior and, of course, not acceptable under real-time constraints.

7.3.4 Synchronization/Priorities

The possibilities for thread scheduling and prioritization offered by Java are insufficient. Real-time systems require a reasonable number of priorities to enable the use of different scheduling strategies. Against this requirement, the number of priorities (the JLS defines 10) is too small. Usually, it is not guaranteed that the Java threads are mapped to threads at the operating system level. However, this is a prerequisite for a correct scheduling in a real-time operating system (RTOS). Although there is a concept for synchronization of threads (keyword `synchronized`) with monitors, no mechanisms for avoidance of priority inversion* are specified by the JLS.

7.3.5 Hardware Access

Finally, as a result of the concept of platform independence realized by the JVM (abstracting the underlining hardware), direct hardware access is not possible in Java. This applies to direct memory access, as well as for specific device functionalities. Although this is no direct requirement for a real-time system, it can be necessary for the I/O-access in control engineering.

All these problems regarding different aspects of standard Java implementations result in the necessity for special modifications or enhancements if Java is to be used with field-level devices under real-time constraints.

---

*Priority inversion occurs if a low-priority thread using a resource is pre-empted by a medium-priority thread. If now a high-priority thread also needs the resource, it is blocked until the low-priority thread is executed again and can free the resource.

7.4 Specifications for Real-Time Java

Although the creation of a uniform specification for real-time behavior in Java is vitally important right against the background of platform independence, it was not possible to find a common approach of all parties thereto. At the moment, there are suggestions for the enhancement/modification of the standard-Java-specification by two consortia. Both of them have a goal to solve the problems stated above and make Java applicable to real-time systems.

7.4.1 Real-Time Specification for Java

The first consortium is the "Real-Time for Java Expert Group," under the leadership of Sun Microsystems, which has developed the "Real-Time Specification for Java" (RTSJ) [2]. The formal scope for the development of this specification is the Java Community Process (JCP), where the RTSJ runs as Java Specification Request (JSR)-000001. The JCP defines a procedure that requires an internal and public review of the draft specifications as well as a reference implementation and a test suite called "Technology Compatibility Kit" (TCK).

The JSR-000001 has reached the state of Final Release on 07 January, 2002, and a reference implementation has been developed by TimeSys Corporation [13].

General principles for the development of the RTSJ were to guarantee the temporally predictable execution of Java programs and to support the current real-time application development practice. For the realization of real-time features, the boundary conditions were set so that no syntactic extension of the Language needed to be introduced, and the backward compatibility was to be maintained by mapping Java Language Semantics (JLS) to appropriate entities providing the required behavior under real-time conditions. Furthermore, the RTSJ should be appropriate for any Java Platform.

The more general WORA concept of standard Java is replaced by the so-called Write Once Carefully Run Anywhere Conditionally (WOCRAC) principle. This is necessary because the influence of the platform has to be taken into account. For instance, if the performance of the platform is not fast enough to meet some deadlines (this should normally not be the case, but may be, if the original platform was much faster), this is a serious problem that cannot be ignored.

In this context, the goal of the RTSJ is not to optimize the general performance of a JVM, but to improve the features causing problems for real-time systems like the GC, synchronization mechanisms, and the handling of asynchronous events.

To achieve this, the RTSJ enhances the Java specification in seven areas. All these additions improve critical aspects in the behavior of Java, or add typical programming features for real-time system development [10]. In the following, these enhancements will be explained more in detail.

7.4.1.1 Thread Scheduling and Dispatching

The RTSJ introduces real-time threads with scheduling attributes to improve the scheduling possibilities compared to Standard-Java. Every real-time thread (implementing the interface `schedulable`) has a reference to a scheduler. Although the RTSJ is open for implementations of various scheduling principles, the only required version is (as default scheduler) a strict fixed-priority pre-emptive scheduling for real-time threads. For these threads, at least 28 unique priority levels are required (plus the ten priorities of the JLS). The assignment of thread-priorities is, as usual in most real-time systems, left to the programmer. The algorithm for the determination of the next thread to run (feasibility algorithm) requires an assignment of priorities following the Rate-Monotonic Analysis.

In the RTSJ, the interface `scheduleable` is implemented by `RealTimeThreads` and `AsyncEventHandlers`. It contains several parameters to describe relevant data for the scheduling. For instance, the abstract class `SchedulingParameters` is used as the base class for the `PriorityParameters` describing the priority of a thread. Besides the basic RealTimeThread, which may use memory areas other than the normal heap, the `NoHeapRealtimeThread` must be created with a scoped memory area (see next paragraph) and is able to preempt the GC. Hence, there are strong restrictions regarding the interaction with objects on the heap (no reading or writing, nor manipulations of references except objects in the ImmortalMemory).

7.4.1.2 Memory Management

The RTSJ provides different kinds of memory to allocate objects outside the memory area controlled by the garbage collector. As the JLS does not require a GC (although nearly every JVM-implementation has one), the RTSJ also does not define the necessity for a GC. ScopedMemory is used to define the lifetime of objects depending on the syntactical scope. If such a scoped memory area is entered, every use of a new statement results in an allocation of memory exclusively within this area. These objects are not garbage collected, but live until the control flows out of this scope. Then the whole memory area will be reclaimed. A scope (real-time thread or closure) can be associated with more than one memory area and, also, a scoped memory area can be associated with one or more scopes. There are two types of this kind of memory distinguishable by the relation of allocation time to object size: linear for LTMemory and variable for VTMemory.

Objects created within the third kind of memory, the ImmortalMemory, will never be affected by any GC and will exist until the Java runtime terminates. For each JVM, there is only one ImmortalMemory, which is used by all real-time threads together. The RTSJ provides a restricted support for memory allocation budgets, which can be used for defining the maximal memory consumption of a thread.

7.4.1.3 Synchronization and Resource Sharing

The RTSJ specifies an implementation of the synchronized primitive, which avoids an unbound priority inversion by using the priority inheritance protocol.[*] Using this method, problems can occur when a RealTimeThread is synchronized with a normal thread. A NoHeapRealTimeThread generally has a higher priority than the garbage collector, and a normal thread always has a priority below. Thus, a Java thread cannot have the same priority as a RealTimeThread. To solve this problem, the RTSJ introduces special wait-free queue classes with unidirectional data flow and nonblocking read/write methods.

7.4.1.4 Asynchronous Event Handling

The RTSJ introduces a mechanism to react on events that occur asynchronously to the program execution. Therefore, a schedulable object (AsyncEventHandler) is bound to an event (represented by an instance of the class AsyncEvent). If an event occurs, the event handler changes its state to "ready" and the event will then be scheduled like any other schedulable object (implementing the interface Schedulable).

7.4.1.5 Asynchronous Transfer of Control

Similar to the normal exception handling in Java, the RTSJ defines the possibility to transfer the flow of control to a predetermined point in the program. It is important to note that the possibility of an interruption has to be declared explicitly before the execution (interface interruptible).

7.4.1.6 Asynchronous Thread Termination

As the current JLS does not provide a mechanism to terminate a thread (the existing method thread.stop is marked as deprecated as it can cause inconsistencies), the RTSJ provides such a capability. It is realized by using asynchronous event handling and asynchronous transfer of control, and typically is deployed to terminate a thread when external events occur.

7.4.1.7 Physical Memory Access

The RTSJ defines two low-level mechanisms for a direct memory access. RawMemoryAccess represents a direct addressable physical memory. The content of this memory can be interpreted, for example, as byte, integer, or short (RawMemoryAccessFloat can be used for floating point numbers). The classes

---

[*] If a low-priority task holds a resource and therefore blocks a high-priority task, its priority is increased to the same as the high-priority task.

ImmortalPhysicalMemory and ScopedPhysicalMemory provide the possibility to allocate Java objects in the physical memory.

For practical use, the RTSJ has several powerful easy-to-use mechanisms. Other features like the immortal memory are potentially dangerous and have to be used very carefully. As objects in the immortal memory never free their allocated memory (until the runtime is shutdown), continued or periodical object creation can easily lead to a situation where the system runs out of memory.

Combining the RTSJ with a real-time GC [5], as explained in [7], can overcome some limitations. In doing so, it is possible to access the heap from the real-time part without limitations and to directly synchronize real-time and non-real-time parts of an application. Unfortunately, the resulting application will not be executable on each RTSJ-compliant JVM.

Also, some conceptual problems of the RTSJ have to be noticed. While the RTSJ provides a defined API, some aspects of the behavior can depend on the underlying RTOS. The API is the same, but the semantics may change! This particularly applies to scheduling and the possibility of implementing scheduling strategies other than the primary scheduler.

The reference implementation (RI) of the RTSJ is provided by TimeSys Corporation. The RI runs on all Linux versions, but the priority inheritance protocol is only supported by TimeSys Linux — a special real-time-capable Linux version. The JVM is based on the JTime-system VM (corresponding to the J2ME-CDC). As the JVM works in an interpreting mode, it is a good system for testing and getting experiences. For product development, one of the available commercial implementations of the RTSJ would be a better choice. Meanwhile, several companies have developed RTSJ-compliant platforms or implemented at least some of the basic concepts of the RTSJ [7, 13, 14].

7.4.2 Real-Time Core Extensions

The second consortium working on real-time specifications for Java is the JConsortium. One group within the JConsortium is the Real-Time Java Working Group (RTJWG). This working group has created the Real-Time Core Extensions (RTCE), finished in September 2000 [2].

The general goal of this specification can be characterized by reaching real-time behavior for Java applications with a performance regarding throughput and memory footprint comparable to compiled C++ code on conventional RTOS. Thus, this specification aims at providing a direct alternative to the existing real-time technologies. It was assumed that for typical applications in this field, the cooperation between the real-time part and the non-real-time part is limited.

As a result of these requirements, the general idea of the RTJWG was to define a set of standard real-time operating system services wrapped by a Java API (the "Core"). In doing so, the standard JVM was not changed, but extended by a Real-Time Core Execution Engine. The components of the core are portable, can be dynamically loaded, and may be extended by profiles providing specialized functionalities. Consequently, all objects implementing real-time behavior are derived from org.rtjwg.Core-Object, which is similar to Java.lang.object. This means the separation between real-time ("Core") and non-real-time ("baseline Java") is type based.

Following the concept of limited cooperation between core components and baseline Java, the core objects have some special characteristics regarding memory management, synchronization, or scheduling. These will be described in the following subsections in more detail.

7.4.2.1 Memory Management

All real-time objects are allocated the core-memory area and are not affected by the GC until the appropriate task* terminates. The RTCE also specifies a stack allocation of objects, which provides a more efficient allocation and better reliability and has a performance that is easier to predict. To allocate an object on the stack, it has to be declared as "stackable" and some restrictions have to be observed; for example, a value of a stackable variable cannot be copied to a nonstackable variable.

---

*In the RTCE, mainly the term "task" is used instead of "thread," which is not unusual for real-time systems.

If a task terminates, all objects of this allocation context may be eligible for reclamation; but as there might still be references from the baseline objects, the core engine has to verify if this is the case or not. If not, the objects can be reclaimed.

7.4.2.2   Synchronization

To avoid priority inversion, the RTCE supports two protocols: the typical priority inheritance protocol and the priority ceiling for "protected objects." The problem of synchronization between baseline and core components (as described for real-time tasks in the RTSJ section) occurs here in the form that the synchronization protocols must not cause unpredictable delays in the real-time tasks. The mechanism to avoid this is the usage of a BufferPair (write/read buffer) in combination with the priority ceiling protocol.

7.4.2.3   Scheduling

The RTCE defines 128 priorities, which are all above the normal Java priorities, so that the GC cannot preempt core tasks. The assignment of priorities is done by the programmer and typically a preemptive priority-based scheduling is used.

7.4.2.4   Cooperation with Baseline Java Objects

As core objects are not garbage collected before release, they must not access baseline Java objects. In contrast, objects in the baseline heap may invoke some special methods to access core objects (although the most fields and methods of core objects are not visible for baseline objects).

Besides this basic behavior, the RTCE provides a set of further features, which are useful for the programming of real-time systems, like signaling and counting semaphores as well as interrupts and I/O ports with integrity. For the practical handling of the RTCE, it has to be noticed that the specification was developed under the assumption that core programmers are "trusted experts" (in contrast to baseline Java programmers) who are aware of the typical problems and pitfalls of real-time systems and can benefit from the functionality provided by the RTCE.

7.4.3   Real-Time Data Access

Another working group within the JConsortium is the RTAWG. The focus of this group is the application field of industrial automation. Hence, the resulting "Real-Time Data Access" (RTDA)-specification focuses on an API for accessing I/O-data in typical industrial and embedded applications rather than on features supporting hard real-time requirements.

Analyzing the typical requirements for the usage of Java in industrial applications, the general idea behind the RTDA is that the support of real time is a basic requirement, but that hard real time with sophisticated features is only needed in a few cases. More important is a concept for a common access to I/O data for real time and non-real-time parts of an application as well as the support of typical traditional procedures regarding configuration and event handling in this domain.

Following these conditions, the RTDA considers real-time capabilities as a prerequisite, assuming that the real-time and non-real-time applications run on a real-time-capable JVM.

7.4.3.1   Real-Time Aspects

The RTDA supports real time by using permanent objects and creating an execution context for these objects. Permanent objects are not garbage collected and have to implement the interface Permantent-MemoryInterface, or the "permanent memory creation" has to be enabled (by invoking the method enable() of the class PermantentMemory). As the main intention is to create all these objects in a setup phase, the permanent objects can also be created by a non-real-time thread. This principle simplifies real-time programming, as the problems of memory allocation for object creation under time constraints and availability of memory are not critical for such a procedure.

The RTDA requires 32 priorities and a system priority for Interrupt Service Routines (ISR) and non-Java threads. The ranges are defined by the constants MIN_PRIORITY – MAX_PRIORITY (typically,

1–10) for the non-real-time threads and MIN_RT_PRIORITY – MAX_RT_PRIORITY for real-time threads (12–32); the GC runs typically with priority 11.

The existing `synchronized` keyword will be used, but with different behavior, depending on the context: usage of the priority inheritance protocol for real-time threads and permanent objects and raising the priority for permanent objects in normal threads to block out the GC.

7.4.3.2 Event Handling and I/O-Data Access

As the event handling and the I/O-access are the core components of the RTDA, it defines a dynamic execution model supporting asynchronous as well as synchronous access to I/O -data. The main instances responsible for creation of appropriate objects are the Event-Managers (asynchronous) and the Channel-Managers (synchronous). Every Event-Manager is created out of a `DataAccessThreadGroup`, which controls the priority of this Event-Manager. Depending on the type of event, there are different Event-Managers like:

- IOInterruptEventManager   (InteruptEvent)
- IOTimerEventManager   (PeriodicEvent or OneShotEvent)
- IOSporadicEventManager   (SporadicEvent)
- IOGenericEventManager   (all types of events, more than one event)

For the realization of a hardware-independent I/O-data access, three components are important: "Device-Descriptions," the classes `IONode`/`IONodeLeaf`, and the I/O-Proxy classes.

A DeviceDescription specifies a hardware component (static in a native DLL or as Java class). As this description does not represent a concrete instance of a hardware, but describes the type of a device, the concrete instantiation is realized by an `IONodeLeaf`-object.

The complete overall (hierarchical) structure of a given I/O-system is described by a tree of `IONodes`. It represents the concrete configuration of a system where the leaves of this tree are objects of the type `IONodeLeaf`. The configuration realizes a mapping on memory addresses as well as on I/O-space. The access path of an `IONodeLeaf` can be described in a way starting from the root node down to the `IONodeLeaf`. During the configuration, which is executed in the setup phase of the Java system, the instances of the `IONodes` (describing a device node) are created and a mapping regarding the addresses is executed.

The entity to access the I/Os in the application is the I/O-proxy, representing physical or software entities. These proxies are generated out of a `IOChannelManager` and can be identified by a physical or a symbolic name. The mapping between both these names takes place in a name map table of the appropriate `IOChannelManager`. Thereby, the design of hardware-independent applications is enabled if only symbolic names are used. Hence, changes in the hardware require only changes within the map table and not within the application.

`IOChannel` is the superclass for all types of I/O-proxies. Every I/O-proxy has a cache for holding the appropriate I/O-data and provides methods for updating the cache from the input channel as well as flushing the cache to the output channel. Furthermore, there is a common error handling.

As the concept of using I/O-proxies in the control application is an important part of the RTDA, there are different types of proxies. Hardware proxies represent a concrete hardware -I/O-device and therefore have a physical name corresponding to the respective entry in an `IODeviceDescription`. Furthermore, there are two different types of software proxies: Empty I/O-channels and generic Software I/O-channels. Empty I/O-channels give a generic possibility for I/O-data access and can be used by manifold non-RT Java components (e.g., HTTP-server). The generic Software I/O-channels provide the possibility to extend the existing I/O-classes by new functionalities without changing the RTDA implementation. Examples for such possible extensions, given in the specification [3], are, for instance, I/O-proxies for remote access, simulation, or accessing legacy software.

As the superclass `IONode` is the base for all I/O-channels and events, the basic concept for the handling of events is similar to the concept of I/O-handling; a similar mechanism for the handling of events are specified.

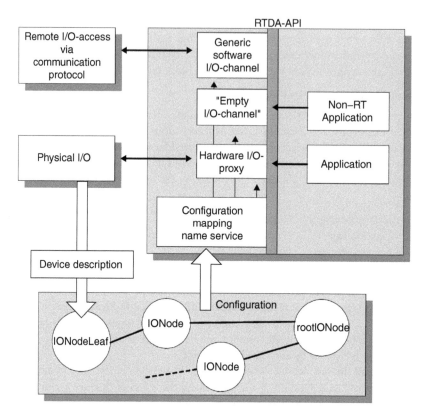

FIGURE 7.2 RTDA — overall archirecture.

Although there is no direct reference implementation for the RTDA, the Java package for SICOMP industrial-PCs by Siemens closely follows this specification of the J-Consortium. This particularly applies to the JFPC system (Java for process control) in relation to the I/O concept of the RTDA [3, 8] (the resulting structure of RTDA is given in Figure 7.2).

7.4.4   Comparison

Comparing all three specifications, it can be stated that all of them follow different approaches but also focus on different application areas. The RTSJ and RTCE have the goal to provide a more general way of enabling Java to become usable in the real-time domain, while the RTDA focuses on the special application field of industrial automation and understands real-time capabilities only as one of several requirements, which are typical for this area.

The RTSJ is the most general approach, following the idea of "making Java more real-time." Therefore, the RTSJ can be used well by experienced Java programmers, as all enhancements are close to common Java concepts. Considering the number of products implementing the specification, currently, the RTSJ has the highest acceptance.

In contrast, the RTCE follows other premises, focusing on providing a performance comparable to the state of the art (e.g., C++) solutions on commercial RTOS for hard-real-time requirements. The JVM was not changed, but extended by a separated "core," which realizes the real-time features. The RTCE provides real-time functionality following the typical concepts of today's real-time application development and therefore it is assumed that it will be used by experienced real-time programmers. Hence, the RTCE is the first choice for problems dealing with hard-real-time constraints and high-performance requirements, although, at the moment, the lack of a sufficient number of available products reduces the applicability.

The RTDA provides a complete concept for the application development in industrial control engineering. Real-time capabilities are seen only as a needed prerequisite for such applications, but the main part deals with the handling of I/O-data and interrupts. The concept is very similar to typical concepts used in conventional control systems and is therefore easy to understand for programmers working in that domain. Nevertheless, the RTDA is a very "closed" world and therefore it is not easy to adapt all ideas of using reusable components when programming control applications in Java according to the RTDA. Besides the JFPC system, which is available only for very powerful and expensive hardware (a Linux implementation will be developed and opens here new possibilities [12]), the RTDA also lacks a wide range of products.

Currently, it is still not clear if there ever will be *the* one specification, or if each of these specifications will find acceptance in a certain area.

7.5   Java Real-Time Systems

To ensure the real-time capability of a Java application, independent from the used JVM, the underlying operating system (OS) has to be real-time capable as well. Therefore, in particular in the area of embedded systems, different possibilities of combining JVM and real-time-OS exist. In general, there are four different categories:

1. *Real-time JVM on conventional RTOS.* In this case, a real-time JVM is running on a "normal" RTOS. This is typical for systems where the possibility to execute Java is an additional feature (of course, it can be the only feature that is used) and where the resources allow one to run a JVM with sufficient execution time. Normally, this is the case for systems such as IPCs and Soft PLCs, less for conventional PLCs because the operating systems and processors of conventional PLCs make the integration of a real-time JVM difficult. Examples of such systems are the SICOMP-IPC with JavaPackage [8] and the Jamaica-VM [7].

2. *JVM and RTOS as one integrated system.* If the only function of the RTOS is to run a JVM (this means the whole RT-application is written in Java), it can be useful to integrate both JVM and RTOS as one product and tailor the RTOS exactly to the needs of the JVM. Such systems can be a basis for full Java-PLCs.

3. *Conventional RTOS with Java co-processor.* If the resources of the system are limited with respect to the requirements of a JVM, it can be useful to have a special Java co-processor for the execution of the Java-Bytecode. This results in an increased execution speed, and also in an increased complexity of the system hardware and software.

4. *Java processors.* Finally, there are systems with a processor that executes Java Bytecode directly as native processor code. Here, the JVM is realized in "hardware," so that the conventional OS level can be dropped. This results in a very efficient (and therefore fast) Bytecode execution, but the flexibility of such systems is reduced due to the fact that non-Java applications cannot be executed. The typical application for the last two combinations are intelligent I/O devices.

7.6   Control Programming in Java

7.6.1   Requirements of Control Applications and New Possibilities in Java

Today, IEC 61131-compliant languages are typical for programming control applications. Although there are concepts for encapsulation of functionalities and data (e.g., function blocks provided in IEC61131-5), the main advantage of Java, compared to these languages, is the possibility to use object-oriented features like encapsulation and inheritance. Of course, networking abilities and stability also play an important role, but object orientation enables completely new ways for code reusability and increases the efficiency of application developments in control programming.

As it is not efficient to implement the whole application from scratch for every new project, it is important to encapsulate functionalities in classes for reasons of reuse. Depending on the concrete device, by means of these classes (or interfaces), generic functions like specific communication protocols or easy access to specific devices can be realized. These existing classes can (if necessary) be modified or extended and then be integrated into the application.

Hence, notable potentials result for industrial automation. Based on the platform independence of Java, parts of applications can be easily reused. Manufacturers of components like I/O-devices or intelligent actors/sensors can deliver classes (representing their components) together with the hardware. Instances of these classes can then be used in the concrete application and provide methods for using the device functionality, concealing the concrete hardware access. In doing so, the manufacturer can provide a clear simplification for the programmer and reduce the danger of improper use of his own hardware.

For complex applications, the concrete hardware access (I/O), communication functions, or other basic functions can be abstracted on different levels. Here, special tailored classes are imaginable, for instance, the usage of IEC 61499 function blocks. A simple example for such a concept shall be explained now more in detail.

7.6.2   Structure of a Control Application in Java — An Example

The goal of the following is to give recommendations for the structure of an application to support efficient programming. This can be ensured on the one hand by as much as possible device-independency, and on the other, by using a consistent procedure for the handling of I/O-variables, independently if these variables represent local I/Os or remote I/O-devices, which are accessible over different communication protocols. The communication can be realized by conventional bus-systems (e.g., CANBus) or via Ethernet (e.g., Modbus/TCP or EtherNet/IP). How should an application for these requirements be structured? In general, it is advantageous to use several levels for the encapsulation of functionalities in Java classes (see Figure 7.4). To explain the structure, a simple example shall be used. The hardware structure of this example is shown in Figure 7.3. A conveyer system consisting of several belts and turntables is controlled by a PC-based Java-PLC. As I/Os, two Modbus/TCP-Ethernet-couplers are used.

Objects of the lower level, called here communication level, realize the access to the concrete hardware. For local I/O access (e.g., memory access or GPIOs of a processor), these classes are device-specific. In contrast to this, for the communication with remote I/O -devices or other controls, generic classes (availability of appropriate hardware assumed) can be used on different platforms. Today, such Java classes are available for the Ethernet-based protocols Modbus/TCP and EtherNet/IP. In our example, the Ethernet coupler (and its I/Os) are represented by two instances of a class Modbus. These objects ensure the read- and write-access to all I/Os of the appropriate coupler and make it possible to read and write the I/Os. These objects, basically realizing a communication, can be used by classes of the next level (I/O-level), which represents the logical I/O-variables (e.g., digital or analog inputs or outputs). Here, a mapping from physical variables (representing physical I/Os) to logical variables is realized. If this mapping is flexible by using configuration information, for example, stored in a configuration file or as an input from an IDE, the control application can be implemented device-independently. In the example, every input or output is represented by an object of the class DigitalIn or DigitalOut, which is linked with the help of a configuration file to the appropriate Modbus-objects.

On the third level, complex control functionalities (e.g., emergency-stop) as well as representations of parts of the plant control system (e.g., components like conveyors, drives, or generically usable actors) can be encapsulated in classes. Objects on this level, as well as groups of I/O-variables on the I/O-level, typically run in their own real-time thread. In the example, there are objects for each element of the

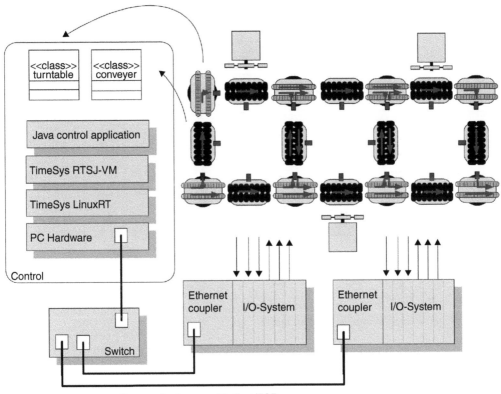

Communication over Modbus/TCP

FIGURE 7.3 Hardware structure of the example.

transport system as instances of the classes `Conveyor` and `Turntable`. These objects provide methods for controlling these elements (e.g., `start()`, `stop()`, `turnleft()`, `turnright()`,...).

This structure simplifies the reusability of huge parts of the control application. Thus, on the highest level (application level), an application can be implemented in a more abstracted, function-related way.

7.6.3 Integration of Advanced Technologies

The use of Java for control programming at the field level shows advantages, especially in systems where Java is also used for non-real-time applications. This applies for visualization or remote access for specific functions like maintenance and code download/distribution. In this context, new technologies, also in the area of industrial automation such as agent-systems or plug-and-participate technologies, play an important role.

For implementing such systems, special attention has to be paid to restrictions regarding the mutual interactions of the control application running under real-time conditions and the non-real-time application. If the real-time Java platform provides support in this respect (like the software I/O proxy of the RTDA), this can be realized very easily. Unfortunately, most real-time Java products and specifications do not follow the requirements of industrial control applications; hence, special attention has to be paid to these aspects.

As a general rule, it can be said that time-critical parts of the application will run with a higher priority, while non-real-time parts have priorities below the GC and will be executed as normal Java applications and, of course, restrictions like synchronous calls from real-time to non-real-time part have to be observed.

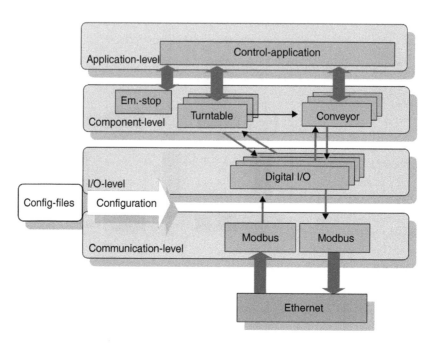

FIGURE 7.4 Structure for control applications in Java.

For this case, an architecture is reasonable, which decouples the control part and allows a synchronization at certain states of the system. The loose coupling of both parts, allowing the access to the control only in exact defined states, can be implemented by a connection layer using, for example, a finite state machine. This allows one to load, parameterize, and start control applications.

As an example for such a system, the so-called "Co-operative Manufacturing Unit" (CMU) shall be mentioned. It was developed within the international research project PABADIS [11]. This project aims at creating a highly flexible structure for automated production systems, replacing parts of the traditional MES-layer by concepts using technologies like mobile agents and plug-and-participate technologies. The CMU is the entity in the system providing the functionalities of automated devices (e.g., welding, drilling) to the PABADIS-system. Although it is also possible to connect conventional controls (e.g., PLCs) to the system, the fully Java-based CMU is the most advanced concept. It avoids the additional communication effort from the object-oriented Java-world to IEC 61131-compliant languages and provides all the advantages of Java stated before. The outcomes of the PABADIS-project give an idea of the possibilities that technologies like Java can bring to future automation systems.

7.6.4   Migration Path for the Step from Conventional Programming to Java Programming

As mentioned before, it is necessary to provide a possibility to migrate in a certain way from the conventional IEC 61131-based programming to Java programming. This way can be opened by applying the ideas of IEC 61499 and the definition of special function blocks for the application parts given in the component level, I/O level, and communication level of Figure 7.4. Based on these predefined structures and self-designed control application blocks programmed in IEC 61131-compliant dialects (which can be automatically translated to Java), a new way of programming can be established.

7.7   Conclusion

Summarizing, it has to be stated that Java technology has reached a status where the technical prerequisites regarding use in industrial automation even in the area of control applications on limited devices are fulfilled. There are Java2 Editions allowing an adaptation to the requirements of the platform, and several products using different approaches providing real-time capabilities in Java are available. This allows for the use of Java on nearly all types of devices at the field level. A common standard for real-time-Java is important against the background of retaining platform independence of Java also for real-time devices. Now the development of concepts regarding how the advantages of object-oriented, high-level languages can best be used for increasing the efficiency of application development (by supporting reusability and providing abstract views on the Java application) is momentous. Besides this, Java opens new possibilities for an easy integration of technologies like XML, Web-Services, (mobile) agents, and plug-and-participate technologies [11].

References

1. Bollella, G., B. Brosgol, P. Dibble, S. Furr, J. Gosling, D. Hardin, M. Turnbull, and R. Belliardi, *The Real-Time Specification for Java™* (First Public Release), Addison-Wesley, Reading, MA, 2001.
2. Real-Time Core Extentions, International JConsortium Specification, 1999, 2000.
3. Real-Time Data Access, International JConsortium Specification 1.0, November 2001.
4. Dibble, P., *Real-Time Java Platform Programming*, Sun Microsystems Press, Prentice-Hall, March 2002.
5. Siebert, F., *Hard Realtime Garbage Collection in Modern Object Oriented Programming Languages*, BoD GmbH, Norderstedt, 2002.
6. Pilsan, H. and R. Amann, *Echtzeit-Java in der Fertigungsautomation Tagungsband SPS/IPC/Drives 2002*, Hüthig Verlag, Heidelberg, 2002.
7. Siebert, F., Bringing the Full Power of Java Technology to Embedded Realtime Applications, *Proceedings of the 2nd Mechatronic Systems International Conference*, Witherthur, Switzerland, October 2002.
8. Hartmann, W., Java-Echtzeit-Datenverarbeitung mit Real-Time Data Access, Java™ SPEKTRUM, 3, 2001.
9. Brich, P., G. Hinsken, and K.-H. Krause, *Echtzeitprogrammierung in JAVA*, Publicis MCD Verlag, München und Erlangen, 2001.
10. Shipkowitz, V., D. Hardin, and G. Borella, The Future of Developing Applications with the Real-Time Specification for Java APIs, JavaOne Session, San Francisco, June 2001.
11. The PABADIS project homepage, www.pabadis.org, 2003.
12. Kleines, H., P. Wüstner, K. Settke, and K. Zwoll, Using Java for the access to industrial process periphery — a case study with JFPC (Java For Process Control), *IEEE Transactions on Nuclear Science*, 49, pp. 465–469, 2002.
13. TimeSys, Real-Time Specification for Java Reference Implementation, www.timesys.com, 2003.
14. Hardin, D., aJ-100: A Low-Power Java Processor, Presentation at the *Embedded Processor Forum*, June 2000, www.ajile.com/Documents/ajile-epf2000.pdf
15. Schwab, C. and K. Lorentz, *Ethernet & Factory, PRAXIS Profiline — Visions of Automation*, Vogel-Verlag, Wuerzburg, 2002.

Section 3.6

Standards for System Design

8

Achieving Reconfigurability of Automation Systems by using the New International Standard IEC 61499: A Developer's View

Hans-Michael Hanisch
University of Halle–Wittenberg

Valeriy Vyatkin
University of Auckland

8.1 Reasons for a New Standard

The development of the IEC 61499 has been stimulated by new requirements coming mainly from the manufacturing industry and by new concepts and capabilities of control software and hardware engineering.

To survive growing competition in more and more internationalized global markets, the production systems need to be more flexible and reconfigurable. This demand is especially strong in highly developed countries with high labor costs and individual customer demands. Acting successfully in such markets means decreased lot sizes and therefore frequent changes of manufacturing orders that may require changes of the manufacturing system itself. This means either a change of parts of the machinery or a change of the interconnection of subsystems by flow of material.

Each change corresponds to a partial or complete redesign of the corresponding control system. The more frequent the changes are, the more time and effort has to be spent on the redesign. As a consequence, there is a growing economic need to minimize these costs by application of appropriate methodologies of control system engineering. The desire of control engineers is to reach so-called "plug-and-play" integration and reconfiguration.

When the production systems are built from automated units, the control engineers naturally try to reuse the software components of the units. This is especially attractive in case of a reconfiguration, when the changes of units' functionality may seem to be minor. However, the obstacles come from the software side.

The current dominating International Standard IEC 61131 for programming of Programmable Logic Controllers (PLCs) [2] is reaching the end of its technological life cycle, and its execution semantics does not fit well into the new requirements for distributed, flexible automation systems. The IEC 61331 systems rely on the centralized programmable control model with cyclically scanned program execution. Integration of this kind of system via communication networks may require quite complex synchronization procedures. Thus, overheads of the integration may be as complex as the startover system development. Furthermore, even different implementations of the same programming language of IEC 61131 may have different execution semantics. Moreover, PLCs of different vendors are not interoperable and require vendor-dependent configuration tools.

For dealing with distributed nonhomogeneous systems, it would be convenient to define control applications in a way that is independent from a particular hardware architecture. However, the architectural concept for such a definition is missing.

These are severe obstacles to the plug-and-play integration and reconfiguration of flexible automation systems.

The newly emerging International Standard IEC 61499 [1] is an attempt to provide the missing architectural concept. The standard defines a reference architecture for open, distributed control systems. This provides the means for real compatibility between automation systems of different vendors. The standard incorporates advanced software technologies, such as encapsulation of functionality, component-based design, event-driven execution, and distribution. As a result, specific implementations of different providers of field devices, controller hardware, human–machine interfaces, communication networks, etc., can be integrated in component-based, heterogeneous systems. The IEC 61499 standard stimulates the development of new engineering technologies that are intended to reduce the design efforts and to enable fast and easy reconfiguration.

This chapter provides an overview of the concepts and design principles of the standard. For the internal details of the standard, the reader is referred to Reference [1], for a more systematic introduction into the subject to Reference [4], and to Reference [7] for the evolution of the idea. Since the design principles differ considerably from those of the well-known IEC 61131, the general issues are introduced in a rather intuitive way. To illustrate the design principles and new features, an example is presented in this chapter. Real applications — although sparse — do exist but are too complex to be explained here in detail. The interested reader is referred to Reference [9]. This chapter is based on the Publicly Available Specification (PAS) of the IEC TC65 from March 2000. As of July 2004, the parts 61499-1 "Architecture" and 61499-2 "Software Tools Requirements" were voted and approved as IEC Standards. The texts are expected to be published by the end of 2004.

8.2   Basic Concepts of IEC 61499

The standard defines several basic *functional* and *structural* entities. They can be used for specification, modeling, and implementation of distributed control applications. Some of them stand for pure software components. Others represent logical abstractions of a mixed software/hardware nature. The main functional entities are function block types, resource types, device types, and applications that are composed of function block instances. System configurations include instances of device types and applications that are placed (mapped) into the devices. The following subsections will briefly discuss all these entities.

The specification of functionality of a control application and the specification of a system architecture can be performed independently. An implementation of an application on a given system architecture is done by mapping of its function blocks onto the devices and their resources. Each function block must be mapped to a single resource and cannot be distributed over different resources.

Thus, the next section discusses the means to define the functionality of a distributed control application, and the section "Specification of the System Architecture" shows how a particular system configuration can be defined.

8.2.1 Describing the Functionality of Control Applications

8.2.1.1 Overview of the Function Block Concept

The basic entity of a portable software component in IEC 61499 is a so-called *function block*. The new standard defines several types of blocks: basic function blocks, service interface function blocks, and composite function blocks. Although the term "function block" is known from IEC 61131, a function block of IEC 61499 is different from IEC 61131 function blocks. Figure 8.1 shows a *function block interface* following the standard (note that the graphical appearance is not normative). The upper part is often referred to as the "head" and the lower as the "body" of the function block. A block may have inputs and outputs. There is a clear distinction between data and event input/output signals. Events serve for synchronization and interactions among the execution control mechanisms in interconnected networks of function blocks. The concept of data types is adopted as in any programming language. In particular, the standard refers to the data types of IEC 61131. In the graphical representation in Figure 8.1, the event inputs and outputs are associated with the head of the function block, and the data inputs and outputs are associated with the body.

The IEC 61499 defines a number of standard function blocks for manipulations with events, such as splitting or merging events, generation of events with delays or cyclically, etc.

The definition of the function block external interface also includes an association between the events and relevant data. It is denoted by vertical lines connecting an event and its associated data inputs/outputs in the graphical representation. The association literally means that the values of the variables associated with a particular event (and only these!) will be updated when the event occurs.

The IEC 61499 standard uses the typing concept of function blocks that is similar to that in IEC 61131 and in the object-oriented programming. Once a function block type is defined, it can be stored in the library of function block types and later instantiated in applications over and over again.

The standard does not determine in detail the languages to define the internal functionality of a function block. However, for each type of blocks, the ways of structuring the functionality are identified. These will be discussed in the following subsections.

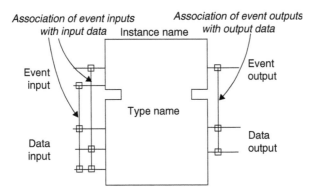

FIGURE 8.1 External interface of a function block.

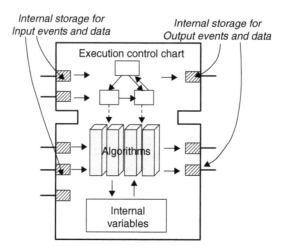

FIGURE 8.2 A basic function block.

8.2.1.2 Basic Function Blocks

Basic function blocks are software structures intended to implement basic functions of distributed control applications. The standard says that in addition to inputs and outputs, the internal structure of a basic function block may include *internal variables*, one or several *algorithms,* and an *execution control chart*.

An algorithm is a structure of finest granularity. It represents a piece of software code operating on the common input, output, and internal data of the function block. The algorithms can be specified, for example, in languages defined in IEC 61131. But, in general, they can be given in any form supported by the implementation platform. It is important to note that an algorithm has to be implemented in one programming language. The internal data of a function block cannot be accessed from outside and can only be used by the internal algorithms of the particular function block (Figure 8.2).

The execution control function specifies the algorithm that must be invoked after a certain input event in a certain state of the execution control function. It is specified by means of Execution Control Charts (ECCs for short). Figure 8.3 shows an example. An ECC is a finite state machine with a designated initial state. An ECC consists of states with associated *actions* (designated by ovals in Figure 8.3) and of state transitions (designated by arrows). The actions contain algorithms to invoke and output events to issue upon the completion of the algorithms' execution.

Each state transition is labeled with a BOOLEAN condition that is a logic expression utilizing one or more event input variables, output variables, or internal variables of the function block. The event inputs are represented in the conditions as BOOLEAN variables that are set to TRUE upon an event and cleared after all possible state transitions (initiated by a single input event) are exhausted.

An input event causes the *invocation of the execution control function* that in more detail is as follows (see Figure 8.4):

Step 1: The input variable values relevant to the input event are made available.
Step 2: The input event occurs, the corresponding BOOLEAN variable is set, and the execution control of the function block is triggered.
Step 3: The execution control function evaluates the ECC as follows. All the transition conditions going out of the current ECC state are evaluated. If no transition is enabled, then the procedure goes to the Step 8. Otherwise, if one or several state transitions are enabled (i.e., if the corresponding conditions are evaluated to TRUE), a single state transition takes place.* The current state is substituted by the following one. The algorithms associated with the

---

*The standard does not determine a rule regarding how to choose a state transition if several are simultaneously enabled.

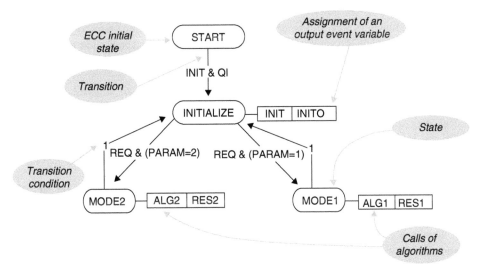

FIGURE 8.3 An example of the ECC [1].

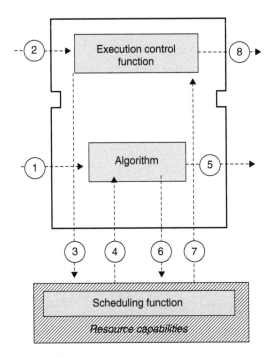

FIGURE 8.4 Execution of a function block.

new current state will be scheduled for execution. The execution control function notifies the resource scheduling function to schedule an algorithm for execution.

Step 4: Algorithm execution begins.

Step 5: The algorithm completes the establishment of values for the output variables.

Step 6: The resource scheduling function is notified that algorithm execution has ended.

Step 7: The scheduling function invokes the execution control function. The procedure resumes from Step 3.

Step 8:   When some of the output event variables were set during this invocation of the execution
control function, then the execution control function signals the corresponding event output
and clears the BOOLEAN variables corresponding to the triggered input and output events.

First conclusions can be drawn at this point:

- A basic function block is an abstraction of a software component adjusted for the needs of measurement and control systems. Its execution semantics is event-driven and platform independent. Basic function blocks are intended to be main instruments of an application developer.
- The standard implies separation of the functions, implemented by algorithms, from the execution control. The algorithms encapsulated in a function block can be programmed in different programming languages.
- The execution of function blocks is event-driven. This means that algorithms are executed only if there is a need to execute them in contrast to the cyclically scanned execution in IEC 61131. The need has to be indicated by events. The source of events can be other function blocks. Some of them may encapsulate interfaces to the environment (controlled process, communication networks, hardware of a particular computational device).
- The execution function of a basic function block is defined in the form of a state machine that is available for documentation and specification purposes even if the algorithms are hidden.
- The function block abstracts from a physical platform (the resource) on which it is located. This means that the specification of the function block can be made without any knowledge of the particular hardware on which it will be later executed.

8.2.1.3   Composite Function Blocks

The standard also defines *composite function blocks*, the functionality of which, in contrast to basic function blocks, is determined by a network of interconnected function blocks inside. Figure 8.5 shows the principle.

More precisely, members of the network are instances of function block types. These can be either basic function blocks or other composite function blocks. Therefore, hierarchical applications can be built.

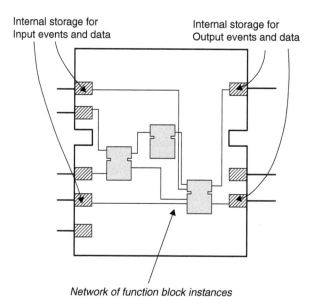

Internal storage for
Input events and data

Internal storage for
Output events and data

Network of function block instances

FIGURE 8.5   Composite function block.

It is important to note that composite function blocks have no internal variables, except for those storing the values of input and output events and data. Thus, the functionality of composite function blocks completely depends on the behavior of the constituent function blocks and their interconnections by events and data.

Along with basic function blocks, the composite function blocks are intended to be main instruments of an application developer.

8.2.1.4 Service Interface Function Blocks

In contrast to basic and composite function blocks, *service interface function blocks* are not intended to be developed by an application developer. These have to be provided by vendors of the corresponding equipment, for example, controllers, field buses, remote input/output modules, intelligent sensors, etc. The application scope determines the differences of this kind of function blocks from the previously considered ones.

To conceal the implementation-specific details of the system from the application, the IEC 61499 defines the concept of services that the system provides to an application. A service is a functional capability of a resource that is made available to an application. A service is specified by a sequence of service primitives that defines properties of the interaction between an application and a resource during the service. The service primitives are specified in a graphical form of the time-sequence diagrams described in Technical Report 8509 of the International Standard Organization (ISO) [3]. This is rather a qualitative specification form, as it does not specify exact timing requirements to the services. An example of time-sequence diagrams is presented in Figure 8.7 and will be briefly discussed later in this section.

A *service interface function block* is an abstraction of a software component implementing the services. Figure 8.6 shows an example of a service interface function block REQUESTER that provides some service upon request to an application (examples of possible services: read the values of sensors, increase the memory used by a resource, shut down a resource, send a message, access remote database, etc.). The standard predefines some names for input/output parameters of service interface function blocks such as INIT for initialization, INITO for confirmation of the initialization, QI for input qualifier, etc.

Some of the services provided by this block are specified in the form of time-sequence diagrams in Figure 8.7. These are "normal establishment" of the service, "normal service," and "application initiated termination" of the service.

The input event INIT serves for initialization/termination of the service, depending on whether the BOOLEAN input QI is true or false. The notation INIT+ means the occurrence of the event INIT with the qualifier value QI=true, and INIT-correspondingly with QI=false.

The input parameter PARAMS stands for the service parameters that have to be taken into account during the service initialization. At the end of the initialization/termination procedure, the service interface function block responds by event INITO its completion and indicates by the BOOLEAN data output QO whether initialization/termination was successful (QO=true) or not (QO=false).

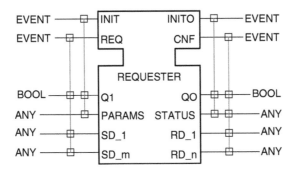

FIGURE 8.6 Generic REQUESTER [1].

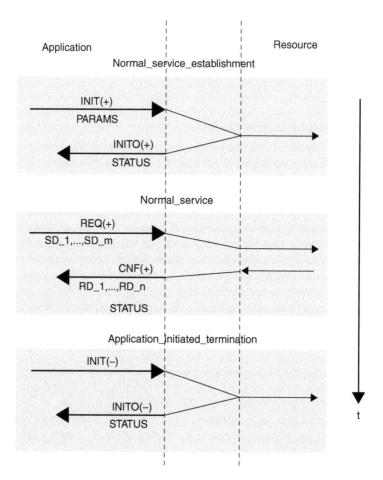

FIGURE 8.7 Diagrams of service sequences for application-initiated interactions.

The input data needed to perform the service are denoted as inputs SD_1 ... SD_m. Note that these data are associated with the event input REQ. The data outputs RD_1 ... RD_n stand for the data computed as a result of the service. These data are associated with event CNF, which represents confirmation of the service. The output STATUS provides information about the status of the service upon the occurrence of an event output.

The execution of service interface function blocks is initiated by input events. The internal structure of service interface function blocks is not specified as firmly as for basic function blocks. For example, a programming implementation of a service interface function block can be done in the form of several encapsulated algorithms (methods, procedures) that are invoked upon a particular event (say, algorithm init stands for the event INIT). The algorithms may check the value of the qualifier QI and then call either the subroutine responsible for the "normal service establishment" (if QI=true) or the one responsible for the "application initiated termination" (if QI=false). Note that the concept of service interface function blocks does not presume the need for internal variables — the conditions for initiating services are described by input events and data (qualifiers).

A particular case of service interface function blocks are *communication interface function blocks*. The standard explicitly defines two generic communication patterns: PUBLISH/SUBSCRIBE for unidirectional transactions and CLIENT/SERVER for bidirectional communication. These patterns can be adjusted to a particular networking or communication mechanism of a particular implementation. Otherwise, a provider of communication hardware/software can specify his own patterns if they differ

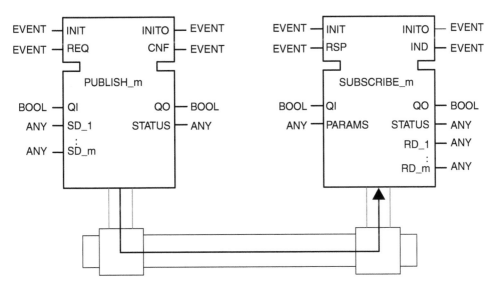

FIGURE 8.8　PUBLISH and SUBSCRIBE communication interface blocks.

from the above-mentioned ones. Figure 8.8 illustrates the generic PUBLISH and SUBSCRIBE blocks performing unidirectional data transfer via a network.

The PUBLISHER serves for publishing data SD_1 ... SD_m that come from one or more function blocks in the application. It is therefore initialized/terminated by the application in the same way as described above.

Upon the request-event REQ from the application, the data that need to be published are sent by the PUBLISHER via an implementation-dependent network. When this is done, the PUBLISHER informs the publishing application via event output CNF.

The SUBSCRIBER function block is initialized by the application that is supposed to read the data RD_1 ... R_m. Normal data transfer is initiated by the sending application via the REQ input event to the PUBLISHER. This is illustrated in Figure 8.9 by means of time-sequence diagrams. The PUBLISHER sends the data and triggers the IND-event at the outputs of the SUBSCRIBER to notify the reading applications that new values of data are available at RD_1 ... RD_m outputs of the SUBSCRIBER. The reading application notifies the SUBSCRIBER by the RSP-event that the data are read.

In summarizing this subsection, one can see that service interface function blocks implement the interface between an application and the specific functionality that is provided by control hardware or system software. The content of service interface function blocks can be concealed, but the means are reserved to specify their functionality in a visual form.

8.2.1.5　Application

An *application* following IEC 61499 is a network of function block instances whose data inputs and outputs and event inputs and outputs are interconnected (see Figure 8.10).

An application can be considered as an intermediate step in the system development. It already defines the desired functionality of the system completely, but it does not specify the system's structure in terms of computational devices where the function blocks can be executed. The next step in the engineering process is to define a particular set of devices and to "cut" the application, assigning the blocks to the devices as illustrated in Figure 8.11.

The way in which the separated parts of the distributed applications communicate with each other has to be explicitly defined. This can be done by adding *communication function blocks* in the places where the "cut" took place (see Figure 8.12).

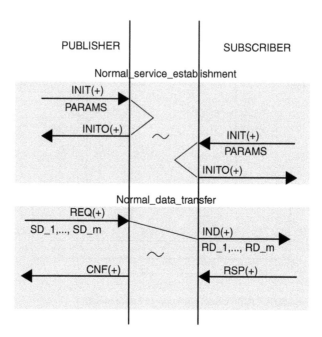

FIGURE 8.9 Communication establishment and normal data transfer sequence.

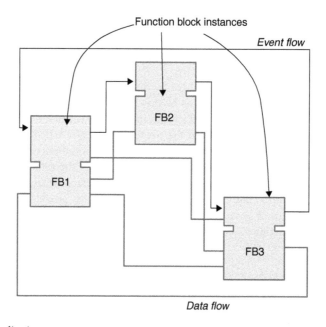

FIGURE 8.10 An application.

A network of function blocks (forming the application) can be encapsulated in a composite function block if needed. In this case, however, it could not be distributed across several devices or resources as a function block can be executed only in a single resource.

In fact, the standard provides a structure (called *subapplication*) that combines features of composite function blocks and of applications. The content of a subapplication can be distributed across several devices. However, the practical applicability of this structure is quite questionable.

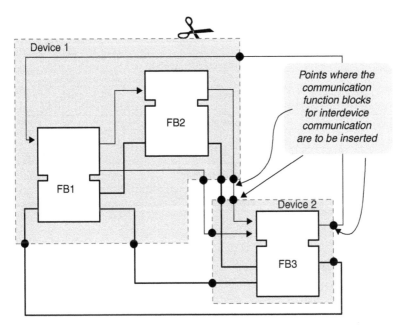

FIGURE 8.11 The application distributed onto two devices.

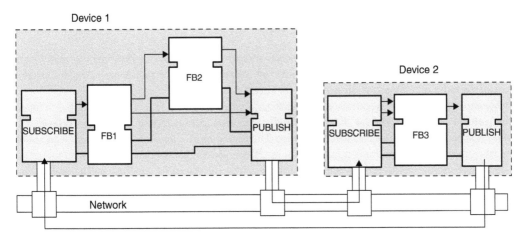

FIGURE 8.12 Communication function blocks explicitly connecting parts of the distributed application.

The following subsection will show the concepts and specifications of a system following IEC 61499 as a platform for implementation and execution of an application.

8.2.2 Specification of the System Architecture

8.2.2.1 Resources and Devices

A *device* in IEC 61499 is an atomic element of a system configuration. The standard provides architectural frames for creating models of devices, including their subdivision in computationally independent resources.

A *device type* (Figure 8.13) is specified by its process interface and communication interfaces. A device can contain zero or more resources (see the description below) and function block networks (this option is reserved for the devices having no resources).

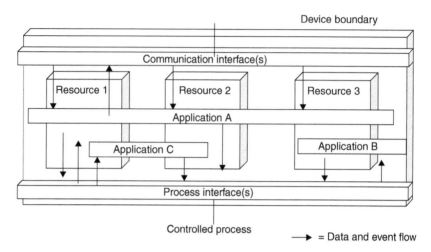

FIGURE 8.13 A device model.

A "process interface" provides a mapping between the physical process (analog measurements, discrete I/O, etc.) and the resources. Information exchanged with the physical process is presented to the resource as data or events, or both.

Communication interfaces provide a mapping between resources and the information exchanged via a communication network. In particular, services provided by communication interfaces may include presentation of communicated information to the resource and additional services to support programming, configuration, diagnostics, etc.

The interfaces are implemented by the libraries of corresponding service interface function blocks. The libraries of these blocks form the "identity" of a device. A device that contains no resources is considered to be functionally equivalent to a resource.

A *resource* (Figure 8.14) is considered to be a functional unit, contained in a device, that has independent control of its operation. It may be created, configured, parameterized, started up, deleted, etc., without affecting other resources within a device. The functions of a resource are to accept data and/or events from the process and/or communication interfaces, process the data and/or events, and to return data and/or events to the process and/or communication interfaces, as specified by the applications utilizing the resource.

Furthermore, a resource provides physical means for running algorithms. This means storage for data, algorithms, execution control, events, etc. It also has to provide software capabilities for managing and controlling the function blocks' behavior, scheduling its algorithms (scheduling-function), etc.

8.2.2.2 System Configuration

A *system* is a collection of one or more devices (Figure 8.15). The devices communicate with each other over communication networks. Also, the devices are linked with the controlled process via sensor and actuator signals.

Applications are mapped onto the devices. This means that their function blocks are assigned to the resources of the corresponding devices. In this way, a *system configuration* is formed. A system configuration is feasible if each device in it supports the function block types that are mapped on it. Otherwise, the block would not be instantiated, and the system will not run.

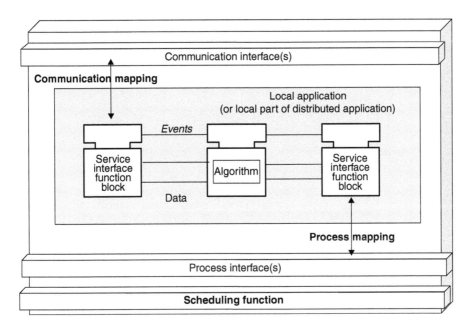

FIGURE 8.14 A model of resource.

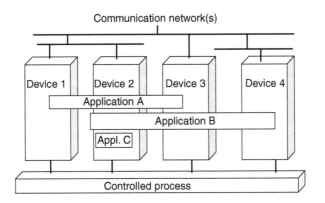

FIGURE 8.15 A system configuration.

8.3 Illustrative Example

8.3.1 Desired Application Functionality

The example "FLASHER" that is used in this section was borrowed from the set of samples provided with the Function Block Development Kit (FBDK). This is the first software tool supporting IEC 61499 system development. It was developed by Rockwell Automation, U.S.A. The toolset can be downloaded from Reference [5]. The example represents an abstraction of an automation system. It is supposed to make four lamps blinking according to a preprogrammed mode of operation.

The system consists of human–machine interface components and of a core functional component. The core component generates the output signals determined by input parameters. The output values are then delivered to the visualization device (lamps). In the form presented here, the system is completely simulated in a computer. All human–machine interface components and lamps exist only on the computer screen. The example visually shows how easily each of the software components that interact with a simulated object can be replaced by components that would interact with the real physical equipment

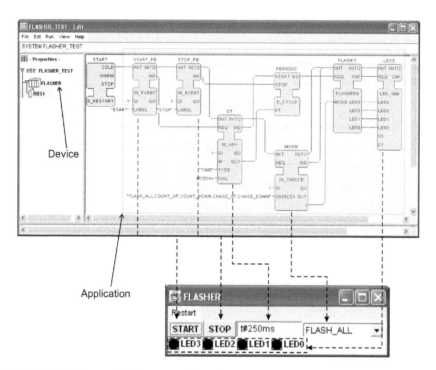

FIGURE 8.16 FLASHER in centralized system configuration.

(e.g., buttons, switchers, knobs). After such a reconfiguration, the whole system will then show the same functionality without changing its structure and without redesigning the other components.

Figure 8.16 shows a screenshot of the FBDK containing a system configuration. As a result of its execution, the output frame is produced. The output frame is located in the figure below the FBDK screen. The arrows connect the function blocks with the screen objects created by them.

The system configuration includes one application. It is placed in one device with only one resource. An instance of the device type FRAME_DEVICE is used in this example. This type of device creates a windows frame on the computer screen. The resource of type PANEL_RESOURCE creates a rectangular panel within this frame. If a function block creates an output to the screen, it will be placed in the corresponding panel.

In this particular case, the system configuration implements the application in a centralized manner. The application includes all the blocks shown in Figure 8.16 in the shaded rectangular area. The only block that falls out of the application is the block START of type E_RESTART. It belongs to the resource type PANEL_RESOURCE.

The application creates the following models of human–machine interface primitives. Buttons START and STOP are created by the blocks START_PB and STOP_PB. Both are instances of the type IN_EVENT. Block DT creates the input field for the TIME parameter. It is an instance of type IN_ANY. Block MODE creates the pull-down menu to select a desired mode of operation.

The block that produces the output combination is FLASHIT. It is an instance of type FLASHER4. The output values are then visualized by the block LEDS of type LED_HMI. The FLASHER4 generates the output values at every pulse of the event input REQ. The pulses are generated by the block PERIODIC with the frequency determined by the value in the field DT that is received from the block DT.

The operation of the system configuration is started when the resource is initialized, for example, when the device is created (or switched on for more realistic devices). At that moment, the service interface function block START produces the event COLD (cold restart). It is connected to the input event INIT of the block START_PB. This input event causes the block of type IN_EVENT to place a button image in the resource's panel. The caption of the button is given by the input parameter "LABEL" of the block, that is, "START" in our case. After that, an output event INITO is generated that is connected to the input INIT of the block STOP_PB. It leads to the creation of the button "STOP" and so forth in the chain until the whole picture is created on the resource panel as shown in Figure 8.16.

Once the button START is pressed (e.g., by a mouse click), it ignites the event output IND. This event propagates through the chain of blocks DT and PERIODIC. It enables the latter to generate the output event EO with the desired frequency. Every moment the event comes to the input REQ of the FLASHIT, the output combination of LED0..LED3 is created, and the output event CNF notifies the block LEDS, which updates the picture.

If the operating mode is changed during the operation, the corresponding event IND of the block MODE would notify the FLASHIT. Then, FLASHIT will change the pattern according to which its outputs are generated.

A more detailed description of the function block type FLASHER4 is given now. Figure 8.17 shows the Execution Control Chart of this block. The state machine either switches to the desired algorithm corresponding to the selected MODE of operation (number in the interval 1..5) at input event REQ, or to the initialization algorithm INIT at the input INIT.

Note that one half of the algorithms encapsulated in the FLASHER4 is programmed in Structured Text (ST), while the other half is programmed in Ladder Logic Diagrams (LD). This is shown in Figure 8.18 and Figure 8.19. This illustrates the opportunities the IEC 61499 function blocks provide to reuse the legacy code and even to combine different programming languages in a single software component.

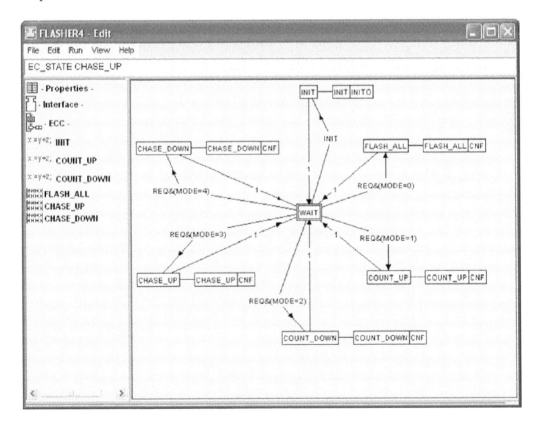

FIGURE 8.17 FLASHER4: execution control chart.

```
ALGORITHM COUNT_UP IN ST                          X

LED0:=NOT LED0;
IF NOT LED0 THEN
  LED1:=NOT LED1;
  IF NOT LED1 THEN
    LED2:=NOT LED2;
    IF NOT LED2 THEN LED3:=NOT LED3;END_IF;
  END_IF;
END_IF;


              Update   Close   Help
```

FIGURE 8.18 Algorithm COUNT_UP programmed in structured text.

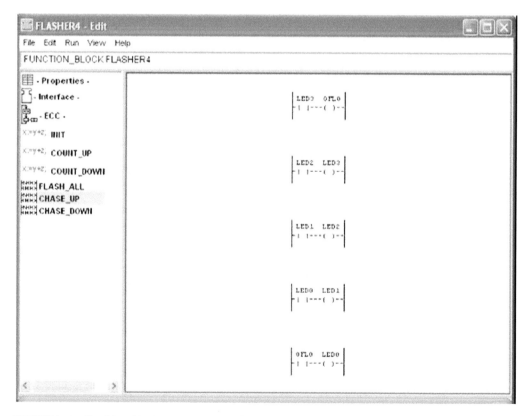

FIGURE 8.19 Algorithm CHASE_UP programmed in ladder logic.

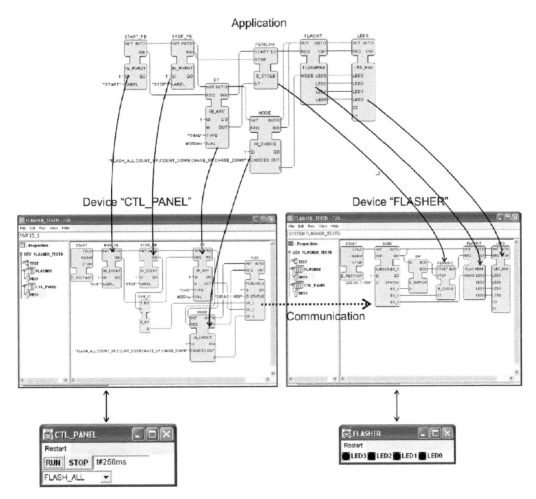

FIGURE 8.20 Application FLASHER distributed across two devices.

8.3.2 Distribution

The distributed version of the same application is shown in Figure 8.20. The system configuration includes two devices: CTL_PANEL and FLASHER. The function blocks of the application are mapped to either of these devices. In addition, the communication function blocks PUBLISH and SUBSCRIBE are added to connect the parts of the application. The devices may be started or shut down completely independently from each other. As soon as the part located in CTL_PANEL produces any changes in the operation parameters, it will notify the FLASHER's part via the communication channel.

Figure 8.21 shows a distribution of the same application across three devices. Device "FLASHER" (on the right) produces no picture. It produces the output values and sends them to the device DISPLAY given the input parameters received from the CTL_PANEL.

8.4 Engineering Methods and Further Development

In contrast to the present practice of IEC 61131, the specification of a control system following IEC 61499 is a complete change of paradigms. Execution of control code in IEC 61131 is sequential and time-driven. The control engineer who uses the IEC 61499 needs to think in terms of event-driven execution of encapsulated pieces of code. The execution is distributed and concurrent. This requires new methodologies for control system design, verification/validation, and implementation.

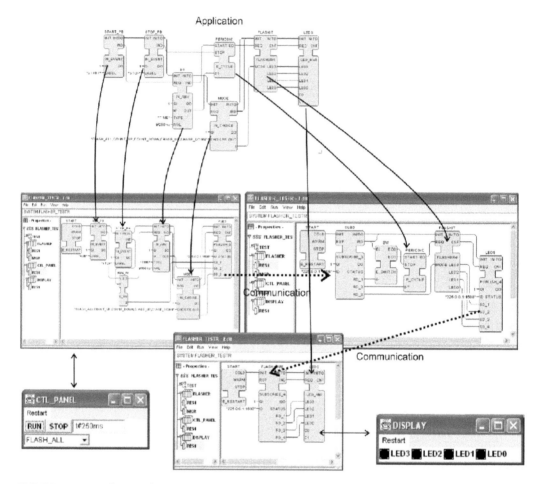

FIGURE 8.21 Distribution of the FLASHER application across three devices.

A natural way of system engineering using IEC 61499 is a method with the following steps:

1. Identification of the functionality of system components.
2. Encapsulation of basic functionality in these components. This gives the basic function blocks or even composite function blocks.
3. Interconnection of the function blocks to build an application. External coordination of the network of function blocks may be needed and added.
4. Mapping of the application into a control system architecture.

At least the following questions must be answered to make such a methodology applicable.

- How to encapsulate the existing controllers (sometimes implementing very sophisticated *ad hoc* algorithms) into new event-driven capsules?
- How to specify the component controllers in a way allowing (semi-)automatic integration into distributed systems?
- Conversely, given a desired behavior of the integrated system, how to decompose it to the control actions of the component controllers?

This requires further research and development, in particular, revolving around the concept of automation objects [11]. The concept of automation objects is understood as a framework for encapsulation and handling of the diverse knowledge relevant to automation systems. This includes operation semantics as well as layouts, CAD data, circuitry, etc. As the scope of IEC 61499 cannot cover all these issues, it probably needs to be combined with ideas arising from other developments, in particular, with IEC 68104; see Reference [12] for a more detailed description of device types.

Conducting such a work requires integration of development activities among leading vendors and users of automation technologies on a global scale. Currently, such an activity is being organized under the support of the Intelligent Manufacturing Systems Research and Development program (www.ims.org) in the form of OOONEIDA project [13].

Currently available engineering methodologies can be found in References [6, 9]. Also, the idea of combining Unified Modeling Language (UML) with the IEC 61499 promises to provide a consistent engineering methodology for control system engineering. Some details on this can be found in Reference [10].

Another major future issue is how to bring the ideas of reconfiguration to practical applications. The standard itself provides the following means.

1. Basic function blocks do not depend on a particular execution platform. They will show equivalent execution semantics on different platforms, regardless of, for example, in which sequence they are listed. This is not the case in the current implementations based on IEC 61131.
2. The functionality of whole applications that are represented as hierarchical networks of function blocks also does not depend on a particular number and topology of computational resources. Thus, system engineering can be done in an implementation-independent way.
3. Models of various devices are represented as device types. This will allow anticipation of the system's behavior after reconfiguration. On the one hand, the way of modeling of devices and resources is very modular. On the other, it uses a very limited number of constituent instruments. It allows modeling of a great variety of system configurations without going into unnecessary details.
4. There are also some other, more technical means provided in IEC 61499 to support reconfiguration. One of these is the use of adapter interfaces to minimize interblock connections by means of predefined patterns. Another one includes standard function blocks for manipulations with events.

Another open question for future research and development is how the correctness of the system can be validated. This requires more formal models for simulation or even for formal analysis of the behavior. The concept of encapsulation is extremely useful for embedding such models in the design. In principle, it allows modeling of the whole measurement and control systems with all their diverse components, such as sensors and actuators, networks, computational resources, etc. The models could reproduce the execution semantics of the real system also, taking into account such factors as communication delays of particular networks. Moreover, models of the controlled objects can be encapsulated in function blocks as well. This allows one to study the behavior of the controller and of the controlled object in the closed loop. The desired properties could be validated by simulation or could even be formally verified. For the first time, this approach was applied for formal modeling and verification of IEC 61499-compliant designs in Reference [8]. This approach can also be used for prototyping of the distributed control systems based on their formal models.

It is obvious that all these future developments must be supported by appropriate tools, compliant devices, runtime systems, etc.

Last but not the least, there is a considerable need for training and education of engineering staff to understand and to apply the new concepts of the standard.

Nonetheless, despite the amount of development that still lies ahead, the potential benefits of using IEC 61499 are very clear, and the ideas and concepts define cornerstones of future development and design of distributed control systems.

Acknowledgments

The authors thank Sirko Karras for help in managing the graphic material of this contribution. The authors furthermore express their gratitude to Rockwell Automation and personally to Dr. James Christensen for providing the FLASHER example, for the permission of using Figure 8.8 and Figure 8.9, and for the fruitful ideas expressed in the personal communication.

References

1. Function Blocks for Industrial Process Measurement and Control Systems, Publicly Available Specification, International Electrotechnical Commission, Part 1: Architecture, Technical Committee 65, Working Group 6, Geneva, 2000.
2. International Standard IEC 1131-3, Programmable Controllers — Part 3, International Electrotechnical Commission, Geneva, Switzerland, 1993.
3. ISO TR 8509-1987, Information processing systems — Open Systems Interconnection — Service Conventions, 1987.
4. Lewis, R., Modeling Distributed Control Systems Using IEC 61499, Institution of Electrical Engineers, London, 2001.
5. www.holobloc.com — Web site devoted to IEC 61499.
6. Christensen, J.H., IEC 61499 Architecture, Engineering, Methodologies and Software Tools, 5th IFIP International Conference BASYS'02, Proceedings, Cancun, Mexico, September 2002.
7. Schoop, R. and H.-D. Ferling, Function Blocks for Distributed Control Systems, DCCS'97, Proceedings, 1997, 145–150.
8. Vyatkin, V. and H.-M. Hanisch, Verification of distributed control systems in intelligent manufacturing, Journal of Intelligent Manufacturing, 14, 123–136, 2003.
9. Vyatkin, V., Intelligent Mechatronic Components: Control System Engineering Using an Open Distributed Architecture, IEEE Conference on Emerging Technologies and Factory Automation (ETFA'03), Proceedings, Vol. II, Lisbon, September 2003, pp. 277–284.
10. Tranoris, C. and K. Thramboulidis, Integrating UML and the Function Block Concept for the Development of Distributed Applications, IEEE Conference on Emerging Technologies and Factory Automation (ETFA'03), Proceedings, Vol. II, Lisbon, September 2003, pp. 87–94.
11. IEC 61804 Function Blocks for Process Control, Part 1 — General Requirement; Part 2 — Specification, Publicly Available Specification, International Electrotechnical Commission, Working Group 6, Geneva, 2002.
12. Automation Objects for Industrial-Process Measurement and Control Systems, Working Draft, International Electrotechnical Commission, Technical Committee No. 65, 2002.
13. OOONEIDA: Open Object-Oriented Knowledge Economy in Intelligent Industrial Automation. Official Web site: http://www.oooneida.info

Section 3.7

Integration Solutions

9

Integration between Production and Business Systems

Claus Vetter
*ABB Corporate Research Center,
Switzerland*

Thomas Werner
*ABB Corporate Research Center,
Switzerland*

9.1 Introduction

More efficient use of production equipment, best possible scheduling, and optimized production processes are the challenges in today's process and manufacturing industries. These trends become apparent for a wide range of industries such as electric utilities [1] and production facilities in batch and chemical operations [2, 3].

With numerous software systems already in place in most production facilities, the challenges ahead lie in seamless integration of the IT landscape, with one of the focus points in manufacturing and process industries being the connectivity between plant floor execution and business systems. Effective and accurate exchange of information is the means to meet those challenges. Existing IT integration projects focused on delivering data from one system to the other, but usually the integration methodology has not concentrated on reusable solutions.

The goal is to shift efforts from point-to-point solutions, which come with a high coding effort when connecting single data sets, toward reusable components, which form building blocks of an enterprise — plant floor integration architecture. The challenge is to leverage existing mainstream technologies such as enterprise application integration (EAI) [4], web services [5] or XML [6], and emerging industry standards (e.g., ISA 95 [7] or CIM [8]) and apply them in the context of the usage scenarios found in today's production environments.

This chapter presents the concepts of an integration study between production and business systems. A software architecture combines a set of functional components for implementing interfaces between manufacturing execution systems and business systems, containing all the elements necessary to develop, execute, and operate the integration in all likely scenarios, including real-time data exchange from shop floor to board room, near-real-time message and event-oriented, and periodic data exchange that extracts and imports data in bulk.

The benefits of the integration concept include:

- Reduced overall effort by developing a single architecture for all connectivity scenarios.
- Faster time-to-market for new functionality, since the focus lies on developing this add-on components instead of developing "infrastructure."
- Reduced interface development effort by utilizing tools and templates to jumpstart development.
- Reduced maintenance costs since each interface is implemented using the same technology concepts.

The study presents thoughts on features to be considered in the architecture and technologies that should be used to implement the functional building blocks. The concepts presented are based upon the experiences of the authors and research of recently available technologies. Key elements of the concept are confirmed in proof-of-concept and prototyping activities.

9.1.1   Objectives and Scope

The objectives of the document are to provide the reader with an understanding of the following:

- Initial functional and technical requirements driving the development of the architecture.
- Technology guiding principles used as the basis for making decisions on the selected technology.
- Available integration technologies and approaches.
- High-level design for the integration components.

This chapter is technical in nature and is intended to be read by (technical) project managers and software architects. It assumes basic knowledge regarding functionality performed by the key applications in the integration context — Manufacturing Execution Systems (MES) and Enterprise Resource Planning (ERP).

9.1.2   Chapter Organization

The chapter is organized as follows: Section 9.2 presents example functional scenarios that provide fundamental requirements for the overall software architecture. Section 9.3 presents integration types, features, and the technology guiding principles that are required to fulfill the base requirements outlined. Integration types define the different technical integration scenarios (view, functional, and data). Required features address the characteristics of architectures common to each integration type. The guiding principles provide high-level criteria that are used in selecting between the various technologies available in constructing the architectures. Additionally, the main components of the presented prototypes — an ABB production system based on the Industrial[IT] framework and SAP's R/3 system — are outlined with a focus on its main interfacing points and structuring concepts relevant for integration. Sections 9.4 through 9.6 outline the available technologies being considered for the integration architectures. Each of them addresses an exemplary use case, the architectural concept, and the prototype design of a specific integration type as defined in Section 9.3.

9.2   Integration Scenarios in a Production Environment

9.2.1   Functional Interaction

In a production environment, three generic scenarios (Figure 9.1) can be distinguished:

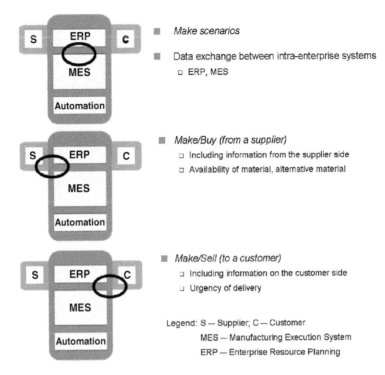

FIGURE 9.1 Integration usage: three main scenarios.

1. *Make (intra-enterprise integration):* This scenario includes data exchange between business and execution systems within one plant (or company) in order to automate data synchronization, and to optimize production and maintenance schedules.
2. *Make/buy (business-to-business):* Information from the supplier side is included for optimization reasons, as, for example, scheduling the production triggers placement of orders to a supplier, or production scheduling depending on the availability of raw material from a supplier.
3. *Make/sell (business-to-consumer):* Production information is available toward customers, for example, availability or production capacity information. Production can be optimized toward certain customers (urgency), or customers can track the production progress of the placed orders.

In the following subsections, the focus lies on the intra-enterprise integration between MES and business systems. It has broad applicability across several industries, such as chemical, food, primary pharmaceutical, and to some extent discrete manufacturing. It is also of importance in the context of process industries and utilities, where the trend of interaction between production and back-office becomes more and more apparent.

9.2.2 Inter-Enterprise Integration Scenarios

Table 9.1 summarizes the transactional "data" interfaces between production planning and materials management modules of an ERP system and the MES systems that typically result from the above-outlined application scenario of ERP and MES functionality, showing the many interfacing points for optimizing and automating data exchange.

Out of the above, the most generic patterns of the functional interaction between production and business systems can be extracted and are outlined in the following subsections.

TABLE 9.1   Sample Functional Integration Requirements for the "Make" Process

From	To	Description
Production planning	MES	Released production orders — Production orders created in production planning or scheduling are released and then sent to MES.
Production planning	MES	Production order changes (dates, quantity, released bill-of-materials [BOMs], recipes and production parameters, routings, and cancellations) — Production order changes are sent from ERP to MES.
Production planning	MES	Master data (BOMs, recipes, routings) and changes are sent to MES.
MES	Production planning	Production order start/stop (typically included in confirmation transactions) — MES users input command: START/STOP and data are sent to ERP system.
MES	Production planning	Production confirmations (in-process and final confirmations) — MES sends production confirmations to ERP.
MES	Production planning	Production order and schedule changes (quantities, dates, cuts, adds) — MES input changes and data are sent to ERP.
MES	Production planning	Material and process deviations (changes to released BOM, recipe and routings) — MES users input command: CHANGES and data is sent to ERP.
MES	Production planning	Real capacity information — Data sent from MES to ERP for improving scheduling.
MES	Production planning	Execution events (batch history) — Data are sent from MES to ERP.
MES	Material management	Inventory transactions (moves, issues, receipts) — Data from MES is sent to ERP.
Plant maintenance	MES	Maintenance time is synchronized with production plan and spare parts.
Plant maintenance	MES	Maintenance schedule of a device in order to chose alternate production routings.
MES	Plant maintenance	Detection that an asset is performing poorly. Send a work order notification to plant maintenance.

9.2.2.1   Download of Production-Relevant Information to an Execution System

This step typically involves scheduled production orders, which are released from the business system (production planning) and are transferred to a production execution system. During this process, a number of data mappings take place: the order recipe is mapped to a production recipe and recipe parameters are resolved (e.g., quantity, production line, equipment, and time and date). If unplanned, late changes occur, information must be communicated to the production system, as, for example, order size or order quality since such changes may impact the current production schedule.

9.2.2.2   Upload of Status Information from a Production System

Upload of status information from the production system includes deviations of the produced order from the actual production status, by, for example, order size, quality, material locations, and material usage. Furthermore, if the status information is continuously sent to the business system, the data can be used for detailed tracking of the production progress.

9.2.2.3   Data Exchange between a Production System and a Maintenance Management System

Automating the process of creating work orders due to equipment malfunctions and performing rescheduling of production due to lines in service can be achieved by synchronizing events from maintenance and production planning modules. Maintenance planning can be taken into account in order to calculate actual and planned capacity of production lines. This allows timely shifting production to alternate lines if maintenance actions are necessary in parts of the production facilities. Users can access equipment maintenance reports to identify causes for degrading production quality and schedule an order on alternate production lines already on the planning level.

9.2.2.4 Data Exchange between a Production and Scheduling System

Using history information from production (e.g., planned vs. actual material used or time needed to perform a production recipe in planning systems) allows more accurate forecasts and optimized production schedules. A planning system allocates all equipment resources for a specific recipe during the execution of the order. If the production system additionally provides detailed information on the equipment usage (e.g., when and how long an equipment is needed), capacity forecasts and scheduling can be optimized.

Summarizing, the integration scenarios range from simple data exchange to complex data mapping and routing functionality. To each of these, one or several appropriate technical integration solutions can be mapped.

9.3 Technical Integration

9.3.1 Integration Options

Depending on the functional integration requirements of data exchange, synchronization, and user access, four technical integration options (Figure 9.2) can be distinguished. They differ with respect to the number of involved systems and therefore the level of complexity, which in turn has a direct effect on the customization effort and adaptations that have to be made. While less complex systems with fewer numbers of interfacing points can be productized quite easily, larger systems with a high number of intersystems communications are usually tailored toward customer requirements.

View integration allows access to a target system, for example, ERP through its Graphical User Interface (GUI), which is embedded in the calling system. This can either be realized by linking the application GUI by simple command calls or through web-based navigation, which effectively depends on the display capabilities of the target system. In other words, for example, a user has access to transaction screens in an ERP system from his MES workplace and therefore can update order changes or material consumptions directly. View integration becomes more powerful if context-based information is shared between the source and target system, as, for example, a device or batch identifier is shared between the systems, which in turn eases the usage for users since it minimizes navigation effort to the requested information. View integration is a real-time user interaction with another system — accessing ERP transaction from within MES environment — and by thus blocking other activities while a transaction is in process. Also, changes to data do not become visible until the transaction has completed.

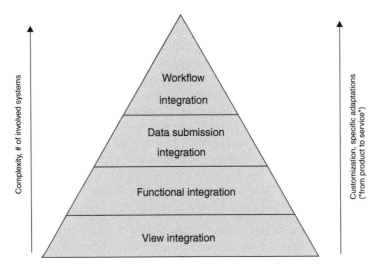

FIGURE 9.2 Technical integration options.

The *functional integration* option realizes interaction between plant-floor and business systems through application programming interfaces (APIs). Data are exchanged through programmable services at two sides, the production and the business system, by thus separating user input (GUI) and data exchange points. View access to the data can be fit to specific requirements of certain user groups such as operators or production engineers. Similar to the view integration, the functional integration also requires context information to be shared between the systems.

An *enterprise application integration* (EAI) component provides the integration layer for the *data submission* scenario. The component provides data transformation and message-based routing as key functionalities, since both involved end-points (business and production systems) do not track the published information, such as releases of production orders (business system) or the status of the production itself (production system). Data submission requires APIs on all involved systems.

Data submission functionality is either triggered from *events* or scheduled in the form of *bulk data transfer*. Since EAI components, in most cases, work through message queues, both the business and the production systems must not be connected all the time — the information is stored in queues and delivered as soon as the message channel is available.

- Event-based functionality is primarily used to monitor for events on one system, such as a data change, that immediately initiates a transaction on the other system. A context (message identification) is required to decide how to handle the information received.
- Bulk data transfer is used for interfaces that are either time dependent or involve large volumes of data. It allows the scheduled extract of multiple records from one system an intermediate transformation and the scheduled import into the other system. The architecture is developed to address interfaces originating in either ERP or MES.

Workflow processing components manage their own data sources and computation states and update this information while processing data flows. A simple workflow (e.g., based on the content of the data) can create, modify, or delete order entries in the production or business system. Examples for advanced workflow processing are advanced scheduling and balancing algorithms or order management, which might not be available as functional modules in either business or production systems themselves.

All components — view and functional integration, as well as data submission — are needed for workflows as the functionality is a mix of user-driven and event-based business process steps.

In general, two concurrent concepts can be distinguished in terms of workflow definition, administration, and execution.

1. A *centralized* solution using standard EAI tools. They provide means to define workflows through an orchestration engine, which supports the users in graphically defining the steps involved in a business workflow, the specification of data exchange maps for the single steps of the workflow, and the integration of the components from, in our case, MES and ERP system.
2. A *decentralized* solution where each participating component administers its own workflow part (function, engine, data translation) but does not know of the overall business scenario. The entry point of the function is responsible for orchestrating the different subfunctions, which are registered in a lookup service (e.g., UDDI) on a domain basis (e.g., each participating system registers a function supporting "outage management"). A component manager on both systems supports the translation from the service lookup to the individual services and functions on the systems that are included in the workflow. Each participating function will only know of its subworkflow and performs the individual services lookup for the following step.

However, both concepts still have to be tested and evaluated with respect to both their applicability in a production environment and their performance for the general characteristics such as reliability, transaction safety, or scalability, which are of utmost importance for a guaranteed execution of workflows. Therefore, they will not be covered in detail in this chapter.

9.3.2 Guiding Principles

The following principles shall be used to guide the development of the technical architecture and product selection decisions throughout the design process:

- Utilize standards to ensure compatibility with current and future IT developments and encapsulate existing APIs of MES and ERP;
- Design for large-scale, complex enterprise environments and provide an infrastructure for secure, reliable, and predictable execution;
- Facilitate the rapid generation of new functional interfaces by leveraging "off-the-shelf" technology and tools from known vendors; and
- Create an extensible integration architecture that can eventually be used to integrate MES with other applications in addition to ERP.

The following characteristics were taken into account for the design of the prototype architectures:

- *Reliability*: Guaranteed transaction processing, logging, and audit trails.
- *Scalability*: Support for multiple processor machines and server farms; distribution across multiple tiers.
- *Interoperability*: Support for multiple platforms and message transport protocols; support for web service technologies.
- *Adaptability*: Highly configurable allowing for fast adoptions; support the ability to adopt new client/user interfaces, and modular to isolate changes within a tier.
- *Availability*: No single points of failure; support for clustering, load balancing, backup, and restore; and monitoring services.
- *Maintainability*: Tools that assist in interface development, usage of widely available technologies, and support for templates and programming standard guidelines.

9.3.3 Integration Approach and Typical Use Cases

The integration concept supports large-scale, high-volume integration scenarios typically encountered in enterprise environments.

Table 9.2 describes real-life examples in a business context of different types of integration that are usually found across a production environment. These include a subset of the user interface and "data"-oriented interfaces. Subsequent sections will refer back to this table.

In the following subsections, the main technical concepts for the integration scenarios are described, covering in detail the *view, functional*, and *data submission (event-based and bulk data)* approach.

For each scenario, a typical use case is presented from both a manufacturing- and process industry-related scenario or from an energy production and delivery scenario of electric utilities. The main technical concept behind the integration is presented and a prototype integration that uses specific technologies is outlined.

9.3.4 Prototype Components

9.3.4.1 MES

Today's MES systems are of a monolithic architecture exposing a system-level interface only to exchange production-relevant information, such as product information (recipes, parameters) or production status information. Scheduling is most often provided from the ERP or advanced planning level; however, the MES application will often perform the detailed scheduling of operations that are highly dependent on the manufacturing process (e.g., creating a detailed schedule to consume a roll of paper to produce an end product, where the production scheduling optimization depends on the quality of the paper as it relates to the location on the roll) [9].

TABLE 9.2　Example Scenarios

Description	Real-Life Example	Illustrated Approach
Production orders and confirmations — Production orders from ERP are downloaded into MES. These orders include recipes stored in ERP.	Production orders created in ERP production planning or scheduling are released and then sent to MES. Recipes are copied at the time of a production order release and sent to MES. The MES creates line-specific versions of the recipes, prepares, and initiates production.	Data submission (event)
	Under control of the control system, a box is filled with a product. After packaging, one unit of production is recorded in the MES system. Once an hour, material production confirmations for a production order are summarized, by material, and sent to ERP as a single production confirmation.	Data submission (bulk)
	Alternatively, the single unit of production is sent immediately from the MES to ERP.	Data submission (event)
Detailed status and schedule changes — Dates, quantities, and cancellations are input by a scheduler into ERP or by a shop floor supervisor into MES.	A master scheduler receives a phone call from customer service with an urgent request to increase a released production order by 10%. After ensuring that the order has not started and that there are sufficient components, the scheduler increases the order by 10%. This change is sent to the MES for execution.	Data submission (event)
	Alternatively, at the beginning of a production order the supervisor discovers that a key component is 50% short and that additional components cannot be expedited into the shop. The supervisor navigates to the ERP production order screen and changes the quantity expected by 50% giving visibility to customer service and production scheduling.	View or functional (UI)
Materials — A shop floor supervisor may make a material supply inquiry or a material demand inquiry.	The material pick list in the MES system is calling for more material than is physically present. The shop floor supervisor navigates to the material supply screen in the ERP system to investigate alternate locations for the required materials.	View or functional (UI)
	Additionally, a shop floor supervisor must decide which of two different high-priority component orders to run. From the MES system, the supervisor navigates to the material demand function in the ERP system to view the "pegged" demand for that component.	View or functional (UI)

Throughout this chapter, the MES referenced to is ABB's batch production system, based on ABB's Industrial^{IT} architecture. Production equipment and recipes can be configured and managed from the framework. Users can view progress of production orders or details on the execution.

The following paragraphs summarize the architectural building blocks as realized in ABB's Aspect Integrator Platform (AIP) being the central component of ABB Industrial IT architecture (Figure 9.3). AIP constitutes the glue between all components of the control system and the MES system.

9.3.4.1.1　Aspect Objects and Aspect Object Types

Concepts, actors, and entities that are relevant within the enterprise and plant context (for example, motors, robots, productions cells, valves, pumps, products, processes, customers, or locations) are represented in the control and production system as aspect objects. In the following, we will refer to it shortly as objects.

Each object in the system is an instance of an object type. Object types, which are related by inheritance hierarchies, determine which aspects and aspect systems (see below) are associated with each object instance.

Instances of objects are organized hierarchically in so-called structures — means to define groupings from a user perspective, such as a location or functional structure. Structures also describe the dependencies between real objects in a certain navigation context. An object can exist in multiple structures (Figure 9.4), for example, a motor object in both a functional and location unit structure. Each structure

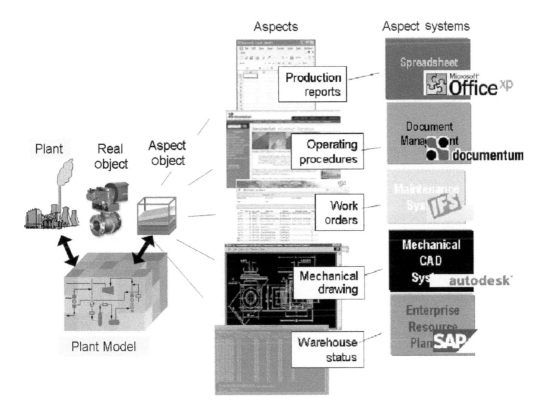

FIGURE 9.3 Aspect systems and aspect objects.

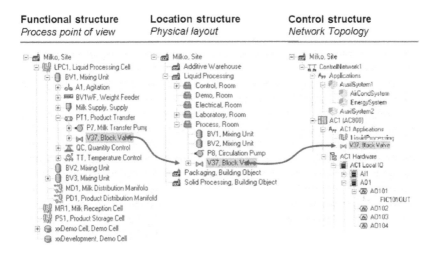

FIGURE 9.4 Structuring concept of aspect object.

represents one meaningful context and notion of relatedness between objects. For example, an asset is (a) a technical equipment, (b) a member of an organizational unit such as a production cell, (c) located at a certain place, and (d) participates in a production process.

9.3.4.1.2 Aspects and Aspect Systems

Each object carries a number of so-called aspects. An *aspect* is a component encapsulating a subset of data (attributes) and corresponding methods, associated with the object and relating to a common context

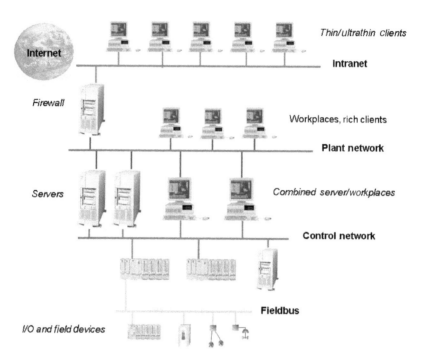

FIGURE 9.5 Industrial IT network architecture.

or purpose (e.g., all maintenance attributes of the equipment objects of a production cell object are part of the consolidated maintenance data aspect of the, e.g., pump object).

Aspects also serve as ports to external data sources. Data are exchanged via protocols like OPC, HTTP, or specific connectors to various ERP systems. *Aspect systems* are the information access applications with or without user interface, which are used to view, edit, maintain, store, and process the information contained in the aspects of an object.

Within AIP, aspect systems can also transparently access any other aspects that allow navigating between aspects in the same context. From the process-related information displayed as process graphics, the user can thus navigate seamlessly to the maintenance related information (e.g., a work order) without losing the context of the object (e.g., object name, object identifier).

Its *aspect object architecture* assumes a system of computers and devices that communicate with each other over different types of communication networks. This layered network (Figure 9.5) consists of

- The *Intranet,* used for communication with thin clients such as mobile devices or browser-only work panels, but also access to third-party nonproduction systems, such as ERP.
- The *plant network,* used for communication between *servers,* and between servers and *workplaces.* Servers run software that provides system functionality, and workplaces run software that provides various forms of user interaction, such as process graphics, alarming, or trending.
- The *control network,* a local area network (LAN) that is optimized for high performance and reliable communication with predictable response times in real time. It is used to connect *controllers* to the servers. Controllers are nodes that run control software.
- *Fieldbuses,* used to interconnect field devices, such as I/O modules, smart sensors and actuators, variable speed drives, or small single loop devices. These devices are connected to the system, either via a controller or directly to a server, through, for example, OLE for process control (OPC).

9.3.4.2 ERP System

SAP's Production Planning and Plant Maintenance modules act as the respective ERP components in the outlined prototype scenarios.

Calling specific functionality in SAP is available through different mechanisms. In this scope, *the business application programming interfaces* (BAPIs) and *intermediate documents* (IDocs) mechanisms are covered. While the first one is a functional interface, IDocs utilize message queues.

An object-oriented approach for access to SAP systems has been introduced by SAP providing business processes and their relevant data as business objects. External applications can access the SAP business objects via standard interfaces called BAPIs. BAPIs offer an abstract, object-oriented view and keep implementation details well concealed. As a consequence, BAPIs represent the building blocks for the construction of interacting components, with the outside world being accessible through a variety of technology connectors (e.g., JAVA, COM, .NET).

IDocs are used in an SAP context to exchange messages with other SAP systems or third-party systems. Messages are sent asynchronously, possibly even as batches. A message carries data, which typically are split into multiple semantically meaningful segments. These segments may be structured hierarchically. Together with the data, a message exchange is accompanied by a control record, which specifies the source and routing of the message, and a status record, which tracks the life cycle of the message.

The IDoc and partner definition process consists of the following sequence of activities:

- IDoc definition (segment definition and message type definition).
- Linkage of IDoc to message type.
- Definition of ports and RFC destination.
- Definition of partner profile.
- Linkage of IDoc and message type to application object.

Figure 9.6 graphically describes an inbound IDoc message exchange. Typically, two partners (two connected software systems) exchange a message, which in this case is a request for a list of active work orders. The message is called GetActiveWorkOrders. Both the external partner, BIZTALKCH, and the message type, GetActiveWorkOrders, have to be registered in the SAP system and configured with appropriate settings. As many partners as needed may be configured in an SAP system.

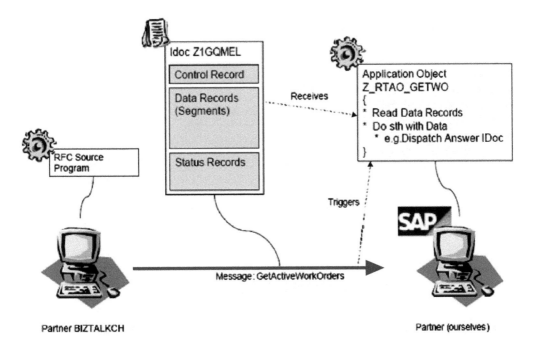

FIGURE 9.6 IDoc exchange between business partners.

9.4   View Integration

The *view integration architecture* is intended to provide a solution to user interface access to all business system functionality from the MES user environment. The architecture allows a production system user direct access to all user-relevant ERP transactions with context information and allows them to view, update, create, or delete data. For example, a user would be able to go directly to the list of material associated with a specific production order.

9.4.1   Use Case

The responsible engineer of a paint shop production line notices that the output quality of a production cell is poor; in fact, looking at color quality trends, he notices that the number of repaintings has increased steadily over the last 24 h. With integrated ERP-CMMS (computerized maintenance management system) connectivity into MES workplaces, he can access the maintenance history for the paint shop cell and its components in order to identify possible causes for the poor quality. Selecting the paint shop cell on the workplace screen, he can invoke "show maintenance history" as an action.

Last planned maintenance on the cell has been done 3 weeks ago, and the report summary indicates that a "normal" maintenance service has been carried out. Also, the performance history of the paint shop cell components does not reveal any unusual behavior.

Navigating on a GUI, the workplace, the user can select a piece of equipment and invoke, through interaction with the UI, one of the available CMMS functions for accessing maintenance information, which is stored in the CMMS system for the selected equipment. The CMMS data are displayed through the proprietary CMMS user interface (Figure 9.7).

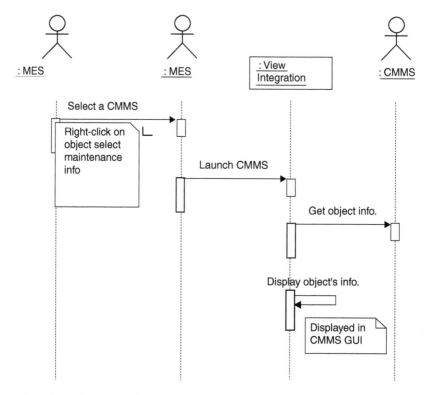

FIGURE 9.7   Accessing maintenance information.

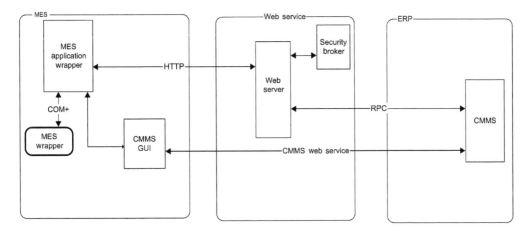

FIGURE 9.8 View integration concept.

9.4.2 Technical Concept

The user is given the possibility to invoke an ERP transaction screen from an MES workstation. While this is a relatively easy solution to implement, it does not provide navigation from the ERP screen to other production system applications, nor does it provide for data translation between the systems.

The architecture utilizes the "wrapper" capabilities of available software components such as Internet Explorer or other to encapsulate the client's transactions (ERP GUI). Through these containers, a variety of target components in the form of ActiveX controls, HTML pages, or Java applets can be accessed. To alleviate the manual configuration required, "smart wrappers" can be developed for specific ERP transactions, which automatically collect relevant context data associated with the entity when the function is initiated and embed it into the ERP GUI invocation. The equipment identifier and ERP transaction number are examples of context data that are passed. The concept of embedded context information allows navigating directly to the required transaction screens and avoiding that the user has to step through a number of menus. The architecture also provides a single sign-on capability to enable access to ERP transactions without having to provide credentials for each transaction. The primary components and message flow required for this architecture are illustrated in Figure 9.8.

9.4.3 Prototype Realization

The prototype builds upon the concept of representing a real-world object through the AIP Aspect Object container. By attaching relevant context information, such as an equipment identifier from the different systems (MES, ERP) as aspects, this concept allows to map and share this information when invoking the view functionality.

The prototype architecture with specific technologies is illustrated in Figure 9.9. The focus of the implementation builds on the functionality of the SAP Internet Transaction Server (ITS), which allows accessing any SAP transaction through a URL as an HTML page. Therefore, a distribution of client software to each MES workstation is not required, which in turn increases the likelihood of supporting direct access to specific transactions. Simple HTML wrappers are used to provide application access to SAP transactions with static or hard-coded context data, whereas the "smart" wrappers are able to collect the required context data to navigate to a specific instance of a transaction.

The user is able to enter the original SAP screen and to navigate in the transaction context with all SAP functionality being available (Figure 9.10). The main disadvantage of this concept is that the screens typically are not customized toward their specific needs by, for example, limiting the information presented, but present the ERP transaction screen that was usually built on requirements from other user groups.

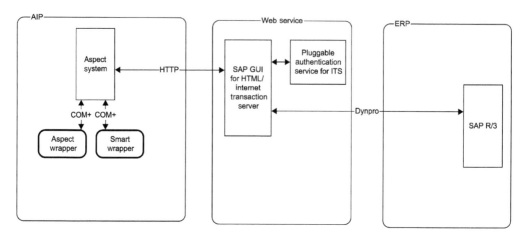

FIGURE 9.9   Prototype: view integration.

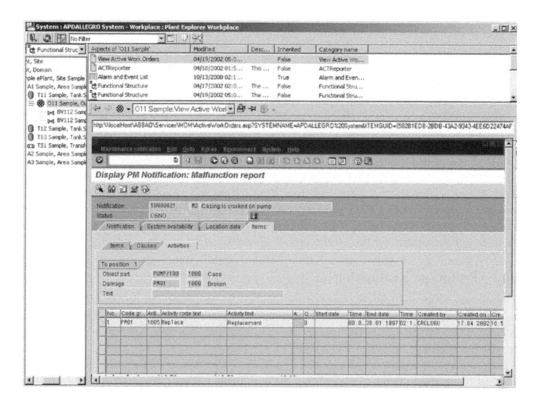

FIGURE 9.10   View integration: example screenshot.

9.5   Functional Integration

This architecture provides interface integration from MES to specific transactions in ERP, and vice versa, through APIs. This is enabled by defining a set of interfaces exposed toward either MES or ERP. Data exchange through interfaces and graphical representation of the data is separated, allowing developing custom GUI depending on user requirements while reducing the interface development efforts. Through definition of domain interfaces, for example, for maintenance or production, only the adapters to the various ERP systems have to be developed, while the exposure of interfaces toward the integration system remains consistent. Thus, adapting to other systems is eased, since only the adapter needs to be exchanged.

9.5.1 Use Case

In the production scenario from the paint shop production line, the user still is not satisfied with the quality of the cell. The next planned maintenance shutdown is scheduled in 1 week. However, since the engineer could not identify the root cause of the quality degradation, he decides to issue a maintenance request for the paint shop cell with severity "high."

Creating a maintenance request with integrated CMMS connectivity is simply done by selecting the paint shop cell on his workplace screen, and invoking "maintenance request" as action. A data screen is shown to the engineer, which already has data entered specific to the selected cell, such as cell identifier, date and time, name of the user, etc. The engineer selects priority "high," and attaches links to the quality trend diagrams. As soon as the work order is submitted, a new notification is generated within the CMMS system and the corresponding data sets are stored. Now, the maintenance planning department can analyze the attached information and decide if an unplanned shutdown is necessary (Figure 9.11).

In addition to retaining a common context as already described in the previous section on view integration, the data semantics between the systems have to be defined, ideally through a common application data model that abstracts specific application functions calls into domain APIs, such as managing maintenance requests or scheduling orders. Only such domain interfaces guarantee interoperability between systems from various vendors.

9.5.2 Technical Concept

Functional integration offers the ability to separate data representation and the communication between involved systems through defined APIs. In this architecture, the MES initiates a call to a web service that exposes domain interfaces (e.g., for maintenance) and in turn uses specific ERP APIs to implement the functionality. The web service brokers access either directly to ERP through an ERP proxy component

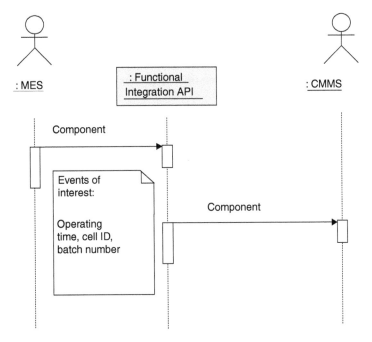

FIGURE 9.11 Transfer of quality information from MES to CMMS.

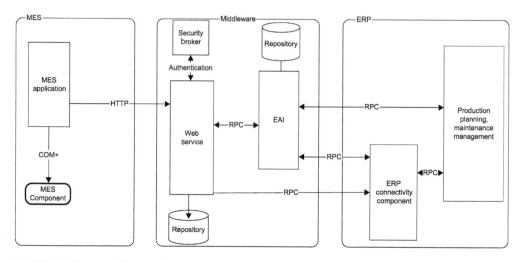

FIGURE 9.12 Functional integration concept.

or through an EAI component and works across firewalls. The latter provides functionality such as context-based data mapping and translation. User interface components use the web services to request or post data; the transaction itself is executed with the call to a specific method on the web service itself.

The primary components and message flow required for this architecture are illustrated in Figure 9.12.

As with the view integration, this architecture could be utilized for integration points as described in Table 9.2.

9.5.3   Prototype Realization

The prototype architecture with specific technologies is illustrated in Figure 9.13. The user interface component gathers the required information necessary to request or process information in SAP and package it into a Simple Object Access Protocol (SOAP) packet. The SOAP packet is then submitted via HTTP to a web service, where the information from within the SOAP message is used to identify the required request for SAP. As a next step, the web service queries the repository for the necessary information needed to call the SAP Microsoft Distributed Component Model (DCOM) generated proxy, instantiate the proxy, and pass the necessary information gathered from the SOAP message.

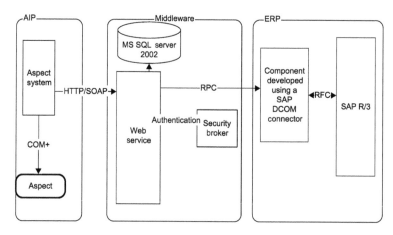

FIGURE 9.13 Prototype: functional integration.

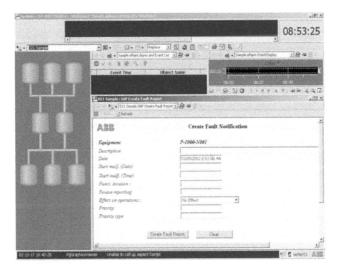

FIGURE 9.14 Functional integration — example screenshot.

In the functional integration architecture, GUIs are typically customized (Figure 9.14) and only context-relevant information is shown in order to perform a specific task. Changing requirements result in coding efforts to adapt data mapping and API calls. Information from the ERP on the selected object is retrieved and displayed to the user through a customized screen that is tailored to his needs and omitting information which is more specific for, for example, maintenance planning. For example, a user would be able to go directly to the fault notification site associated with a specific piece of plant equipment. However, the interface definitions and the ERP specific code do not change.

9.6 Data Submission

Data submission requires functional interfaces as described in the preceding section on both business and plant-floor systems. An EAI component provides the integration layer for the data submission scenario. This component provides data transformation and message-based routing as key functionalities, since both involved end-points (business and production systems) do not track the published information, such as releases of production orders (business system) or the status of the production itself (production system). The EAI component uses functions, which query or listen for information from one system and convert data from the format delivered from the source to the format required by the target system. Since EAI components, in most cases, work through message queues, both the business and the production systems must not be connected all the time — the information is stored in queues and delivered as soon as the message channel is available. A context (message identification) is required to decide how to handle the information received through the queues.

9.6.1 Event-Based Data Submission

9.6.1.1 Use Case

As a precondition to the described scenario, a customer order is received by a dispatcher, who enters the order into the planning system as specified by the customer: product type, quantity, and delivery date. The manufacturing plant is manually chosen by the dispatcher. After the order has been entered into the planning system, verification is carried out against available material in stock. The order can be released to the production system, and the relevant data are transferred.

As soon as the order is completed, status information (e.g., material used) is transferred back to the production planning system, which triggers an update of the material and inventory stocks (planned vs.

actual). The dispatcher can look up the inventories and check if the order has been executed against the given customer order. As soon as the information is updated in the inventory, the customer can be notified and the delivery can be initiated.

9.6.1.2 Technical Concept

The presented architecture provides a loosely coupled integration between the MES and a production planning system. This will be beneficial in the cases where the transaction has the potential to "block" the user's workstation while they wait for a response from ERP. Instead, a user can make the asynchronous request and perform other tasks while waiting for a message to return, indicating that the transaction has been processed. Requests are submitted automatically through system-generated events originating in either ERP or production system.

The implementation involves an EAI component, which interacts both with the ERP system as well as with the production system. It is the central component of the integration architecture. Many of the other integration architecture components are determined by the EAI application selected. Utilizing an off-the-shelf EAI application greatly reduces the effort involved in developing and maintaining the integration architecture between production systems and ERP. EAI applications typically provide the following functionality:

- *Process flow integration and management*: GUI tools for process integration and management, workflow, and state management across applications and enterprise boundaries.
- *Development tools*: Including configuration management, source control, debugging tools, and general coding environment.
- *Technology architecture*: Reliability, scalability, availability, adaptability, as well as operations support.
- *Transformation and formatting*: Transformation, translation, mapping, and formatting for integration purposes (i.e., to reconcile the differences between data from multiple systems and data sources).
- *Business-to-business capabilities*: Integration with trading partners, partner management, and Internet standards support — XML, HTTP, SMTP, FTP, SSL.

Utilizing an EAI application to develop interfaces helps to reduce the long-term maintenance effort by providing a central repository for data mapping between data sources and targets. The repository allows developers to reuse mappings and translations consistently across multiple interfaces. EAI also provides customers with a platform to integrate the MES with other applications in addition to production planning.

Interaction with the production system is realized by two custom services (Figure 9.15), which hook up to the proprietary APIs provided by the production system. One service listens to production orders from the EAI component, while the other polls the status interface of the production system.

Production orders are sent from ERP to the EAI component as ERP messages and are placed in an EAI inbound message queue. The information contained in the message is read and transformed according to the defined mapping and placed in an outbound message queue. The custom service listens to

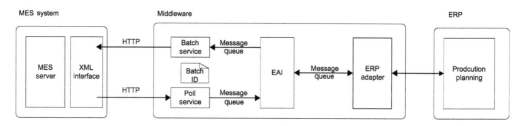

FIGURE 9.15 Event-based data submission concept.

outbound messages, reads the message, and calls the appropriate CreateOrder function on the production system and passes the corresponding parameters. At the same time, a log file is created containing the batch identifier number that allows the second service to correlate the status to active orders.

The second service continuously polls the status interface of the production system and checks the status of the order number that is logged. As soon as the production status changes from "running" to "completed," the service reads the status information (actual material used) from the production system, composes a message, and places the message to an EAI inbound message queue. Again, the message content is read and mapped according to the PollStatus mapping definition, and the resulting message is placed in an outbound message queue. Finally, the EAI architecture for ERP reads the message, composes an ERP message, and routes it into the ERP system.

9.6.1.3   Prototype Realization

Microsoft BizTalk Server 2002 has been chosen as an EAI component, providing the server, tools, and plug-ins needed to integrate and automate the business between the production system and ERP. A key benefit of BizTalk Server is its ability to integrate XML web services and to supply a central repository for mappings and transformations, which are stored natively in XML. BizTalk Server comes with Software Development Kits (SDKs) available for transports, document types, and application architectures. Custom BizTalk mapping functions (functoids) can be developed and reused for multiple interfaces to accommodate the specific transformations of batch data to ERP data and vice versa.

Development is performed using wizard-based design tools such as BizTalk Mapper, Orchestration Designer, and Message Manager. If needed, C# and scripting languages (as, e.g., VBScript) supporting COM or the .NET framework can be used as well.

BizTalk supports Remote Function Call (RFC), BAPI, and iDOC integration with SAP through plug-ins. There are a number of different connectors available, both from Microsoft and independent system vendors that provide ERP-specific integration capabilities for BizTalk. The chosen Microsoft connector provides the following:

- retrieves iDoc structure, and generates XML schemas for iDoc automatically;
- defines routes for documents within BizTalk Server environment;
- guarantees successful delivery of an iDoc both into and out of ERP; and
- supports both BizTalk Orchestration Services using a COM component and the BizTalk Messaging Manager using a BizTalk Application Integration Component (AIC).

The architecture makes extensive use of the data mapping and transformation capabilities of EAI. Figure 9.16 depicts how BizTalk Server as one example of an EAI integration product can be utilized to define data mappings between batch and ERP, while highlighting BizTalk Server's Message Mapper tool.

Two mappings have been defined for the architecture: CreateOrder and PollStatus. The first one defines the relationship between ERP's message format and a custom message format in order to deliver the necessary information with corresponding parameters to schedule and initiate a production order. The second mapping contains status information definition for updating ERP's inventories.

The prototype architecture provides two means of submitting asynchronous requests. Requests can be submitted either through an MES user interface or automatically through system-generated events originating in either SAP or MES.

Custom UI control that collect the required input data from the user and submit or receive requests are developed through the web services features of the .NET framework. The user will receive a message indicating that the transaction has been processed as the response corresponding to the initial transaction.

An MES service monitor receives system generated events from the production system. When an event is received, the MES service formats the request into a SOAP packet and submits the packet into an outbound message queue. The architecture utilizes Microsoft Message Queue (MSMQ) as the transport mechanism between MES and BizTalk Server to enable HTTP as the transport protocol. The data transported using MSMQ are received by BizTalk Server, transforms, or maps, the XML stream received from the MSMQ into an IDoc format to be processed by SAP. BizTalk Server, and then invokes BizTalk

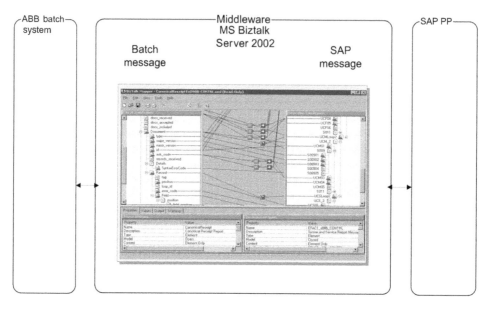

FIGURE 9.16 Message mapping.

Adapter for SAP connector to submit the IDoc into SAP. BizTalk Adapter utilizes the DCOM Connector
to initiate the receipt of the IDoc from SAP.

Requests from SAP to MES travel a similar route. SAP initiates the Remote Function Call (RFC) Server,
COM4ABAP, through a transactional RFC (tRFC) call. COM4ABAP deposits the IDoc into an MSMQ.
Once the IDoc is in MSMQ, BizTalk Server can initiate any transformations needed for the MES inte-
gration. The transformed message is dropped on an outgoing MSMQ to be delivered to an MES inbound
MSMQ. The MES service reads the MSMQ, and based on the information supplied within the message,
the appropriate production system API is called.

The prototype architecture with specific technologies is illustrated in Figure 9.17.

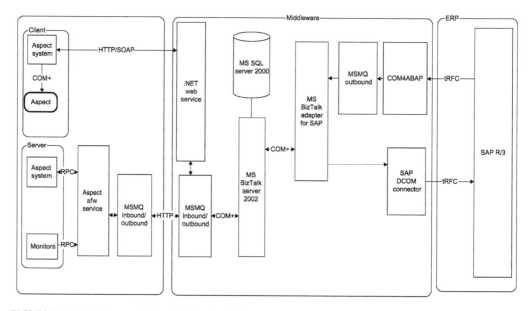

FIGURE 9.17 Prototype event-based data submission.

9.6.2 Data Submission Using Bulk Data Transfer

Bulk data integration architecture provides the ability to extract data from either MES or ERP, apply a custom data transformation, and import it into the other system. The architecture is used for moving large volumes of data and for transactions that are time dependent. An example would be the periodic update of selected master data records from ERP to MES. Another example would be the scheduled summarizing of material production confirmations.

9.6.2.1 Use Case

Production schedules are released to the MES according to the incoming orders from the customer. While the MES performs a detailed scheduling of the production orders and distributes the schedules among the available production lines, the ERP system has to be notified of finished production in order to initiate delivery to the customer and fill up material stocks for remaining production. Therefore, finished production batches are collected at the MES and sent in bulk to the ERP. Additionally, every 24 h, the material consumption, or the various production lines, is recorded at the MES and sent to the ERP for updating stock lists and initiating further material orders (Figure 9.18).

9.6.2.2 Technical Concept

Bulk data integration architecture is driven by an external scheduling system that initiates jobs on each system through a scheduling agent.

A scheduling agent on the MES system triggers export and import jobs through a custom data manager, which determines the appropriate import and export routines to execute, log audit data, and initiate outbound data transfers. The scheduling agent on the EAI server initiates the data transformation routines and outbound data transfers. On the ERP system side, the scheduling agent initiates export and import jobs through a data manager that determines the appropriate import and export objects to execute, log audit data, and initiate outbound data transfers.

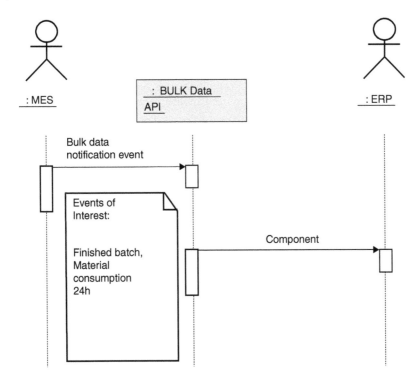

FIGURE 9.18 Transfer of quality information from MES to ERP.

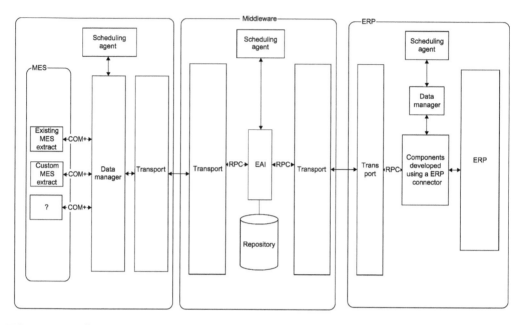

FIGURE 9.19 Bulk data submission concept.

The primary components and message flow required for this architecture are illustrated in Figure 9.19.

The architecture relies on the availability of an enterprise job scheduling application that can initiate jobs across multiple systems and platforms. The scheduling agents are provided by the enterprise scheduler, required on the MES and ERP servers where extracts and imports are running. The agent is also required on the EAI server. The agents initiate jobs on the local servers and report status back to the enterprise scheduler.

Providing the correct containers for moving large data volumes is of equal importance as the scheduling architecture. The following protocols are being considered for transporting files from server to server.

- *File Transfer Protocol (FTP)* — FTP is a commonly used client/server protocol that allows a user on one computer to transfer files to and from another computer over a TCP/IP network. FTP is often used as a reliable data transfer method between dissimilar systems. Utilities to ensure the complete transfer of files could easily be developed or purchased at a nominal cost.
- *File System* — The files could be transferred by copying files between file system shares. This would most likely be supported in a Windows-based environment, but would require additional utilities if the ERP server was on a platform other than Windows.
- *HTTP* — Utilizing HTTP would ensure access through firewalls; however, HTTP has been shown to have poor performance in transferring larger files.
- *Message Queues* — MQ support the guaranteed delivery of messages. Using message queues for this architecture would require the development of a utility to break large files into smaller pieces depending on the maximum message size.

9.6.2.3 Prototype Realization

The prototype architecture utilizes the customer's enterprise scheduling system and agents to control the execution of the bulk data integration interfaces. The scheduler provides agents for the MES, BizTalk Server, and the SAP system.

The scheduling agent on the MES calls the import and export programs directly. Upon successful completion of an export, the scheduling agent initiates an FTP transfer of the export file to BizTalk Server. The scheduling agent then calls a component on the BizTalk Server to audit the file, confirms a successful transfer, and initiates an XLANG schedule in BizTalk Server. XLANG is an extension of the Web Service

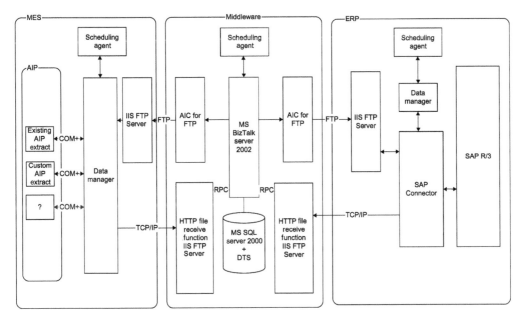

FIGURE 9.20 Prototype bulk data submission.

Definition Language (WSDL) and provides both the model of an orchestration of services as well as collaboration contracts between orchestrations. The XLANG schedule performs any required transformations and mappings, adds new audit information to the file, and transfers the file to the SAP R/3 server. BizTalk Server utilizes a connector component for FTP to send the file to the SAP server. On the SAP side, the scheduling agent calls the file transfer audit program to confirm the successful file transfer, and then initiates the data import and logs the results. A similar process is implemented for interfaces from SAP to MES.

The realized architecture with specific technologies is illustrated in Figure 9.20.

9.7 Conclusions and Outlook

We have presented integration scenarios in an intra-enterprise production environment, and have outlined technical integration options as building blocks for reusable integration architectures. Prototype concepts were outlined for each of the use cases presented.

The following conclusions can be drawn regarding integration between plant-floor and business systems:

- point-to-point data exchange solutions restrict integration flexibility;
- an integration solution is composed of different technical components;
- for a reusable solution, domain interfaces have to be standardized and developed, which allow connecting different systems without the need to redevelop the whole architecture;
- systems with fewer numbers of interfacing points can be productized quite easily; larger systems with a high number of intersystems communications are usually tailored toward customer requirements;
- standards serving cross-system integration, such as ISA S95, will further strengthen the role of EAI as a bridging layer between automation and business systems through their data mapping and orchestration capabilities; and
- using standard technologies such as XML and web services eases the task of application integration across vendors and platforms.

However, the full solution potential can only be achieved if the following integration issues are carefully taken into account:

- *Data consistency and engineering*: Data entry must be facilitated and promoted across the various subsystems and must, by all means, be kept synchronized across the various subsystems; this implies detecting changes (insert, modify, delete) in systems on objects (e.g., equipment, batches) and their attributes (e.g., status data) and replicate changes to "connected" systems according to replication rules and in a defined sequence.
- *Data exchange*: Common sets of functionality for various subsystems (such as production planning), high-level APIs to facilitate the access to these systems, and common data description to overcome semantic differences of the involved systems must be developed. A uniform data access for users and applications, hiding the origin of data from different systems, can be achieved by combining functional and data integration capabilities as described in this document. To the outside (user or application), a uniform access interface allows requesting information of a specified "category" — for example, operational data, maintenance-related information, performance-related information, etc, which are composed of attributes of object instances from different systems.

In order to achieve this functionality, additional concepts, such as data and attribute views (i.e., typed concepts) should be introduced with the assignment of object instances to types, and the ability to define relations between the content of types.

Transparent access to object attribute information will be the functionality that enables applications and users to access information according to defined "data views" independent of the information sources. The origin of the information (source systems) is hidden. Benefits of this functionality include a better access management to applications, simplified aggregation of data across systems, and the synchronization of objects and their attributes between systems.

References

1. Deliang, Z., L. Jizhen, and L. Yanquan, Control-Decision and Integrated Information System of Power Plant, in *Proceedings of Powercon '98*, Beijing, China, Vol. 2, 1998, pp. 1241–1245.
2. Canton, J., D. Ruiz, C. Benqlilou, J.M Nougues, and L. Puigjaner, Integrated Information System for Monitoring, Scheduling and Control Applied to Batch Chemical Processes, in *Proceedings of IEEE International Conference on Emerging Technologies and Factory Automation*, Vol. 1, 1999, pp. 211–217.
3. Herzog, U., R. Hantikainen, and M. Buysse, A Real World Innovative Concept for Plant Information Integration, in *Proceedings of Cement Industry Technical Conference, IEEE-IAS/PCA 44th*, 2002, pp. 323–334.
4. Wells, D. et al., Enterprise Application Integration, Report, Ovum Ltd., 2002.
5. Axtor, C. et al., Web Services for the Enterprise, Opportunities and Challenges, Report, Ovum Ltd., 2002.
6. XML resources, http://www.w3c.org/xml
7. Enterprise-Control System Integration, ANSI/ISA-95.00.01-2000, Part 1: Models and Terminology, ANSI/ISA-95.00.02-2001, Part 2: Object Model Attributes.
8. De Vos, A., S. Widergren, and J. Zhu, XML for CIM Model Exchange, in *Proceedings of IEEE Conference for Power Industry Computer Applications, PICA 2001*, Sydney, Australia, 2001, pp. 31–37.
9. McClellan, Michael, *Applying Manufacturing Execution Systems*, APICS Series on Resource Management, St. Lucie Press, Boca Raton, FL, 1997.

Part 4

Network-Based Integration Technologies in Industrial Automated Systems

Section 4.1

Field Devices — Technologies and Standards

10

A Smart Transducer Interface Standard for Sensors and Actuators

Kang Lee
National Institute of Standards and Technology

10.1 Introduction

Sensors are used in many devices and systems to provide information on the parameters being measured or to identify the states of control. They are good candidates for increased built-in intelligence. Microprocessors can make smart sensors or devices a reality. With this added capability, it is possible for a smart sensor to directly communicate measurements to an instrument or a system. In recent years, the concept of computer networking has gradually migrated into the sensor community. Networking of transducers (sensors or actuators) in a system and communicating transducer information via digital means vs. analog cabling facilitates easy distributed measurements and control. In other words, intelligence and control, which were traditionally centralized, are gradually migrating to the sensor level. They can provide flexibility, improve system performance, and ease system installation, upgrade, and maintenance. Thus, the trend in industry is moving toward distributed control with intelligent sensing architecture. These enabling technologies can be applied to aerospace, automotive, industrial automation, military and homeland defenses, manufacturing process control, smart buildings and homes, and smart toys and appliances for consumers. As examples: (1) in order to reduce the number of personnel to run a naval ship from 400 to less than 100 as required by the reduced-manning program, the U.S. Navy needs tens of thousands of networked sensors per vessel to enhance automation, and (2) Boeing needs to network hundreds of sensors for monitoring and characterizing airplane performance.

Sensors are used across industries and are going global [1]. The sensor market is extremely diverse, and it is expected to grow to $43 billion by 2008. The rapid development and emergence of smart sensor and field network technologies have made the networking of smart transducers a very economical and attractive solution for a broad range of measurement and control applications. However, with the existence of a multitude of incompatible networks and protocols, the number of sensor interfaces and amount of hardware and software development efforts required to support this variety of networks are enormous for both sensor producers and users alike. The reason is that a sensor interface customized for a particular network will not necessarily work with another network. It seems that a variety of networks will coexist to serve their specific industries. The sensor manufacturers are uncertain of which network(s) to support and are restrained from full-scale smart sensor product development. Hence, this condition has impeded the widespread adoption of the smart sensor and networking technologies despite a great desire to build and use them. Clearly, a sensor interface standard is needed to help alleviate this problem [2].

10.2 A Smart Transducer Model

In order to develop a sensor interface standard, a smart transducer model should first be defined. As defined in the IEEE Std 1451.2-1997 [3]:

> a smart transducer is a transducer that provides functions beyond those necessary for generating a correct representation of a sensed or controlled quantity. This functionality typically simplifies the integration of the transducer into applications in a networked environment.

Thus, let us consider the functional capability of a smart transducer. A smart transducer should have:

- integrated intelligence closer to the point of measurement and control,
- basic computation capability, and
- capability to communicate data and information in a standardized digital format.

Based on this premise, a smart transducer model is shown in Figure 10.1. It applies to both sensors and actuators. The output of a sensor is conditioned and scaled, then converted to a digital format through an analog-to-digital (A/D) converter. The digitized sensor signal can then be easily processed by a microprocessor using a digital application control algorithm. The output, after being converted to an analog signal via a digital-to-analog (D/A) converter, can then be used to control an actuator. Any of the measured or calculated parameters can be passed on to any device or host in a network by means of network communication protocol.

The different modules of the smart transducer model can be grouped into functional units as shown in Figure 10.2. The transducers and signal conditioning and conversion modules can be grouped into a building block called a *smart transducer interface module* (STIM). Likewise, the application algorithm and network communication modules can be combined into a single entity called a *network-capable application processor* (NCAP). With this functional partitioning, transducer-to-network interoperability can be achieved in the following manner:

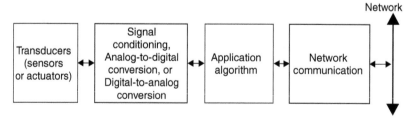

FIGURE 10.1 A smart transducer model.

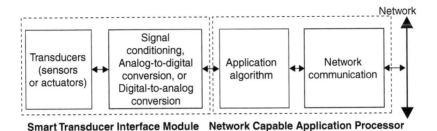

FIGURE 10.2 Functional partitioning.

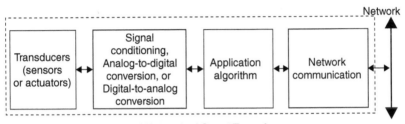

FIGURE 10.3 An integrated networked smart transducer.

1. STIMs from different sensor manufacturers can "plug-and-play" with NCAPs from a particular sensor network supplier,
2. STIMs from a sensor manufacturer can "plug-and-play" with NCAPs supplied by different sensor or field network vendors, and
3. STIMs from different manufacturers can be interoperable with NCAPs from different field network suppliers.

Using this partitioning approach, a migration path is provided to those sensor manufacturers who want to build STIMs with their sensors, but do not intend to become field network providers. Similarly, it applies to those sensor network builders who do not want to become sensor manufacturers.

As technology becomes more advanced and microcontrollers become smaller relative to the size of the transducer, integrated networked smart transducers that are economically feasible to implement will emerge in the marketplace. In this case, all the modules are incorporated into a single unit as shown in Figure 10.3. Thus, the interface between the STIM and NCAP is not exposed for external access and separation. The only connection to the integrated transducer is through the network connector. The integrated smart transducer approach simplifies the use of transducers by merely plugging the device into a sensor network.

10.3 Networking Smart Transducers

Not until recently have sensors been connected to instruments or computer systems by means of a point-to-point or multiplexing scheme. These techniques involve a large amount of cabling, which is very bulky and costly to implement and maintain. With the emergence of computer networking technology, transducer manufacturers and users alike are finding ways to apply this networking technology to their transducers for monitoring, measurement, and control applications [4]. Networking smart sensors provides the following features and benefits:

- enable peer-to-peer communication and distributed sensing and control,
- significantly lower the total system cost by simplified wiring,

- use prefabricated cables instead of custom laying of cables for ease of installation and maintenance,
- facilitate expansion and reconfiguration,
- allow time-stamping of sensor data,
- enable sharing of sensor measurement and control data, and
- provide Internet connectivity, meaning *global* or *anywhere,* access of sensor information.

10.4   Establishment of the IEEE 1451 Standards

As discussed earlier, a smart sensor interface standard is needed in industry. In view of this situation, the Technical Committee on Sensor Technology of the Institute of Electrical and Electronics Engineers (IEEE)'s Instrumentation and Measurement Society sponsored a series of projects for establishing a family of IEEE 1451 Standards [5]. These standards specify a set of common interfaces for connecting transducers to instruments, microprocessors, or field networks. They cover digital, mixed-mode, distributed multi-drop, and wireless interfaces to address the needs of different sectors of industry. A key concept in the IEEE 1451 standards is the Transducer Electronic Data Sheets (TEDS), which contain manufacture-related information about the sensor such as manufacturer name, sensor types, serial number, and calibration data and standardized data format for the TEDS. The TEDS has many benefits:

- *Enable self-identification of sensors or actuators* — a sensor or actuator equipped with the IEEE 1451 TEDS can identify and describe itself to the host or network via the sending of the TEDS.
- *Provide long-term self-documentation* — the TEDS in the sensor can be updated and stored with information such as location of the sensor, recalibration date, repair record, and many maintenance-related data.
- *Reduce human error* — automatic transfer of TEDS data to the network or system eliminates the entering of sensor parameters by hands, which could induce errors due to various conditions.
- *Ease field installation, upgrade, and maintenance of sensors* — this helps to reduce life cycle costs because only a less skilled person is needed to perform the task by simply using "plug-and-play."

IEEE 1451, designated as Standard for a Smart Transducer Interface for Sensors and Actuators, consists of six document standards. The current status of their development are as follows:

1. IEEE P1451.0,* Common Functions, Communication Protocols, and TEDS Formats — *In progress.*
2. IEEE Std 1451.1-1999, NCAP Information Model for Smart Transducers [6] — *Published standard.*
3. IEEE std 1451.2-1997, Transducer to Microprocessor Communication Protocols and TEDS Formats — *published standard.*
4. IEEE std 1451.3-2003, Digital Communication and TEDS Formats for Distributed Multidrop Systems — *Published standard.*
5. IEEE std 1451.4-2004, Mixed-mode Communication Protocols and TEDS Formats — *Published standard.*
6. IEEE P1451.5, Wireless Communication and TEDS Formats — *In progress.*

10.5   Goals of IEEE 1451

The goals of the IEEE 1451 standards are to:

- develop network- and vendor-independent transducer interfaces,
- define TEDS and standardized data formats,
- support general transducer data, control, timing, configuration, and calibration models,

---

*P1451.0 — the "P" designation means that P1341.0 is a draft standard development project. Once the draft document is approved as a standard, "P" will be dropped.

- allow transducers to be installed, upgraded, replaced, and moved with minimum effort by simple "plug-and-play,"
- eliminate error prone, manual entering of data, and system configuration steps, and
- ease the connection of sensors and actuators by wireline or wireless means.

10.6 The IEEE 1451 Standards

10.6.1 The IEEE 1451 Smart Transducer Model

The IEEE 1451 smart transducer model parallels the smart transducer model discussed in Figure 10.2. In addition, the IEEE 1451 model includes the TEDS. The model for each of the IEEE 1451.X standards is discussed in the following.

10.6.1.1 IEEE P1451.0 Common Functionality

Several standards in the IEEE 1451 family share certain characteristics, but there is no common set of functions, communications protocols, and TEDS formats that facilitate interoperability among these standards. The IEEE P1451.0 standard provides that commonality and simplifies the creation of future standards with different physical layers that will facilitate interoperability in the family.

This project defines a set of common functionalities for the family of IEEE P1451 smart transducer interface standards. This functionality is independent of the physical communications media. It includes the basic functions required to control and manage smart transducers, common communications protocols, and media-independent TEDS formats. The block diagram for IEEE P1451.0 is shown in Figure 10.4. IEEE P1451.0 defines functional characteristics, but it does not define any physical interface.

10.6.1.2 IEEE 1451.1 Smart Transducer Information Model

The IEEE 1451.1 Standard defines a common object model for the components of a networked smart transducer and the software interface specifications to these components [7]. Some of the components are the NCAP block, function block, and transducer block.

The networked smart transducer object model provides two interfaces.

1. The interface to the transducer block, which encapsulates the details of the transducer hardware implementation within a simple programming model. This makes the sensor or actuator hardware interface resemble an input/output (I/O)-driver.

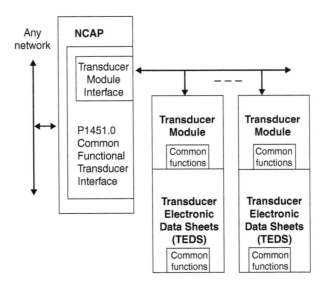

FIGURE 10.4 The block diagram for IEEE P1451.0.

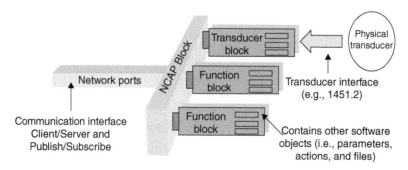

FIGURE 10.5 Conceptual view of IEEE 1451.1.

2. The interface to the NCAP block and ports encapsulate the details of the different network protocol implementations behind a small set of communications methods.

Application-specific behavior is modeled by function blocks. To produce the desired behavior, the function blocks communicate with other blocks both on and off the smart transducer. This common network-independent application model has the following two advantages:

1. Establishment of a high degree of interoperability between sensors/actuators and networks, thus enabling "plug-and-play" capability.
2. Simplification of the support of multiple sensor/actuator control network protocols.

A conceptual view of IEEE 1451.1 NCAP is shown in Figure 10.5, which uses the idea of a "backplane" or "card cage" to explain the functionality of the NCAP. The NCAP centralizes all system and communications facilities. Network communication can be viewed as a port through the NCAP, and communication Interfaces support both client–server and publish–subscribe communication models. Client–server is a tightly coupled, point-to-point communication model, where a specific object, the client, communicates in a one-to-one manner with a specific server object, the server. On the other hand, the publish–subscribe communication model provides a loosely coupled mechanism for network communications between objects, where the sending object, the publisher object, does not need to be aware of the receiving objects, the subscriber objects. The loosely coupled, publish–subscribe model is used for one-to-many and many-to-many communications. A function block containing application code or control algorithm is "plugged" in as needed. Physical transducers are mapped into the NCAP using transducer block objects via the hardware Interface, for example, the IEEE 1451.2 interface.

The IEEE 1451 logical interfaces are illustrated in Figure 10.6. The transducer logical interface specification defines how the transducers communicate with the NCAP block object via the transducer block. The network protocol logical interface specification defines how the NCAP block object communicates with any network protocol via the ports.

10.6.1.3 IEEE 1451.2 Transducer-to-Microprocessor Interface

The IEEE 1451.2 standard defines a TEDS, its data format, and the digital interface and communication protocols between the STIM and NCAP [8]. A block diagram and detailed system diagram of IEEE 1451 are shown in Figure 10.7 and Figure 10.8, respectively. The STIM contains the transducer(s) and the TEDS, which is stored in a nonvolatile memory attached to a transducer. The TEDS contains fields that describe the type, attributes, operation, and calibration of the transducer. The mandatory requirement for the TEDS is only 179 bytes. The rest of the TEDS specification is optional. A transducer integrated with the TEDS provides a very unique feature that makes possible the self-description of transducers to the system or network. Since the manufacture-related data in the TEDS always go with the transducer, and this information is electronically transferred to an NCAP or host, human errors associated with manual entering of sensor parameters into the host are eliminated. Because of this distinctive feature of

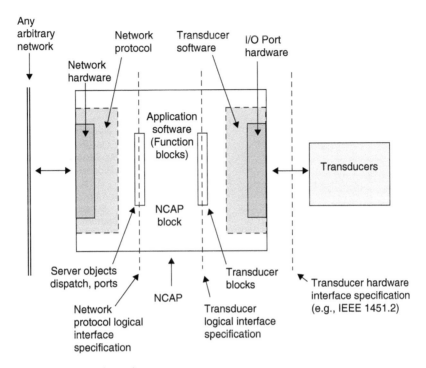

FIGURE 10.6 IEEE 1451 logical interfaces.

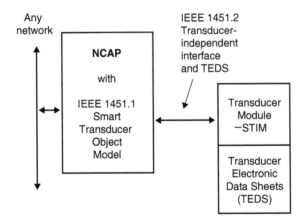

FIGURE 10.7 Block diagram of IEEE 1451.

the TEDS, upgrading transducers with a higher accuracy and enhanced capability or replacing transducers for maintenance purpose is simply considered "plug-and-play."

Eight different types of TEDS are defined in the standard. Two of them are mandatory and six are optional. They are listed in Table 10.1. The TEDS are divided into two categories. The first category of TEDS contains data in a machine-readable form, which is intended for use by the NCAP. The second category of TEDS contains data in a human-readable form. The human-readable TEDS may be represented in multiple languages using different encoding for each language.

The Meta TEDS contains the data that describe the whole STIM. It contains the revision of the standard, the version number of the TEDS, the number of channels in the STIM, and the worst-case timing required to access these channels. This information will allow the NCAP to access the channel information. In

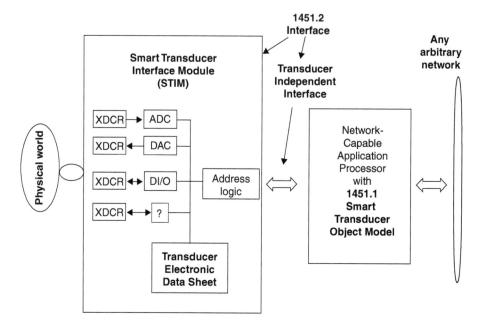

FIGURE 10.8 Detailed system block diagram of an IEEE 1451 smart transducer interface.

TABLE 10.1 Different Types of TEDS

TEDS Name	Type	Optional/Mandatory
Meta TEDS	Machine readable	Mandatory
Channel TEDS	Machine readable	Mandatory
Calibration TEDS	Machine readable	Optional
Generic extension TEDS	Machine readable	Optional
Meta-identification TEDS	Human readable	Optional
Channel identification TEDS	Human readable	Optional
Calibration identification TEDS	Human readable	Optional
End-user application-Specific TEDS	Human readable	Optional

addition, the Meta TEDS includes the channel groupings that describe the relationships between channels. Each transducer is represented by a channel.

Each channel in the STIM contains a Channel TEDS. The Channel TEDS lists the actual timing parameters for each individual channel. It also lists the type of transducer, the format of the data word being output by the channel, the physical units, the upper and lower range limits, the uncertainty or accuracy, whether or not a calibration TEDS is provided, and where the calibration is to be performed.

The Calibration TEDS contains all the necessary information for the sensor data to be converted from the analog-to-digital converter raw output into the physical units specified in the Channel TEDS. If actuators are included in the STIM, it also contains the parameters that convert data in the physical units into the proper output format to drive the actuators. It also contains the calibration interval and last calibration date and time. This allows the system to determine when a calibration is needed. A general calibration algorithm is specified in the standard.

The Generic Extension TEDS is provided to allow industry groups to provide additional TEDS in a machine-readable format.

The Meta Identification TEDS is human-readable data that the system can retrieve from the STIM for display purposes. This TEDS contains fields for the manufacturer's name, the model number and serial number of the STIM, and a date code.

The Channel Identification TEDS is similar to the Meta Identification TEDS. When transducers from different manufacturers are built into an STIM, this information will be very useful for the identification of channels. The Channel Identification TEDS provides information about each channel, whereas the Meta Identification TEDS provides information for the STIM.

The Calibration Identification TEDS provides details of the calibration in the STIM. This information includes who performed the calibration and what standards were used.

The End-User Application-Specific TEDS is not defined in detail by the standard. It allows the user to insert information such as installation location, the time it was installed, or any other desired text.

The STIM module can contain a combination of sensors and actuators of up to 255 channels, signal conditioning/processing, A/D converter, D/A converter, and digital logics to support the transducer-independent interface (TII). Currently, the P1451.2 working group is considering an update to the standard to include a popular serial interface, such as RS232, in addition to the TII for connecting sensors and actuators.

10.6.1.4   IEEE 1451.3 Distributed Multidrop Systems

The IEEE 1451.3 defines a transducer bus for connecting transducer modules to an NCAP in a distributed multidrop manner. A block diagram is shown in Figure 10.9. The physical interface for the transducer bus is based on Home Phoneline Networking Alliance (HomePNA) specification. Both power and data run on a twisted pair of wires. Multiple transducer modules, called transducer bus interface modules (TBIMs), can be connected to an NCAP via the bus. Each TBIM contains transducers, signal conditioning/processing, A/D, D/A, and digital logics to support the bus, and can accommodate large arrays of transducers for synchronized access at up to 128 Mbps with HomePNA 3.0 and up to 240 Mbps with extensions. The TEDS is defined in the eXtensible Markup Language (XML).

10.6.1.5   IEEE 1451.4 Mixed-Mode Transducer Interface

The IEEE 1451.4 defines a mixed-mode transducer interface (MMI), which is used for connecting transducer modules, mixed-mode transducers (MMTs), to an instrument, a computer, or an NCAP. The block diagram of the system is shown in Figure 10.10. The physical transducer interface is based on the Maxim/Dallas Semiconductor's one-wire protocol, but it also supports up to 4 wires for bridge-type sensors. It is a simple, low-cost connectivity for analog sensors with a very small TEDS — 64 bits mandatory and 256 bits optional. The mixed-mode interface supports a digital interface for reading and writing the TEDS by the instrument or NCAP. After the TEDS transaction is completed, the interface switches into analog mode, where the analog sensor signal is sent straight to the instrument and NCAP, which is equipped with A/D to read the sensor data.

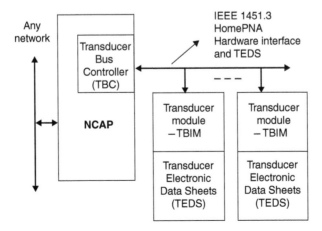

FIGURE 10.9   Block diagram of IEEE 1451.3.

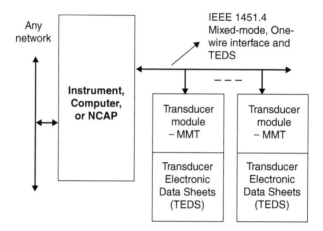

FIGURE 10.10   Block diagram of IEEE 1451.4.

10.6.1.6   IEEE P1451.5 Wireless Transducer Interface

Wireless communication is emerging, and low-cost wireless technology is on the horizon. Wireless communication links could replace the costly cabling for sensor connectivity. It could also greatly reduce sensor installation cost. Industry would like to apply the wireless technology for sensors; however, there is a need to solve the interoperability problem among wireless sensors, equipment, and data. In response to this need, the IEEE P1451.5 working group is working to define a wireless sensor communication interface standard that will leverage existing wireless communication technologies and protocols [9]. A block diagram of IEEE P1451.5 is shown in Figure 10.11. The working group seeks to define the wireless message formats, data/control model, security model, and TEDS that are scalable to meet the needs of low cost to sophisticated sensor or device manufacturers. It allows for a minimum of 64 sensors per access point. Intrinsic safety is not required but the standard would allow for it. The physical communication protocol(s) being considered by the working group are: (1) IEEE 802.11 (WiFi), (2) IEEE 802.15.1 (Bluetooth), and (3) IEEE 802.15.4 (ZigBee).

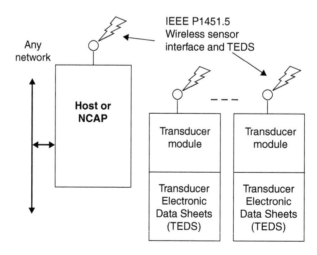

FIGURE 10.11   Block diagram of IEEE P1451.5 wireless transducer.

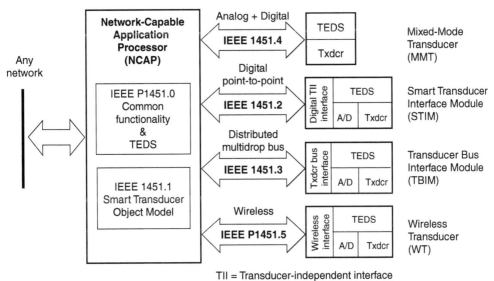

TII = Transducer-independent interface

Txdcr = Transducer (sensor or actuator)

FIGURE 10.12 Family of IEEE 1451 Standards.

10.6.2 IEEE 1451 Family

Figure 10.12 summarizes the family of IEEE 1451 Standards. Each of the IEEE P1451.X is designed to work with the other. However, they can also stand on their own. For example, IEEE 1451.1 can work without any IEEE 1451.X hardware interface. Likewise, IEEE 1451.X can also be used without IEEE 1451.1, but software with similar functionality should provide sensor data/information to each network.

10.6.3 Benefits of IEEE 1451

IEEE 1451 defines a set of common transducer interfaces, which will help to lower the cost of designing smart sensors and actuators because designers would only have to design to a single set of standardized digital interfaces. Thus, the overall cost to make networked sensors will decrease.

Incorporating the TEDS with the sensors will enable self-description of sensors and actuators, eliminating error-prone, manual configuration.

10.6.3.1 Sensor Manufacturers

Sensor manufacturers can benefit from the standard because they only have to design a single standard physical interface. Standard calibration specification and data format can help in the design and development of multilevel products based on TEDS with a minimum effort.

10.6.3.2 Application Software Developers

Applications can benefit from the standard as well because standard transducer models for control and data can support and facilitate distributed measurement and control applications. The standard also provides support for multiple languages — which is good for international developers.

10.6.3.3 System Integrators

Sensor system integrators can benefit from IEEE 1451 because sensor systems become easier to install, maintain, modify, and upgrade. Quick and efficient transducer replacement results by simple "plug-and-play." It can also provide a means to store installation details in the TEDS. Self-documentation of hardware and software is done via the TEDS. Best of all is the ability to choose sensors and networks based on merit.

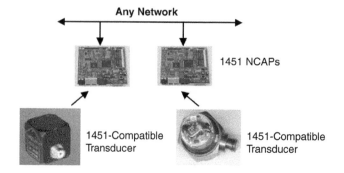

FIGURE 10.13 IEEE 1451 enables "plug-and-play" of transducers to a network.

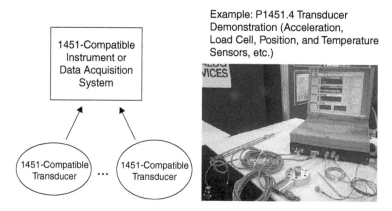

FIGURE 10.14 IEEE 1451 enables "plug-and-play" of transducers to data acquisition/instrumentation system.

10.6.3.4 End Users

End users can benefit from a standard interface because sensors will be easy to use by simple "plug-and-play." Based on the information provided in the TEDS, software can automatically provide the physical units, readings with significant digits as defined in the TEDS, and installation details such as instruction, identification, and location of the sensor.

10.6.3.5 "Plug-and-Play" of Sensors

IEEE 1451 enables "plug-and-play" of transducers to a network as illustrated in Figure 10.13. In this example, IEEE 1451.4-compatible transducers from different companies are shown to work with a sensor network. IEEE 1451 also enables "plug-and-play" of transducers to a data acquisition system/instrumentation system as shown in Figure 10.14. In this example, various IEEE 1451.4-compatible transducers such as an accelerometer, a thermistor, a load cell, and a linear variable differential transformer (LVDT) are shown to work with a LabVIEW-based system.[2]

10.7 Example Application of IEEE 1451.2

IEEE 1451-based sensor networks consisting of sensors, STIM, and NCAP are designed and built into a cabinet as shown in Figure 10.15. There were a total of four STIM and NCAP network nodes as shown in Figure 10.15. Thermistor sensors were used for temperature measurements. They were calibrated in the laboratory to generate IEEE 1451.2-compliance calibration TEDS for all four STIMs and NCAPs.

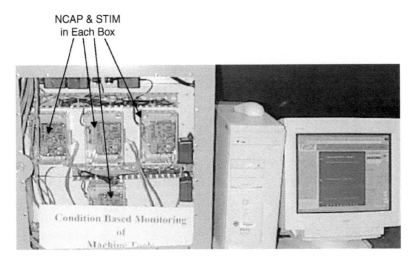

FIGURE 10.15 NCAP-based condition monitoring system.

FIGURE 10.16 Three-axis vertical machining center.

The thermistors were mounted on the spindle motor housing, bearing, and axis drive motors of a 3-axis vertical machining center, which is shown in Figure 10.16. Since each NCAP has a built-in micro web server, a custom web page was constructed using the web tool provided with the NCAP. Thus, remote monitoring of the machine thermal condition was easily achieved via the Ethernet network and the Internet using a readily available common web browser. The daily trend chart of the temperature of the spindle motor (top trace) and the temperature of the Z-axis drive motor (bottom trace) in the machine is shown in Figure 10.17. The temperature rise tracks the working of the machine during the day, and the temperature fall indicates that the machine is cooling off after the machine shop is closed.

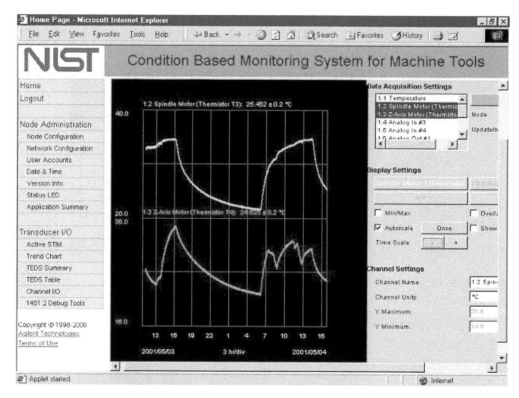

FIGURE 10.17 Temperature trend chart.

10.8   Application of IEEE 1451-Based Sensor Network

A distributed measurement and control system can be easily implemented based on the IEEE 1451 standards [10]. An application model of IEEE 1451 is shown in Figure 10.18. Three NCAP/STIMs are used to illustrate the distributed control, remote sensing or monitoring, and remote actuating. In the first scenario, a sensor and actuator are connected to the STIM of NCAP #1, and an application software running in the NCAP can perform a locally distributed control function, such as maintaining a constant temperature for a bath. The NCAP reports measurement data, process information, and control status to a remote monitoring station or host. It frees the host from the processor-intensive, closed-loop control operation. In the second scenario, only sensors are connected to NCAP #2, which can perform remote process or condition monitoring functions, such as monitoring the vibration level of a set of bearings in a turbine. In the third scenario, based on the broadcast data received from NCAP #2, NCAP #3 activates an alarm when the vibration level of the bearings exceeds a critical set point. As illustrated in these examples, an IEEE 1451-based sensor network can easily facilitate peer-to-peer communications and distributed control functions.

10.9   Summary

The IEEE 1451 smart transducer interface standards are defined to allow a transducer manufacturer to build transducers of various performance capabilities that are interoperable within a networking system. The IEEE 1451 family of standards has provided the common interface and enabling technology for the connectivity of transducers to microprocessors, field networks, and instrumentation systems using wired and wireless means. The standardized TEDS allows the self-description of sensors, which turns out to be a very valuable tool for condition-based maintenance. The expanding Internet market has created a

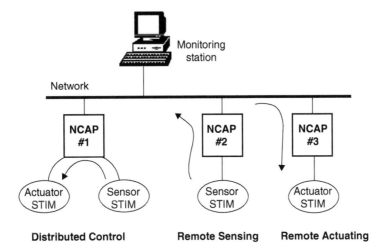

FIGURE 10.18 Application model of IEEE 1451.

good opportunity for sensor and network manufacturers to exploit web-based and smart sensor technologies. As a result, users will greatly benefit from many innovations and new applications.

Acknowledgments

The author sincerely thanks the IEEE 1451 working groups for the use of the materials in this chapter. Through its program in Smart Machine Tools, the Manufacturing Engineering Laboratory of the National Institute of Standards and Technology has contributed to the development of the IEEE 1451 standards.

References

1. Amos, Kenna, Sensor market goes global, *InTech — The International Journal for Measurement and Control*, June, 40–43, 1999.
2. Bryzek, Janusz, Summary Report, *Proceedings of the IEEE/NIST First Smart Sensor Interface Standard Workshop*, NIST, Gaithersburg, MD, March 31, 1994, pp. 5–12.
3. IEEE Std 1451.2-1997, *Standard for a Smart Transducer Interface for Sensors and Actuators — Transducer to Microprocessor Communication Protocols and Transducer Electronic Data Sheet (TEDS) Formats*, Institute of Electrical and Electronics Engineers, Inc., Piscataway, NJ, 1997.
4. Eidson, J. and S. Woods, A Research Prototype of a Networked Smart Sensor System, *Proceedings Sensors Expo*, Boston, Helmers Publishing, May 1995.
5. URL http://ieee1451.nist.gov
6. IEEE Std 1451.1-1999, *Standard for a Smart Transducer Interface for Sensors and Actuators — Network Capable Application Processor (NCAP) Information Model*, Institute of Electrical and Electronics Engineers, Inc., Piscataway, NJ, 1999.
7. Warrior, Jay. IEEE-P1451 Network Capable Application Processor Information Model, *Proceedings Sensors Expo*, Anaheim, Helmers Publishing, April 1996, pp. 15–21.
8. Woods, Stan et al., IEEE-P1451.2 Smart Transducer Interface Module, *Proceedings Sensors Expo*, Philadelphia, Helmers Publishing, October 1996, pp. 25–38.
9. Lee, K.B., J.D. Gilsinn, R.D. Schneeman, and H.M. Huang, *First Workshop on Wireless Sensing*, National Institute of Standards and Technology, NISTIR 02-6823, February 2002.
10. Lee, Kang and Richard Schneeman, Distributed Measurement and Control Based on the IEEE 1451 Smart Transducer Interface Standards, *Instrumentation and Measurement Technical Conference 1999*, Venice, Italy, May 24–26, 1999.

11

Integration Technologies of Field Devices in Distributed Control and Engineering Systems

Christian Diedrich
Institut für Automation und
Kommunikation eV
Germany

11.1 Introduction

Control system engineering and the instrumentation of industrial automation belong to a very innovative industrial area with a scope for cost reduction. That is why the costs for equipment and devices are already at a relatively low level. This trend is accompanied by a paradigm shift to digital processing in devices and digital communication among them.

This chapter describes the historical development steps from analog electronic devices connected via 4–20 mA, or 24 V technology, to digital devices with industrial communication connections such as fieldbus and Ethernet/TCP/IP. From these steps, the changes of the commissioning and system integration requirements are derived, and the arising technologies to support device manufacturers and system integrators are introduced. All these integration technologies support the instrumentation tasks of field devices. Examples of these technologies are the Device Description Languages (provided, e.g., by PROFI-BUS, Device Net, Fieldbus Foundation), standardized interfaces such as OPC and Field Device Architecture (FDT), control application integration using proxy function blocks written in PLC languages (e.g., in IEC 61131-3 languages), and vertical communication from field devices to SCADA, Decentral

Control Systems (DCS), and Manufacturing Execution Systems (MES) by means of XML. The reader will become familiar with up-to-date technologies used in field device engineering and instrumentation tools. The chapter concludes with a formal modeling approach, specifically with a device model. This model provides an abstract view of the described technologies and makes the relations among them visible.

11.2   History of Smart Devices

Digital information computation in field devices and digital fieldbus communication leads to a change in the handling of automation systems in manufacturing and process control. The field devices now contain much more information than the 4–20 mA signal. In addition, they carry out some functions that are/were originally programmed within the PLC or DCS. These field devices are also known as *smart devices.*

The consequence is a distributed system. The tools for the design and programming of control applications, commissioning, as well as maintenance need both access to the data of the field device and/ or an exact machine-readable product description of the field devices including their data and functions. These device descriptions, from the design of the description to its use, are called device description technology. Using these device product descriptions, it is necessary to integrate standard interface specifications and standardized industrial communication systems.

The following example of a transmitter shows the transformation from an analog 4–20 mA device to a smart fieldbus device. Other types of automation devices were or are moving in a similar direction. Even digital, discrete, and I/O field devices are much simpler; for example, the 24 V technology will also be replaced by fieldbus connected devices.

The transmitter in Figure 11.1 is composed of electronics, which detect the specific measurement value (e.g., mV, mA) and transform the detected signal into the standardized 4–20 mA. The adjustment to the specific sensor and wiring is done by trim resistors. Each transmitter is connected to the PLC by its own wires.

Digital signal computation provides a higher accuracy. Therefore, the signal processing is carried out by microprocessors (Figure 11.2). An analog/digital and a digital/analog unit transform the signals two times. The signal processing may be influenced by several parameters, which make the transmitter more flexible. These parameters have to be accessed by the operator. For this purpose, the manufacturer provides a local operator panel at the transmitter, which consists of a display and very few buttons. PC

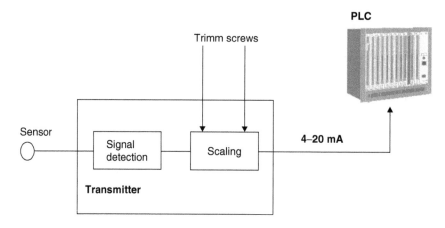

FIGURE 11.1   Structure of analog transmitter.

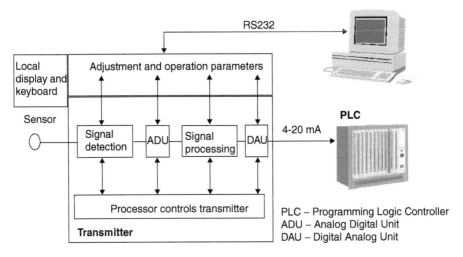

FIGURE 11.2 Structure of smart 4–20 mA transmitters.

tools provide more ergonomic solutions for the commissioning of such smart transmitters, which is necessary if they are more complex.

In principle, in smart fieldbus transmitters, the digital/analog unit at the end of the signal chain in the transmitter is replaced by the fieldbus controller. This increases the accuracy of the devices again. In addition, devices according to Figure 11.3 need specific communication commissioning using additional fieldbus configuration tools.

The smart field device parameterization has moved from manual screwing over local device terminals to PC application software. Additionally, the communication system configuration has to be managed during the commissioning of the instrumentation.

The field device is typically integrated as a component in an industrial automation system. The automation system performs the automation-related part of the complete application. The components of an industrial automation system may be arranged in multiple hierarchical levels connected by communication systems, as illustrated in Figure 11.4.

The field devices are components in the process level connected via inputs and outputs to the process or the physical or logical subnetworks (IEC Guideline, 2002). This also includes programmable devices

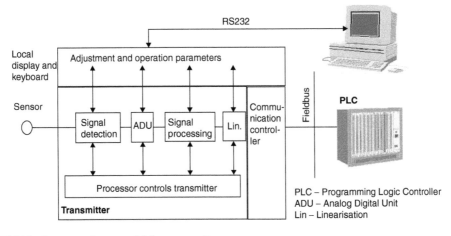

FIGURE 11.3 Structure of a smart fieldbus transmitter.

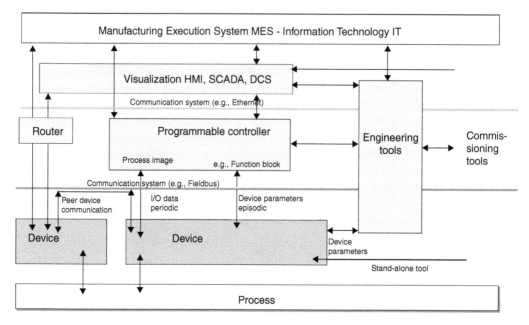

FIGURE 11.4   Typical automation configuration (IEC Guideline, 2002).

and router or gateways. A communication system (e.g., a fieldbus) connects the field devices to the upper-level controllers, which are typically programmable controllers or DCS or even MES. Since the engineering tools and the commissioning tools should have access to the field devices and to the controllers, these tools are also located in the controller level. The field devices may directly communicate via the fieldbus or the controller (Programmable Controller). In larger automation systems, another higher level may exist connected via a communication system such as LAN or Ethernet. In these higher-level visualization systems (HMI), DCS, central engineering tools, and SCADA are located. Multiple clusters of field devices with or without a controller as described above may be connected over the LAN with each other or to the higher-level systems. MES, Enterprise Resource Planning (ERP), and other Information Technology (IT) systems can have access indirectly to field devices via the LAN and the controllers or direct via routers.

The result of this development is a large variety of PC-based tools for planning, design, commissioning configuration, and maintenance (Figure 11.4). This problem is also known in the home and consumer electronics. TV, CD player, video recorder, and record player are bought gradually by a family, and may be produced by different manufacturers. Each device is purchased with remote controllers. The user has to deal with different styles, programming approaches, and outlines. This is unacceptable both in consumer electronics and in the industrial environment. Of course, manufacturers desire to be unique regarding functions and features of their products and provide special approaches to interact with the device. The end user gets products very suitable to specific requirements, but he or she is faced with a broad variety of device interfaces.

The field devices are an integrated part of the entire life cycle of control systems, starting with planning/design, over purchase, system integration/commissioning, operation, and ending with maintenance. During each phase of the life cycle, specific information represented in different formats is used from the tools. Planning and design define the requirements coming from the process that defines the type and the properties of the field device. The results are the pipe and instrumentation diagram (P&ID) in process control and electrotechnical drawings (E-CAD) in facturing automation combined with device lists. Purchase is made by requests of offers or by web market places directly. Commissioning carries out the parameterization and configuration of field devices in connection with the programming of the programmable controllers. During operation, there are mostly interactions between the controllers and the field devices. The field devices interact with special tools during maintenance. The view to field

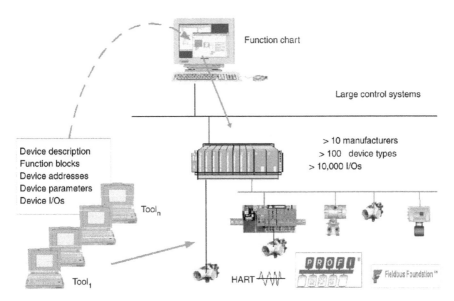

FIGURE 11.5 Different tools and multiple data input have determined field device integration to date (Bruns et al., 1999).

devices, that is, the used subset of the entire information range, the format of the information (text in manuals, files, data base entries, HTML pages), and the source of the information (paper, machine readable in the system, online from the device), differs between the lifecycle phases. Therefore, integration technologies deal with multiple technologies. B2B and MES are not within the scope of these collections of technologies. The main reason is that this chapter concentrates on the functional aspects of the field device integration. Even planning and design are only mentioned in certain sections because product data management does not yet have tight connections to the functional design of the control system.

11.3 Field Device Instrumentation

Today, the large number of different device types and suppliers within a control system project makes the field device parameterization and configuration task difficult and time-consuming. Different tools must be mastered and data must be exchanged between these tools. The data exchange is not standardized; therefore, data conversions are often necessary, requiring detailed specialist knowledge. In the end, the consistency of data, documentation, and configurations can only be guaranteed by an intensive system test.

The central workplace for service and diagnostic tasks in the control system does not fully cover the functional capabilities of the field devices. Furthermore, the different device-specific tools cannot be integrated into the system's software tools. Typically, device-specific tools can only be connected directly to a fieldbus line or directly to the field device (Figure 11.5).

In order to maintain the continuity and operational reliability of process control technology, it is necessary to fully integrate field devices as a subcomponent of process automation.

Field devices often have to be adjusted to their concrete application purpose; therefore, additional software components are necessary for parameterization reasons. These components are necessary because local operator keyboards, with only a few bottoms and small displays, are not suitable to provide the complex parameterization issues to the operator. In principle, the following cases of application can be found (Diedrich et al., 2001):

- Devices with fixed functionality and without parameterization of its device application. These devices have to provide their communication properties, which is only done according to descriptions such as PROFIBUS GSD and CAN EDS. A communication configuration tool verifies all properties of all devices that are connected at one communication segment and generate the communication configuration in terms of the communication parameters such as baud rate and device addresses.

- Devices with only little parameter data have to be fixed only once during the commissioning phase (e.g., minimum and maximum speed of a drive, calibration for a transmitter). This device should receive the data via fieldbus from the control station. Therefore, many fieldbus systems have added parameterization keywords in their communication-related descriptions, such as PROFIBUS GSD and CAN EDS. Fieldbus configuration tools provide the possibility to edit the parameter values; the controller (e.g., PLC) ensures the persistence of the parameter data, for example, for a restart. This case of application does not need any additional parameterization tools. The replacement of a device after failure can be made easily.
- Devices with many parameters and complex parameterization means (e.g., transmitters, actuators in the process control field) have, for a long time, had local terminals and/or their own commissioning tools. Tools that are able to parameterize many different types of devices from different manufacturers have been established in the market. Well-known languages are HART DDL (Device Description Language [HART, 1995]), Fieldbus Foundation DDL (FF, 1996), and the PROFIBUS EDD (Electronic Device Description [PNO, 2001b]), which belong to the same language family (IEC 61804, 2003). These languages are characterized by off-line parameterization, device-specific guidance by the operators, and extensive consistency checks. The management of the persistence data is directly related to parameterization tools; therefore, a replacement of these devices without these tools is not possible.
- Devices with complex parameterization sequences and complex data types (e.g., graphical information representation or video representation) cannot be described by the languages mentioned. Therefore, these devices need device-specific commissioning tools (e.g., Laser scanner).

11.3.1 Fieldbus Communication Configuration

11.3.1.1 GSD Language

A communication segment has to be configured for a certain device configuration. The baud rate, device addresses, and special communication timing parameters have to be adjusted according to the collection of devices and their properties. Therefore, fieldbuses provide communication feature lists in a machine-readable format. Each device is purchased with such a list, which is used by a communication configuration tool to generate the possible or optimal communication configuration.

One example is the PROFIBUS GSD. The abbreviation GSD stands for German Device Data Base (GeräteStammDaten). There is no translation of the abbreviation at all. The GSD describes the communication features and the cyclic data of simple devices such as binary and analog I/O. Communication features are potential baud rates, communication protocol time parameters, and the support of PROFIBUS services and functions (e.g., remote address assignment).

Simple I/O devices (also known as Remote I/O or intelligent clip) are mostly modular. A basic device with power consumption and a central processing unit has several slots, offered to plug in modules with different signal qualities (analog, binary) and quantities (e.g., 2, 4, 8, 16 channels). According to the chosen modules, a different set of data has to be transferred in the cyclic transfer between the PROFIBUS master and slave devices. In other words, every module contributes with a specific set of data to the cyclic telegram. If we consider that the configuration tools of the PROFIBUS master devices (PLC mostly) have to be used device manufacturer-independent, an unambiguous and manufacturer-independent description of the modules is necessary. This is the second main task of the GSD description.

The GSD language is a list of keywords accompanied by their value assignment for each feature of the PROFIBUS DP protocol, which is configurable, and the module specification. This list is accomplished by a device and GSD file identification. The following example shows what a GSD file looks like (Figure 11.6).

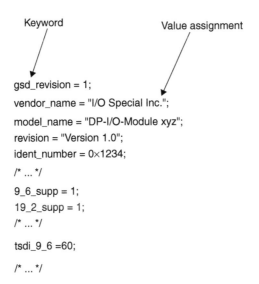

FIGURE 11.6 Example GSD file subset.

Modules are represented in the GSD by the keyword "Module." Between the Module and the End_Module keyword, the declaration of the contribution of the module to the cyclic data telegram is specified in terms of binary codes, the so-called Identifier Bytes. Each module has a name that is printed at the configuration tool screen. If the module is chosen, the specified Identifier Bytes will be concatenated to the configuration string, which is transferred from the master to the slave during the startup of the cyclic data transfer (CFG service of PROFIBUS-DP [2]). In addition, the range and the default value of variables can be configured, which is supported by the variable declaration possibility of GSD. These data specified in the module GSD construct are transferred with the so-called PRM service (PRM from parameterization) of PROFIBUS DP ([2]).

11.3.1.2 GSD Tool Set

All PROFIBUS DP manufacturers are obliged to sell these devices with a GSD file, which is checked by the PROFIBUS User Organisation (PNO). The PNO offers a special GSD editor, which guides the engineer during the GSD development (Figure 11.7). The engineer chooses the features of the device with a mouse click from the superset of all existing PROFIBUS-DP features. The editor generates the syntactic right GSD ASCII file. This file is not compiled.

The configuration tools read these GSD files and present the specified device features for communication configuration and adjustment of some application parameters. The results of this configuration process are the configuration and parameterization strings for the PROFIBUS DP services (Figure 11.8).

The handling of GSD is very simple and needs no specific training. It matches exactly the needs of simple I/O with no or less application parameterization.

11.3.2 Field Device Application Parameterization — Device Description Languages

Two kinds of users can be distinguished for the task of parameterization:

1. the end user or operator of a plant or machine and
2. the system integrator.

For the end user of a distributed control system, the most important thing concerning device description is its transparency. The end user wishes to see only the graphical user interface of a device represented in the SCADA-software or other Human–Machine Interfaces (HMIs). Therefore, the electronic device

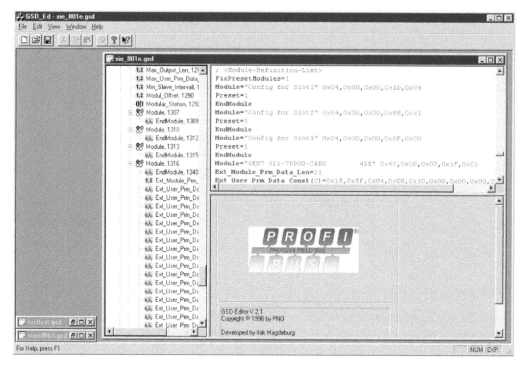

FIGURE 11.7   GSD editor snapshot.

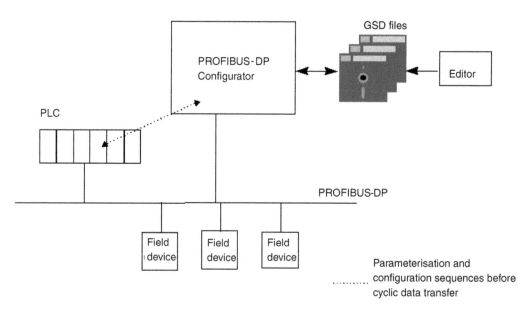

FIGURE 11.8   Tools and handling of GSD.

description has to be constructed following the plug-and-play concept. Furthermore, exchanging a device in an application is required not to lead to a big engineering task, but is required to be done simply and securely by a technician. The device description has to support this.

In contrast to the end user, the system integrator, who carries out the engineering, instrumentation, and documenting, has different targets to meet. His goal is to reduce the engineering time for solving interoperability problems but to enhance the engineering effort to design the optimal application with regard to the quality of the product to be produced with the process control system. The effectiveness of his work, comprising all preproducing phases of the plant/machine life cycle, depends directly on the support of a well-defined, standardized, and completely machine-readable description of the devices he has to deal with.

The conclusions from this analysis are that the Device Description Language has to be designed mainly from the system integrator's point of view, because he has to deal directly with the electronic device description. The fieldbus community developed some approaches to exchange data by electronic means. This includes DDL (e.g., for HART, FF, and Electronic Device Description for PROFIBUS).

The main features of this description technology, including the according language, can be summarized as follows:

- An EDD (in terms of a file) is delivered by the device vendor, together with the device.
- EDD is used in the engineering process of the distributed control system, supporting planning, commissioning, operation, diagnostics, and maintenance.
- There are different EDD presentations possible, source in a human-readable way and binary format.
- The EDD is mostly stored on disk and can additionally be stored within the device (transport via fieldbus).
- The EDD is nearly independent from the underlying fieldbus system.
- The EDD is used to describe information identifying each item and defining relationships between them (hierarchical, relational).
- The EDD offers language elements for presentation within an HMI and for communication access.
- An EDD file represents a static description, that is, the declarative part of the device; therefore, only the external interface/behavior of the device is described (no interest in internal code).

A parameterization tool needs special adaptation to functions and parameters of each device, which should be commissioned or visualized (Figure 11.9). The EDD contains all device functions and parameters. The functions and parameters differ between the devices, but they are described with a defined language. The tools understand this language and adapt themselves to the described functionality. Adaptation means modification of the screen outline (e.g., content of the menus and bars) and the interactions with the devices. Generally speaking, the devices will be purchased with disks containing their own device

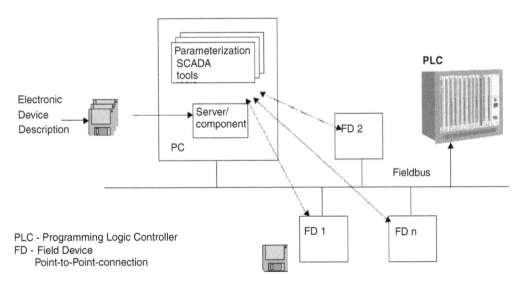

FIGURE 11.9 Using EDD in parameterization tools.

description, and the tools get their specific configuration by reading and interpreting this device description. The main parts of the EDD are as follows:

- Description of variables and parameters, including their attributes (Process Variable — PV with label for screen prints, e.g., Level hydrostatic, data type floating point, access right read/only, default value = 0).
- Presentation of variables in HMI tools (e.g., PV with label "Temperature").
- Guidance of the operator during commissioning (e.g., order of menu entries at hand-held terminals or PC tools).
- Table of content of all visible elements of the device.
- Device configuration parameter (e.g., identifier for plug-in modules of modular devices).
- Communication configuration parameters (e.g., baud rate = 500 kBaud).

The following example gives a short look-and-feel of the language.

EDD is a formal language used to describe completely and unambiguously what a field instrument looks like when it is seen through the "window" of its digital communication link. EDD includes descriptions of accessible variables, the instrument's command set, and operating procedures such as calibration. It also includes a description of a menu structure that a host device can use for a human operator. The EDD, written in a readable text format, consists of a list of items ("objects") with a description of the features ("attributes" or "properties") of each. Some example fragments from an (imaginary) flowmeter EDD are shown in Figure 11.10.

The major benefit of DDL for suppliers is that it decouples the development of host and field devices. Each designer can complete product development with the assurance that the new product will interoperate correctly with current and older devices, as well as with future devices not yet invented. In addition, a simulation program can be used to "test" the user interface of the EDD, allowing iterative evaluation and improvement, even before the device is built.

For the user, the major benefit is the ability to mix products from different suppliers, with the confidence that each can be used to its full capacity. Easy field upgrades allow host devices to accept new field devices. Innovation in new field devices is encouraged. The EDD is restricted for the description of a single device and use in a mostly stand-alone tool.

Software tools for automation are very complex and implement a lot of know-how. The number of sold products is relatively small in comparison with office packets. The definition of standardized DDL increases the potential users of the tools and speeds up the use of fieldbus-based automation.

11.3.3 Programming of the Control Applications with Integrated Field Device Functions

The Proxy-Concept organizes the functional and data relations between field devices and programmable controllers (e.g., PLCs; Figure 11.11). Field devices implement some signal processing, such as scaling or limit check, which was a part of the programmable controller libraries using the 4–20 mA technology. The consequence of this change is an interruption of the sequence of functions, which have already been carried out in one compact PLC program. This program has been developed with one software tool. Now different devices with different resources, which are scheduled in an asynchronous manner, carry out the functions and it is unclear if the sequence order works in the necessary way. Additionally, PLC and field devices are programmed with different tools. The IEC 61499 (IEC 61499, 2001), PROFInet (PROFInet, 2002) and IDA (IDA, 2001) standards provide means to address the problems; however, these standards are not yet available on the market.

It would be favorable if the PLC programmer can use the decentralized functions in its normal programming environment with all the characteristic features. This would be possible if there is a device function block in the PLC library. This device function block represents the field device as its data input and output interface. The internals of this device function block are manufacturer specific. This concept of device function blocks of remote functions is known as the proxy concept. The communication between

```
VARIABLE v_low_flow_cutoff
{
LABEL LowFlowCutoff,
HELP „The value below which the process variable will indicate zero,
to prevent noise or a small zero error begin interpreted as a real flow rate".
TYPE FLOAT
{
IF v_precision = high
DISPLAY_FORMAT „4.61",
ELSE
DISPLAY_FORMAT „4.21",
}
CONSTANT_UNIT „%",
HANDLING READ & WRITE;
}

MENU m_configuration
{
LABEL „configuration"
ITEMS
{
v-flow_units,                      /*variable*/
range, /*edit-display*/
v_low_flow_cutoff,                 /*variable*/
m_flow_tube_config,                /*menu*/
m_pulse_output_config             /*menu*/
}
}

COMMAND write_low_flow_cutoff
{
NUMBER 137;
OPERATION WRITE;
TRANSACTION
{
REQUEST
{
low_flow_cutoff
}
REPLY
{
response_code,
device_status,
low_flow_cutoff
}
}
RESPONSE_CODES
{
0. SUCESS,                  [no_command_specific_errors];
3. DATA_ENTRY_ERROR,        [passed_parameter_too_large];
4. DATA_ENTRY_ERROR,        [passed_parameter_too_small];
6. MSC_ERROR,               [too_few_data_bytes_received];
7. MODE_ERROR,              [in_write_protect_mode];
}
}
```

FIGURE 11.10 Device Description example.

the proxy and the field device is carried out by cyclic (for process variables) and acyclic (for device parameterization, e.g., batch) means, that is, communication function blocks of the PLC. If these communication function blocks are standardized and the PLC is IEC 61131-3 [11] compliant, the device function blocks are portable in an easy manner. Portable device function blocks mean that the

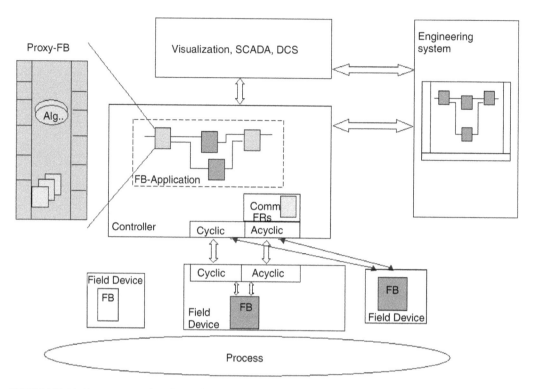

FIGURE 11.11   Proxy-concept based on a function block.

communication function blocks are PLC manufacturer independent at its interface and main behavior. The specification of IEC 61131-5 that specifies such communication function blocks does not fulfill the requirements for an interoperable data transfer and is not well accepted in the market. Therefore, PROFIBUS has built its own specification, which is based on the IEC.

Proxies, that is, device function blocks, are additional components that have to be provided by a field device manufacturer. This is a consequence of the increase of device functionality, the decentralization of the system and its engineering, and of the resulting distributed system. Proxies are a very good basis for manufacturer- and user-specific functionalities, which can be implemented as PLC programs instead of embedded software in the field device. Therefore, more flexibility of the manufacturer satisfying user requests is possible. This flexibility can be implemented using different platforms of different PLC manufacturers.

11.3.4   Field Device System Integration

Common for the technologies described in the previous subsections is the limited scope of data in so-called "stand-alone" tools. The engineering and supervisory system (Figure 11.4) needs data from all devices and components in order to provide an uninterrupted information flow between the tools.

In order to maintain the continuity and operational reliability of process control technology, it is therefore necessary to fully integrate field devices as a subcomponent of process automation [2]. To resolve the situation, the German Electrical and Electronic Manufacturers' Association (ZVEI) initiated a working group in 1998 to define a vendor-independent FDT architecture. This FDT concept defines interfaces between device-specific software components (DTM, Device Type Manager) supplied by device manufacturers, and engineering systems supplied by control system manufacturers. The device manufacturers are responsible for the functionality and quality of the DTMs, which are integrated into engineering systems via the FDT interface. With DTMs integrated into engineering systems in a unified way, the connection between engineering systems (e.g., PLC applications) and inconsistent field devices

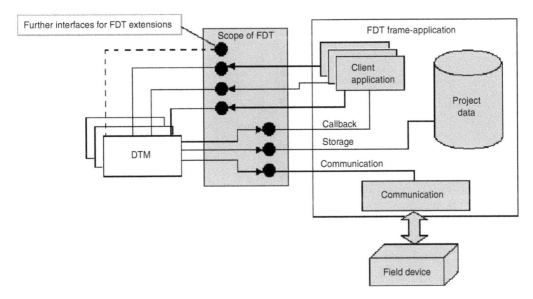

FIGURE 11.12 The architecture of Field Device Tools (PNO 2001a).

becomes available. The FDT specification specifies what the interfaces are, not the implementation of these interfaces [2]. Figure 11.12 shows the FDT architecture. From the figure, we can see that DTMs act as bridges between the frame-application and field devices.

11.3.4.1 FDT Frame-Application

As one component of the FDT structure, the frame-application is supposed to manage data and communication with the device and embedded DTMs. According to the environment a DTM is running in, the frame-application can be an engineering tool, or even a web page. But here, we consider frame-application as an engineering environment that integrates field devices and their configuration tools and controls DTMs within a project. The project is a logical object to describe the management and controls, at least in the lifetime of device instances in terms of the DTM within a frame-application (PNO, 2001a). In view of a DTM, interfaces such as IPersistPropertyBag (Connection Storage in Figure 11.12) and IfdtCommunication (Connection Communication in Figure 11.12) specified in PNO (2001a) determine what the frame-application can do.

11.3.4.2 DTM

The DTM is the key concept of field device integration, with each DTM representing one field device. In addition, more than one device of a common type can also be represented by a single DTM object. Moreover, a device can be subdivided into modules and submodules, for example, a device with its remote I/O modules. If a device supports remote I/O channels (in the FDT concept, the I/O of a device is called a channel), it also acts as a gateway. For example, a gateway from Profibus to HART is logically represented by a gateway DTM that implements the operations for channel access. Each time a user wants to communicate with a certain device connected via an interface card between the communication section of the frame-application and the device, an instance of the corresponding DTM is asked to be built. During the user's operation on the device, the whole lifetime of this DTM object is controlled by the frame-application. This is the concept of COM technology. In fact, DTM is implemented as a Microsoft COM object, which will be dynamically loaded when the user intends to obtain information of the device or set parameters to the corresponding field device, and then released after the operations. All these interactions are carried out via interfaces specified in the FDT specification (Figure 11.13).

The interfaces IDtm and IDtmInformation specified in PNO (2001a) provide basic functions used by the frame-application to obtain information for the DTMs and execute certain operations on the

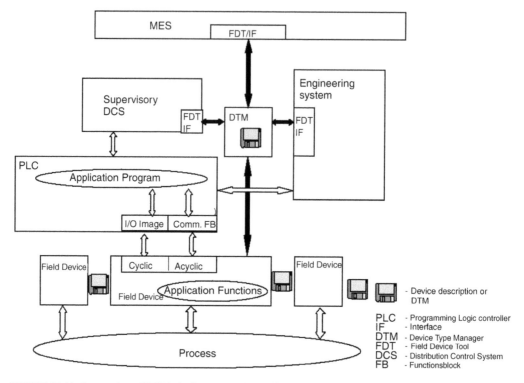

FIGURE 11.13 Integration of DTMs in the automation architecture using FDT interfaces.

corresponding devices. If the device represented by a DTM has its process values, another interface specified in PNO (2001a), IFdtChannel, acts as the gateway to deal with the connection to substructure, such as remote devices connected via Remote I/O channel to the master device. For some special tasks, such as documentation and communication, a DTM must implement the task-related interfaces if it supports the tasks.

The device manufacturer has multiple choices for the development of these DTMs. The DTM can be written manually in a high-level language, for example, in C++ or VB. It is also possible to generate DTMs out of existing device descriptions or provide DTM interpreting existing DTMs. Parameterization tools can be transformed to DTMs developing a wrapper around the tools that are conformed with the FDT interfaces.

11.3.4.3 FDT Interfaces

From the above parts, interfaces of frame-application and DTMs are mentioned and shown as the bridges between frame-application and DTMs. From the outside of an object, we can see only those interfaces that are specified according to their special functionalities but their implementation remains invisible and encapsulated.

With the help of FDT specifications and the corresponding interface technology, the user (engineering system manufacturer) will be able to handle devices and their integration into engineering tools and other frameworks in a consistent manner.

11.4 Fieldbus Profiles

Profiles are functional agreements between field device manufacturers to provide interoperable device functions of a certain device class to other field devices, controllers or DCS, SCADA, and engineering systems. Device classes have common functional kernels accompanied by the variables and parameters. These common set of variables, parameters, and functions are called profiles. Profiles reduce the degrees

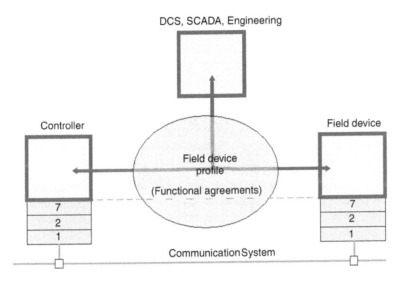

Agreement between device manufacturer on how to use communication means

Quelle: PNO TC3 profile guideline

FIGURE 11.14 Profile in the automation architecture.

of freedom using the variety of the communication system and the choice of application functions of the device classes (Figure 11.14).

There are certain degrees of compatibility and, accordingly, degrees of cooperation between profile-based devices (compatibility levels). Compatibility levels are applicable for various roles of a device, for example, control, diagnosis, parameterization/configuration, and even applicable to subsets of its functionality. This means that one device can have different compatibility levels regarding different interfaces to the system. The levels are dependent on well-defined communication and application device features (Figure 11.15).

Device Feature	Incompatible	Coexistent	Interconnectable	Interworkable	Interoperable	Interchangeable
Dynamic behavior						X
Application functionality					X	X
Parameter semantics					X	X
Data types				X	X	X
Data access			X	X	X	X
Communication interface			X	X	X	X
Communication protocol		X	X	X	X	X

Device Profile Application Part — Dynamic behavior, Application functionality, Parameter semantics

Device Profile Communication Part — Data types, Data access, Communication interface, Communication protocol

FIGURE 11.15 Levels of functional compatibility (IEC 61804, 2003).

The device features are either related to the communication system as it is specified in the standard (e.g., protocols, service interfaces, data access, data types specified in IEC 61185, and IEC 61784) or related to the device application such as data types and semantic of the parameters, application functions, and dynamic behavior of the application. Profiles usually provide a mixture of compatibility levels regarding parts of the profile and their different users. For example, the main measurement value of a device is defined very precisely regarding data type, semantic, including dynamic behavior; hence, devices are interchangeable for this measurement value. The same profile may skip to specify parameters that are used in the function chain from the electrical signal at the process attachment to the measurement value. Then, devices are not fully interoperable regarding their parameterization.

The main benefits of profiles are:

- The state-of-the-art functionality of device classes are specified, including their parameter semantic. This makes it possible that human device users as well as tools find the same functions and parameters with the same names and behavior in devices from different manufacturers.
- It is possible to provide communication feature lists (e.g., GSD), EDDs, DTMs, and proxies for device classes with profiles.

11.5   Model for Engineering and Instrumentation

The handling of the life cycle of DCS is a complex process that can only be done using sophisticated tools (hardware, software). Here, it is very important to design a noninterrupted life cycle, that is, to achieve a information transfer from one step to another step without losing information, and to ensure a single-source principle while putting information into the system. This information is used in each step to create a special view for the user. This cannot be done using paper documents for storing and transporting information. However, the usage of database management systems is not sufficient as long as open technologies are not applied. It is necessary to use databases and communication systems that follow a commonly agreed transfer syntax and standardized information models that ensure the meaning of the information (semantics). Basically, a connection between all tools in a DCS must be created (Diedrich, Ch. and P. Neumann, 1998a).

There are several possibilities to classify the life cycle/the engineering of DCS (Alznauer, 1998). Such classification can be made using:

- the hierarchy of the control functions,
- their timing sequence, and
- their logical dependencies.

All life cycle phases that are connected to field devices should be united under the rubric "instrumentation" and described as use cases. Instrumentation is defined as follows:

Instrumentation comprises of all activities within the life cycle of the distributed control system where handling of the field devices (logically or physically) is necessary.

Therefore, instrumentation can be considered as the intersection of the life cycle of the distributed control system and the field device. Figure 11.16 shows the instrumentation steps as a UML use case diagram. There is only one actor, who is not described in detail.

11.5.1   Device Model

Field devices are linked both with process, via I/O hardware/software, and with other devices via communication controllers/transmission media. The center of our attention is on the field device as the computational power is increasing rapidly, as mentioned above. Thus, the applications are run more and more on these devices, and the application processes are becoming more and more distributed. We have to solve the problem of configuring and parameterization of these field devices during the operation for

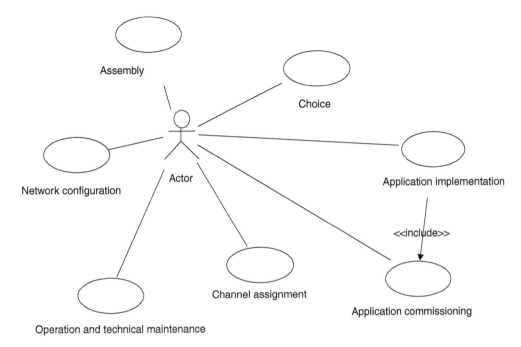

FIGURE 11.16 Use case Diagram Instrumentation (Simon, 2001).

real-time data processing purposes, diagnosis, parameter tuning, etc. Therefore, there is a need to model such field devices (Diedrich and Neumann, 1998a). A field device can be characterized by:

- internal data management (process I/O image, communication parameters, application parameters),
- process interface,
- information processing (e.g., Function Blocks),
- communication interface (Fieldbus, Ethernet-TCP/IP, etc.),
- (optional) man/machine interface (local display, buttons, switches, LEDs), and
- (optional) persistent memory and others.

We can define a device model as shown in Figure 11.17 (represented by UML packages), which supports the data exchange between instrumentation steps. This is a very abstract presentation of a device model.

The packages DIFunction, DIHardwareArchitecture, DISoftwareArchitecture, DIProcess, DICommunication, DIManagement, and DIOperation (DI stands for Device Instrumentation) are important. The packages depicted in Figure 11.17 contain the detailed model represented by UML class diagrams modeling the different views on a device. The details of the packages are beyond the scope of this chapter. However, the figures show the internal structure of elements (i.e., classes), for example., variable, function, and function block are described in detail in terms of its attributes and methods. The described relations between the classes and the attributes and methods are the basis for the development of the tools for the noninterrupted engineering and instrumentation. Figure 11.18 depicts the class diagram of the package DIFunction.

Figure 11.19 depicts the class diagram of the package DICommunication.

Figure 11.20 depicts the class diagram of the package DIOperation.

Simon (2001) contains all other class diagrams needed for modeling the semantics of field devices as well as further explanations. This model has to be described by description languages to generate the basic information for an uninterrupted tool chain.

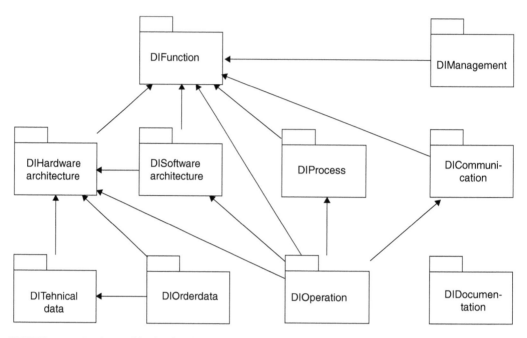

FIGURE 11.17 Device model related to the instrumentation (Simon, 2001).

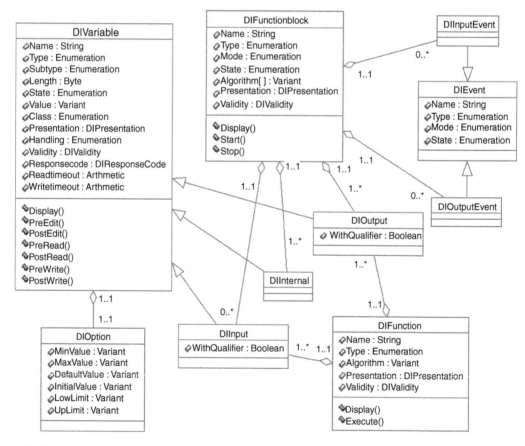

FIGURE 11.18 Package DIFunction (Simon, 2001).

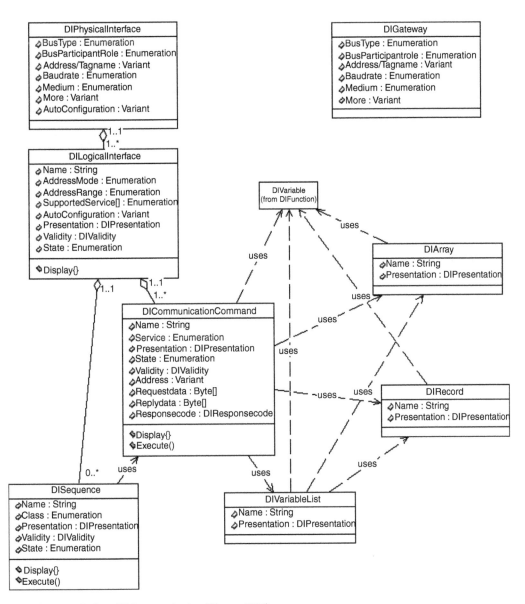

FIGURE 11.19 Package DICommunication (Simon, 2001).

11.5.2 Description and Realization Opportunities

The device model can be implemented (realized) in several ways. It is possible to derive the FDT/DTM structure and interfaces, EDD, field device proxies, function blocks, and other technologies from this device model. For this chapter, we chose EDD and an Extended Markup Language (XML)-based language. For the computable description of device parameters for automation systems components, the so-called Electronic Device Description Language (EDDL) has been specified (NOAH, 1999); (PNO, 2001b; Simon and Demartini, 1999). EDD is used to describe the configuration and operational behavior of a device and covers the following aspects (Neumann et al., 2001):

- description of the device parameters, semantically defined by the field device model mentioned above,
- support of parameter dependencies,

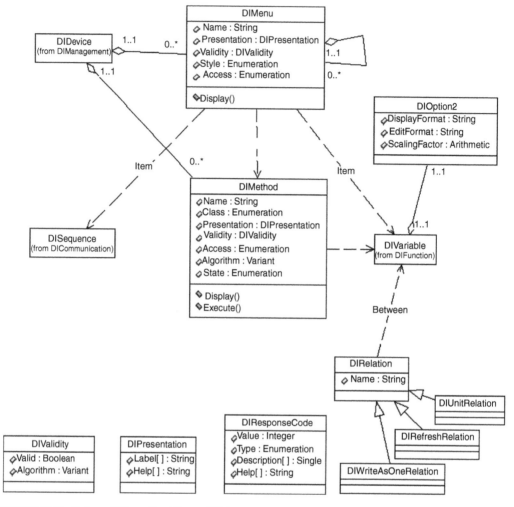

FIGURE 11.20 Package DIOperation (Simon, 2001).

- logical grouping of the device parameters,
- selection and execution of supported device functions, and
- description of the device parameter access method.

11.5.3 Overall Example Using EDDL

EDD is based on the ASCII standard. XML could be a promising approach for the future, especially because of its use in other areas. Both approaches contain definitions for the exchange of device descriptions using files. These definitions are not given here; however, a small example is used to show that different realizations can and must be based on the same solid foundation — the device model.

The example comprises of a variable (package DIFunction), which is described by name, data type, label, and help.

The class DIVariableExample as shown in Figure 11.21 is the starting point.

The realization as an EDDL is described using language production rules shown in Figure 11.22.

A sentence created according to these rules may resemble Figure 11.23.

Using this definition of a variable, a commissioning tool provides the human–machine interface (Figure 11.24):

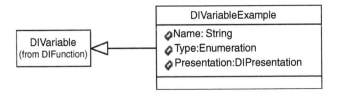

FIGURE 11.21 Class DIVariableExample.

```
variable
    =        'VARIABLE'        Identifier        '{'
variable_attribute_list '}'

variable_attribute_list
    = variable_attribute_listR

variable_attribute_listR
    = variable_attribute
    = variable_attribute_listR variable_attribute

variable_attribute
    = help
    = label
    = type
```

FIGURE 11.22 Language production rules (EDD).

```
VARIABLE    Temperature
{
    LABEL      "Temperatur";
    TYPE       DOUBLE;
    HELP       "Temperatur";
}
```

FIGURE 11.23 A variable definition (EDD).

Temperatur 42.0   K

FIGURE 11.24 Human–machine interface showing a variable.

11.5.4 The XML Approach

The XML (Bray et al., 1998) expands the description language HTML with user-defined tags, data types, and structures. In addition, a clear separation between the data descriptions, the data, and their representation in a browser have been introduced. Furthermore, declaring syntactical and semantical information in a separate file (Document Type Definition, DTD) allows reusing the description structure in different contexts. This provides a number of benefits when using the same XML description file for different tasks. Different views can be implemented on top of the same data. The description can be hierarchically organized. Depending on the functions to be performed, the XML data can be filtered and associated to software components (controls, Java beans, etc.). The selection of the necessary information and the definition of their presentation details can be performed by means of scripts and style sheets. The style sheets are a part of the development of XML (Boumphrey, 1998). In most cases, they are implemented using the extensible Style Language (XSL). The XML file, the scripts, and the different style sheets can be used to generate HTML pages, special text files, and binary files (components, applets) necessary to build the certain functions of the software tools. The distribution of the generated HTML pages and associated software components is done following the concepts used in an Internet environment. The major benefit of this solution is a unique, reusable description with an excellent consistency and reduced efforts of the description process.

For the realization using the XML approach, the specification of a schema is necessary. Figure 11.25 shows part of it describing the element variable.

An instance of this schema may look as shown in Figure 11.26.

A standard web browser creates interface as in Figure 11.27.

The unit (Kelvin) is not supported by the example model. The help text is not visible, and the name of the variable is not used. Based on the data type, the value provided by the device is shown.

The presentation of the small example underlines the objectives targeted by the modeling approach. If the internal structures of different realizations are similar, that is, they follow the same field device model, then it is possible to build translators from one realization to another and to secure investments already done. Similar presentations with the same contents provide the opportunity to simplify training and education through previous recognition.

```
<ElementType name="DIVariable" content="mixed" model="closed">
        <attribute type ="Name" required="yes"/>
        <element type="operation:DIPresentation" minOccurs="1"  maxOccurs="1"/>
        <element type="Type" minOccurs="1"  maxOccurs="1"/>
    </Element Type=>
```

FIGURE 11.25 Schema specification.

```
<DIVariable
            Name="Temperature"
            <operation:DIPresentation>
                <operation:Label
                        String="Temperatur">
                </operation:Label>
                </operation:Help
                        String="Temperature">
                </operation:Help>
            </operation:DIPresentation>
            <Type>
                <Arithmetic>
                    <Integer>
                    </Integer>
                </Arithmetic>
            </Type>
        </DIVariable>
```

FIGURE 11.26 A variable definition (XML).

Temperatur   42.0   K

FIGURE 11.27 Web browser showing a variable.

Device Descriptions are necessary for the integration of intelligent field devices in commissioning tools, maintenance tools, engineering systems, or MES/ERP systems. Device Descriptions comprise device models and presentations based on the models (e.g., ASCII files). The XML concept may help to extend the application scope of Device Description because it is becoming one of the basic technologies of the fast-growing Internet and various e-engineering activities. At present, we can observe a transition to XML-based Device Descriptions, which is characterized by the following issues:

- XML- and ASCII-based Device Descriptions have to use the same device model to retain the semantics already developed by automation industry,
- new application functions supporting different phases of the life cycle of an automation system have to be defined using different views on the Device Description, and
- consequently, a new chain of tools for creating and using Device Descriptions has to be developed.

A necessary precondition is the international standardization in this area, which is currently done in the European standardization CENELEC and the IEC.

11.6 Summary

Field devices are integrated parts of the entire automation system. Therefore, new technologies arise to provide means for the device manufacturer, system integrators, and system suppliers to design, commissioning, operate, and maintain systems with a large variety of field device classes (Figure 11.28). These technologies are named integration/instrumentation technologies.

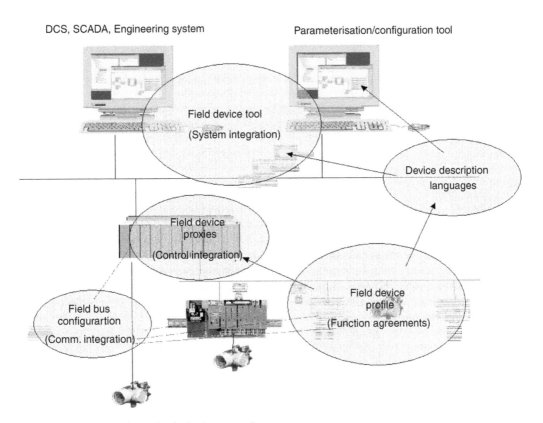

FIGURE 11.28 Integration technologies in automation systems.

Field device profiles define common sets of functionality that are provided by devices of different manufacturers in an interoperable way. This is the semantic basis for the other technologies. The Bases for all integration are configured communication systems, which use feature list languages. The EDDL, the field device proxies, and FDT/DTMs provide means to integrate device-specific and innovative features in the system. All these are integration/instrumentation technologies of field devices in distributed control and engineering systems.

References

Boumphrey, F., *Professional Style Sheets for HTML and XML*, Wrox Press, Birmingham, U.K., 1998.

Bray, T., Paoli, J., and Sperberg-McQueen, C.M., Extensible Markup Language (XML) 1.0., http://www.w3.org/TR/REC-xml, 1998.

Bruns, H., Hempen, U., Ott, W., and Vahldieck, R., Ein Konzept für eine COM/DCOM-basierte herstellerunabhängige Integration von Feldgeräten in Engineeringsysteme, *Atp-Automatisier ungstechnische Praxis*, 41, 10, S.34ff, 1999.

Diedrich, Ch. and Neumann, P., Field device integration in DCS engineering using a device model, *IECON'98, IEEE Conference, Proceedings*, Aachen, 1998a, pp. 164–168.

Diedrich, Ch., Wollschlaeger, M., Riedl, M., and Simon, R., Three Component Model for Field Device Integration in Control Systems, IFAC/FET 2001, Nancy, France, Proceedings, 15–16 Nov., 2001.

FF, Foundation Specification, Device Description Language, Fieldbus Foundation, Austin, TX, 1996.

HART, Device Description Language Specification: HCF, Austin, Texas, 1995.

IDA, IDA — The Internet of Automation Technology, White paper V1.0, April 2001, www.ida-group. org

IEC 61499, Function Blocks for Industrial-Process Measurement and Control Systems — Part 1 and 2, Public Available Specification (PAS), Geneva, 2001.

IEC 61804, Function Blocks for Process Control — Part 2, Committee Draft for Vote (CDV), Geneva, 2003.

IEC Guideline, Preliminary Joint Working Group: Device Profile Guideline, IEC Document for Comment, 65/290/DC, 2002.

Neumann, P., Simon R., Diedrich, Ch., and Riedl, M., Field Device Integration, 8th IEEE International Conference on Emerging Technologies and Factory Automation, ETFA 2001, Proceedings, Antibes, 2001, pp. 63–68.

NOAH, Language Specification of Electronic Device Description, Deliverable 321, Network Oriented Application Harmonisation (NOAH), 1999.

PNO, Specification for PROFIBUS Device Description and Device Integration, Vol. 3, Field Device Tool, Version 1.2, PROFIBUS Guideline, PROFIBUS User Organisation, 2001a.

PNO, Specification for PROFIBUS Device Description and Device Integration, Vol. 2, Electronic Device Description, Version 1.0., PROFIBUS Guideline, PROFIBUS User Organisation, 2001b.

PROFInet, PROFInet — More than just Ethernet, Brochure, www.profibus.com, 2002.

Simon, R., Methods for Field Instrumentation of Distributed Computer Control Systems, Ph.D. thesis, Otto-von-Guericke University Magdeburg, 2001 (in German).

Simon, R. and Demartini, C., Electronic Device Description, *Proceedings of FET'99*, Magdeburg, 23–24 Sept. 1999, Springer-Verlag, Wien, 1999, pp. 429–436.

12

Open Controller Enabled by an Advanced Real-Time Network (OCEAN)

Fabrizio Meo

FIDIA S.p.A., Italy

12.1 OCEAN Background

Open Controller Enabled by an Advanced real-time Network (OCEAN) is a project funded by the European Commission under the Information Society Technologies (IST) priority. The activities are conducted through a consortium of ten partners, as of the end of 2003. The project started in August 2002 and was completed by July 2005.

OCEAN is realizing a real-time-capable platform for distributed control applications. This platform will enable a dynamic integration of control components that are based on an open specification. Through this, a flexible and application-specific configuration of control systems will be possible.

The basic idea is to continue the approach of OSACA by adopting its benefits and realizing a fundamental improvement on the basis of newly available technologies. This strategy promises a broad industrial acceptance of open control technologies based on the DCRF.

12.2   Objectives

The two main objectives of the project are:

1. The development of a "Distributed Control System Real-Time Framework" (DCRF) for numerical controls based on standardized communication systems and delivered as an open source. The DCRF is envisaged to host control components in distributed open platforms and provide a real-time communication application programming interface (API). This framework is to be coupled with standardized interfaces, enabling the integration of external real-time-critical and non-real-time-critical control components. The DCRF is based on open source components (e.g., RT-Linux as operating system, RT-Common Object Request Broker Architecture (CORBA) for standardized communication mechanisms) and will be available, as an open source, at the end of the project. Figure 12.1 shows the structure of the DCRF and motion control components.

   A. The communication within a complex, dynamic environment requires a maximum flexibility for the data transfer. Hence, the deployment of transparent communication standards is essential. Nowadays, reliable open communication systems like CORBA are available, which follow the object-oriented concept for data exchange. This property guarantees a transparent handling of data objects and furthermore makes use of the network transparency, which hides from the software developer the problems caused by the communication between distributed components. Embracing this philosophy means bringing a component-based view of the software to a network distributed environment. Applications use object interfaces without the need to know whether they use a local or a remote object. In terms of communication, this approach is the backbone of the DCRF that closes the gap between the distributed control components. With respect to communications, a main focus of the OCEAN project is on the real-time capability of the DCRF. Real-time communication is crucial for distributed control systems because it is a prerequisite for a productive, high-quality, and fail-safe production. The DCRF will be real-time capable in contrast to former approaches like OSACA, OPC, etc. This capability is crucial to distributed control systems.

   B. The DCRF will enable a flexible composition and reconfiguration of control systems with manufacturing task-specific functionality on the basis of a common communication platform. Suitable lower-level layers (hardware, bus system, RTOS) and the communication platform

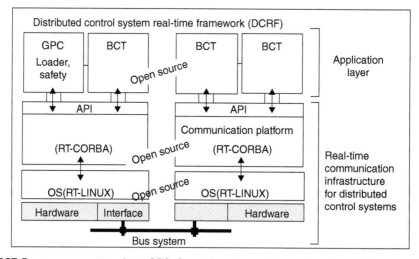

BCT, Base component template; GPC, General purpose component; OS, Operating system;

FIGURE 12.1 Distributed Control System Real-Time Framework (DCRF).

(RT-CORBA) have to be extended and integrated so that they can serve as a distributed real-time framework for the whole control system with all its single components.

    C. General-purpose components (GPC in Figure 12.1) for configuration, loader, and safety aspects are also a very important part of the DCRF. In particular, safety functions will be designed and implemented, which can assure a fail-safe operation of component-based control systems in industrial production.

2. The definition and realization of a component-based open numerical control reference architecture for machine tool; in order to take advantage of the open control systems, it is necessary to extend the existing reference architecture and to decompose the monolithic blocks into components with clearly defined interfaces, described in a standardized format. The open control reference architecture will not be delivered as an open source but with publicly available, new standardized interfaces for motion control components of machine tools, which will make use of the DCRF. These interfaces will be published for further comments and implementations by users in the field of control techniques (the Base Component Templates (BCT) visible in Figure 12.1). Thus, it will become possible to integrate additional functionality and third-party software by just using the standardized interface description without a need to adapt interfaces. The component-based reference architecture, which will be developed in the OCEAN project, has to be seen as a starting point for the specification of further control system applications in the future. The first step is the design and implementation of motion control base components that are mandatory to operate a machine such as HMI, motion control kernel, PLC, and kinematics components (Figure 12.2).

    A. The second step aims at the design and implementation of motion control extension components that are not yet covered by any reference architecture, such as filter and process control components. Particular focus will be put on a new safety component that is specifically control related and includes mechanisms like real-time comparators for a multichannel environment. The component-based concept will be proved with tests and simulations of the single components as well as of the complete control system on different machine tool demonstrators.

Clearly, the OCEAN project has a potential to substantially influence the practices of major vendors by demonstrating that the adoption of the OCEAN open source-based architecture (Linux, CORBA, etc.), combined with standards like POSIX, IEC 61131, and IEC 61508, will stimulate the development of competitive products. In particular, the OCEAN project will allow one to integrate additional functionality and third-party software by just using the standardized interface description, and to merge components provided by different CNC manufacturers. Two examples will clarify the importance of the

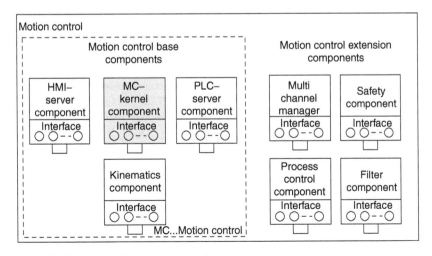

FIGURE 12.2 Subdivision of complex motion control structure based on components.

results; it will become possible and easy for the machine tool builder to implement his or her own kinematics component or filter component and even reuse it for different CNCs from different control manufacturers. Moreover, the possibility to substitute the user interface, leaving the rest of the numerical control unchanged (hardware and software), will overcome one of the major problems that machine tool builders have to face today, namely the obvious desire by end users to have uniform user interfaces for all machine tools in the same plant.

12.3 Status of the Project

12.3.1 Analysis of Communication Systems, Platforms, and Tools

The first part of the project dealt with the analysis of Communication Systems, Platforms, and Tools. The main objective was to find an appropriate communication infrastructure for a "DCRF" that would suit the technical and cost aspects, and could be provided to the end user as an open source.

CORBA implementations, as well as available Linux real-time extensions, were analyzed and compared. The two main criteria for this analysis were the real-time capability and the Open Source availability of the investigated software. As a result, TAO was found to be the most suitable CORBA Object Request Broker (ORB), and RTAI the preferable Linux real-time extension.

TAO has been developed at the Washington University of St. Louis with a special focus on its real-time capability. TAO offers a variety of CORBA services as well as the quality-of-service (QoS) functionality that can be applied to the supervision of modular software systems. Additionally, an implementation of the CORBA Component Model (CCM) is currently being developed for TAO. The CCM allows for a better software reuse and a dynamic configuration of CORBA applications. RTAI is preferred as the Linux real-time extension rather than RTLinux because RTAI offers the best real-time support, which is crucial for the realization of the DCRF. RTAI allows to run hard real-time processes in kernel space simultaneously with soft real-time processes in user space, for instance. Furthermore, RTAI offers the best C++ support and a more complete feature set. An additional criteria on is the availability of LXRT for RTAI that enables the operation of hard real-time tasks in user space. This allows the use of various standard operating system functionalities in real-time applications.

The consortium has performed analysis of the benefits of existing open controller approaches that can be considered for the design of the DCRF. Due to the fact that DCRF will be based on CORBA, the main focus of the analysis was on the comparison between CORBA as a middleware and the communication infrastructure of existing open controller approaches. The following open controller solutions were investigated:

- OSACA (Europe)
- OMAC (U.S.A.)
- OSEC and FAOP (Japan)

With respect to the communication infrastructure and the communication system APIs, all benefits of the different open controller approaches are fulfilled by CORBA. Compared with OSACA, the total functional range of CORBA Services will even have to be reduced to the minimum functionality necessary for a control platform. This will result in a smaller footprint of DCRF. The modularization of control functionality achieved by the existing open controller approaches, as well as the configuration and start-up of the software modules, can be mapped into the respective functions of the CORBA Component Model. Finally, the OSACA reference architecture that specifies the software interfaces of control modules can be mapped into a reference architecture based on the Object Management Group Component Interface Definition Language (OMG CIDL).

The requirements for the infrastructure and the general-purpose components of DCRF were compiled. These requirements address on the one hand the data exchange between different real-time components and on the other the communication between hard real-time and less time-critical software components. Additionally, requirements for safety functions in DCRF were addressed.

Requirements were compiled for the following categories:

- Functional requirements
- Latency requirements
- Task scheduling jitter
- Clock synchronization
- Task interruption
- Software safety aspects

The first part of the project was concluded once all information and data needed for designing the DCRF were collected. As a result of the investigations, the TAO–RTAI solution has been selected as the best combination to fulfill the requirements of DCRF. It allows a distribution and cross-platform deployment of components in a real-time environment. The benefits of existing open controller approaches have been identified and can be mapped into the CORBA-based development of DCRF.

12.3.2 Design, Implementation, and Validation of a DCRF

The second phase of the project, currently in progress, is the design, implementation, and validation of a DCRF. The first step for this implementation is porting of TAO to RTAI, done by the consortium and OCI. After this phase, the consortium will verify if the services of the TAO implementation satisfy the real-time approach. In this case, it will adapt the result to be the basis for the DCRF. One of the main aims of the DCRF is to have CNC components from different vendors interoperable with the DCRF framework, which means they have to be hardware independent or at least easily adaptable to the hardware of different control. To solve this problem, hardware abstractions have to be investigated and evaluated so that an exchange of components will be easy and feasible.

Moreover, it is mandatory for the real time components that their communication channels meet their time constraints (QoS). Therefore, suitable transport mechanisms have to be chosen. The lower level CORBA layer, which is called the "pluggable protocol framework" in TAO, has to be extended to meet the needs of DCRF.

Another parallel activity is the design and implementation of general-purpose components (for loading, configuration, and safety).

12.3.3 Design, Implementation, and Validation of Motion Control Base Components for an Open Numerical Control System

As of 2003, the detailed functionality of the components is another focus area. The aim here is to achieve a definition of a reference architecture for the motion control components of the open control systems by extending existing architectures (OSACA, OMAC, JOP) and new proposals. The goal is to provide standardized and published interface descriptions for components that are up to now part of monolithic control blocks and that will be extracted from there; this way, machine tool builders or end users will be able to integrate additional functionalities independent of control manufacturers. The components and their interfaces will be defined with formal methods like IDL or XML.

12.4 Foundations of the OCEAN Project, and Results

12.4.1 OSACA

The first goal of the OCEAN project is the development of DCRF based on Open Source Software, which provides a communication infrastructure for distributed real-time control applications. OCEAN draws considerably from the results and experience of the OSACA project.

The OSACA system platform is composed of three main parts: reference architecture, configuration system, and communication system. The reference architecture specifies a data model for numerical controls. The configuration system coordinates the setup of control systems. The communication system handles data exchange.

In the OSACA reference architecture, five main functional units of numerical control have been specified:

1. Motion Control
2. Motion Control Manager
3. Axis Control
4. Spindle Control
5. Logic Control

The reference architecture that was developed in OSACA mainly deals with non-real-time control tasks. It was intended to be a specification for uniform interfaces between the NC kernel and the human–machine interface (HMI).

The OCEAN project aims at the development of real-time capable control tasks. Therefore, data models for components within the NC kernel have to be defined and implemented. The development is split into two steps. In the first step, the basic components that are mandatory to operate an NC are designed and implemented, followed by the development of extension components that will enhance the functionality of numerical control. The base components are Motion Control Kernel Component, Kinematics Component, PLC-Server Component, and HMIServer Component. The benefits of the OSACA Motion Control, Axis Control, and Spindle Control units are taken into account for specifying the OCEAN Motion Control Kernel and HMI-Server Components. Furthermore, the OSACA Logic Control unit, which is very basic and only offers three data items, is taken into consideration for the OCEAN PLC-Server Component. The specification for the OCEAN Kinematics Component was developed from scratch. With respect to the Extension Components, all specifications were developed from scratch, except for the Multi-Channel Manager Component. This component can partly be based on the OSACA Motion Control Manager unit. The experiences from the OSACA configuration system are used for the development of the General-Purpose Components. In OCEAN, these are components for the configuration of distributed control systems, for supervision concerning safety, and for dynamically loading additional control components. For this purpose, the OSACA configuration system is extended by the CCM specification, if applicable. The scope of the CCM and its impact in "resource management," "QoS," and usability has to be considered and adapted for the use in DCRF.

The OSACA communication system was completely replaced by RT-CORBA. In this context, migration from OSACA to RT-CORBA of the three different types of communication objects developed in OSACA (Variable Objects for data exchange, Process Objects for commands, and Event Objects for notifications about data updates) was required. This exchange of the communication systems also affects the development of the General-Purpose Components. For instance, the OSACA configuration system had to be further redeveloped to allow for the use of RT-CORBA.

12.4.2 Linux

Linux is an operating system created by Linus Torvalds at the University of Helsinki in Finland. The first version, 1.0, of the Linux Kernel was released in 1994. At present, Linux is developed under the GNU General Public License and its source code is freely available to everyone. This, however, does not mean that Linux and its assorted distributions are free. Companies and developers may charge money for it as long as the source code remains available. Linux may be used for a wide variety of purposes, including networking, software development, and as an end-user platform. Linux is often considered an excellent, low-cost alternative to other, more expensive operating systems.

Due to the very nature of Linux's functionality and availability, it has become quite popular worldwide and a vast number of software programmers have taken Linux's source code and adapted it to meet their individual needs.

12.4.3   Linux Real-Time Extensions

For real-time extensions, a distinction between soft real-time extensions and hard real-time extensions can be made. Linux real-time extensions can be defined as *hard* if they guarantee deterministic worst-case latencies of interrupt servicing and scheduling. They can be defined as *soft* if a fast response in terms of the hard real-time requirements cannot be guaranteed. In the case of Numerical Control, a hard real-time extension that guarantees fast response is mandatory.

12.4.3.1   Available Implementations

The two available implementations of real-time Linux are RTAI and RTLinux. RTAI and RTLinux are both patches (additions, but patch is the real technical term) to the standard Linux kernel. The multi-threaded real-time kernel runs standard Linux as the lowest priority thread that is always preemptible. This makes hard real-time functionality as well as most features of the non-real-time Linux kernel available to programmers. RTLinux is a small POSIX 1003.13/PSE51 compatible hard real-time operating system. RTAI is a variant of RTLinux that has gone off on its own path. It provides compatibility with POSIX 1003.1c and 1003.1b (Threads and Queues).

Some characteristics of RTL are:

- Majority of APIs are not portable.
- Dynamic thread creation is dangerous (uses kmalloc non-atomic).
- Support for Mutexes and Condition Variables.
- POSIX calls are part of the scheduler core.

Characteristics of RTAI:

- Modeled after standard LinuxThreads (by Xavier Leroy).
- Supports dynamic thread creation (from preallocated pool).
- Supports mutexes with priority inheritance.
- Supports condition variables.
- POSIX API provided by an optional module.
- Good basis for extending to meet the EL/IX specification.
- Supports user mode real-time tasks.
- Provided through a GPL License.

RTAI now has two separate approaches toward user space real-time: one extends the microkernel with system calls from the user space (via the trap mechanism), and the other approach extends the Linux task API. The former is meant to work with ADEOS, as a completely Linux-independent real-time operating system, if desired. The latter can never work without Linux. The development in this issue is ongoing and not yet finished.

Based on the comparison, it can be judged that RTAI provides better RT support, meaning that:

- The modification required to install the patch on the Linux Kernel is smaller.
- It allows soft real-time in user space along with hard real time in kernel space.
- It provides excellent performance in terms of low jitter and low latency.
- It provides better C++ support and a more complete feature set than RTLinux.
- RTAI has the better open source approach with lots of feedback from developers.

Furthermore, RTAI has evolved by gaining a good support for real-time access to system peripherals (e.g., real-time drivers for serial ports are now available).

RTAI now offers development of interrupt handlers in user mode and an enhanced scheduling policy.

12.4.3.2   License Agreements

The basic mechanism that stands at the basis of RTLinux is protected under U.S. Patent 5,995,745 and licensed without a fee to RTLinux users for commercial and noncommercial purposes. This means that RTLinux can be used royalty-free under two conditions:

1. Under the terms of the GPL-Licence. It can be used, and even modified, if it is used with any software that also runs under the terms of the GPL-License.
2. *Together with software that uses "Open RTLinux Execution Environment."* In this case, any software can be run together with RTLinux even if the software is not open or not complying with the terms of GPL (e.g., LGPL software). "Open RTLinux Execution Environment" means a computer hardware system where the interrupt control hardware of processors and system boards is under the direct control of an unmodified Open RTLinux Software in binary form. A configuration done by using the options of a configuration tool is not considered a modification.

RTAI is a real-time extension that has partly been changed from LGPL conditions to GPL conditions. It is covered by the terms of Version 2 of the Open RTLinux Patent License, which says: "The Patented Process may be used, without any payment of a royalty."

12.4.4   CORBA

CORBA is a standard architecture for distributed object systems (Figure 12.3). It allows a distributed, heterogeneous collection of objects to interoperate. The definition of CORBA is managed by the Object Management Group (OMG). The OMG comprises over 700 companies and organizations, including almost all the major vendors and developers of distributed object technology, such as platform, database, and application vendors, as well as software tool and corporate developers.

12.4.5   Comparison of Available CORBA Implementations

For the comparison of available CORBA implementations, an Internet research was performed. In the scope of this research, open source ORBs as well as commercial solutions were investigated.

The following ORBs were included in the investigation:

- TAO, Distributed Object Group at Washington University, St. Louis, U.S.A.
- ROFES, Aachen University of Technology, Germany
- MICO, University of Frankfurt, Germany
- ORBacus, IONA, Dublin, Ireland
- ORBExpress, Object Interface Systems, Herndon, VA, U.S.A.
- e*ORB, Vertel, Woodland Hills, CA, U.S.A.
- ORBit, GNOME Project
- omniORB2, AT&T Laboratories, Cambridge, U.K.
- ORBIX E2A, IONA, Dublin, Ireland
- VisiBroker, Borland, Scotts Valley, CA, U.S.A.
- JacORB, Software Engineering and Software Group, FU Berlin

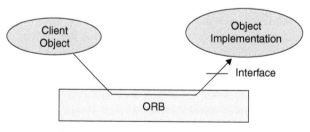

FIGURE 12.3 Basic CORBA concept.

The research was based on the following criteria:

- Open Source availability
- CORBA version implemented
- real-time capability and implemented RT CORBA features
- available language bindings
- implemented CORBA services
- supported multithreading concept
- supported protocols
- tested Linux and compiler support

Except for the real-time version of VisiBroker and JacORB, all investigated ORBs have been ported into and tested on Linux platforms. All ORBs (except the Java-based JacORB) offer language bindings for C++. Real-time capability is provided by the Open Source ORBs TAO and ROFES. Further real-time versions of ORBExpress, e*ORB, and VisiBroker are available but only in commercial distributions. The range of CORBA services that is implemented for each ORB is quite different. The basic services like Naming, Event, and Trading are provided by almost all ORBs. Additional CORBA Services like Notification, Time, Scheduling, etc. vary among implementations. The largest range of implemented CORBA Services offer TAO and ORBit.

12.4.6 Interoperability between Different ORBs

The interoperability between different ORBs is an essential feature for the realization of distributed control systems using CORBA as middleware. The OMG specification covers the aspect of ORB interoperability since CORBA 2.0. This revision of the specification introduced a general ORB interoperability architecture, which is called General Inter-ORB Protocol (GIOP). GIOP is an abstract protocol that specifies the transfer syntax and a standard set of message formats to allow independently developed ORBs to communicate over any connection-oriented transport. The Internet Inter-ORB Protocol (IIOP) specifies how GIOP is implemented over TCP/IP (Figure 12.4).

ORB interoperability also requires standardized object reference formats. Object references are opaque to applications but they contain information that ORBs need in order to establish communications between clients and target objects. The standard object reference format, called the Interoperable Object reference (IOR), is flexible enough to store information for almost any inter-ORB protocol imaginable. An IOR identifies one or more supported protocols and contains information specific to that protocol. For IIOP, an IOR contains a host name, a TCP/IP port number, and an object key that identifies the target object at the given host name and port combination.

All investigated ORBs were implemented in conformance with the CORBA 2.0 specification. Therefore, all of them support IIOP, which is the basis for interoperability. Statements for successful interoperability tests were found for the combination of the three ORBs: TAO, MICO, and JacORB.

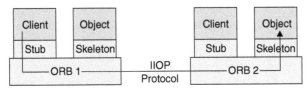

IIOP:Internet inter–ORB protocol

FIGURE 12.4 CORBA interoperability concept.

To ensure and to test the interoperability between different ORBs, research projects have been launched for the development of test suits. One of them is the EC-funded project CORVAL2 (Enhanced Techniques for CORBA Validation; IST-1999-11131) and a second one is the open source projects COST (CORBA Open Source Testing). The main objective of these projects is to investigate new mechanisms and tools for validating the implementation of object-oriented technology. The resulting test suits can be applied by users and providers of CORBA technology to test products against the CORBA specifications, and hence to ensure product interoperability and application portability.

12.4.7   Investigation of the CCM

12.4.7.1   Introduction to the CCM

To provide higher-level reusable components, the OMG specifies a set of CORBA Object Services that defines standard interfaces to access common distribution services, such as naming, trading, and event notification. By using CORBA and its Object Services, system developers can integrate and assemble large, complex distributed applications, and systems using features and services from different providers.

Unfortunately, the traditional CORBA object model, as defined by CORBA 2.4, has the following limitations.

1. *No standard way to deploy object implementations.* The earlier CORBA specification did not define a standard for deployment of object implementations in server processes. Deployment involves distributing object implementations, installing those implementations in their execution contexts, and activating the implementations in an ORB. Thus, system designers developed ad hoc strategies to instantiate all objects in a system. Moreover, since objects may depend on one another, the deployment and instantiation of objects in a large-scale distributed system are complicated and nonportable.

2. *Limited standard support for common CORBA server programming patterns.* The CORBA family of specifications provides a rich set of features to implement servers. For example, the CORBA 2.2 specification introduced *Portable Object Adapter* (POA), which is the ORB mechanism that forwards client requests to concrete object implementations. The POA specification provides standard APIs to register object implementations with the ORB, to deactivate those objects, and to activate object implementations on demand. The POA is flexible and provides numerous policies to configure its behavior. In many application domains, however, only a limited subset of these features is ever used repeatedly; yet, server developers face a steep learning curve to understand how to configure POA policies *selectively* to obtain their desired behavior.

3. *Limited extension of object functionality.* In the traditional CORBA object model, objects can be extended only via inheritance. To support new interfaces, therefore, application developers must: (1) use CORBA´s Interface Definition Language (IDL) to define a new interface that inherits from all the required interfaces; (2) implement the new interface; and (3) deploy the new implementation across all their servers. Multiple inheritance in CORBA IDL is fragile because overloading is not supported in CORBA; therefore, multiple inheritance has limited applicability. Moreover, applications may need to expose the same IDL interface multiple times to allow developers to either provide multiple implementations or multiple instances of the service through a single access point. Unfortunately, multiple inheritance cannot expose the same interface more than once, nor can it alone determine which interface should be exported to clients.

4. *Availability of* CORBA Object *Services not defined in advance.* The CORBA specification does not mandate which Object Services are available at runtime. Thus, object developers used ad hoc strategies to configure and activate these services when deploying a system.

5. *No standard object life-cycle management.* Although the CORBA Object Service defines a Life Cycle Service, its use is not mandated. Therefore, clients often manage the life cycle of an object explicitly in ad hoc ways. Moreover, the developers of CORBA objects controlled through the life cycle service must define auxiliary interfaces to control the object life cycle. Defining these interfaces is

tedious and should be automated when possible, but earlier CORBA specifications lacked the capabilities required to implement such automation.

In summary, the inadequacies, outlined above, of the CORBA specification, prior to and including version 2.4, often yield tightly coupled, ad hoc implementations of objects that are hard to design, reuse, deploy, maintain, and extend. The CCM extends the CORBA object model by defining features and services that enable application developers to implement, manage, configure, and deploy components that integrate commonly used CORBA services — such as transaction, security, persistent state, and event notification services — in a standard environment. In addition, the CCM standard allows greater software reuse for servers and provides greater flexibility for dynamic configuration of CORBA applications.

Figure 12.5 shows an example CCM component with IDL specification, and Figure 12.6 shows the tool chain using the Component Implementation Definition Language.

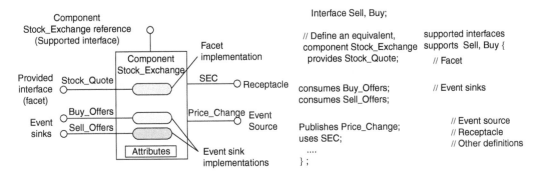

FIGURE 12.5　Example CCM component.

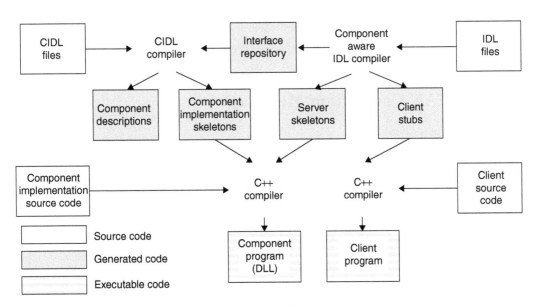

FIGURE 12.6　Tool chain using the Component Implementation Definition Language.

12.5   Conclusion

The CORBA object model is increasingly gaining acceptance as the industry standard, cross-platform, cross-language distributed object computing model. The recent addition of the CCM integrates a successful component programming model from EJB, while maintaining the interoperability and language-neutrality of CORBA. The CCM programming model is thus suitable for leveraging proven technologies and existing services to develop the next generation of highly scalable distributed applications. However, the CCM specification is large and complex. Therefore, ORB providers have only started implementing the specification recently. As with first-generation CORBA implementations several years ago, it is still hard to evaluate the quality and performance of CCM implementations. Moreover, the interoperability of components and containers from different providers is not yet well understood.

CCM providers are now implementing the complete specification, as well as support value-added enhancements to their implementations, just as operating system and ORB providers have done historically. In particular, containers provided by the CCM component model implementation provide QoS capabilities for CCM components, and can be extended to provide more services to components to relieve components from implementing these functionalities in an ad hoc way. These container QoS extensions provide services that can monitor and control certain aspects of components behaviors that crosscut different programming layers or require close interaction among components, containers, and operating systems. As CORBA and the CCM evolve, we expect that some of these enhancements will be incorporated into the CCM specification. CIAO will be the first implementation available with QoS RT-CCM enhancements and it is worth taking it into account for DCRF.

12.5.1   Selection of Suitable ORB

For the development of DCRF, TAO is favored as ORB. This selection is based on the following arguments:

- TAO is available in open source,
- offers the majority of features defined in the RT CORBA 1.0 specification,
- comprises the main CORBA services, and
- a release including the CORBA Component Model is planned for 2003.

The only drawback of using TAO is its size, which should not be a problem in most cases, since a large amount of memory is needed to host a complete CNC.

12.6   OCEAN Consortium Members

As of 2003, the consortium involves:

1. FIDIA S.p.A., coordinator: manufacturer of numerical controls, high speed milling machines and CAM products (I)
2. USTUTT: The Numerical Controls Institute of the University of Stuttgart (D)
3. RWTH: The Machine Tool Institute of the Aachen University (D)
4. KUL: The Department of Mechanical Engineering of the Katholieke Universiteit Leuven (B)
5. HOMAG: Manufacturer of numerical controls and wood working machines (D)
6. FATRONIK: Research Institute for Machine Tool and Production Techniques (E)
7. FAGOR: Numerical controls manufacturer (E)
8. OSAI: Numerical controls manufacturer (I)
9. GORATU: Machine-tool (lathes, milling and High speed milling machine) manufacturer, (E)
10. ITIA: The Industrial Technologies and Automation Institute, of the Italian National Research Council (I)

Section 4.2

Fieldbus Technology

13

Fieldbus Systems: History and Evolution

Thilo Sauter
Austrian Academy of Sciences

13.1   What Is a Fieldbus?

Throughout the history of automation, many inventions and developments have influenced the face of manufacturing and information processing. But few novelties have had such a radical effect as the introduction of fieldbus systems, and no single achievement was so heavily disputed as these industrial networks. And yet, they have made automation what it is today. But even after some 20 years of fieldbus development, there exists no clear-cut definition for the term. The "definition" given in the IEC 61158 fieldbus standard is more a programmatic declaration, or a least common multiple compromise, than a concise formulation [1]: "A fieldbus is a digital, serial, multidrop, data bus for communication with industrial control and instrumentation devices such as — but not limited to — transducers, actuators and local controllers." It comprises some important characteristics, but is far from being complete. On the other hand, it is a bit too restrictive.

A more elaborate explanation is given by the Fieldbus Foundation, the user organization supporting one of the major fieldbus systems [2]: "A Fieldbus is a digital, two-way, multi-drop communication link among intelligent measurement and control devices. It serves as a Local Area Network (LAN) for advanced process control, remote input/output and high speed factory automation applications." Again, this is a bit restrictive, for it limits the application to process and factory automation, the primary areas where the Foundation Fieldbus is used.

The lack of a clear definition is mostly due to the complex evolutionary history of fieldbuses. A look at today's situation reveals that fieldbus systems are employed in all automation domains ranging from the aforementioned process and factory areas to building and home automation, machine building, automotive and railway applications, and avionics. In all those fields, bus systems emerged primarily to break up the conventional star-type point-to-point wiring schemes connecting simple digital and analog input and output devices to central controllers, thereby laying the grounds for the implementation of really distributed systems with more intelligent devices. As was declared in the original mission statement of the International Electrotechnical Commission (IEC) work, "the Field Bus will be a serial digital communication standard which can replace present signalling techniques such as 4-20 mA ... so that more information can flow in both directions between intelligent field devices and the higher level control systems over a shared communication medium ..." [3, 4].

But even though the replacement of especially the traditional 4–20 mA current loop by a digital interface is bequeathed as the sole impetus of fieldbus development, still in contemporary publications [5], there is much more to the idea of the fieldbus:

- **Flexibility and modularity**: A fieldbus installation like any other network can be extended much more easily than a centralized system, provided the limitations of addressing space, cable length, etc., are not exceeded.
- **Configureability**: A network — other than an analog interface — permits the parameterization and configuration of complex field devices, which facilitates system setup and commissioning and is the primary requirement for the usability of intelligent devices.
- **Maintainability**: Monitoring of devices, applying updates, and other maintenance tasks are easier, if at all possible, via a network.
- **Distribution**: A network is the prerequisite of distributed systems; many data processing tasks can be removed from a central controller and placed directly in the field devices if the interface can handle reasonably complex ways of communication.

These aspects are not just theoretical contemplations but actual user demands that influenced the development from the beginning [4]. However, as the application requirements in the various automation domains were quite different, so were the solutions, and that makes it difficult to find a comprehensive definition.

The purpose of this contribution is not to find the one and only precise definition for what constitutes a fieldbus. The vast literature on this topic shows that this is a futile attempt. Furthermore, such a definition would be mostly of an academic nature and is not necessary either. Instead, the following sections will treat the fieldbus as a given phenomenon in automation and look at it from different sides. Typical characteristics will be discussed as well as the role of fieldbus systems in a globally networked automation world. The major part of this chapter will be devoted to the historical evolution and the standardization processes, which will enlighten the current situation. Finally, future aspects and evolutionary potential are briefly discussed.

13.2 Notions of a Fieldbus

Fieldbus systems have to be seen as an integrative part of a comprehensive automation concept and not as stand-alone solutions. The name is therefore programmatic and evocative. It seems to give an indication of the intentions the developers had in mind and thus deserves special attention.

13.2.1 The Origin of the Word

Interestingly enough, not even the etymology of the term itself is fully clear. The English word *fieldbus* is definitely not the original one. It appeared around 1985 when the fieldbus standardization project within IEC TC65 was launched [4] and seems to be a straightforward literal translation of the German term *Feldbus*, which can be traced back until about 1980 [6]. Indeed, the overwhelming majority of early publications in the area are available only in German. The word itself was coined in process industry and primarily refers to the process field, designating the area in a plant where lots of distributed field devices, mostly sensors and actuators, are in direct contact with the process to be controlled. Slightly after the German expression and sharing its etymological root, the French word *réseau de terrain* (or *réseau d'instrumentation*, instrumentation network) emerged. This term was not specifically targeted at the process industry, but refers also to large areas with scattered devices. The connection of such devices to the central control room was traditionally made via point-to-point links, which resulted in a significant and expensive cabling need. The logical idea, powered by the advances of microelectronics in the late 1970s, was to replace this star-like cabling in the field by a party-line, bus-like installation connecting all devices via a shared medium — the fieldbus [7, 8].

Given the large dimensions of process automation plants, the benefits of a bus are particularly evident. However, the concept was not undisputed when it was introduced. The fieldbus approach was an ambitious concept: a step toward decentralization, including the preprocessing of data in the field devices, which both increases the quality of process control and reduces the computing burden for the centralized controllers [9]. Along with it came the possibility to configure and parameterize the field devices remotely via the bus. This advanced concept, on the other hand, demanded increased communication between the devices that goes far beyond a simple data exchange. This seemed infeasible to many developers, and still in the mid-1980s, one could read statements like the following [10]: "The idea of the fieldbus concept seems promising. However, with reasonable effort it is not realizable at present."

The alternative and somewhat more conservative approach was the development of so-called *field multiplexers*, devices that collect process signals in the field, serialize them, and transfer them via one single cable to a remote location where a corresponding device de-multiplexes them again [11]. For quite some time, the two concepts competed and coexisted [12], but ultimately the field multiplexers mostly disappeared, except for niches in process automation, where many users still prefer such *remote input/ output* (I/O) systems despite the advantages of fieldbus solutions [13]. The central field multiplexer concept of sampling I/O points and transferring their values in simple data frames also survived in some fieldbus protocols, especially designed for low-level applications.

The desire to cope with the wiring problem getting out of hand in large installations was certainly the main impetus for the development of fieldbus systems. Other obvious and appealing advantages of the concept are modularity, the possibility to easily extend installations, and the possibility to have much more intelligent field devices that can communicate not just for the sake of process data transfer, but also for maintenance and configuration purposes [14, 15]. A somewhat different viewpoint that led to different design approaches was to regard bus systems in process control as the spine of distributed real-time systems [16]. While the wiring optimization concepts were in many cases rather simple bottom-up approaches, these distributed real-time ideas resulted in sophisticated and usually well-investigated top-down designs.

13.2.2 Fieldbuses as Part of a Networking Concept

An important role in the fieldbus evolution has been played by the so-called automation pyramid. This hierarchical model was defined to structure the information flow required for factory and process automation. The idea was to create a transparent, multilevel network — the basis for computer-integrated manufacturing (CIM). The numbers vary, but typically this model is composed of up to five levels [7, 8]. While the networks for the upper levels already existed by the time the pyramid was defined, the field level was still governed by point-to-point connections. Fieldbus systems were therefore developed also with the aim of finally bridging this gap. The actual integration of field-level networks into the rest of the hierarchy was in fact considered in early standardization [4]; for most of the proprietary developments, however, it was never the primary intention.

In the automation pyramid, fieldbuses actually populate two levels: the field level and the cell/process level. For this reason, they are sometimes further differentiated into two classes:

- Sensor–actuator buses or device buses have very limited capabilities and serve to connect very simple devices with, e.g., programmable logic controllers (PLCs). They can be found exclusively on the field level.
- Fieldbuses connect control equipment like PLCs and PCs as well as more intelligent devices. They are found on the cell level and are closer to computer networks.

Depending on the point of view, there may even be a third sublevel [17]. This distinction may seem reasonable but is in fact problematic. There are only few fieldbus systems that can immediately be allocated to one of the groups; most of them are used in both levels. Therefore, it should be preferable to abandon this arbitrary differentiation.

How do fieldbus systems compare to computer networks? The classical distinction of the different network types used in the automation pyramid hinges on the distances the networks span. From top down, the hierarchy starts with global area networks (GANs), which cover long, preferably intercontinental distances and nowadays mostly use satellite links. On the second level are wide area networks (WANs). They are commonly associated with telephone networks (no matter if analog or digital). Next come the well-known local area networks (LANs), with Ethernet as the most widely used specimen today. They are the classical networks for office automation and cover only short distances. The highest level of the model shown in Figure 13.1 is beyond the scope of the original definition, but is gaining importance with the availability of the Internet. In fact, Internet technology is penetrating all levels of this pyramid all the way down to the process level.

From GANs to LANs, the classification according to the spatial extension is evident. One step below, on the field level, this criterion fails because fieldbus systems or field area networks (FANs) can cover even larger distances than LANs. Yet, as LANs and FANs evolved nearly in parallel, some clear distinction between the two network types seemed necessary. As length is inappropriate, the classical border line drawn between LANs and FANs relies mostly on the characteristics of the data transported over these networks. Local area networks have high data rates and carry large amounts of data in large packets. Timeliness is not a primary concern, and real-time behavior is not required. Fieldbus systems, by contrast, have low data rates. Since they transport mainly process data, the size of the data packets is small, and real-time capabilities are important.

For some time, these distinction criteria between LANs and FANs were sufficient and fairly described the actual situation. Recently, however, drawing the line according to data rates and packet sizes is no longer applicable. In fact, the boundaries between LANs and fieldbus systems have faded. Today, there are fieldbus systems with data rates well above 10 Mbit/s, which is still standard in older LAN installations. In addition, more and more applications require the transmission of video or voice data, which results in large data packets.

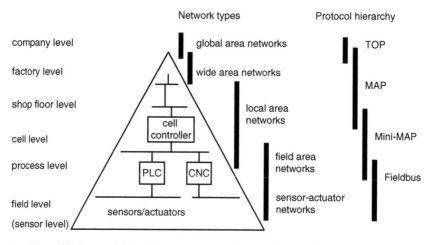

FIGURE 13.1 Hierarchical network levels in automation and protocols originally devised for them.

On the other hand, Ethernet as *the* LAN technology is becoming more and more popular in automation and is bound to replace some of today's widely used midlevel fieldbus systems. The real-time extensions under development tackle its most important drawback and will ultimately permit the use of Ethernet in low-level control applications. At least for the next 5 years, however, it seems that Industrial Ethernet will not make the lower-level fieldbuses fully obsolete. They are much better optimized for their specific automation tasks than the general-purpose network Ethernet. But the growing use of Ethernet results in a reduction of the levels in the automation hierarchy. Hence the pyramid gradually turns into a flat structure with at most three, maybe even only two, levels.

Consequently, a more appropriate distinction between LANs and FANs should be based on the functionality and the application area of these networks. According to this argumentation, a fieldbus is simply a network used in automation, irrespective of topology, data rates, protocols, or real-time require-ments. Consequently, it need not be confined to the classical field level; it can be found on higher levels (provided they still exist) as well. A LAN, on the other hand, belongs to the office area. This definition is loose, but mirrors the actual situation. Only one thing seems strange at first: following this definition, the Industrial Ethernet changes into a fieldbus, even though many people are inclined to associate it with LANs. However, this is just another evidence that the boundaries between LANs and FANs are fading.

13.3 History

The question of what constitutes a fieldbus is closely linked to the evolution of these industrial networks. The best approach to understanding the essence of the concepts is to review the history and intentions of the developers. This review will also falsify one of the common errors frequently purported by marketing divisions of automation vendors: that fieldbus systems were a revolutionary invention. They may have revolutionized automation — there is hardly any doubt about it. However, they were only a straightforward evolution that built on preexisting ideas and concepts.

13.3.1 The Roots of Industrial Networks

Although the term *fieldbus* appeared only about 20 years ago, the basic idea of field-level networks is much older. Still, the roots of modern fieldbus technology are mixed. Both classical electrical engineering and computer science have contributed their share to the evolution, and we can identify three major sources of influence:

- Communication engineering with large-scale telephone networks
- Instrumentation and measurement systems with parallel buses and real-time requirements
- Computer science with the introduction of high-level protocol design

This early stage is depicted in Figure 13.2. One foundation of automation data transfer has to be seen in the classic telex networks and also in standards for data transmission over telephone lines. Large distances called for serial data transmission, and many of these comparatively early standards still exist, like V.21 (data transmission over telephone lines) and X.21 (data transmission over special data lines). Various protocols have been defined, mostly described in state machine diagrams and rather simple because of the limited computing power of the devices available at that time. Of course, these commu-nication systems have a point-to-point nature and therefore lack the multidrop characteristic of modern fieldbus systems, but nevertheless, they were the origin of serial data transmission. Talking about serial data communication, one should notice that the engineers who defined the first protocols often had a different understanding of the terms *serial* and *parallel* than we have today. For example, the serial interface V.24 transmits the application data serially, but the control data in a parallel way over separate control lines.

In parallel to the development of data transmission in the telecommunication sector, hardware engi-neers defined interfaces for stand-alone computer systems to connect peripheral devices such as printers. The basic idea of having standardized interfaces for external devices was soon extended to process control and instrumentation equipment. The particular problems to be solved were the synchronization of

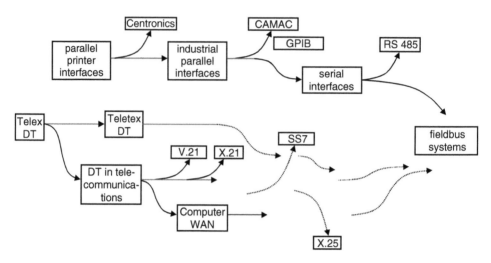

FIGURE 13.2   Roots of fieldbus systems.

spatially distributed measurement devices and the collection of measurement data from multiple devices in large-scale experimental setups. This led to the development of standards like CAMAC (computer-automated measurement and control, mostly used in nuclear science) and GPIB (general purpose interface bus, later also known as IEEE 488). To account for the limited data processing speed and real-time requirements for synchronization, these bus systems had parallel data and control lines, which is also not characteristic for fieldbus systems. However, they were using the typical multidrop structure.

Later on, with higher integration density of integrated circuits and thus increased functionality and processing capability of microcontrollers, devices became smaller and portable. The connectors of parallel bus systems were now too big and clumsy, and alternatives were sought [18]. The underlying idea of developments like I²C [19] was to extend the already existing serial point-to-point connections of computer peripherals (based on RS 232) to support longer distances and finally also multidrop arrangements. The capability of having a bus structure with more than just two connections together with an increased noise immunity due to differential signal coding eventually made RS 485 a cornerstone of fieldbus technology up to the present day.

Historically the youngest root of fieldbus systems, but certainly the one that left the deepest mark, was the influence of computer science. Its actual contribution was a structured approach to the design of high-level communication systems, contrary to the mostly monolithic design approaches that had been sufficient until then. This change in methodology had been necessitated by the growing number of computers used worldwide and the resulting complexity of communication networks. Conventional telephone networks were no longer sufficient to satisfy the interconnection requirements of modern computer systems. As a consequence, the big communication backbones of the national telephone companies gradually changed from analog to digital systems. This opened the possibility to transfer large amounts of data from one point to another. Together with an improved physical layer, the first really powerful data transmission protocols for wide area networks were defined, such as X.25 (packet switching) or SS7 (common channel signaling). In parallel to this evolution on the telecommunications sector, local area networks were devised for the local interconnection of computers, which soon led to a multitude of solutions. It took nearly a decade until Ethernet and the Internet Protocol (IP) suite finally gained the dominating position they have today.

13.3.2   The Evolution of Fieldbuses

The preceding section gave only a very superficial overview of the roots of networking, which laid the foundations not only of modern computer networks, but also of those on the field level. But let us now look more closely at the actual evolution of the fieldbus systems. Here again, we have to consider the different influences of computer science and electrical engineering. First and foremost, the key contribution undoubtedly came from the networking of computer systems, when the International Organization for Standardization (ISO) introduced the Open Systems Interconnection (OSI) model [20, 21]. This seven-layer reference model was (and still is) the starting point for the development of many complex communication protocols.

The first application of the OSI model to the domain of automation was the definition of the Manufacturing Automation Protocol (MAP) in the wake of the CIM idea [22]. MAP was intended to be a framework for the comprehensive control of industrial processes covering all automation levels, and the result of the definition was a powerful and flexible protocol [23]. Its complexity, however, made implementations extremely costly and hardly justifiable for general-purpose use. As a consequence, a tightened version called MiniMAP, using a reduced model based on OSI layers 1, 2, and 7, was proposed to better address the problems of the lower automation layers [24]. Unfortunately, it did not have the anticipated success either. What did have success was Manufacturing Message Specification (MMS). It defined the cooperation of various automation components by means of abstract objects and services and was later used as a starting point for many other fieldbus definitions [25]. The missing acceptance of MiniMAP and the inapplicability of the original MAP/MMS standard to time-critical systems [26] were finally the reasons for the IEC to launch the development of a fieldbus based on the MiniMAP model, but tailored to the needs of the field level. According to the original objectives, the higher levels of the automation hierarchy should be covered by MAP or PROWAY (process data highway) [22].

Independent of this development in computer science, the progress in microelectronics brought forward many different integrated controllers (ICs), and new interfaces were needed to interconnect the ICs in an efficient and cheap way. The driving force was the reduction of both the interconnect wires on the printed circuit boards and the number of package pins on the ICs. Consequently, electrical engineers — without knowledge of the ISO/OSI model or similar architectures — defined simple buses like the I²C. Being interfaces rather than fully fledged bus systems, they have very simple protocols, but they were and still are widely used in various electronic devices.

Long before the invention of board-level buses, the demand for a reduction of cabling weight in avionics and space technology had led to the development of the MIL-STD-1553 bus, which can be regarded as the first real fieldbus. Introduced in 1970, it showed many characteristic properties of modern fieldbus systems: serial transmission of control and data information over the same line, master–slave structure, the possibility to cover longer distances, and integrated controllers. It is still used today. Later on, similar thoughts (reduction of cabling weight and costs) resulted in the development of several bus systems in the automotive industry, but also in the automation area. A characteristic property of these fieldbuses is that they were defined in the spirit of classical interfaces, with a focus on the lower two protocol layers, and no or nearly no application layer definitions. With time, these definitions were added to make the system applicable to other areas as well. Controller Area Network (CAN) is a good example of this evolution: for the originally targeted automotive market, the definition of the lowest two OSI layers was sufficient. Even today, automotive applications of CAN typically use only these low-level communication features because they are easy to use and the in-vehicle networks are usually closed. For applications in industrial automation, however, where extensibility and interoperability are important issues, higher-level functions are important. So, when CAN was found to be interesting also for other application domains, a special application layer was added. The lack of such a layer in the original definition is the reason why there are many different fieldbus systems (like CANopen, Smart Distributed System (SDS), DeviceNet) using CAN as a low-level interface.

From today's point of view, it can be stated that all fieldbuses that still have some relevance were developed using the top-down or computer science-driven approach, i.e., a proper protocol design with abstract high-level programming interfaces to facilitate usage and integration in complex systems. The fieldbuses that followed the bottom-up or electrical engineering-driven approach, i.e., that were understood as low-level computer interfaces, did not survive due to their inflexibility and incompatibility with modern software engineering, unless some application layer functions were included in the course of the evolution.

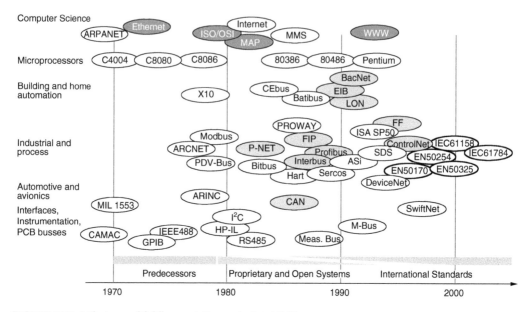

FIGURE 13.3 Milestones of fieldbus evolution and related fields.

From the early 1980s on, when automation made a great leap forward with PLCs and more intelligent sensors and actuators, something like a gold rush set in. The increasing number of devices used in many application areas called for reduced cabling, and microelectronics had grown mature enough to support the development of elaborated communication protocols. This was also the birth date for the fieldbus as an individual term. Different application requirements generated different solutions, and from today's point of view, it seems that creating new fieldbus systems was a trendy and fashionable occupation for many companies in the automation business. Those mostly proprietary concepts never had a real future, because the number of produced nodes could never justify the development and maintenance costs. Figure 13.3 depicts the evolution timeline of fieldbus systems and their environments. The list of examples is of course not comprehensive; only systems that still have some significance have been selected. Details about the individual solutions are summarized in the tables in the appendix.

As the development of fieldbus systems was a typical technology push activity driven by the device vendors, the users first had to be convinced of the new concepts. Even though the benefits were quite obvious, the overwhelming number of different systems appalled rather than attracted the customers, who were used to perfectly compatible current-loop or simple digital inputs and outputs as interfaces between field devices and controllers and were reluctant to use new concepts that would bind them to one vendor. What followed was a fierce selection process where not always the fittest survived, but often those with the highest marketing power behind them. Consequently, most of the newly developed systems vanished or remained restricted to small niches. After a few years of struggle and confusion on the user's side, it became apparent that proprietary fieldbus systems would always have only limited success and that more benefit lies in creating open specifications, so that different vendors may produce compatible devices, which gives the customers back their freedom of choice [8, 27]. As a consequence, user organizations were founded to carry on the definition and promotion of the fieldbus systems independent of individual companies [28]. It was this idea of open systems that finally paved the way for the breakthrough of the fieldbus concept.

13.4 Fieldbus Standardization

From creating an open specification to the standardization of a fieldbus system it is only a small step. The basic idea behind it is that a standard establishes a specification in a very rigid and formal way, ruling out the possibility of quick changes. This attaches a notion of reliability and stability to the specification, which in turn secures the trust of the customers and, consequently, also the market position. Furthermore, a standard is vendor independent, which guarantees openness. Finally, in many countries standards have a legally binding position, which means that when a standard can be applied (e.g., in connection with a public tender), it *has* to be applied. Hence, a standardized system gains a competitive edge over its nonstandardized rivals. This position is typical for, e.g., Europe (see [29] for an interesting U.S.-centric comment). It is therefore no wonder that after the race for fieldbus developments, a race for standardization was launched. This was quite easy on a national level, and most of today's relevant fieldbus systems soon became national standards. Troubles started when international solutions were sought.

One problem of fieldbus standardization is that the activities are scattered among a multitude of committees and working groups according to the application fields. This reflects the historical evolution and underpins the previous statement that the fieldbus is not a unique and revolutionary technology, but emerged independently in many different areas. Interestingly enough, the standardization activities are not even confined to the electrotechnical standardization bodies. Inside the IEC, committees concerned are:

- IEC TC65/SC65C: Industrial-Process Measurement and Control/Digital Communications
- IEC TC17/SC17B: Switchgear and Controlgear/Low-Voltage Switchgear and Controlgear

In the ISO, work is being done in:

- ISO TC22/SC3: Road Vehicles/Electrical and Electronic Equipment
- ISO TC184/SC5: Industrial Automation Systems and Integration/Architecture, Communications and Integration Frameworks
- ISO TC205/WG3: Building Environment Design/Building Control Systems Design

The second player in the international standardization arena is the European standardization bodies CENELEC and CEN.[*] They are not mirrors of the IEC and ISO; the committees work independently, even though much work is being done in parallel. In recent years, cooperation agreements were established with the aim of facilitating the harmonization of international standardization. The cooperation of ISO and CEN is governed by the Vienna Agreement [30] (1990), and that of IEC and CENELEC by the Dresden Agreement [31] (1996). Roughly, these documents define procedures to carry out parallel votings and to simultaneously adopt standards on both the international and European levels. In practice, this comes down to international standards always superseding European ones, even though there is the theoretical possibility of European work being adopted on an international level. Hence, European committees are today much more closely connected to their worldwide counterparts than they were at the beginning of the fieldbus era. Within CENELEC, such relevant committees are:

- CLC TC65CX: Fieldbus
- CLC TC17B: Low-voltage Switchgear and Controlgear Including Dimensional Standardization
- CLC TC205: Home and Building Electronic Systems (HBES)

In CEN, fieldbuses are defined in:

- CEN TC247: Building Automation, Controls and Building Management

The committee with the longest track record in fieldbus standardization is IEC SC65C, which in May 1985 started the ambitious endeavor of defining an international and uniform fieldbus standard for process and industrial automation. This initiative came relatively early, soon after the trend toward field-level networking and the inability of MAP to fully cover it became apparent. With the background of several industry-driven solutions emerging all around, however, this project caused heavy turbulences and opened a battlefield for politics that gradually left the ground of technical discussion. Table 13.1 shows the overall timeline of these fieldbus wars, which form an essential and obscure chapter in the fieldbus history and thus deserve special attention.

---

[*]CENELAC, Comité Européen de Normalisation Electrotechnique (European Committee for Electrotechnical Standardization); CEN, Comité Européen de Normalisation (European Committee for Standardization).

TABLE 13.1   Fieldbus Standardization Timeline from the Viewpoint of IEC 61158

Period	Status of Standards	Major Activities
1985–1990	The claims are staked	Start of the IEC fieldbus project; selection of various national standards — German Profibus and French FIP are the main candidates; first attempts to combine the two approaches
1990–1994	German–French fieldbus war	Attempt of a general specification based on WorldFIP and the Interoperable System Project (ISP)
1995–1998	Standardization locked in stalemate	Development of the American Foundation Fieldbus (FF) in response to the European approach and formation of the CENELEC standards comprising several fieldbus systems in one standard; deadlock of the international standard through obstructive minorities
1999–2000	The compromise	The eight-type specification becomes a standard
2000–2002	Amendments to reach maturity for the market	The standard is enhanced by more types and the necessary profiles are specified in IEC 61784

13.4.1 The German–French Fieldbus War

The actual starting point of international fieldbus standardization in IEC SC65C was a new work item proposed by the German national mirror committee [32]. The task was allocated to the already existing working group 6 dealing with the definition of PROWAY, another fieldbus predecessor. At that time, the development of fieldbus systems was mainly a European endeavor, thrust forward by research projects that still had a strong academic background as well as many proprietary developments. The European activities — at least those on the nonproprietary level — also have to be seen as a response to MAP, where the U.S. had a dominating position. Hence, the two big European fieldbus projects at that time, Factory Instrumentation Protocol (FIP) and Profibus, were intended to be counterweights for the international automation world.

The IEC work started with a definition of requirements a fieldbus must meet [4]. In parallel, the ISA SP50 committee started its own fieldbus project on the U.S. level and defined a slightly different set of requirements [24, 33]. Work was coordinated between the two committees, with ISA taking the more active part. It launched a call for proposals to evaluate existing solutions. In response to this call, the following systems were identified as possible candidates [34]:

- FIP (*Flux Information Processus*, Factory Instrumentation Protocol), a French development started around 1982
- Profibus (derived from process field), a German project started around 1984
- A proposal from Rosemount based on the ISO 8802.4 token-passing bus
- A proposal from Foxboro based on the high-level data link control (HDLC) protocol
- The IEEE 1118 project, in fact an extension of Bitbus
- An extension of MIL-STD-1553B defined by a U.K. consortium

All these proposals were evaluated, and finally the two most promising projects retained for further consideration were the French FIP and the German Profibus. Unfortunately, the approaches of the two systems were completely different. Profibus was based on a distributed control idea and in its original form supported an object-oriented vertical communication according to the client–server model in the spirit of the MAP/MMS specification, with the lower two layers taken from the exiting PROWAY project. FIP, on the other hand, was designed with a central, but strictly real-time-capable control scheme and with the newly developed producer–consumer (producer–distributor–consumer) model for horizontal communication. In fact, the idea behind FIP was to have a distributed operating system; a communication protocol was just one building block.

Different as they were, the two systems were well suited for complementary application areas [35]. Evidently, a universal fieldbus had to combine the benefits of both, and so the following years saw strong efforts to find a viable compromise and a convergence between the two approaches. The most problematic part was the data link layer, where Profibus supported a token-passing scheme, while FIP relied on a central scheduling approach. The suggestion to standardize both in parallel was not supported, and so it came that two different proposals were put to vote: a token-passing approach and a new proposal defined by an expert group with the aim of reconciling the two worlds [32]. The latter was more FIP oriented and finally prevailed [36], but it was very complex and left many Profibus supporters skeptical about its practical usability.

In the meantime, the leading role in the standardization efforts on the IEC level had been taken not by the Europeans, but by the work of the SP50 committee of the Instrumentation, Systems and Automation Society (ISA, at that time still standing for Instrument Society of America). Owing to its mandatory composition involving manufacturers *and* users, it had taken a more pragmatic view and had been much more efficient during the late 1980s [37]. Actually, the committee had defined and issued (as a U.S. standard in 1993) a solution on its own. The results of this work exerted an important influence on the layer structure of the standard as we have it today [8, 38]. Finally, ISA and IEC decided to have joint meetings [35], and from that point onward the actual technical work was done within ISA SP50, while IEC restricted its activities to organizing the voting process.

By the mid-1990s, the IEC committee was still struggling to overcome the differences between Profibus and FIP in what was sarcastically called the two-headed monster. With respect to its goal of defining a uniform fieldbus solution, it had not produced any substantial outcome for more than 8 years. The only exception was the definition of the physical layer, which was adopted as IEC 61158-2 in 1993. This part is the one that has since been used very successfully, mainly in the process automation area.

On top of the physical layer, however, the standardization drafts became more and more comprehensive and overloaded with all kinds of communication and control principles imported from the different systems. In the data link layer specification, for example, three different types of tokens were introduced: the scheduler token, which determines which station controls the timing on the bus; the delegated token, with which another station can temporarily gain control over the bus; and the circulated token, which is passed from station to station for bus access. The problem with these all-inclusive approaches was that a full implementation of the standard was too expensive, whereas a partial implementation would have resulted in incompatible and not interoperable devices (a problem that was encountered also in the early implementations of, e.g., Profibus-FMS (fieldbus message specification), where significant parts of the standard are optional and not mandatory).

Outside the international standardization framework, but alerted by the inability of the committees to reach a resolution, the big vendors of automation systems launched two additional initiatives to find a compromise. The foundation of the international WorldFIP project in 1993 had the goal of adding the functionality of the client–server model to FIP [39]. On the other side, the Interoperable System Project (ISP) attempted to demonstrate from 1992 onward how Profibus could be enhanced with the publisher–subscriber communication model, which is about the same as the producer–consumer model of FIP. Strangely enough, the ISP was abandoned in 1994, before reaching a mature state, for strategic reasons [40].

13.4.2 The International Fieldbus War

In 1994, after long years of struggle between German and French experts to combine the FIP and Profibus approaches, several, mainly American, companies decided to no longer watch the endless discussions. With the end of the ISP project, several former project members joined forces with the WorldFIP North America organization and formed the Fieldbus Foundation. This new association began the definition of a new fieldbus optimized for the process industry: the Foundation Fieldbus (FF). The work was done outside the IEC committees within the ISA, and for some time, the IEC work seemed to doze off.

Meanwhile in Europe, disillusion had run rampant [3]. Following the failure to find an acceptable IEC draft for a universal fieldbus, several players deemed it necessary to make a new start at least on a European level. Therefore, the CENELEC committee TC65CX was established in 1993 with the aim of finding an intermediate solution until an agreement was reached within IEC. By that time, the standardization issue had ceased to be a merely technical question. Fieldbus systems had already made their way into the

TABLE 13.2 Contents of the CENELEC Fieldbus Standards and Their Relation to IEC IS 61158

CENELEC Standards Part	Contained in IEC Standard	Brand Name
EN 50170-1 (July 1996)	IS 61158 type 4	P-Net
EN 50170-2 (July 1996)	IS 61158 type 1/3/10	Profibus
EN 50170-3 (July 1996)	IS 61158 type 1/7	WorldFIP
EN 50170-A1 (Apr. 2000)	IS 61158 type 1/9	Foundation Fieldbus
EN 50170-A2 (Apr. 2000)	IS 61158 type 1/3	Profibus-PA
EN 50170-A3 (Aug. 2000)	IS 61158 type 2	ControlNet
EN 50254-2 (Oct. 1998)	IS 61158 type 8	Interbus
EN 50254-3 (Oct. 1998)	(IS 61158 type 3)	Profibus-DP (Monomaster)
EN 50254-4 (Oct. 1998)	(IS 61158 type 7)	WorldFIP (FIPIO)
EN 50325-2 (Jan. 2000)	IS 62026-3 (2000)	DeviceNet
EN 50325-3 (Apr. 2000)	IS 62026-5 (2000)	SDS
EN 50325-4 (July 2002)		CANOpen
EN 50295-2 (Dec. 1998)	IS 62026-2 (2000)	AS-Interface

Note: The dates given in parentheses are the dates of ratification by the CENELEC Technical Board. The parenthetical IEC types denote that the respective fieldbus is contained in a superset definition.

market, much effort and enormous amounts of money had been invested in the development of protocols and devices, and there were already many installations. Nobody could afford to abandon a successful fieldbus; hence it was — from an economical point of view — impossible to start from scratch and create a unified but new standard that was incompatible with the established and widely used national ones.

The emerging market pressure was also a reason that within CENELEC no uniform fieldbus solution could be agreed upon. However, the national committees found after lengthy and controversial discussions [3] a remarkable and unprecedented compromise: all national standards under consideration were simply compiled "as is" to European standards [41]. Every part of such a multipart standard is a copy of the respective national standard, which means that every part is a fully functioning system. Although this approach is very pragmatic and seems easy to carry out once adopted, it took a long time to reach it. After all, with the strict European regulations about the mandatory application of standards, being part of it would ensure competitiveness for the respective system suppliers. As there were mostly representatives of the big players present in the committees, they naturally tried to optimize their own positions. Consequently, the contents of the individual CENELEC standards that were adopted step by step still reflect the strategic alliances that had to be formed by the national committees to get "their" standard into the European ones.

To make the CENELEC collection easier to handle, the various fieldbus systems were bundled according to their primary application areas. EN 50170 contains general-purpose field communication systems, EN 50254 has high-efficiency communication subsystems for small data packages, and EN 50325 is composed of different solutions based on the CAN technology. In the later phases of the European standardization process, the British national committee played the part of an advocate of the American companies and submitted also FF, DeviceNet, and ControlNet for inclusion in the European standards. Table 13.2 shows a compilation of all these standards, as well as their relation to the new IEC standard. For the sake of completeness, it should be noted that a comparable, though much less disputed, standardization process also took place for bus systems used in machine construction (dealt with by ISO), as well as building automation (in CEN and more recently in ISO).

While the Europeans were busy standardizing their national fieldbus systems and simply disregarded what happened in IEC, the Fieldbus Foundation prepared its own specification. This definition was modeled after the bus access scheme of FIP and the application layer protocol of the ISP work (which was in turn based on Profibus-FMS). The FF specification naturally influenced the work in the IEC committee, and consequently, the new draft evolved into a mixture of FF and WorldFIP. By several members of IEC TC65, this was seen as a reasonable compromise able to put an end to the lengthy debate.

However, when the draft was put to a vote in 1996, it was rejected by a very narrow margin, and the actual fieldbus war started. What had happened?

The *casus belli* was that Profibus (specifically the variant PA, which was named after the target application area, process automation, and which had been developed by the Profibus User Organization based on the ideas of the abandoned ISP project) was no longer properly represented in the IEC draft. When the majority of ISP members had teamed up with WorldFIP North America to form the Fieldbus Foundation, the main Profibus supporters had been left out in the rain. The fact that Profibus was already part of a CENELEC standard was no consolation. Given the strict European standardization rules and the Dresden Agreement, according to which international (i.e., IEC) standards supersede opposing CENELEC standards, the Profibus proponents feared that FF might gain a competitive advantage and "their" fieldbus might lose ground. Consequently, the countries where Profibus had a dominant position managed to organize an obstructive minority that prohibited the adoption of the standard. The fact that the IEC voting rules make it easier to cast positive votes (negative votes have to be justified technically) was no particular hindrance, as there were still many inconsistencies and flaws in the draft that could serve as a fig leaf.

The FF empire (as it was seen by the Profibus supporters) could not take this and struck back to save "their" standard. They launched an appeal to cancel negative votes that had, in their opinion, no sufficient technical justification. The minority of votes against the draft was very small, so the cancellation of a few votes would have been enough to turn the voting result upside down. Because this idea of using rather sophisticated legal arguments to achieve the desired goal was rather delicate, they proposed that the members of the IEC committee (i.e., the respective national mirror committees) should decide about the (non-)acceptance of the incriminated votes — a procedure that is not in conformance with the IEC rules and caused substantial exasperation. The discredited countries filed a complaint to the Committee of Action (CoA) of the IEC and asked it to resolve the situation. Owing to the infrequent meetings and rather formal procedures, the controversy sketched here carried on for several months.

In the meantime, a new draft had been prepared with most of the editorial errors removed. The main discussion point was again the data link layer draft. But now the question was whether the draft in its present form could really be implemented to yield a functioning fieldbus. The Profibus supporters claimed it was not possible, and they envisioned — especially in Europe — a dreary scenario of a nonfunctional IEC fieldbus standard replacing the market-proven European counterparts. The FF proponents maintained it was possible. Their argument was that the Foundation Fieldbus was implemented according to the draft and that products were already being sold. The debate waved to and fro, and Figure 13.4 tries to depict why it was so difficult to judge what was right. Over the years of development, several different versions of the data link layer specification had been submitted to the various standardization committees or implemented as products. Hence, both sides could find ample evidence for their claims.

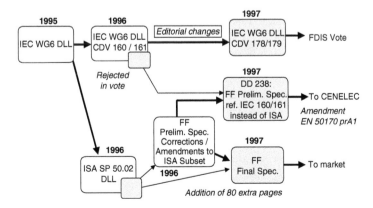

FIGURE 13.4 Evolution of the IEC 61158 data link layer and the Foundation Fieldbus (FF) demonstrating the various inconsistent flavors of the document.

In the course of subsequent voting processes, the battle raged and things grew worse. There were countries voting — both in favor and against — that had never cast a vote before or that according to their status in the IEC were not even allowed to vote. There were votes not being counted because they were received on a fax machine different from that designated at the IEC and thus considered late (because the error was allegedly discovered only after the submission deadline and it took several days to carry the vote to the room next door). Finally, there were rumors about presidents of national committees who high-handedly changed the conclusions of their committee experts. Throughout this entire hot phase of voting, the meetings of the national committees burst of representatives of leading companies trying to convince the committees of one or the other position. Never before or afterwards was the interest in fieldbus standardization so high, and never were the lobbying efforts so immense — including mobilization of the media, who had difficulties getting an objective overview of the situation [42]. The spiral kept turning faster and faster, but by and large, the obstruction of the standard draft remained unchanged, and the standardization process had degenerated to a playground for company tactics, to an economical and political battle that was apt to severely damage the reputation of standardization as a whole.

13.4.3   The Compromise

On June 15, 1999, the Committee of Action of the IEC decided to go a completely new way to break the stalemate. One month later, on July 16, the representatives of the main contenders in the debate (Fieldbus Foundation, Fisher Rosemount, ControlNet International, Rockwell Automation, Profibus User Organization, and Siemens) signed a "Memorandum of Understanding," which was intended to put an end to the fieldbus war. The Solomonic resolution was to create a large and comprehensive IEC 61158 standard accommodating all fieldbus systems — a move that left unhappy many of those who had been part of the IEC fieldbus project from the beginning [36, 43].

However, other than CENELEC, where complete specifications had been copied into the standard, the IEC decided to retain the original layer structure of the draft with physical, data-link, and application layers, each separated into services and protocols parts (Table 13.3). The individual fieldbus system specifications had to be adapted to so-called types to fit into this modular structure. In a great effort and under substantial time pressure, the draft was compiled and submitted for vote. The demand of the CoA was clear-cut: either this new draft would finally be accepted, or the old draft would be adopted without further discussion. Hence it was no wonder that the new document passed the vote, and the international fieldbus was released as a standard on the carefully chosen date of December 31, 2000.

It was evident that the collection of fieldbus specification modules in the IEC 61158 standard was useless for any practicable implementation. What was needed was a manual for the practical use showing which parts can be compiled to a functioning system and how this can be accomplished. This guideline was compiled later on as IEC 61784-1 as a definition of so-called communication profiles [44]. At the same time, the specifications of IEC 61158 were corrected and amended. The collection of profiles shows

TABLE 13.3   Structure of the IEC 61158 Fieldbus for Industrial Control Systems

Standards Part	Contents	Contents and Meaning
IEC 61158-1	Introduction	Only technical report
IEC 61158-2	PhL: Physical Layer	8 types of data transmission
IEC 61158-3	DLL: Data Link Layer Services	8 types
IEC 61158-4	DLL: Data Link Layer Protocols	8 types
IEC 61158-5	AL: Application Layer Services	10 types
IEC 61158-6	AL: Application Layer Protocols	10 types
IEC 61158-7	Network Management	Must be completely revised
IEC 61158-8	Conformance Testing	Work has been canceled

TABLE 13.4 Profiles and Protocols according to IEC 61784-1 and IEC 61158

IEC 61784 Profile	IEC 61158 Protocols			CENELEC Standard	Brand Names
	Phy	DLL	AL		
CPF-1/1	Type 1	Type 1	Type 9	EN 50170-A1 (Apr. 2000)	Foundation Fieldbus (H1)
CPF-1/2	Ethernet	TCP/UDP/IP	Type 5	—	Foundation Fieldbus (HSE)
CPF-1/3	Type 1	Type 1	Type 9	EN 50170-A1 (Apr. 2000)	Foundation Fieldbus (H2)
CPF-2/1	Type 2	Type 2	Type 2	EN 50170-A3 (Aug. 2000)	ControlNet
CPF-2/2	Ethernet	TCP/UDP/IP	Type 2	—	EtherNet/IP
CPF-3/1	Type 3	Type 3	Type 3	EN 50254-3 (Oct. 1998)	Profibus-DP
CPF-3/2	Type 1	Type 3	Type 3	EN 50170-A2 (Oct. 1998)	Profibus-PA
CPF-3/3	Ethernet	TCP/UDP/IP	Type 10	—	PROFInet
CPF-4/1	Type 4	Type 4	Type 4	EN 50170-1 (July 1996)	P-Net RS-485
CPF-4/2	Type 4	Type 4	Type 4	EN 50170-1 (July 1996)	P-Net RS-232
CPF-5/1	Type 1	Type 7	Type 7	EN 50170-3 (July 1996)	WorldFIP (MPS, MCS)
CPF-5/2	Type 1	Type 7	Type 7	EN 50170-3 (July 1996)	WorldFIP (MPS, MCS, SubMMS)
CPF-5/3	Type 1	Type 7	Type 7	EN 50170-3 (July 1996)	WorldFIP (MPS)
CPF-6/1	Type 8	Type 8	Type 8	EN 50254-2 (Oct. 1998)	Interbus
CPF-6/2	Type 8	Type 8	Type 8	EN 50254-2 (Oct. 1998)	Interbus TCP/IP
CPF-6/3	Type 8	Type 8	Type 8	EN 50254-2 (Oct. 1998)	Interbus subset
CPF-7/1	Type 6	Type 6	—	—	Swiftnet transport
CPF-7/2	Type 6	Type 6	Type 6	—	Swiftnet full stack

that the international fieldbus today consists of seven different main systems (communication profile families) that in turn can be subdivided (see Table 13.4). All important fieldbuses from industrial and process automation are listed here, and the world's biggest automation companies are represented with their developments.

Foundation Fieldbus consists of three profiles. The H1 bus is used in process automation, whereas high-speed Ethernet (HSE) is planned as an Ethernet backbone and for industrial automation. H2 is a remainder of the old draft. It allows for a migration of the WorldFIP solution toward FF, but in the profile description it is explicitly noted that there are no products available. From the Profibus side, the two profiles DP (decentralized periphery) and PA are present (even the new PROFInet has been included). Interestingly, the experts did not consider it worthwhile to list the original version of Profibus, the FMS, which is a strong sign for the diminishing importance, if not abandonment, of this hard-to-engineer fieldbus that is currently only contained in the EN 50170-2. The Danish fieldbus P-Net was taken over like all definitions and variants of WorldFIP and Interbus. In the latter case, the extensions for the tunneling of TCP/IP traffic have also been foreseen in the standard. A newcomer in the fieldbus arena is Swiftnet, which is widely used in airplane construction. The correct designation of an IEC fieldbus profile is shown for the example of Profibus-DP: compliance to IEC 61784 Ed.1:2002 CPF 3/1. Table 13.5 shows some technical characteristics and the main fields of application for the different systems. Low-level fieldbus systems for simple inputs/outputs (I/Os) such as the ones based on CAN or the AS-Interface are not part of IEC 61158; it is planned to combine them in IEC 62026.

13.5 Fieldbus Characteristics

The application areas of fieldbus systems are manifold; hence, many different solutions have been developed in the past. Nevertheless, there is one characteristic and common starting point for all those efforts. Fieldbus systems were always designed for efficiency, with two main aspects:

- Efficiency concerning data transfer, meaning that messages are rather short according to the limited size of process data that must be transmitted at a time
- Efficiency concerning protocol design and implementation, in the sense that typical field devices do not provide ample computing resources

TABLE 13.5 Technical Characteristics and Application Domains of the Different Profiles

Profile	Name	Industry	Special Features	Processing	Bus Access	Nodes per Segment
CPF-1/1	FF (H1)	Process	Function blocks	Centralized	Producer–consumer	Max. 32
CPF-1/2	FF (HSE)	Factory	for decentralized	Decentralized	with distributor	
		Process	control	Decentralized	CSMA/CD	Max. 30
CPF-1/3	FF (H2)	Factory		Centralized	Producer–consumer	Max. 32
				Decentralized	with distributor	
CPF-2/1	ControlNet	Factory	Optimized for	Centralized	Producer–consumer	Max. 99
CPF-2/2	EtherNet/IP	Factory	factory applications			Max. 30
CPF-3/1	Profibus-DP	Factory	Optimized for	Centralized	Master–slave with	Max. 126
			remote I/O		token passing	Max. 32
CPF-3/2	Profibus-PA	Process	Optimized for	Centralized		
			process control			
CPF-3/3	PROFInet	Factory	Distributed	Decentralized	Producer–consumer	Max. 30
			automation objects			
CPF-4/1	P-Net RS-485	Factory	Multinet capability	Centralized	Master–slave with	Max. 32
CPF-4/2	P-Net RS-232	Shipbuilding			token passing	
CPF-5/1	WorldFIP	Factory	Distributed real-time	Centralized	Producer–consumer	Max. 256
CPF-5/2			database	Decentralized	with distributor	
CPF-5/3						
CPF-6/1	Interbus	Factory	Optimized for	Centralized	Single master with	Max. 256
CPF-6/2	Interbus TCP/IP		remote I/O		synchronized shift	
CPF-6/3	Interbus Subset				register	
CPF-7/1	Swiftnet transport	Aircraft	Optimized for	Decentralized	Producer–consumer	Max. 1024
CPF-7/2	Swiftnet full stack		aircraft		with distributor	

These two aspects, together with characteristic application requirements in the individual areas with respect to real-time, topology, and economical constraints, have led to the development of concepts that still are very peculiar of fieldbus systems and present fundamental differences to LANs.

13.5.1 Communication Concepts

One difference to LANs concerns the protocol stack. Like all modern communication systems, fieldbus protocols are modeled according to the ISO/OSI model. However, normally only layers 1, 2, and 7 are actually used [14]. This is in fact a tribute to the lessons learned from the MAP failure, where it was found that a full seven-layer stack requires far too many resources and does not permit an efficient implementation. For this reason, the MiniMAP approach and, based on it, the IEC fieldbus standard explicitly prescribe a three-layer structure consisting of physical, data link, and application layers.

In most cases, this reduced protocol stack reflects the actual situation found in many automation applications anyway. Fieldbuses typically are single-segment networks, and extensions are realized via repeaters or, at most, bridges. Therefore, network and transport layers — which contain routing functionality and end-to-end control — are simply not necessary. If functions of these layers, as well as layers 5 and 6, are still needed, they are frequently included in layer 2 or 7. For the IEC 61158 fieldbus standard, the rule is that layer 3 and 4 functions can be placed in either layer 2 or layer 7, whereas layer 5 and 6 functionalities are always covered in layer 7 (Figure 13.5) [45]. In the building automation domain (LonWorks, EIB/KNX [European installation bus and its successor, Konnex], BacNet), the situation is different. Owing to the possibly high number of nodes, these fieldbus systems must offer the capability of hierarchically structured network topologies, and a reduction to three layers is not sensible.

For typical process control applications, determinism of data transfer is a key issue, and cycle time is a critical parameter. This fact has been the optimization criterion for many different fieldbus protocols and the reason that they are different from conventional LANs. Particularly the physical layer has to meet substantially more demanding requirements like robustness, immunity to electromagnetic disturbances,

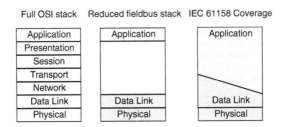

FIGURE 13.5 Layer structure of a typical fieldbus protocol stack as defined by IEC 61158.

intrinsic safety for hazardous areas, and costs. The significance of the physical layer is underpinned by the fact that this area was the first that reached (notably undisputed) consensus in standardization.

On the data link layer, all medium access strategies also known from LANs are used, plus many different subtypes and refinements. Simple master–slave polling (ASi, Profibus-DP) is used as well as token-based mechanisms in either explicit (Profibus, WorldFIP) or implicit (P-Net) form. Carrier-sense multiple access (CSMA) is mostly used in a variant that tries to avoid collisions by either the dynamic adaptation of retry waiting times (LonWorks) or the use of asymmetric signaling strategies (CAN, EIB). Especially for real-time applications, time-division multiple-access (TDMA)-based strategies are employed (TTP [time-triggered protocol], but also Interbus). In many cases, the lower two layers are implemented with application-specific integrated circuits (ASICs) for performance and cost reasons. As a side benefit, the preference of dedicated controllers over software implementations also improves interoperability of devices from different manufacturers.

An essential part of fieldbus protocol stacks is comprehensive application layers. They are indispensable for open systems and form the basis for interoperability. Powerful application layers offering abstract functionalities to the actual applications, however, require a substantial software implementation effort, which can negatively impact the protocol processing time and also the costs for a fieldbus interface. This is why in many cases (like Interbus or CAN) an application layer was originally omitted. While the application areas were often regarded as limited in the beginning, market pressure and the desire for flexibility finally enforced the addition of higher-layer protocols, and the growing performance of controller hardware facilitated their implementation.

Network management inside fieldbus protocols is traditionally not very highly developed. This stems from the fact that a fieldbus normally is not designed for the setup of large, complex networks. There are exceptions, especially in building automation, which consequently needs to provide more elaborated functions for the setup and maintenance of the network. In most cases, however, the flexibility and functionality of network management is adapted to the functionality and application area of the individual fieldbus. There are systems with comparatively simple (ASi, Interbus, P-Net, J1939) and rather complex management functions (Profibus-FMS, WorldFIP, CANopen, LonWorks, EIB). The latter are typically more flexible in their application range but need more efforts for configuration and commissioning. In any case, network management functions are normally not explicitly present (in addition to the protocol stack, as suggested by the OSI model), but rather included in the protocol layers (mostly the application layer).

13.5.2 Communication Paradigms

The characteristic properties of the various data types inside a fieldbus system differ strongly according to the processes that must be automated. Application areas like manufacturing, processing, and building automation pose different *timing* and *consistency* requirements that are not even invariant and consistent within the application areas [46]. Typical examples for different timing parameters are continuous measurement data that are sampled and transmitted in discrete-time fashion and form the basis for continuous process control and monitoring (like temperature, pressure, etc.). Other data are typically event based, i.e., they need transmission only in case of status changes (like switches, limit violations,

TABLE 13.6 Properties of Communication Paradigms

	Client–Server Model	Producer–Consumer Model	Publisher–Subscriber Model
Communication relation	Peer to peer	Broadcast	Multicast
Communication type	Connection oriented	Connectionless	Connectionless
Master–slave relation	Monomaster, multimaster	Multimaster	Multimaster
Communication service	Confirmed, unconfirmed, acknowledged	Unconfirmed, acknowledged	Unconfirmed, acknowledged
Application classes	Parameter transfer, cyclic communication	Event notification, alarms, error, synchronization	State changes, event-oriented signal sources (e.g., switches)

etc.). As far as consistency is concerned, there are on the one hand process data that are continuously updated and on the other hand parameterization data that are transferred only upon demand. In case of error, the former can easily be reconstructed from historical data via interpolation (or simply be updated by new measurements). The systemwide consistency of configuration data, on the other hand, is an important requirement that cannot be met by mechanisms suitable for process data.

These fundamental differences led to the evolution of several communication paradigms that are used either individually or in combination. The applicability in different fieldbus systems is quite different because they require various communication services and media access strategies. The three basic paradigms are:

- Client–server model
- Producer–consumer model
- Publisher–subscriber model

The most relevant properties of these three are summed up in Table 13.6.

The overview shows that processes with mostly event-based communication can get along very well with producer–consumer-type communication systems, especially if the requirements concerning dynamics are not too stringent. The obvious advantage is that all connected devices have direct access to the entire set of information since the broadcasting is based on identification of messages rather than nodes. Reaction times on events can be very short due to the absence of slow polling or token cycles. Generally, producer–consumer-type systems (or subsystems) are necessarily multimaster systems because every information source (producer) must have the possibility to access the bus. The selection of relevant communication relationships is solely based on message filtering at the consumer's side. Such filter tables are typically defined during the planning phase of an installation.

The publisher–subscriber paradigm uses very similar mechanisms; the only difference is that multicast communication services are employed. The subscribers are typically groups of nodes that listen to information sources (publishers). Relating publishers and subscribers can be done online. As both paradigms are message based and therefore connectionless on the application layer, they are not suited for the transmission of sensitive, nonrepetitive data such as parameter and configuration values or commands. Connectionless mechanisms can inform the respective nodes about communication errors on layer 2, but not about errors on the application layer.

The client–server paradigm avoids this problem by using connection-oriented information transfer between two nodes with all necessary control and recovery mechanisms. The communication transfer itself is based on confirmed services with appropriate service primitives (request, indication, response, confirm) as defined in the OSI model. Basically, a client–server-type communication can be implemented in both mono- and multimaster systems. In the latter cases (CSMA- and token-based systems) every master can take on the role of a client, whereas in monomaster systems (polling based) this position is reserved for the bus master. Consequently, the client–server paradigm is used mainly for monomaster systems as well as generally for discrete-time (cyclic) information transfer and for reliable data transfer on the application level (e.g., for parameterization data).

It is a characteristic feature of fieldbus systems that they do not adhere to single communication paradigms, but support a mix of strategies on different levels of sophistication. Examples for typical client–server systems are Interbus, Profibus, P-Net, and ASi. Broadcast services are here only used for special cases like synchronization purposes. Likewise, there are special ways of receiving messages (e.g., direct slave-to-slave communication) that require temporary delegation of certain bus master aspects. The other two paradigms are widely used in systems like CAN, CANopen, DeviceNet, ControlNet, EIB, and LonWorks. Yet, these systems also employ the client–server paradigm for special functions such as node configuration, file transfer, or the like.

13.5.3 Above the OSI Layers: Interoperability and Profiles

A key point for the acceptance of open fieldbus systems was the possibility to interconnect devices of different vendors. Multivendor systems and interoperability are still important arguments in fieldbus marketing. The standardization of fieldbuses was originally thought to be sufficient for interoperable systems, but reality quickly showed that it was not. Standards often leave room for interpretation, and implementations may vary, even if they conform to the standard. Certification of the devices is a suitable way to reduce the problems, but by no means a guarantee. Another reason for troubles is that the semantics of data objects are not precisely defined. This problem has been disregarded in many cases until recently. In fact, it is not a problem of the fieldbus itself, but of the application. Consequently, it must be tackled beyond the ISO/OSI model. The definition of appropriate *profiles* (or companion standards in MMS) addresses this problem.

The creation of profiles originated from the recognition that the definition of the protocol layers alone is not sufficient to allow for the implementation of interoperable products, because there are simply too many degrees of freedom. Therefore, profiles limit the top-level functionality and define specialized subsets for particular application areas [47]. Likewise, they specify communication objects, data types, and their encoding. So they can be seen as an additional layer on top of the ISO/OSI model, which is why they have also been called *layer 8* or *user layer*. One thing to be kept in mind is that nodes using them literally form islands on a fieldbus, which contradicts the philosophy of an integrated, decentralized system. Different profiles may coexist on one fieldbus, but communication between the device groups is normally very limited or impossible.

From a systematic viewpoint, profiles can be distinguished into communication, device, and branch profiles. A bus-specific *communication profile* defines the mapping of communication objects onto the services offered by the fieldbus. A *branch profile* specifies common definitions within an application area concerning terms, data types, and their coding and physical meaning. *Device profiles* build on communication and branch profiles and describe functionality, interfaces, and in general the behavior of entire device classes such as electric drives, hydraulic valves, and simple sensors and actuators.

The work of defining profiles is scattered among different groups. Communication profiles are usually in the hands of fieldbus user groups. They can provide the in-depth know-how of the manufacturers, which is indispensable for bus-specific definitions. Device and branch profiles are increasingly a topic for independent user groups. For them, the fieldbus is just a means to an end — the efficient communication between devices. What counts more in this respect is the finding and modeling of uniform device structures and parameters for a specific application. This forms the basis for a mapping to a communication system that is generic within a given application context.

The ultimate goal is the definition of fieldbus-independent device profiles [47]. This is an attempt to overcome on a high level the still overwhelming variety of systems. Finally, such profiles are also expected to facilitate the employment of fieldbus systems by the end user, who normally is only concerned about the overall functionality of a particular plant — and not about the question of which fieldbus to use. The methods used to define data types, indices, default values, coding and meanings, identification data, and device behavior are based on functional abstractions (most promising are currently function blocks [43, 48]) and universal modeling techniques [49]. A first step in the direction of fieldbus harmonization has been taken by the European research project NOAH (Network-Oriented Application Harmonization [48, 50]), the results of which are currently under standardization by IEC SC65C in project IEC 61804.

13.5.4 Management

Owing to the different capabilities and application areas of fieldbus systems, fieldbus management shows varying complexity and its solutions are more or less convenient for the user. It has already been stated above that the various fieldbuses offer a wide range of management services with grossly varying levels of sophistication. Apart from the functional boundary conditions given by the protocols, fieldbus management always strongly relies on the tool support provided by the manufacturers. This significantly adds to inhomogeneity of the fieldbus world in that entirely different control concepts, user interfaces, and implementation platforms are used. Furthermore, a strict division between communication and application aspects of fieldbus management is usually not drawn.

Typical *communication-related* management functions are bus parameter settings like address information, data rate, or timing parameters. These functions are rather low level and implicitly part of all fieldbus protocols. The user can access them via software tools mostly supplied by the device vendor. *Application-related* management functions concern the definition of communication relations, systemwide timing parameters (such as cycle times), priorities, or synchronization. The mechanisms and services offered by the fieldbus systems to support these functions are very diverse and should be integrated in the management framework for the application itself (e.g., the control system using the fieldbus).

As a matter of fact, a common management approach for various automation networks is still not available today, and vendor-specific solutions are preferred. From the users' point of view (which includes not only the end users, but also system integrators), this entails significantly increased costs for the buildup and maintenance of know-how because they must become acquainted with an unmanageable variety of solutions and tools. This situation actually revives one of the big acceptance problems that fieldbus systems originally had among the community of users: the missing interoperability. Communication interoperability (as ensured by the fieldbus standards) is a necessary but not sufficient precondition. For the user, handling interoperability of devices from different vendors is equally important. What is needed are harmonized concepts for configuration and management tools. As long as such concepts do not exist, fieldbus installations will typically be single-vendor systems, which is naturally a preferable situation for the manufacturers to secure their market position.

With the increasing importance of LAN and Internet technologies in automation, new approaches for fieldbus management appeared that may be apt to introduce at least a common view at various fieldbuses. All these concepts aim at integrating fieldbus management into existing management applications of the higher-level network, which is nowadays typically IP based. One commonly employed high-level network management protocol is the Simple Network Management Protocol (SNMP), which can also be used to access fieldbus data points [51, 52]. Another approach involves the use of Directory Services [53]. These two solutions permit the inclusion of a large number of devices in specialized network management frameworks. An alternative that has become very popular is the use of Web technology, specifically HTTP tunneled over the fieldbus, to control device parameters. This trend is supported by the increasing availability of embedded Web servers and the use of Extensible Markup Language (XML) as a device description language [54]. The appealing feature of this solution is that no special tools are required and a standard Web browser is sufficient. However, Web pages are less suitable for the management of complete networks and rather limited to single-device management. Nevertheless, this approach is meanwhile pursued by many manufacturers.

13.6 New Challenges: Industrial Ethernet

As stated before, Ethernet has become increasingly popular in automation. And like in the early days of fieldbus systems, this boom is driven mainly by the industry — on an academic level, the use of Ethernet had been discussed decades ago. Hence, the initial situation is comparable to that of 15 years ago, and there is enough conflict potential in the various approaches to use Ethernet in automation. After all, a

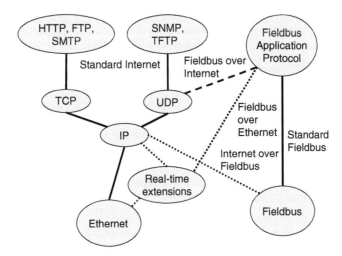

FIGURE 13.6 Structures of Ethernet and fieldbus combinations.

key argument for the introduction of Ethernet was its dominating role in the office world and the resulting status of a uniform network solution. It was exactly this picture of uniqueness that marketing campaigns tried to project also onto the automation world: Ethernet as *the* single, consistent network for all aspects.

A quick look at reality, however, shows that things are different. Ethernet per se is but a solution for the two lower OSI layers, and as fieldbus history already showed, this is not sufficient. Even if the commonly used Internet protocol suite with TCP (Transport Control Protocol) and UDP (User Datagram Protocol) is taken into account, only the lower four layers are covered. Consequently, there are several possibilities to get Ethernet or Internet technologies into the fieldbus domain, all of which are actually used in practice (Figure 13.6):

- Tunneling of a fieldbus protocol over UDP/TCP/IP
- Definition of new real-time-enabled protocols
- Reduction of the free medium access in standard Ethernet
- Tunneling of TCP/IP over an existing fieldbus

The future role of Ethernet in the automation area is not clear. Initially, Ethernet was considered inappropriate because of its lack of real-time capabilities. With the introduction of switched Ethernet and certain modifications of the protocol, however, these problems have been alleviated. And even if there are still doubts about the predictability of Ethernet [55], its penetration into the real-time domain will influence the use of fieldbus-based devices and most likely restrict the future use of fieldbus concepts [56]. Today, Ethernet already takes the place of midlevel fieldbus systems, e.g., for the connection of PLCs. There exist first applications in manufacturing and building automation where no other fieldbuses are installed but Ethernet.

To replace the existing lower-level fieldbuses by Ethernet and TCP/UDP/IP, more efforts are needed. One critical issue is (hard) real time, and there exist already different solutions to make Ethernet and TCP/IP meet the requirements of industrial applications [57]. One step below, on the sensor–actuator level, cost and implementation complexity are the most important factors. At the moment, fieldbus connection circuits for simple devices, often only one ASIC, are still cheaper than Ethernet connections. However, with modifications and simplifications of the controller hardware and the protocol implementations, Ethernet could finally catch up and become an interesting option.

13.6.1 Ethernet in IEC 61158

Only recently has standardization begun to deal with the question of Industrial Ethernet. Still, in the wake of the fieldbus wars, several solutions based on Ethernet and TCP/UDP/IP have made their way into the IEC 61158 standard without much fighting (see also Table 13.4):

- High-speed Ethernet (HSE) of the Foundation Fieldbus
- EtherNet/IP of ControlNet and DeviceNet
- PROFInet defined by Profibus International
- TCP/IP over Interbus

HSE and EtherNet/IP (note that here IP stands for Industrial Protocol) are two solutions with a fieldbus protocol being tunneled over TCP/IP. To be specific, it is no real tunneling, where data packets of a lower fieldbus OSI layer are wrapped in a higher-layer protocol of the transport medium. Instead, the same application layer protocol, which is already defined for the fieldbus, is also used over the TCP/IP or UDP/IP stack. In the case of ControlNet and DeviceNet, this is the Control and Information Protocol [58]. This solution allows the device manufacturers to base their developments on existing and well-known protocols. The implementation is without any risk and can be done fast.

The idea behind PROFInet is more in the direction of implementing a new protocol. For the actual communication, however, it was decided to use the component object model (COM)/distributed component object model (DCOM) mechanism known from the Windows world. This solution opens a wide possibility of interactions with the office IT software available on the market. The possibility to use fieldbus devices like objects in office applications will increase the vertical connectivity. On the other hand, this also includes the risk of other applications overloading the network, which has to be avoided. Basically, the COM/DCOM model defines an interface to use modules as black boxes within other applications. PROFInet offers a collection of automation objects with COM interfaces independent of the internal structure of the device. So the devices can be virtual, and the so-called proxy servers can represent the interfaces of any underlying fieldbus. This encapsulation enables the user to apply different implementations from different vendors. The only thing the user has to know is the structure of the interface. Provided the interfaces of two devices are equal, the devices are at least theoretically interchangeable. Although this proxy mechanism allows the connection of the Ethernet to all types of fieldbus systems, it will not be a simple and real-time-capable solution. A second problem is that in order to achieve portability, the COM/DCOM mechanism has to be reprogrammed for different operating systems. DCOM is tightly connected to the security mechanisms of Windows NT, but there is also the possibility of using WIN95/98 systems or — with restrictions — some UNIX systems. To simplify this, the PROFInet runtime system includes the COM/DCOM functionality, and the standard COM/DCOM functions inside the operating system have to be switched off if PROFInet is used.

The solution of tunneling TCP/IP over a fieldbus requires some minimum performance in terms of throughput from the fieldbus to be acceptable. Normally, throughput of acyclic data (the transport mechanism preferably used in this case) is not the strongest point of fieldbus systems. Nevertheless, Interbus defines the tunneling of TCP/IP over its acyclic communication channel [59]. The benefit of this solution is the parameterization of devices connected to the fieldbus with standard Internet services and well-known tools, e.g., a Web browser. This approach opens the possibility of achieving a new quality of user interaction, as well as a simpler integration of fieldbus management into existing high-level systems. On the downside, however, it forces the manufacturer of the field device to also implement the complete TCP/IP stack, maybe together with a Web server, on the device and the installation personnel to handle the configuration of the IP addressing parameters.

13.6.2   Real-Time Industrial Ethernet

The Industrial Ethernet solutions discussed so far build on Ethernet in its original form, i.e., they use the physical and data link layers of ISO/IEC 8802-3 without any modifications. Furthermore, they assume that Ethernet is low loaded or Fast Ethernet switching technology is used, in order to get a predictable performance. Switching technology does eliminate collisions, but delays inside the switches and lost packages under heavy load conditions are unavoidable with switches [60]. This gets worse if switches are used in a multilevel hierarchy and may result in grossly varying communication delays. The real-time capabilities of native Ethernet are therefore limited and must rely on application-level mechanisms controlling the data throughput. For advanced requirements, like drive controls, this is not sufficient. These known limitations of conventional Ethernet stimulated the development of several alternative solutions that were more than just adaptations of ordinary fieldbus systems. These entirely new approaches were originally outside the IEC standardization process, but are now candidates for inclusion in the real-time Ethernet (RTE) standard, i.e., the second volume of IEC 61784.

The initial and boundary conditions for the standardization work, which started in 2003, are targeted at backward compatibility with existing standards. First of all, RTE is seen as an extension to the Industrial Ethernet solutions already defined in the communication profile families in IEC 61784-1. Furthermore, coexistence with conventional Ethernet is intended. The scope of the working document [61] states that "the RTE shall not change the overall behavior of an ISO/IEC 8802-3 communication network and their related network components or IEEE 1588, but amend those widely used standards for RTE behaviors. Regular ISO/IEC 8802-3 based applications shall be able to run in parallel to RTE in the same network." Reference to the time distribution standard IEEE 1588 [62] is made because it will be the basis for the synchronization of field devices.

The work program of the RTE working group essentially consists of the definition of a classification scheme with RTE performance classes based on actual application requirements [63]. This is a response to market needs that demand scalable solutions for different application domains. One possible classification structure could be based on the reaction time of typical applications in automation:

- A first low-speed class with reaction times around 100 ms. This timing requirement is typical for the case of humans involved in the system observation (10 pictures per second can already be seen as a low-quality movie), for engineering, and for process monitoring. Most processes in process automation and building control fall into this class. This requirement may be fulfilled with a standard system with a TCP/IP communication channel without many problems.
- In a second class the requirement is a reaction time below 10 ms. This is the requirement for most tooling machine control systems like PLCs or PC-based control. To reach this timing behavior, special care has to be taken in the RTE equipment: sufficient computing resources are needed to handle TCP/IP in real-time or the protocol stack must be simplified and reduced to get these reaction times on simple, cheap resources.
- The third and most demanding class is defined by the requirements of motion control: to synchronize several axes over a network, a time precision well below 1 ms is needed. Current approaches to reach this goal rely on modifications of both protocol medium access and hardware structure of the controllers.

These classes will then be the building blocks for additional communication profiles. The intended structural resemblance to the fieldbus profiles is manifested by the fact that the originally attributed document number IEC 62391 was changed to IEC 61784-2. The technological basis for the development will mostly be switched Ethernet.

At the moment there are several systems that have the potential to fulfill at least parts of such an RTE specification and that are already introduced on the market or will be shortly. From these systems, three are extensions to fieldbuses already contained in IEC 61784:

EtherNet/IP: Defined by Rockwell and supported by Open DeviceNet Vendor Association (ODVA) and ControlNet International, EtherNet/IP makes use of the Common Industrial Protocol (CIP), which is common to the networks EtherNet/IP, ControlNet, and DeviceNet. CIP defines objects and their relations in different profiles and fulfills the requirements of class 1 on EtherNet/IP. As such, it is part of IEC 61784-1. With the CIP Sync extensions it is possible to get isochronous communication that satisfies class 2 applications. These extensions use 100 MBit/s networks with the help of IEEE 1588 time synchronization.

PROFInet: Defined mainly by Siemens and supported by Profibus International. Only the first version is currently included in the international fieldbus standard. A second step was the definition of

TABLE 13.7 Industrial Ethernet Profiles Defined in
IEC 61784

IEC 61784 Profile	Volume	Brand Names
CPF-1	1	Foundation Fieldbus
CPF-2	1, 2	EtherNet/IP
CPF-3	1, 2	PROFInet
CPF-6	1, 2	Interbus
CPF-10	2	VNET/IP
CPF-11	2	TCnet
CPF-12	2	EtherCAT
CPF-13	2	EPL (Ethernet Powerlink)
CPF-14	2	EPA
CPF-15	2	Modbus

a soft real-time (SRT) solution for PROFInet IO. In this version class 2 performance is also reached for small and cheap systems by eliminating the TCP/IP stack for process data. I/O data are directly packed into the Ethernet frame with a specialized protocol. Class 3 communication is reached with a special switch ASIC with a short and stable cut-through time and special priority mechanism for real-time data [64]. Synchronization is based on an extension of IEEE 1588 using on-the-fly time stamping, an idea that has been introduced in a different context [65]. The first application planned for PROFInet isochronous real time (IRT) is the PROFIdrive profile for motion control applications.

Interbus: Will also have an RTE extension, which will be identical to PROFInet. Still, it will be listed as a separate profile.

Apart from these approaches that merely extend well-known fieldbus systems, there is a multitude of new concepts collected in IEC 61784-2 (Table 13.7), not all of which were known in detail at the time of this writing:

VNET/IP: Developed by Yokogawa. The real-time extension of this protocol is called RTP (Real-Time and Reliable Datagram Protocol). Like many others, it uses UDP as a transport layer. Characteristic for the approach are an optimized IP stack (with respect to processing times) and a concept for redundant network connections.

TCnet: A proposal from Toshiba. Here, the real-time extension is positioned in the medium access control (MAC) layer. Also, a dual redundant network connection is proposed, based on shared Ethernet.

EtherCAT: Defined by Beckhoff and supported by the Ethercat Technology Group (ETG), EtherCAT uses the Ethernet frames and sends them in a special ring topology [66]. Every station in the net removes and adds its information. This information may be special input/output data or standard TCP/IP frames. To realize such a device, a special ASIC is needed for medium access that basically integrates a two-port switch into the actual device. The performance of this system is very good: it may reach cycle times of 30 μs.

Powerlink: Defined by B&R and now supported by the Ethernet Powerlink Standardization Group (EPSG). It is based on the principle of using a master–slave scheduling system on top of a regular shared Ethernet segment [67]. The master ensures the real-time access to the cyclic data and lets standard TCP/IP frames pass through only in specific time slots. To connect several segments, a synchronization based on IEEE 1588 is used. This solution is the only product available on the market that already fulfills the class 3 requirements today. In the future, the CANopen drive profiles will be supported.

EPA (Ethernet for Process Automation) protocol: A Chinese proposal. It is a distributed approach to realize deterministic communication based on a time-slicing mechanism.

Modbus/TCP: Defined by Schneider Electric and supported by Modbus-IDA,* Modbus/TCP uses the well-known Modbus over a TCP/IP network. This is probably the most widely used Ethernet solution in industrial applications today and fulfills the class 1 requirements without problems. Modbus/TCP was — contrary to all other fieldbus protocols — submitted to Internet Engineering Task Force (IETF) for standardization as an RFC (request for comments) [68]. The real-time extensions use the Real-Time Publisher–Subscriber (RTPS) protocol, which runs on top of UDP.

Originally outside the IEC SC65C was *SERCOS*, well known for its optical ring interface used in drive control applications. SERCOS III, also an Ethernet-based solution, is under development [69]. The ring structure is kept and the framing replaced by Ethernet frames to allow easy mixture of real-time data with TCP/IP frames. In every device a special software or, for higher performance, an application-specific integrated circuit will be needed that separates the real-time time slot from the TCP/IP time slot with a switch function. Recently, cooperation between the committee working on SERCOS and SC65C has been established to integrate SERCOS in the RTE standard.

The recent activities of IEC SC65C show that there is a substantial interest, especially from industry, in the standardization of real-time Ethernet. This situation closely resembles fieldbus standardization at the beginning of the 1990s, which ultimately led to the fieldbus wars. Given the comparable initial situation, will history repeat itself? Most likely not, because the structure of the intended standard documents already anticipates a multipart solution. So, the compromise that in former days needed so long to be found is already foreseen this time. Furthermore, the big automation vendors have learned their lessons to allow them to avoid time- and resource-consuming struggles that eventually end up in compromises anyway. Finally, the IEC itself cannot afford a new standardization war that would damage its image. Hence, all parties involved should have sufficient interest for the standardization process to be smooth and fast without too much noise inside the committees. Another evidence for this attitude is that the CENELEC committee TC65CX explicitly decided not to carry out standardization on the European level, but to wait for the outcome of the IEC work. The final standard is expected in 2007.

13.7 Aspects for Future Evolution

Even though fieldbus systems have reached a mature state, applications have become more demanding, which in turn creates new problems. Much work is still being done to improve the fieldbus itself, in particular concerning transmission speed and the large area of real-time capabilities [46, 70]. Another subject receiving considerable attention is the extension of fieldbuses to wireless physical layers [71, 72]. Apart from such low-level aspects, other problems are lurking on the system and application levels.

13.7.1 Driving Forces

Historically, the most important driving forces behind the development of fieldbus systems were the reduction of cabling and the desire to integrate more intelligence into the field devices. At least in Europe, the general need for automation, of which fieldbus systems are an integral part, also had a socioeconomic reason. Raising production costs due to comparatively high wages required a higher degree of automation to stay competitive in an increasingly globalized market. The enabling technology for automation was, of course, microelectronics. Without the availability of highly integrated controllers, the development of fieldbus systems would have never been possible.

---

*IDA, Interface for Distributed Automation, a consortium that originally worked on an independent solution, but finally merged with Modbus.

Today's driving forces for further evolution mainly come from the application fields that will be reviewed. Nevertheless, there are also technology push factors that promote the application of new technologies, mainly at the lower layers of communication (e.g., Ethernet). It must not be overlooked, however, that these factors are to a certain extent marketing driven and aim at the development of new market segments or the redistribution of already existing ones.

One important factor is what has recently become known as *vertical integration*. It concerns the possibly seamless interconnection between the traditional fieldbus islands and higher-level networks. The driving force behind this development is that people have become used to the possibility of accessing any information at any time over the Internet. Computer networks in the office area have reached a high level of maturity. Moreover, they are (quasi) standards that permitted worldwide interconnectivity and — even more important — easy access and use for nonspecialists. Hence, it is not astonishing that the anytime–anywhere concept is also extended to fieldbuses and automation systems in general. A common solution today is to have the coexistence of real-time fieldbus traffic and not time-critical tasks like configuration and parameterization based on, e.g., user-friendly Web-based services on the same communication medium. This becomes possible by the use of embedded Web servers in the field devices and the tunneling of TCP/IP over the fieldbus. Other approaches employ gateways to translate between the two worlds. In the near future, the increased use of Ethernet on the field level is supposed to further alleviate network integration, even though it will not be able to solve all problems.

Another driving force for the development of new concepts comes from the area of building automation. Although networks in this field emerged relatively late compared with industrial automation, the benefits are evident: the operating costs of a building can be reduced dramatically, if information about the status of the building is available for control purposes. This concerns primarily the energy consumption, but also service and maintenance costs. Energy control is a particularly interesting topic. Provided electrical appliances are interconnected via a fieldbus, they can adjust their energy consumption so as to balance the overall load [73, 74]. This demand-side management avoids peak loads, which in turn is honored by the utility companies with lower energy prices.

Even more important will be the combination of fieldbuses in buildings (and also private homes) with Internet connections. This is a particular aspect of vertical integration and opens a window for entirely new services [75]. External companies could offer monitoring and surveillance services for private houses while the owners are on vacation. Currently, such services already exist, but are limited to company customers (mostly within the context of facility management). A very important topic for utility companies in many countries is remote access to energy meters [76]. Having an appropriate communication link, they can more precisely and with finer granularity monitor the actual energy consumption of their customers, detect possible losses in the network, and better adapt their own productions and distributions. As a side benefit, billing can be automated and tariffs can be made more flexible when load profiles can be recorded. Eventually, if the energy meters support prepayment, even billing is no longer necessary.

An application field that is becoming increasingly relevant for networks is safety-relevant systems. As this domain is subject to very stringent normative regulations, and thus very conservative, it was dominated for a long time (and still is) by point-to-point connections between devices. The first bus system to penetrate this field was the CAN-based safety bus [77]. It took a long time and much effort for this system to pass the costly certification procedures. Nevertheless, it was finally accepted by the users, which was by no means obvious in an area concerned with the protection of human life, given that computer networks usually have the psychological disadvantage of being considered unreliable. After this pioneering work, other approaches like the ProfiSafe profile [78], Interbus safety [79], ASi safety [80], and recently EtherNet/IP safety [81] and WorldFIP [82] readily followed. The next big step is just ahead in car manufacturing, where in-vehicle networks in general and x-by-wire technology in particular will become determining factors [83]. Here, safety is of even more obvious relevance, and the latest developments of fieldbus systems for automotive use clearly address this issue. In the current Industrial Ethernet standardization process, safety considerations also play an important role.

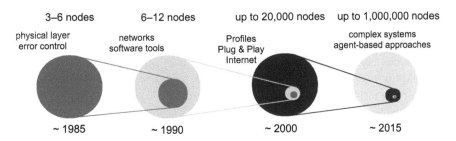

FIGURE 13.7 With increasing complexity of fieldbus installations, the important topics in research and practice change.

Microelectronics will continue to be the primary enabling technology for automation networks. Increasing miniaturization and the possibility to integrate more and more computing power while at the same time reducing energy consumption will be the prerequisite for further evolution. Today, system-on-a-chip (SoC) integration of a complete industrial PC with Ethernet controller, on-chip memory, and a complete IP stack as firmware is available. Of course, the computing resources of such integrated solutions cannot be compared with high-end PCs, but they are sufficient for smart and low-cost sensors and actuators. This evolution is, on the one hand, the foundation of the current boom of Ethernet in automation. On the other hand, it will stimulate more research in the emerging field of sensor networks [84]. Currently, most of the effort in this area is being put into wireless networking approaches, but it can be expected that work on other aspects will gain importance in the future. From an application point of view, other emerging fields like ubiquitous computing [85] or concepts inspired by bionics [86] will also rely on low-level networking as an essential technological cornerstone.

13.7.2 System Complexity

If we consider the evolution of fieldbus systems, we observe a very interesting aspect. Until the mid-1990s, the developers of fieldbus systems concentrated on the definition of efficient protocols. Since the computing resources in the field devices were limited and the developers did not expect fieldbuses to have a complex network structure, most protocols only use the lower two or three layers and the top layer of the OSI/ISO model. In those days, typical applications in industrial automation had only about six nodes on average, so the assumption of not-so-complex structures was justified. With the availability of more fieldbus devices and a growing acceptance of the technology, the number of nodes in a typical installation has also increased. A decade ago, the average application in industrial automation had 6 to 12 nodes.

With time, however, it turned out that the main costs of fieldbus systems were determined not so much by the development of the nodes, but rather by the maintenance of the node software, as well as the software tools necessary to integrate and configure the network. Actually, the development of a fieldbus system means much more than just designing a clever protocol and implementing a few nodes — an aspect that was often underrated in the past. More important for the success of a fieldbus is the fact that a user-friendly configuration and operating environment is available. This was, by the way, a strong argument in favor of open systems, where the development of field devices and software tools can be accomplished by different companies. For proprietary systems, by contrast, the inventor must supply both devices and software, which is likely to overstrain a single company.

Today, the number of nodes per installation is increasing dramatically. The enormous numbers shown in Figure 13.7 are of course not found in industrial automation, but in the area of building automation, where installations with 20,000 or more nodes are nowadays feasible. This evolution goes hand in hand with the advances of sensor networks in general. If we extrapolate the experience from other fields of computer technology, we can try to sketch the future evolution: the prices of the nodes will fall, and at the same time the performance will increase, allowing for the integration of more and more intelligence into the individual node. This way, we can have complex networks with up to 1 million nodes working together. Such complex systems will be the challenge for the next decades.

It is evident that applications in such systems must be structured differently from today's approaches. What is required is a true distribution of the application. A promising concept is holonic systems that have been thoroughly investigated in manufacturing systems [87, 88]. A holonic system consists of distributed, autonomous units (holons) that cooperate to reach a global goal. In artificial intelligence, the same concept is better known as a multiagent system. Such agents could be an interesting way to cope with complex systems [89, 90]. The main problem, however, will be to provide tools that can support the user in creating the distributed application.

A problem directly connected with system complexity is installation and configuration support through some plug-and-play capability. The ultimate meaning here is that new nodes can be attached to an existing network and integrate themselves without further input from the user. Realistically, this will remain only an appealing vision, as the user will always have to define at least the semantics of the information flow (i.e., in the trivial case of building automation, which switch is associated with which lamp), but nodes will have to be much more supportive than they are today. To date, the concepts for plug-and-play or at least plug and participate are at a very early stage. There are exemplary solutions for the automatic configuration of Profibus-DP devices [91] based on a manager–agent model inspired by management protocols like SNMP or the management framework of the ISO/OSI model. Here, a manager controls the status of the fieldbus and initiates the start-up and commissioning of the system in cooperation with the agents on the individual devices. The necessary data are kept in a (distributed) management information base (MIB).

Service broker approaches, such as Jini [92], could also be a suitable approach to tackle the problem of plug-and-play. The goal of Jini is to make distributed resources in a client–server network accessible. The term *resource* has a very abstract meaning and is composed of both hardware and software. To locate the resources in the network, services offered, as well as service requests, are published by the nodes and matched by the service broker [93]. A problem of Jini is that it builds on the relatively complex programming language Java. Hence, all Jini-enabled devices need to have a Java Virtual Machine as an interpreter, which is rather computing intensive. Jini is well developed today; however, hardware support still does not exist, and the breakthrough in smart devices as originally intended is not in sight. Competing approaches like Universal Plug and Play (UPnP) are catching up, but it is also questionable whether they will be suitable for complex systems.

13.7.3 Software Tools and Management

The fieldbus as a simple means to communication is only one part of an automation system. Today, it is the part that is best understood and developed. What becomes increasingly a problem, especially with increasing complexity, is the support through software tools. Historically, such tools are provided by the fieldbus vendors or system integrators and are as diverse as the fieldbuses themselves. Moreover, there are different (and often inconsistent) tool sets for different aspects of the life cycle of a plant, like planning, configuration, commissioning, testing, and diagnosis or maintenance. Such tools typically support only a topological view on the installation, whereas modern complex systems would require rather functionality-oriented, abstract views. A major disadvantage of the tool variety is that they operate in many cases on incompatible databases, which hampers system integration and is likely to produce consistency problems. More advanced concepts build on unified data sets that present consistent views to the individual tools with well-defined interfaces [94, 95]. The data structures are nevertheless still specific for each fieldbus. Unification of the data representations is one of the goals of NOAH [50].

For fieldbus-independent access to the field devices and their data (not necessarily covering the entire life cycle), several solutions have been proposed. They mostly rely on a sort of middleware abstraction layer using object-oriented models. Examples are OPC (OLE for Process Control) [96], Java, and other concepts [97]. Such platforms can ultimately be extended through definition of suitable application frameworks that permit the embedding of generic or proprietary software components in a unified environment spanning all phases of the life cycle. Relevant approaches are, e.g., Open Control [95], Field Device Tool [98], and a universal framework of the ISO [99].

Beyond pure communication management, in the application domain, essential aspects of engineering and management are also not yet universally solved. The ample computing resources of modern field devices, however, allow the introduction of new and largely fieldbus-independent concepts for the modeling of applications. A promising development are function blocks, standardized in IEC 61499 [100]. Historically evolved as an extension to the PLC programming standard IEC 61131, they can be used to create a functional view (rather than a topological one) on distributed applications. The function block concept integrates the models known from PLCs in factory automation, as well as typical functions from process automation that are in many fieldbuses available as proprietary implementations. With its universal approach, it is also a good option for the implementation of fieldbus profiles.

In the context of management and operation frameworks, the unified description of device and system properties becomes of eminent importance. To this end, device description languages were introduced. The descriptions of the fieldbus components are mostly developed by the device manufacturer and are integral parts of the products. Alternatively, they are contained in libraries where they can be downloaded and parsed for further use. Over the years, several mutually incompatible languages and dialects were developed [101, 102]. This is not surprising, as device descriptions are the basis for effective installation and configuration support. Thus, they are a necessary condition for the already discussed plug-and-play concepts. In recent years, the diversity of description languages is being addressed by the increased usage of universal languages like XML [103, 104], which is also the basis for the electronic device description language (EDDL) standardized in IEC 61804 [105, 106].

13.7.4 Network Interconnection and Security

Security has never been a real issue in conventional fieldbus systems. This is understandable insofar as fieldbuses were originally conceived as closed, isolated systems, which raised no need for security concepts. In building automation, where networks are naturally larger and more complex, the situation is different, and at least rudimentary security mechanisms are supported [107]. In factory and process automation, things changed with the introduction of vertical integration and the interconnection of fieldbuses and office-type networks. In such an environment, security is an essential topic on all network levels. Given the lack of appropriate features on the fieldbus level, the development and application of security concepts is typically confined to the actual network interconnection [108, 109].

One important aspect is that popular firewalls are not sufficient to guarantee security. Likewise, encryption is no cure-all, albeit an important element of secure systems. To reach a meaningful security level, a thorough risk analysis is the first step. On this basis, a security strategy needs to be developed detailing all required measures, most of which are organizational in nature. In practice, one will face two major problems: (1) the additional computational effort for security functions on the field devices (e.g., for cryptographic function), which may contradict real-time demands; and (2) the logistical problem of distributing and managing the keys whose secrecy forms the basis of every security policy. Both problems can — to a certain extent — be tackled with the introduction of security tokens such as smart cards [107].

With the introduction of Ethernet in automation, a reconsideration of field-level security is also possible. This is facilitated by the fact that many Industrial Ethernet solutions use IP and the Internet transport protocols UDP and TCP on top of Ethernet, which means that standard security protocols like Transport Layer Security (TLS) [110] can be used. One should recognize, however, that there are other approaches that use proprietary protocols above Ethernet, and that Ethernet per se is not the layer where security features can be reasonably implemented.

The fact that automation networks do not have security features up to now is also reflected in the recent standardization work of IEC SC65C WG13. Unlike other working groups, where the aim of the members is to get concrete proposals of established systems into the standards, no ready-to-use proposals exist. Apart from general considerations, the work has to be started largely from scratch. There is, however, related work in other fields that is being considered:

- IEC 61508: Functional safety of electrical/electronic/programmable electronic safety-related systems, maintained by IEC/SC 65A. Functional safety is in principle covered by the work of WG 12, but the common understanding is that safety-related systems necessarily have security aspects.
- Work being done in IEC TC57/WG15: Power systems management and associated information exchange/data and communication security.
- ISO/IEC 17799: Code of Practice for Information Security Management.
- ISO/IEC 15408: Common Criteria for IT Security Evaluation.
- ISA SP99: Manufacturing and Control Systems Security. It can be expected that this U.S. activity will have significant influence on the WG 13 work.
- AGA/GTI 12: Cryptographic Protection of SCADA Communications.
- NIST PCSRF: Process Control Security Requirements Forum.

13.8 Conclusion and Outlook

Fieldbus systems have come a long way from the very first attempts of industrial networking to contemporary highly specialized automation networks. What is currently at hand — even after the selection process during the last decade — nearly fully covers the complete spectrum of possible applications. Nevertheless, there is enough evolution potential left [70, 86]. On the technological side, the communication medium itself allows further innovations. Up to now, the focus has been on wired links, twisted pair being the dominant solution. Optical media have been used comparatively early for large distances and electromagnetically disturbed environments. Recently, plastic optical fibers have reached a status of maturity that allows longer cable lengths and smaller prices. Another option, especially for building automation, is the use of electrical power distribution lines. This possibility, although tempting in principle, is still impaired by bad communication characteristics of the medium. Substantial research effort will be needed to overcome these limitations, which in fact comes down to a massive use of digital signal processing.

The most promising research field for technological evolution is the wireless domain. The benefits are obvious: no failure-prone and costly cabling and high flexibility, even mobility. The problems, on the other hand, are also obvious: very peculiar properties of the wireless communication channel must be dealt with, such as attenuation, fading, multipath reception, temporarily hidden nodes, and the simple access for intruders [71]. Wireless communication options do exist today for several fieldbuses [72]. Up to now, they have been used just to replace the conventional data cable. A really efficient use of wireless communication, however, would necessitate an entire redefinition of at least the lower fieldbus protocol layers. Evaluation of currently available wireless technologies from the computer world with respect to their applicability in automation is a first step in this direction. Ultimately we can expect completely new automation networks optimized for wireless communication, where maybe only the application layer protocol remains compatible with traditional wired solutions to achieve integration.

Apart from mere technological issues, the currently largest trend is the integration of fieldbus systems in higher-level, heterogeneous networks and process control systems. Internet technologies play a particularly prominent role here, and the penetration of the field level by optimized Ethernet solutions creates additional momentum. The ultimate goal is a simplification and possibly harmonization of fieldbus operation. For the fieldbus itself, this entails increasing complexity in the higher protocol levels. At the same time, more and more field-level applications employ standard PC-based environments and operating systems like Windows or Linux [111]. These two trends together result in a completely new structure of the automation hierarchy. The old multilevel pyramid finally turns into a rather flat structure with two, maybe three levels, as shown in Figure 13.8. Here, functions of the traditional middle layers (like process and cell levels) are transferred into the intelligent field devices (and thus distributed) or into the management level. The traditional levels may persist in the organizational structure of the company, but not in the technical infrastructure.

Does all this mean we have reached the end of the fieldbus era? The old CIM pyramid, which was a starting point for the goal-oriented development of fieldbus systems, ceases to exist, and Ethernet is

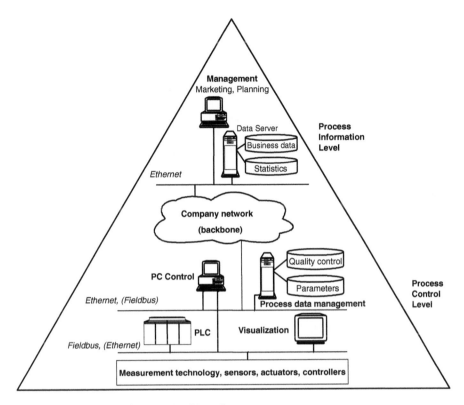

FIGURE 13.8 Flattened, two-level automation hierarchy.

determined to reach down into the field level. This may indeed be the end of the road for the traditional fieldbus as we know it, but certainly not for networking in automation. What we are likely to see in the future are Ethernet- and Internet-based concepts at all levels, probably optimized to meet special performance requirements on the field level but still compatible with the standards in the management area. Below, very close to the technical process, there will be room for highly specialized sensor–actuator networks — new fieldbus systems tailored to meet the demands of high flexibility, energy optimization, small-footprint implementation, or wireless communication. The next evolution step in fieldbus history is just ahead.

Acknowledgments

The author thanks Dietmar Dietrich, Kurt Milian, Eckehardt Klemm, Peter Neumann, and Jean-Pierre Thomesse for the extensive discussions, especially about the historical aspects of fieldbus systems.

References

1. International Electrotechnical Commission, IEC 61158, Digital Data Communications for Measurement and Control: Fieldbus for Use in Industrial Control Systems, 2003.
2. Fieldbus Foundation, What Is Fieldbus? http://www.fieldbus.org/About/FoundationTech/.
3. G.G. Wood, Fieldbus status 1995, *IEE Computing and Control Engineering Journal*, 6, 251–253, 1995.
4. G.G. Wood, Survey of LANs and standards, *Computer Standards and Interfaces*, 6, 27–36, 1987.
5. N.P. Mahalik (Ed.), *Fieldbus Technology: Industrial Network Standards for Real-Time Distributed Control*, Spinger, Heidelberg, 2003.
6. H. Töpfer, W. Kriesel, Zur funktionellen und strukturellen Weiterentwicklung der Automatisierungsanlagentechnik, *Messen Steuern Regeln*, 24, 183–188, 1981.

7. T. Pfeifer, K.-U. Heiler, Ziele und Anwendungen von Feldbussystemen, *Automatisierungstechnische Praxis*, 29, 549–557, 1987.
8. H. Steusloff, Zielsetzungen und Lösungsansätze für eine offene Kommunikation in der Feldebene, *Automatisierungstechnik*, 855, 337–357, 1990.
9. L. Capetta, A. Mella, F. Russo, Intelligent field devices: user expectations, *IEE Coll. on Fieldbus Devices: A Changing Future*, 6/1–6/4, 1994.
10. K. Wanser, Entwicklungen der Feldinstallation und ihre Beurteilung, *Automatisierungstechnische Praxis*, 27, 237–240, 1985.
11. J.A.H. Pfleger, Anforderungen an Feldmultiplexer, *Automatisierungstechnische Praxis*, 29, 205–209, 1987.
12. H. Junginger, H. Wehlan, Der Feldmultiplexer aus Anwendersicht, *Automatisierungstechnische Praxis*, 31, 557–564, 1989.
13. W. Schmieder, T. Tauchnitz, FuRIOS: fieldbus and remote I/O: a system comparison, *Automatisierungstechnische Praxis*, 44, 61–70, 2002.
14. P. Pleinevaux, J.-D. Decotignie, Time critical communication networks: field buses, *IEEE Network*, 2, 55–63, 1988.
15. E.H. Higham, Casting a crystal ball on the future of process instrumentation and process measurements, in *IEEE Instrumentation and Measurement Technology Conference (IMTC '92)*, New York, May 1992, pp. 687–691.
16. J.P. Thomesse, Fieldbuses and interoperability, *Control Engineering Practice*, 7, 81–94, 1999.
17. J.-C. Orsini, *Field Bus: A User Approach*, Cahier Technique Schneider Electric 197, 2000, http://www.schneider-electric.com.tr/ftp/literature/publications/ECT197.pdf.
18. R.D. Quick, S.L. Harper, HP-IL: A Low-Cost Digital Interface for Portable Applications, *Hewlett-Packard Journal*, January 1983, pp. 3–10.
19. Philips Semiconductor, *The I²C-Bus Specification*, 2000, http://www.semiconductors.philips.com/buses/i2c/.
20. H. Zimmermann, OSI reference model: the ISO model of architecture for open system interconnection, *IEEE Transactions on Communications*, 28, 425–432, 1980.
21. J. Day, H. Zimmermann, The OSI reference model, *Proceedings of the IEEE*, 71, 1334–1340, 1983.
22. D.J. Damsker, Asessment of industrial data network standards, *IEEE Trans. Energy Conversion*, 3, 199–204, 1988.
23. H.A. Schutz, The role of MAP in factory integration, *IEEE Transactions on Industrial Electronics*, 35, 6–12, 1988.
24. B. Armitage, G. Dunlop, D. Hutchison, S. Yu, Fieldbus: an emerging communications standard, *Microprocessors and Microsystems*, 12, 555–562, 1988.
25. S.G. Shanmugham, T.G. Beaumariage, C.A. Roberts, D.A. Rollier, Manufacturing communication: the MMS approach, *Computers and Industrial Engineering*, 28, 1–21, 1995.
26. T. Phinney, P. Brett, D. McGovan, Y. Kumeda, FieldBus: real-time comes to OSI, in *International Phoenix Conference on Computers and Communications*, March 1991, pp. 594–599.
27. K. Bender, Offene Kommunikation: Nutzen, Chancen, Perspektiven für die industrielle Kommunikation, in *iNet '92*, 1992, pp. 15–37.
28. T. Sauter and M. Felser, The importance of being competent: the role of competence centres in the fieldbus world, in *FeT '99 Fieldbus Technology*, Magdeburg, Germany, September 1999, pp. 299–306.
29. Gesmer Updegrove LLP, Government Issues and Policy, http://www.consortiuminfo.org/government/.
30. M.A. Smith, Vienna Agreement on Technical Cooperation between ISO and CEN, paper presented at *ISO/IEC Directives Seminar*, Geneva, June 1995, isotc.iso.ch/livelink/livelink/fetch/2000/2123/SDS_WEB/sds_dms/vienna.pdf.
31. International Electrotechnical Commission, IEC-CENELEC Agreement, http://www.iec.ch/about/partners/agreements/cenelec-e.htm.

32. E. Klemm, Der Weg durch die Gremien zur internationalen Feldbusnorm, paper presented at *VDE Seminar Die neue, internationale Feldbusnorm: Vorteile, Erfahrungen, Beispiele, Zukunft*, November 2002, Mannheim.
33. Instrument Society of America Standards and Practices 50, Draft Functional Guidelines, March 10, 1987, document ISA-SP50-1986-17-D.
34. G.G. Wood, Current fieldbus activities, *Computer Communications*, 11, 118–123, 1988.
35. C. Gilson, Digital Data Communications for Industrial Control Systems or How IEC 61158 (Just) Caught the Bus, paper presented at *IEC E-TECH*, March 2004, http://www.iec.ch/online_news/etech/arch_2004/etech_0304/focus.htm#fieldbus.
36. P. Leviti, IEC 61158: an offence to technicians?, in *IFAC International Conference on Fieldbus Systems and Their Applications, FeT 2001*, Nancy, France, November 15–16, 2001, p. 36.
37. T. Phinney, Mopping up from bus wars, *World Bus Journal*, 22–23, December 2001.
38. H. Engel, Feldbus-Normung 1990, *Automatisierungstechnische Praxis*, 32, 271–277, 1990.
39. H. Wölfel, Die Entwicklung der digitalen Prozeßleittechnik: Ein Rückblick (Teil 4), *Automatisierungstechnische Praxis*, 40, S25–S28, 1998.
40. J. Rathje, The fieldbus between dream and reality, *Automatisierungstechnische Praxis*, 39, 52–57, 1997.
41. G.H. Gürtler, Fieldbus standardization, the European approach and experiences, in *Feldbustechnik in Forschung, Entwicklung und Anwendung*, Springer, Heidelberg, 1997, pp. 2–11.
42. S. Bury, Are you on the right bus?, *Advanced Manufacturing*, 1, 26–30, 1999, http://www.advancedmanufacturing.com/October99/fieldbus.htm.
43. G.G. Wood, State of play, *IEE Review*, 46, 26–28, 2000.
44. International Electrotechnical Commission, IEC 61784-1, Digital Data Communications for Measurement and Control: Part 1: Profile Sets for Continuous and Discrete Manufacturing Relative to Fieldbus Use in Industrial Control Systems, 2003.
45. International Electrotechnical Commission, IEC 61158-1, Digital Data Communications for Measurement and Control: Fieldbus for Use in Industrial Control Systems: Part 1: Introduction, 2003.
46. J.-P. Thomesse, M. Leon Chavez, Main paradigms as a basis for current fieldbus concepts, in *Fieldbus Technology*, Springer, Heidelberg, 1999, pp. 2–15.
47. C. Diedrich, Profiles for fieldbuses: scope and dedcription technologies, in *Fieldbus Technology*, Springer, Heidelberg, 1999, pp. 90–97.
48. U. Döbrich, P. Noury, ESPRIT Project NOAH: Introduction, in *Fieldbus Technology*, Springer, Heidelberg, 1999, pp. 414–422.
49. R. Simon, P. Neumann, C. Diedrich, M. Riedl, Field devices-models and their realisations, in *IEEE International Conference on Industrial Technology (ICIT '02)*, Bangkok, December 2002, pp. 307–312.
50. A. di Stefano, L. Lo Bello, T. Bangemann, Harmonized and consistent data management in distributed automation systems: the NOAH approach, in *IEEE International Symposium on Industrial Electronics*, ISIE 2000, Cholula, Mexico, December 2000, pp. 766–771.
51. M. Knizak, M. Kunes, M. Manninger, T. Sauter, Applying Internet management standards to fieldbus systems, in *WFCS '97*, Barcelona, October 1997, pp. 309–315.
52. M. Kunes, T. Sauter, Fieldbus-Internet connectivity: the SNMP approach, *IEEE Transactions on Industrial Electronics*, 48, 1248–1256, 2001.
53. M. Wollschlaeger, Integration of VIGO into Directory Services, paper presented at *6th International P-NET Conference*, Vienna, May 1999.
54. M. Wollschlaeger, Framework for Web integration of factory communication systems, in *IEEE International Conference on Emerging Technologies and Factory Automation (ETFA)*, Antibes Juan-Les-Pins, France, October 2001, pp. 261–265.
55. J.D. Decotignie, A perspective on Ethernet-TCP/IP as a fieldbus, in *IFAC International Conference on Fieldbus Systems and Their Applications, FeT 2001*, Nancy, France, November 15–16, 2001, pp. 138–143.

56. E. Byres, Ethernet to Link Automation Hierarchy, *InTech Magazine*, June 1999, pp. 44–47.

57. M. Felser, Ethernet TCP/IP in automation, a short introduction to real-time requirements, in *Conference on Emerging Technologies and Factory Automation, ETFA 2001*, Antibes Juan-Les-Pins, France, October 15–18, 2001, pp. 501–504.

58. V. Schiffer, The CIP family of fieldbus protocols and its newest member: EtherNet/IP, in *Conference on Emerging Technologies and Factory Automation, ETFA 2001*, Antibes Juan-Les-Pins, France, October 15–18, 2001, pp. 377–384.

59. M. Volz, Quo Vadis Layer 7? *The Industrial Ethernet Book*, No. 5, Spring 2001.

60. K.C. Lee, S. Lee, Performance evaluation of switched Ethernet for real-time industrial communications, *Computer Standards and Interfaces*, 24, 411–423, 2002.

61. TC65/SC65C, New work item proposal, 65C/306/NP, 2003.

62. IEEE 1588, Standard for a Precision Clock Synchronization Protocol for Networked Measurement and Control Systems, 2002.

63. TC65/SC65C, Meeting minutes, 65C/318/INF, 2003.

64. A. Boller, Profinet V3: bringing hard real-time and the IT world together, *Control Engineering Europe*, September 2003, http://www.manufacturing.net/ctl/article/CA318939.

65. R. Höller, G. Gridling, M. Horauer, N. Kerö, U. Schmid, K. Schossmaier, SynUTC: high precision time synchronization over Ethernet networks, in *8th Workshop on Electronics for LHC Experiments (LECC)*, Colmar, France, September 9–13, 2002, pp. 428–432.

66. http://www.ethercat.org/.

67. http://www.ethernet-powerlink.com/.

68. Schneider Automation, Modbus Messaging on TCP/IP Implementation Guide, May 2002, http://www.modbus.org/.

69. E. Schemm, SERCOS to link with ethernet for its third generation, *IEE Computing and Control Engineering Journal*, 15, 30–33, 2004.

70. J.-D. Decotignie, Some future directions in fieldbus research and development, in *Fieldbus Technology*, Springer, Heidelberg, 1999, pp. 308–312.

71. L. Rauchhaupt, J. Hähniche, Opportunities and problems of wireless fieldbus extensions, in *Fieldbus Technology*, Springer, Heidelberg, 1999, pp. 308–312.

72. L. Rauchhaupt, System and device architecture of a radio based fieldbus: the RFieldbus system, in *IEEE Workshop on Factory Communication Systems*, Västerås, Sweden, 2002, pp. 185–192.

73. P. Palensky, Distributed Reactive Energy Management, Ph.D. thesis, Vienna University of Technology, Austria, 2001.

74. G. Gaderer, T. Sauter, Ch. Eckel, What it takes to make a refrigerator smart: a case study, in *IFAC International Conference on Fieldbus Systems and Their Applications (FeT)*, Aveiro, Portugal, July 2003, pp. 85–92.

75. L. Haddon, Home Automation: Research Issues, paper presented at EMTEL Workshop: The European Telecom User, Amsterdam, November 10–11, 1995.

76. M. Lobashov, G. Pratl, T. Sauter, Implications of power-line communication on distributed data acquisition and control systems, in *IEEE International Conference on Emerging Technologies and Factory Automation (ETFA)*, Lisboa, Portugal, September 2003, pp. 607–613.

77. R. Piggin, An introduction to safety-related networking, *IEE Computing and Control Engineering Journal*, 15, 34–39, 2004.

78. PROFIBUS International, Profile for Failsafe with PROFIBUS, DP-Profile for Safety Applications, Version 1.2, October 2002, http://www.profibus.com.

79. INTERBUS Club, INTERBUS Safety, White Paper, 2003.

80. http://as-i-safety.net.

81. ODVA, Safety Networks: Increase Productivity, Reduce Work-Related Accidents and Save Money, Open DeviceNet Vendor Assoc., White Paper, 2003, http://www.odva.org.

82. J.-P. Froidevaux, O. Nick, M. Suzan, Use of fieldbus in safety related systems, an evaluation of WorldFIP according to proven-in-use concept of IEC 61508, *WorldFIP News*, http://www. worldfip. org.

83. G. Leen, D. Heffernan, Expanding automotive electronic systems, *IEEE Computer*, 35, 88–93, 2002.

84. H. Gharavi, S.P. Kumar (Eds.), Special issue on sensor networks and applications, *Proceedings of the IEEE*, 91, 2003.

85. G. Borriello, Key challenges in communication for ubiquitous computing, *IEEE Communications Magazine*, 40, 16–18, 2002.

86. D. Dietrich, T. Sauter, Evolution potentials for fieldbus systems, in *Proceedings of the 3rd IEEE International Workshop on Factory Communication Systems*, Porto, 2000, pp. 343–350.

87. A. Koestler, *The Ghost in the Machine*, Arkana Books, London, 1967.

88. F. Pichler, On the construction of A. Koestler's holarchical networks, in *Cybernetics and Systems 2000*, Austrian Society for Cybernetic Systems, Vienna, 2000.

89. P. Palensky, The convergence of intelligent software agents and field area networks, in *1999 IEEE Conference on Emerging Technologies and Factory Automation*, Barcelona, 1999, pp. 917–922.

90. T. Wagner, An agent-oriented approach to industrial automation systems, in *Agent Technologies, Infrastructures*, R. Kowalczyk et al. (Eds.), Springer-Verlag, Berlin, 2003, pp. 314–328.

91. A. Pöschmann, P. Krogel, Autoconfiguration Management für Feldbusse: PROFIBUS Plug & Play, *Elektrotechnik und Informationstechnik*, 117, 5, 2000.

92. W. Kastner, M. Leupold, How dynamic networks work: a short tutorial on spontaneous networks, in *IEEE Conference on Emerging Technologies and Factory Automation, ETFA 2001*, Antibes Juan-Les-Pins, France, October 15–18, 2001, pp. 295–303.

93. S. Deter, Plug and participate for limited devices in the field of industrial automation, in *IEEE Conference on Emerging Technologies and Factory Automation, ETFA 2001*, Antibes Juan-Les-Pins, France, October 15–18, 2001, pp. 263–268.

94. O. Cramer Nielsen, A real time, object oriented fieldbus management system, in *3rd IEEE International Workshop on Factory Communication Systems*, Porto, 2000, pp. 335–340.

95. A. Baginski, G. Covarrubias, Open control: the standard for PC-based automation technology, in *IEEE International Workshop on Factory Communication Systems*, October 1997, pp. 329–333.

96. OPC Data Access Automation Specification, Version 2.0, OPC Foundation, October 14, 1998.

97. R. Bachmann, M.S. Hoang, P. Rieger, Component-based architecture for integrating fieldbus systems into distributed control applications, in *Fieldbus Technology*, Springer-Verlag, Heidelberg, 1999, pp. 276–283.

98. R. Simon, M. Riedl, C. Diedrich, Integration of field devices using field device tool (fdt) on the basis of electronic device descriptions (EDD), in *IEEE International Symposium on Industrial Electronics, ISIE '03*, June 9–11, Rio de Janeiro, 2003, pp. 189–194.

99. W.H. Moss, Report on ISO TC184/SC5/WG5 open systems application frameworks based on ISO 11898, in *5th International CAN Conference (iCC '98)*, San Jose, CA, 1998, pp. 07-02–07-04.

100. Function Blocks for Industrial-Process Measurement and Control Systems: Committee Draft, IEC TC65/WG6, ftp://ftp.cle.ab.com/stds/iec/sc65bwg7tf3/html/news.htm.

101. GSD Specification for PROFIBUS-FMS (version 1.0), PNO Karlsruhe.

102. Device Description Language specification, HART Communication Foundation, Austin, TX, 1995.

103. T. Bray, J. Paoli, C. M. Sperberg-McQueen, Extensible Markup Language (XML) 1.0, 1998, http://www.w3.org/TR/REC-xml.

104. M. Wollschlaeger, Descriptions of fieldbus components using XML, *Elektrotechnik und Informationstechnik*, 117, 5, 2000.

105. International Electrotechnical Commission, IEC 61804-2, Function Blocks (FB) for Process Control: Part 2: Specification of FB Concept and Electronic Device Description Language (EDDL), 2003.

106. P. Neumann, C. Diedrich, R. Simon, Engineering of field devices using device descriptions, paper presented at IFAC World Congress 2002, Barcelona, 2002.

107. C. Schwaiger, A. Treytl, Smart card based security for fieldbus systems, in *2003 IEEE Conference on Emerging Technologies and Factory Automation*, Lisbon, September 2003, pp. 398–406.

108. T. Sauter, Ch. Schwaiger, Achievement of secure Internet access to fieldbus systems, *Microprocessors and Microsystems*, 26, 331–339, 2002.

109. P. Palensky, T. Sauter, Security considerations for FAN-Internet connections, in *IEEE International Workshop on Factory Communication Systems*, Porto, September 2000, pp. 27–35.

110. E. Rescorla, *SSL and TLS*, Addison-Wesley, Reading, MA, 2000.

111. W. Kastner, C. Csebits, M. Mayer, Linux in factory automation? Internet controlling of fieldbus systems, in *1999 IEEE Conference on Emerging Technologies and Factory Automation*, Barcelona, 1999, pp. 27–31.

112. CAMAC, A Modular Instrumentation System for Data Handling, EUR4100e, March 1969.

113. http://www.hit.bme.hu/people/papay/edu/GPIB/tutor.htm.

114. National Instruments, GPIB Tutorial, www.raunvis.hi.is/~rol/Vefur/%E9r%20Instrupedia/CGP-TUTO.PDF.

115. W. Büsing, Datenkommunikation in der Leittechnik, *Automatisierungstechnische Praxis*, 28, 228–237, 1986.

116. G. Färber, *Bussysteme*, 2nd ed., Oldenbourg-Verlag, Munich, 1987.

117. M-Bus Usergroup, The M-Bus: A Documentation, Version 4.8, November 11, 1997, http://www.m-bus.com/mbusdoc/default.html.

118. G. Leen, D. Heffernan, A. Dunne, Digital networks in the automotive vehicle, *IEE Computer and Control Engineering Journal*, 10, 257–266, 1999.

119. CAN-in-Automation, CAN history, http://www.can-cia.de/can/protocol/history/.

120. Condor Engineering, MIL-STD-1553 tutorial, http://www.condoreng.com/support/downloads/tutorials/MIL-STD-1553Tutorial.PDF.

121. Grid Connect, The Fieldbus Comparison Chart, http://www.synergetic.com/compare.htm.

122. Interbus Club, Interbus Basics, 2001, http://www.interbusclub.com/en/doku/pdf/interbus_basics_en.pdf.

123. H. Kirrmann, Industrial Automation, lecture notes, EPFL, 2004, http://lamspeople.epfl.ch/ kirrmann/IA_slides.htm.

124. H. Wölfel, Die Entwicklung der digitalen Prozeßleittechnik: Ein Rückblick (Teil 3), *Automatisierungstechnische Praxis*, 40, S17–S24, 1998.

125. T. Sauter, D. Dietrich, W. Kastner (Eds.), *EIB Installation Bus System*, Publicis MCD, Erlangen, Germany, 2001.

126. E.B. Driscoll, The History of X10, http://home.planet.nl/~lhendrix/x10_history.htm.

Appendix

The tables presented here give an overview of selected fieldbus systems, categorized by application domain. The list is necessarily incomplete, although care has been taken to include all approaches that either exerted a substantial influence on the evolution of the entire field or are significant still today. The year of introduction refers to the public availability of the specification or first products. This year is also the one used in the timeline in Figure 13.3. Note that despite careful research, the information obtained from various sources was frequently inconsistent, so there may be an uncertainty in the figures. Where respective data could be obtained, the start of the project has been listed as well because there are several cases where much time elapsed between the start of development of the fieldbus and its first release.

TABLE 13.8 Instrumentation and PCB-Level Buses

Fieldbus	Developer (Country)	Introduced in	Standard	References
CAMAC	ESONE (Europe)	1969 (start of development 1966)	IEEE 583 (1970, 1982, 1994) IEEE 595 (1974, 1982) IEEE 596 (1972, 1982) IEEE 758 (1979)	[112]
GPIB (HP-IB)	Hewlett-Packard (U.S.)	1974 (start of development 1965)	ANSI IEEE-488 (1975, 1978) ANSI IEEE-488.2 (1987, 1992) IEC 60625 (1979,1993)	[113, 114, 115]
HP-IL	Hewlett-Packard (U.S.)	1980 (start of development 1976)	—	[18]
I²C	Philips (Netherlands)	1981	—	[116]
M-Bus	University of Paderborn, TI, Techem (Germany)	1992	EN 1434-3 (1997)	[117]
Measurement Bus	Industry consortium (Germany)	1988	DIN 66348-2 (1989) DIN 66348-3 (1996)	

TABLE 13.9 Automotive and Aircraft Fieldbuses

Fieldbus	Developer (Country)	Introduced in	Standard	References
ABUS	Volkswagen (Germany)	1987	—	[118]
ARINC	Aeronautical Radio, Inc. (U.S.)	1978	AEEC ARINC 429 (1978, 1995)	
CAN	Bosch (Germany)	1986 (start of development 1983), CAL 1992	ISO 11898 (1993, 1995) ISO 11519 (1994)	[119]
Flexray	DaimlerChrysler, BMW (Germany)	2002	—	
J1850	Ford, GM, Chrysler (U.S.)	1987	SAE J1850 (1994, 2001) ISO 11519-4	[118]
J1939	SAE (U.S.)	1994	SAE J1939 (1998)	[118]
LIN	Industry consortium	1999	— (open spec)	
MIL-1533	SAE (military and industry consortium, U.S.)	1970 (start of development 1968)	MIL-STD-1553 (1973) MIL-STD-1553A (1975) MIL-STD-1553B (1978)	[120]
VAN	Renault, PSA Peugeot-Citroen (France), ISO TC22	1988	ISO 11519-3 (1994)	[118]
SwiftNet	Ship Star Assoc., Boeing (U.S.)	1997	IEC 61158 (2000)	
TTP	Vienna University of Technology (Austria)	1996	—	[118]

TABLE 13.10 Fieldbuses for Industrial and Process Automation and Their Foundations

Fieldbus	Developer (Country)	Introduced in	Standard	References
ARCNET	Datapoint (U.S.)	1977	ANSI ATA 878 (1999)	[121]
ASi	Industry and university consortium (Germany)	1991	EN 50295-2 (1998, 2002) IEC 62026-2 (2000)	
Bitbus	Intel (U.S.)	1983	ANSI IEEE 1118 (1990)	
CC-Link	Mitsubishi (Japan)	1996	— (open spec)	
CANopen	CAN in Automation (user group, Germany)	1995 (start of development 1993)	EN 50325-4 (2002)	[119]
ControlNet	Allen-Bradley (U.S.)	1996	EN 50170-A3 (2000)	[121]
DeviceNet	Allen-Bradley (U.S.)	1994	EN 50325-2 (2000)	[119]
FF	Fieldbus Foundation (industry consortium, U.S.)	1995 (start of development 1994)	BSI DD 238 (1996) EN 50170-A1 (2000)	[121]
Hart	Rosemount (U.S.)	1986	— (open spec)	
Interbus-S	Phoenix Contact (Germany)	1987 (start of development 1983)	DIN 19258 (1993) EN 50254-2 (1998)	[122]
MAP	General Motors (U.S.)	1982 (start of development 1980)	MAP 1.0 (1982) MAP 2.0 (1985) MAP 3.0 (1988)	[123]
MMS	ISO TC 184	1986	ISO/IEC 9506 (1988, 2000)	
Modbus	Gould, Modicon (U.S.)	1979	— (open spec)	
PDV-Bus	Industry and university consortium (Germany)	1979 (start of development 1972)	DIN 19241 (1982)	[124, 115]
P-NET	PROCES-DATA (Denmark)	1983	DS 21906 (1990) EN 50170-1 (1996)	
PROWAY C	IEC TC 65	1986 (start of development 1975)	ISA S72.01 (1985) IEC 60955 (1989)	[14]
Profibus	Industry and university consortium (Germany)	1989 (start of development 1984)	FMS: DIN 19245-1 and -2 (1991) DP: DIN 19245-3 (1993) PA: DIN 19245-4 (1995) FMS/DP: EN 50170-2 (1996) DP: EN 50254-3 (1998) PA: EN 50170-A2 (2000)	
SDS	Honeywell (U.S.)	1994	EN 50325-3 (2000)	[119]
Sercos	Industry consortium (Germany)	1989 (start of development 1986)	IEC 61491 (1995) EN 61491 (1998)	
Seriplex	APC, Inc. (U.S.)	1990	IEC 62026-6 (2000)	[121]
SINEC L2	Siemens (Germany)	1992	—	
SP50 Fieldbus	ISA SP 50 (U.S.)	1993	ISA SP 50 (1993)	
(World)FIP	Industry and university consortium (France)	1987 (start of development 1982)	AFNOR NF C46601-7 (1989–1992) EN 50170-3 (1996) DWF: AFNOR NF C46638 (1996) DWF: EN 50254-4 (1998)	[16]

TABLE 13.11 Fieldbuses for Building and Home Automation

Fieldbus	Developer (Country)	Introduced in	Standard	References
BACnet	ASHRAE SPC135P (industry consortium, U.S.)	1991	ANSI/ASHRAE 135 (1995) ENV 1805-1 (1998) ENV 13321-1 (1999) ISO 16484-5 (2003)	
Batibus	Industry consortium (France)	1987	AFNOR NF 46621-3 and -9 (1991) ENV 13154-2 (1998)	
CEBus	Industry consortium (U.S.)	1984	ANSI EIA 600 (1992)	
EHS	Industry consortium (Europe)	1987	ENV 13154-2 (1998)	
EIB	Industry consortium (Germany)	1990	AFNOR NFC 46624-8 (1991) DIN V VDE 0829 (1992) ENV 13154-2 (1998)	[125]
HBS	Industry consortium (Japan)	1986 (start of development 1981)	EIAJ/REEA ET2101	
LonWorks	Echelon (U.S.)	1991	ANSI EIA 709 (1999) ENV 13154-2 (1998)	[121, 126]
Sigma I	ABB (Germany)	1983	—	
X10	Pico Electronics (U.K.)	1978 (start of development 1975)	—	[126]

14

PROFIBUS: Open Solutions for the World of Automation

Ulrich Jecht
UJ Process Analytics

Wolfgang Stripf
Siemens AG

Peter Wenzel
PROFIBUS International

14.1 Basics

Fieldbuses are industrial communication systems with bit-serial transmission that use a range of media such as copper cable, fiber optics, or radio transmission to connect distributed field devices (sensors, actuators, drives, transducers, analyzers, etc.) to a central control or management system. Fieldbus technology was developed in the 1980s with the aim of saving cabling costs by replacing the commonly used central parallel wiring and dominating analog signal transmission (4- to 20-mA or ±10-V interface) with digital technology. Due to the different industry-specific demands to sponsored research and development projects or preferred proprietary solutions of large system manufacturers, several bus systems with varying principles and properties were established in the market. The key technologies are now included in the recently adopted standards IEC 61158 and 61784 [1]. *PROFIBUS is an integral part of these standards.*

Fieldbuses create the basic prerequisite for distributed automation systems. Over the years they evolved to instruments for automated processes with high productivity and flexibility compared to conventional technology.

PROFIBUS is an open, digital communication system with a wide range of applications, particularly in the fields of factory and process automation, transportation, and power distribution. PROFIBUS is suitable for both fast, time-critical applications and complex communication tasks (Figure 14.1).

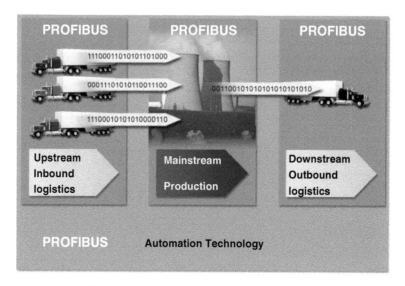

FIGURE 14.1 PROFIBUS suitable for all decentralized applications.

The application and engineering aspects are specified in the generally available guidelines of the PROFIBUS International [2]. This fulfills user demand for standardization, manufacturer independence, and openness; and ensures communication between devices of various manufacturers.

Based on a very efficient and extensible communication protocol, combined with the development of numerous application profiles (communication models for device type families) and a fast-growing number of devices and systems, PROFIBUS began its record of success, initially in factory automation and, since 1995, in process automation. Today, PROFIBUS is the world market leader for fieldbuses with more than a 20% share of the market, approximately 500,000 equipped plants, and more than 12 million nodes. Today, there are more than 2000 PROFIBUS products available from a wide range of manufacturers.

The success of PROFIBUS stems in equal measures from its progressive technology and the strength of its noncommercial PROFIBUS User Organization e.V. (PNO), the trade body of manufacturers and users founded in 1989. Together with the 25 other regional PROFIBUS associations within countries all around the world and the international umbrella organization PROFIBUS International (PI) founded in 1995, this organization now totals more than 1200 members worldwide. Objectives are the continuing further development of PROFIBUS technology and increasing worldwide acceptance.

PROFIBUS has a modular structure (PROFIBUS toolbox) and offers a range of *transmission and communication technologies*, numerous *application and system profiles*, and *device management and integration tools* [8]. Thus, PROFIBUS covers the various and application-specific demands from the field of factory to process automation, from simple to complex applications, by selecting the adequate set of components out of the toolbox (Figure 14.2).

14.2 Transmission Technologies

PROFIBUS features four different transmission technologies, all of which are based on international standards. They all are assigned to PROFIBUS in both IEC 61158 and IEC 61784: RS485, RS485-IS, MBP-IS (IS stands for intrinsic safety protection), and fiber optics.

RS485 transmission technology is simple and cost-effective and primarily used for tasks that require high transmission rates. Shielded, twisted-pair copper cable with one conductor pair is used. No expert knowledge is required for installation of the cable. The bus structure allows addition or removal of stations or the step-by-step commissioning of the system without interfering with other stations. Subsequent expansions (within defined limits) have no effect on stations already in operation.

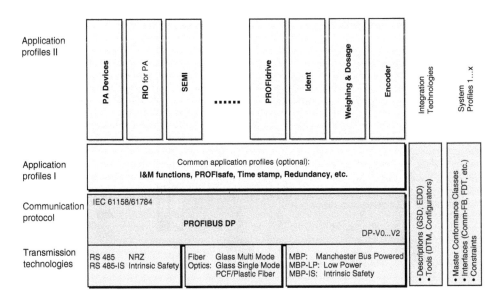

FIGURE 14.2   Structure of PROFIBUS system technology.

Various *transmission rates* can be selected from 9.6 Kbit/s up to 12 Mbit/s. One uniform speed is selected for all devices on the bus when commissioning the system. Up to *32 stations* (master or slaves) can be connected in a single segment. For connecting more than 32 stations, repeaters can be used. The maximum permissible *line length* depends on the transmission rate.

Different cable types (type designation A to D) for different applications are available on the market for connecting devices either to each other or to network elements (segment couplers, links, and repeaters). When using RS485 transmission technology, PI recommends the use of cable type A.

RS485-IS transmission technology responds to an increasing market demand to support the use of RS485 with its fast transmission rates within intrinsically safe areas. A PROFIBUS guideline is available for the configuration of intrinsically safe RS485 solutions with simple device interchangeability. The interface specification details the levels for current and voltage that must be adhered to by all stations in order to ensure safe operation during interconnection. An electric circuit limits currents at a specified voltage level. When connecting active sources, the sum of the currents of all stations must not exceed the maximum permissible current. In contrast to the FISCO model (see below), *all* stations represent active sources. Up to 32 stations may be connected to the intrinsically safe bus circuit.

MBP type transmission technology (Manchester coding and bus powered) is a new term that replaces the previously common terms for intrinsically safe transmission such as physics in accordance with IEC 61158-2, 1158-2, etc. In the meantime, the current version of IEC 61158-2 (physical layer) describes *several different* transmission technologies, MBP technology being just one of them. Thus, differentiation in naming was necessary.

MBP is a synchronous, Manchester-coded transmission with a defined transmission rate of 31.25 Kbit/s. In the *MBP-IS* version, it is frequently used in process automation as it satisfies the key demands of the chemical and petrochemical industries for intrinsic safety and bus powering using two-wire technology.

MBP transmission technology is usually limited to a specific segment (field devices in hazardous areas) of a plant, which is then linked to a RS485 segment via a segment coupler or links (Figure 14.3).

Segment couplers are signal converters that modulate the RS485 signals to the MBP signal level and vice versa. They are transparent from a bus protocol's point of view. In contrast, *links* provide more computing power. They virtually map the entire field devices connected to the MBP segment into the RS485 segment as a single slave. Tree or line structures (and any combination of the two) are network topologies supported by PROFIBUS with MBP transmission with up to 32 stations per segment and a maximum of 126 per network.

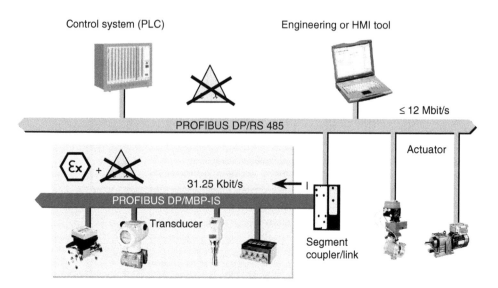

FIGURE 14.3   Intrinsic safety and powering of field devices using MBP-IS.

Fiber-optic transmission technology is used for fieldbus applications with very high electromagnetic interference or that are spread over a large area or distance. The PROFIBUS guideline for fiber-optic transmission [3] specifies the technology available for this purpose, including multimode and single-mode glass fiber, plastic fiber, and hard-clad silica (HCS) fiber. Of course, while developing these specifications, great care was taken to allow problem-free integration of existing PROFIBUS devices in a fiber-optic network without the need to change the protocol behavior of PROFIBUS. This ensures *backward compatibility* with existing PROFIBUS installations.

The internationally recognized *FISCO model* considerably simplifies the planning, installation, and expansion of PROFIBUS networks in potentially explosive areas. FISCO stands for *fieldbus intrinsically safe concept*. It was developed by the German PTB [4]. The model is based on the specification that a network is intrinsically safe and requires no individual intrinsic safety calculations when the relevant four bus components (field devices, cables, segment couplers, and bus terminators) fall within predefined limits with regard to voltage, current, output, inductance, and capacity. The corresponding proof can be provided by certification of the components through authorized accreditation agencies, such as PTB (Germany), UL and FM (U.S.), and others.

If FISCO-approved devices are used, not only is it possible to operate more devices on a single line, but the devices can be replaced during runtime by devices of other manufacturers, or the line can be expanded — all without the need for time-consuming calculations or system certification. So you can simply plug and play, even in hazardous areas.

14.3   Communication Protocol

14.3.1   PROFIBUS DP

At the protocol level, PROFIBUS with decentralized peripherals (DP) and its versions DP-V0 to DP-V2 offer a broad spectrum of optional services, which enable optimum communication between different applications.

DP has been designed for *fast data exchange at the field level*. Data exchange with the distributed devices is primarily cyclic. The communication functions required for this are specified through the DP basic functions (version DP-V0). Geared toward the special demands of the various areas of application, these basic DP functions have been expanded step by step with special functions, so that DP is now available in three versions — DP-V0, DP-V1, and DP-V2 — whereby each version has its own special key features. All versions of DP are specified in detail in IEC 61158 and 61784, respectively.

Version DP-V0 provides the basic functionality of DP, including cyclic data exchange as well as station diagnosis, module diagnosis, and channel-specific diagnosis.

Version DP-V1 contains enhancements geared toward process automation, in particular acyclic data communication for parameter assignment, operation, visualization, and alarm handling of intelligent field devices, in coexistence with cyclic user data communication. This permits online access to stations using engineering tools. In addition, DP-V1 defines alarms. Examples for different types of alarms are status alarm, update alarm, and a manufacturer-specific alarm.

Version DP-V2 contains further enhancements and is geared primarily toward the demands of drive technology. Due to additional functionalities, such as isochronous slave mode and slave-to-slave(s) communication (data exchange broadcast (DXB)), etc., DP-V2 can also be implemented as a drive bus for controlling fast movement sequences in drive axes.

14.3.2 System Configuration and Device Types

DP supports implementation of both monomaster and multimaster systems. This affords a high degree of flexibility during system configuration. A maximum of 126 devices (masters or slaves) can be connected to a bus network. In *monomaster systems*, only *one* master is active on the bus during operation of the bus system. Figure 14.4 shows the system configuration of a monomaster system. In this case, the master is hosted by a programmable logic controller (PLC).

The PLC is the central control component. The slaves are connected to the PLC via the transmission medium. This system configuration enables the shortest bus cycle times. In *multimaster systems* several masters are sharing the same bus. They represent both independent subsystems, comprising masters and their assigned slaves, and additional configuration and diagnostic master devices. The masters are coordinating themselves by passing a token from one to the next. Only the master that holds the token can communicate. PROFIBUS DP differentiates three groups of device types on the bus.

DP master class 1 (DPM1) is a central controller that cyclically exchanges information with the distributed stations (slaves) at a specified message cycle. Typical DPM1 devices are PLCs or PCs. A DPM1 has active bus access with which it can read measurement data (inputs) of the field devices and write the set-point values (outputs) of the actuators at fixed times. This continuously repeating cycle is the basis of the automation function (Figure 14.4).

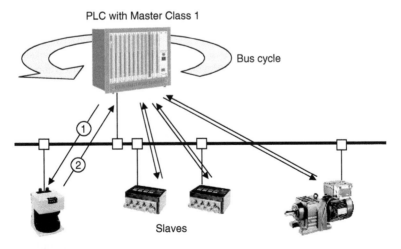

FIGURE 14.4 PROFIBUS DP monomaster system (DP-V0).

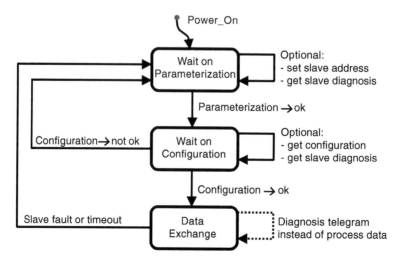

FIGURE 14.5 State machine for slaves.

DP master class 2 (DPM2) consists of engineering, configuration, or operating devices. They are put in operation during commissioning and for maintenance and diagnostics in order to configure connected devices, evaluate measured values and parameters, and request the device status. A DPM2 does not have to be permanently connected to the bus system. The DPM2 also has active bus access.

DP slaves are peripherals (input/output (I/O) devices, drives, human machine interfaces (HMIs), valves, transducers, analyzers), which read in-process information or use output information to intervene in the process. There are also devices that solely provide input information or output information. As far as communication is concerned, slaves are passive devices: they only respond to direct queries (see Figure 14.4, sequences ① and ②). This behavior is simple and cost-effective to implement. In the case of DP-V0, it is already completely included in the Bus-ASIC.

14.3.3 Cyclic and Acyclic Data Communication Protocols

Cyclic data communication between the DPM1 and its assigned slaves is automatically handled by the DPM1 in a defined, recurring sequence (Figure 14.4). The appropriate services are called MS0. The user defines the assignment of the slave(s) to the DPM1 when configuring the bus system. The user also defines which slaves are to be included/excluded in the cyclic user data communication. DPM1 and the slaves are passing three phases during start-up: parameterization, configuration, and cyclic data exchange (Figure 14.5).

Before entering the cyclic data exchange state, the master first sends information about the transmission rate, the data structures within a PDU, and other slave-relevant parameters. In a second step, it checks whether the user-defined configuration matches the actual device configuration. Within any state the master is enabled to request slave diagnosis in order to indicate faults to the user.

An example for the telegram structure for the transmission of information between master and slave is shown in Figure 14.6. The telegram starts with some synchronization bits, the type (SD) and length (LE) of the telegram, the source and destination addresses, and a function code (FC). The function code indicates the type of message or content of the load (processing data unit) and serves as a guard to control the state machine of the master. The PDU, which may carry up to 244 bytes, is followed by a safeguard mechanism frame-checking sequence (FCS) and a delimiter.

One example for the usage of the function code is the indication of a fault situation on the slave side. In this case, the master sends a special diagnosis request instead of the normal process data exchange that the slave replies to with a diagnosis message. It comprises 6 bytes of fixed information and user-definable device and module- or channel-related diagnosis information [1, 7].

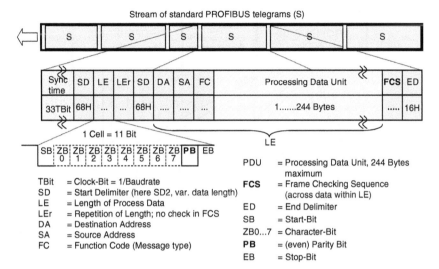

FIGURE 14.6 PROFIBUS DP telegram structure (example).

In addition to the single station-related user data communication, which is automatically handled by the DPM1, the master can also send control commands to all slaves or a group of slaves simultaneously. These control commands are transmitted as multicast messages and enable sync and freeze modes for *event-controlled synchronization* of the slaves [1, 7].

For safety reasons, it is necessary to ensure that DP has effective *protective functions* against incorrect parameterization or failure of transmission equipment. For this purpose, the DP master and the slaves are fitted with monitoring mechanisms in the form of time monitors. The monitoring interval is defined during configuration.

Acyclic data communication is the key feature of version DP-V1. This forms the requirement for parameterization and calibration of the field devices over the bus during runtime and for the introduction of confirmed alarm messages. Transmission of acyclic data is executed parallel to cyclic data communication, but with lower priority. Figure 14.7 shows some sample communication sequences for a master class 2, which is using MS2 services. In using MS1 services, a master class 1 is also able to execute acyclic communications.

Slave-to-slave communications (DP-V2) enable direct and timesaving communication between slaves using broadcast communication without the detour over a master. In this case, the slaves act as publisher; i.e., the slave response does not go through the coordinating master, but directly to other slaves embedded in the sequence, the so-called subscribers (Figure 14.8). This enables slaves to directly read data from other slaves and use them as their own input. This opens up the possibility of completely new applications; it also reduces response times on the bus by up to 90%.

The *isochronous mode* (DP-V2) enables clock synchronous control in masters and slaves, irrespective of the busload.

The function enables highly precise positioning processes with clock deviations of <1 μs. All participating device cycles are synchronized to the bus master cycle through a global control broadcast message. A special sign of life (consecutive number) allows monitoring of the synchronization.

Clock control (DP-V2) via a new master slave service synchronizes all stations to a system time with a deviation of <1 ms. This allows the precise tracking of events. This is particularly useful for the acquisition of timing functions in networks with numerous masters. This facilitates the diagnosis of faults as well as the chronological planning of events.

Upload and download (DP-V2) allows the loading of any size of data area in a field device with the help of a few commands. Within IEC 61158 these services are called *load region*. This enables, for example, programs to be updated or devices replaced without the need for manually loading processes.

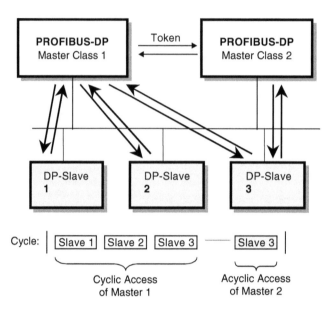

FIGURE 14.7 Cyclic and acyclic data communication with DP-V1.

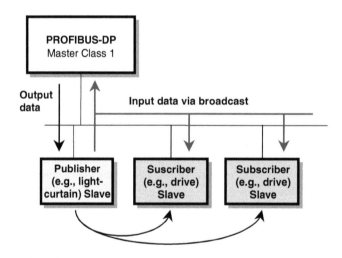

FIGURE 14.8 Slave-to-slaves data exchange.

Addressing with slot and index is used for both cyclic and acyclic communication services (Figure 14.9).
When addressing data, PROFIBUS assumes that the physical structure of the slaves is *modular* or can be structured internally in logical functional units, so-called *modules* (Figure 14.9). The slot number addresses the module and the index addresses the data records assigned to a module. Each data record can be up to 244 bytes. The modules begin at slot 1 and are numbered in ascending contiguous sequence. The slot number 0 is for the device itself. Compact devices are regarded as units of virtual modules. These can also be addressed with slot number and index.

14.4 Application Profiles

Profiles are used in automation technology to define specific properties and behavior for devices, device families, or entire systems. Only devices and systems using a vendor-independent profile provide interoperability on a fieldbus, thereby fully exploiting the advantages of a fieldbus. Profiles take into account

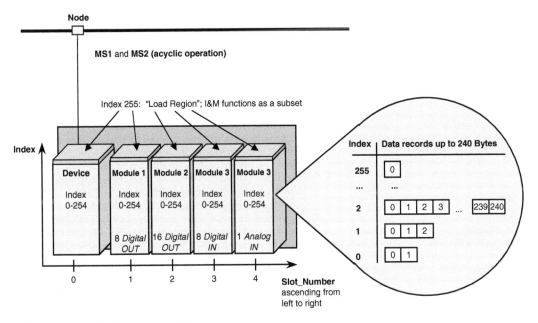

FIGURE 14.9 Slot/index address model of a slave.

application- and type-specific special features of field devices, controls, and methods of integration (engineering). The term *profile* ranges from just a few specifications for a specific device class to comprehensive specifications for applications in a specific industry. The generic term for all profiles is *application profiles*.

A distinction is then drawn between *general application profiles* with implementation options for different applications (this includes, for example, the profiles identification and maintenance (I&M) functions, PROFIsafe, redundancy, and time stamp); *specific application profiles*, which are developed for a specific application, such as PROFIdrive, SEMI (Semiconductor Equipment and Materials International), or process automation (PA) devices; and *system and master profiles*, which describe specific system performance that is available to field devices.

PROFIBUS offers a wide range of such application profiles, which allow application-oriented implementation.

14.4.1 General Application Profiles

Identification and maintenance functions are mandatory for all PROFIBUS devices with MS1 or MS2 services. The main purpose of the I&M functions described hereinafter is to support the end user during various scenarios of a device's life cycle, be it configuration, commissioning, parameterization, diagnosis, repair, firmware update, asset management, audit trailing, etc. It is kind of a type plate or boiler plate. The corresponding parameters all are stored at the same address space within the PROFIBUS slot/index address model.

Using the call mechanism of the load region services [1] opens up an additional subindex address space of 65535 data records. The I&M functions are assigned a space between 65000 and 65199 for basic, profile-specific, and manufacturer-specific items. Table 14.1 itemizes the individual parameters.

The usage of IDs may not be very helpful for the user if a tool is displaying the information directly out of the device. However, nowadays laptops or engineering tools normally have access to the Internet, at least temporarily. Thus, it is quite easy to reference a central database (e.g., on the PROFIBUS Web server) and retrieve comprehensive and always actual information, even in a desired language (Figure 14.10).

PROFIsafe is a comprehensive, open fieldbus solution for safety-relevant applications without the use of a second relay-based layer or proprietary safety buses. PROFIsafe defines how fail-safe devices

TABLE 14.1 Basic Identification and Maintenance Functions

I&M Function	Data Format	Notes
MANUFACTURER_ID	2 Octets	I&M functions are using company IDs instead of names. These IDs are harmonized with the list of the HART Foundation and comprise extensions for additional companies.
ORDER_ID	20 Octets	Order number for a particular device type. For virtual modular devices the root or highest level of the basic device.
SERIAL_NUMBER	16 Octets	Unique identifier for a particular device (counter).
HARDWARE_REVISION	2 Octets	The content of this parameter characterizes the edition of the hardware only.
SOFTWARE_REVISION	4 Octets	The content of this parameter characterizes the edition of the software or firmware of a device or module. The structure supports coarse and detailed differentiation that may be defined by the manufacturer: Vx.y.z.
REV_COUNTER	2 Octets	Indicates unplugging of modules or write access.
PROFILE_ID	2 Octets	Device or module corresponds to a particular PROFIBUS profile.
PROFILE_SPECIFIC_TYPE	2 Octets	This identifier references a device class defined within a PROFIBUS profile.
IM_VERSION	2 Octets	Version of the I&M functions implemented within a device or module.
IM_SUPPORTED	2 Octets	Directory for the subset of I&M functions implemented within a device or module.
TAG_FUNCTION	32 Octets	User definable information about the "role" of the device within a plant facility.
TAG_LOCATION	22 Octets	User definable information about the "location coordinates" of the device within a plant facility.
INSTALLATION_DATE	16 Octets	Indicates the date of installation or commissioning of a device or module, e.g., 1995-02-04 16:23.
DESCRIPTOR	54 Octets	User defined comments.
SIGNATURE	54 Octets	Allows parameterization tools to store a "security" code as a reference for a particular parameterization session and audit trail tools to retrieve the code for integrity checks. Used for safety applications according 21 CFR 11 [6] or hazardous machinery (PROFIsafe).

(emergency stop push buttons, light curtains, level switches, etc.) can communicate over PROFIBUS with fail-safe controllers in such a manner that they can be used for safety-relevant automation tasks up to category 4 compliance with EN 954 (ISO 13849) or SIL3 (safety integrity level) according to IEC 61508. It implements safe communications over a profile, i.e., over a special PROFIsafe data frame and a special protocol. PROFIsafe is a single-channel *software solution* that is implemented in the devices as an additional layer above layer 7 (Figure 14.11); the standard PROFIBUS components, such as lines, application-specific integrated circuits (ASICs), or protocols, remain unchanged. This ensures redundancy mode and retrofit capability. Devices with the PROFIsafe profile can be operated in coexistence with standard devices without restriction on the same bus (cable).

PROFIsafe takes advantage of the acyclic communication (DP-V1) for full maintenance support of the devices and can be used with RS485, fiber-optic, or MBP transmission technology. This ensures both fast response times (important for the manufacturing industry) and low power consumption with intrinsically safe operation (important for process automation).

HART on PROFIBUS DP integrates HART devices installed in the field, in existing or new PROFIBUS systems. It includes the benefits of the PROFIBUS communication mechanisms without any changes required to the PROFIBUS protocol and services, the PROFIBUS, or the state machines and functional characteristics. This profile is implemented in the master and slave above layer 7, thus enabling mapping of the HART client–master–server model on PROFIBUS. The cooperation of the HART Foundation on the specification work ensures complete conformity with HART specifications.

The HART client application is integrated in a PROFIBUS master and the HART master in a PROFIBUS slave (Figure 14.12), whereby the latter serves as a multiplexer and handles communication to the HART devices.

The *time-stamp* application profile describes mechanisms for supplying certain events and actions with a time stamp, which enables precise time assignment. Precondition is a clock control in the slaves through

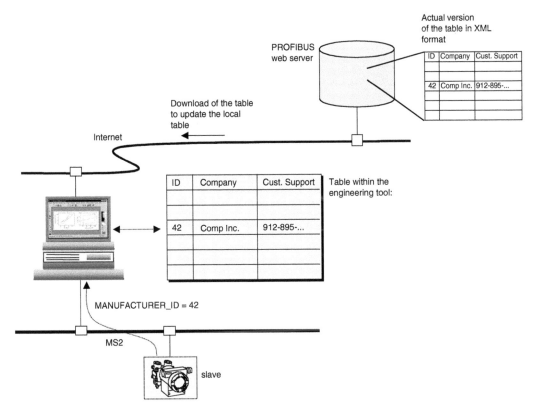

FIGURE 14.10 Referencing IDs via the Internet.

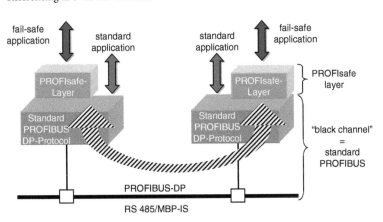

FIGURE 14.11 Safe communication on PROFIBUS DP.

a clock master via special services. An event can be given a precise system time stamp and read out accordingly. The concept of graded messages is used. The message types are summarized under the term *alerts* and are divided into high-priority *alarms* (these transmit a diagnostics message) and low-priority *events*. In both cases, the master acyclically reads the time-stamped process values and alarm messages from the alarm and event buffer of the field device (Figure 14.13).

The *slave redundancy* application profile provides a slave redundancy mechanism (Figure 14.14): slave devices contain two different PROFIBUS interfaces that are called *primary* and *backup* (slave interface). These may be in a single device or distributed over two devices.

The devices are equipped with two independent protocol stacks with a special *redundancy expansion*.

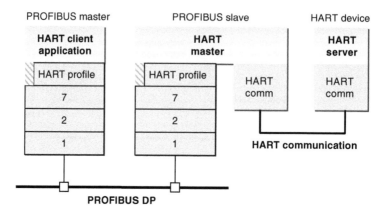

FIGURE 14.12 Operating HART devices over PROFIBUS DP.

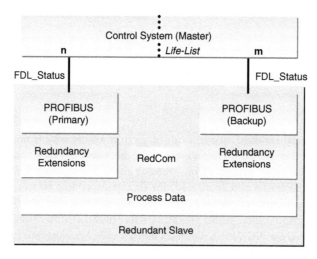

FIGURE 14.13 Time stamping and alarm messages.

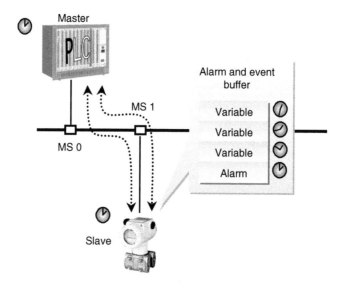

FIGURE 14.14 Slave redundancy in PROFIBUS.

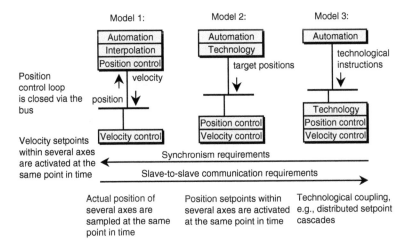

FIGURE 14.15 Different requirements for distributed drive applications.

A redundancy communication (RedCom) runs between the protocol stacks, i.e., within a device or between two devices. It is independent of PROFIBUS and its performance capability is largely determined by the redundancy reversing times.

Only one device version is required to implement different redundancy structures, and no additional configuration of the backup slave is necessary. The redundancy of PROFIBUS slave devices provides high availability, short reversing times, and no data loss and ensures fault tolerance.

14.4.2 Specific Application Profiles

The *PROFIdrive* application profile defines device behavior and the access procedure to drive data for electric drives on PROFIBUS, from simple frequency converters to highly dynamic servo-controls.

The method of integrating drives in automation solutions is highly dependent on the task of the drives (Figure 14.15). The more the drives are acting independently from central host controllers, the more they require slave-to-slave communication capabilities. On the other hand, the more the central host controllers are taking over the computing tasks, the more that synchronization of the involved drives is required.

For this reason, PROFIdrive defines six classes covering the majority of applications (Figure 14.16):

Standard drives (class 1): The drive is controlled by means of a main set-point value (e.g., rotational speed), whereby speed control is carried out in the drive controller.

Standard drives with technological function (class 2): The automation process is broken down into several subprocesses and some of the automation functions are shifted from the central programmable controller to the drive controllers. PROFIBUS serves as the technology interface in this case. Slave-to-slave communication between the individual drive controls is a requirement for this solution.

Positioning drive (class 3): Integrates an additional position controller in the drive, thus covering an extremely broad spectrum of applications (e.g., the twisting on and off of bottle tops). The positioning tasks are passed to the drive controller over PROFIBUS and started.

Central motion control (classes 4 and 5): Enables the coordinated motion sequence of multiple drives. The motion is primarily controlled over a central numeric control (CNC). PROFIBUS serves to close the position control loop as well as synchronize the clock (Figure 14.17). The position control concept (dynamic servo-control) of this solution also supports extremely sophisticated applications with linear motors.

Distributed automation by means of clocked processes and electronic shafts (class 6): Can be implemented using slave-to-slave communication and isochronous slaves. Sample applications include electrical gears, curve discs, and angular synchronous processes.

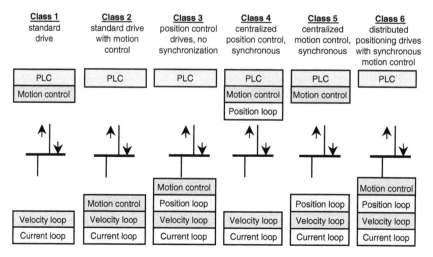

FIGURE 14.16 PROFIdrive defines six application classes.

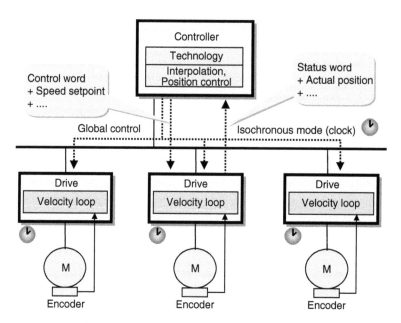

FIGURE 14.17 Positioning with central interpolation and position control.

In contrast to other drive profiles, PROFIdrive only defines the access mechanisms to the parameters and a subset of approximately 30 *profile parameters*, which includes fault buffers, drive controllers, device identification, etc.

All other parameters (which may number more than 1000 in complex devices) are *manufacturer specific*; they provide drive manufacturers great flexibility when implementing control functions.

The profile for *PA devices* defines all functions and parameters for different classes of devices for process automation with local intelligence. They can execute part of the information processing or even take over the overall functionality in automation systems. The profile includes all steps of a typical signal flow — from process sensor signals to the preprocessed process value that is communicated to the control system together with a value qualifier (Figure 14.18).

The profile for PA devices is documented in a *general model description* containing the currently valid specifications for all device types and in *device data sheets* containing the agreed additional specifications

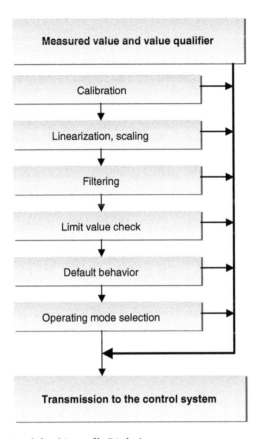

FIGURE 14.18 Signal processing defined in profile PA devices.

for individual device classes. Version 3.0 of the profile for PA devices includes device data sheets for quantity measurement of pressure and differential pressure, level, temperature, and flow rate, and data sheets for valves, actuators, analyzers, analog, and digital inputs and outputs.

In process engineering it is common to use *blocks* for describing the characteristics and functions of a measuring point or manipulating point at a certain control point and to represent an automation application through a combination of these blocks. Therefore, the specification for PA devices uses a *function block model* according to IEC 61804 to represent functional sequences, as shown in Figure 14.19. The blocks are implemented by the manufacturers as software in the field devices and, taken as a whole, represent the functionality of the device.

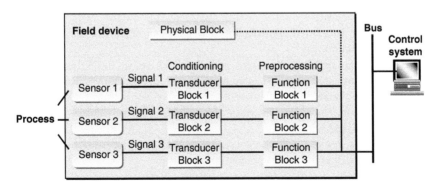

FIGURE 14.19 Block structure of a PA field device.

The following three block types are used:

A *physical block* (PB) contains the characteristic data of a device, such as device name, manufacturer, version, serial number, etc. There can only be one physical block in each device.

A *transducer block* (TB) contains all the data required for processing an unconditioned signal delivered from a sensor for passing on to a function block. If no processing is required, the TB can be omitted. Multifunctional devices with two or more sensors have a corresponding number of TBs.

A *function block* (FB) contains all data for final processing of a measured value prior to transmission to the control system or, on the other hand, for processing of a setting before the setting process. Different FBs are available:

Analog input (AI) — Delivers the measured value from the sensor/TB to the control system

Analog output (AO) — Provides the device with the value specified by the control system

Digital input (DI) — Provides the control system with a digital value from the device

Digital output (DO) — Provides the device with the value specified by the control system

The profile for *ident systems* defines complete communication and processing models for bar code readers and transponder systems. These are primarily intended for extensive use with DP-V1 functionality. While the cyclic data transmission channel is used for small data volumes to transfer status/control information, the acyclic channel serves the transmission of large data volumes that result from the information in the bar code reader or transponder. The definition of standard PROXY function blocks [5] according to IEC 61131-3 has facilitated the use of these systems and paves the way for the application of open solutions on completion of international standards, such as ISO/IEC 15962 and ISO/IEC 18000.

The profile for *weighing and dosage systems* follows approaches similar to those of the ident systems. Communication and processing models are defined for four classes of devices or systems:

- Simple scale
- Comfort scale
- Continuous scale
- Batch scale

These new types of profiles will dramatically reduce the engineering costs and improve the bidding process during project execution.

14.4.2.1 Summary of Specific Application Profiles

PROFIdrive: The profile specifies the behavior of devices and the access procedure to parameters for variable-speed electrical drives on PROFIBUS DP.

PA devices: The profile specifies the data formats for cyclic data exchange and the characteristics for *process engineering* of devices for process automation.

Robots/NC: The profile describes how *handling and assembly robots* are controlled via PROFIBUS.

Panel devices: The profile describes the interfacing of simple *human machine interface devices* (HMIs) to control components.

Encoders: This profile describes the interfacing of rotary, angle, and linear *encoders* with single-turn or multiturn resolution.

Fluid power: The profile describes the control of hydraulic drives via PROFIBUS.

SEMI: The profile defines models of devices for semiconductor production such that they comply with the PA model and the SEMI model.

Low-voltage switchgear: The profile describes data exchange for low-voltage switchgear like circuit breakers, switches, and starters.

Weighing/dosage: The profile describes the communication and processing models for simple and comfort scales as well as for batch and continuous scales.

Ident systems: The profile describes the communication and processing models for bar code readers and transponders.

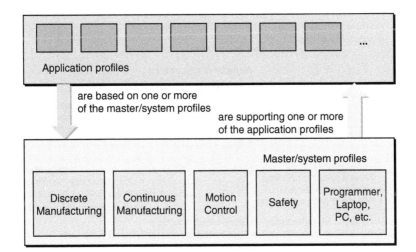

FIGURE 14.20   Master and system profiles for PROFIBUS DP.

Remote IO for PA devices: This profile takes into account the special conditions of physically modular slaves like limited communication resources and extreme cost sensitivity. The profile follows the model of PA devices as much as possible but has some simplifications.

14.4.3   Master and System Profiles

Master profiles for PROFIBUS describe classes of controllers, each of which supports a specific subset of all the possible master functionalities, such as cyclic and acyclic communications, diagnostics, alarm handling, clock control, slave-to-slaves communication, isochronous mode, and safety.

System profiles for PROFIBUS go a step further and describe classes of systems, including the master functionality, the possible functionality of *standard program interfaces* (FB in accordance with IEC 61131-3, safety layer, and field device tool (FDT)), and *integration options* (general station description (GSD), electronic device description (EDD), and device type manager (DTM)). Figure 14.20 shows the standard platforms available today.

In the PROFIBUS DP system, the master and system profiles provide the much needed counterpart to the application profiles: master and system profiles describe specific system parameters that are made available to the field devices; application profiles require specific system parameters in order to simplify their defined characteristics.

By using these profiles the *device manufacturers* can focus on existing or specified system profiles and the system manufacturers can provide the platforms required by the existing or specified device application profiles.

14.5   Integration Technologies

Modern field devices in both factory and process automation provide a wide range of information and also execute functions that were previously executed in PLCs and control systems. To execute these tasks, the tools for commissioning, maintenance, engineering, and parameterization of these devices require an exact and complete description of device data and functions, such as the type of application function, configuration parameters, range of values, units of measurement, default values, limit values, identification, etc. The same applies to the controller/control system, whose device-specific parameters and data formats must also be made known (integrated) to ensure error-free data exchange with the field devices.

PROFIBUS has developed a number of methods and tools (integration technologies) for this type of device description, which enables standardization of device management. The performance range of these

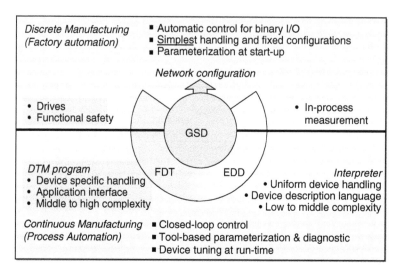

FIGURE 14.21   Integration technologies for PROFIBUS DP.

tools is optimized to specific tasks (Figure 14.21), which has given rise to the term *scalable device integration*. GSD and EDD are both a sort of electronic device data sheet, developed with different languages according to the special scope, while DTM is a software component containing specific field device functions for parameterization, configuration, diagnostics, and maintenance, generated by mapping and to be used together with the universal software interface FDT, which is able to implement software components.

A GSD is an electronically readable ASCII text file that contains both general and device-specific specifications for *communication* and network configuration. Each of the entries describes a feature that is supported by a device. By means of keywords, a configuration tool reads the device identification (ID number), the adjustable parameters, the corresponding data type, and the permitted limit values for the configuration of the device from the GSD. Some of the keywords are *mandatory*, for example, Vendor_Name. Others are *optional*, for example, Sync_Mode_supported. A GSD replaces the previously conventional manuals and supports automatic checks for input errors and data consistency, even during the configuration phase.

Distinction is made between a *device GSD* (for an individual device only) and a *profile GSD*, which may be used for devices that comply exactly with a profile such as PROFIdrive version 3 or PA devices version 3.

GSD for *compact devices*, whose block configuration is already known on delivery, can be created completely by the device manufacturer. GSD for *modular devices*, whose block configuration is not yet conclusively specified on delivery, must be configured by the user in accordance with the actual module configuration using the configuration tool.

The device manufacturers are responsible for the scope and quality of the GSD of their devices. Submission of a profile GSD (contains the information from the profile of a device family) or an individual device GSD (device specific) is essential for certification of a device.

Like a GSD, an EDD is an electronic device data sheet, but it is developed by using a more powerful and universal language, the electronic device description language (EDDL). An EDD typically describes the application-related parameters and functions of a field device, such as configuration parameters, ranges of values, units of measurement, default values, etc. An EDD is a versatile source of information for engineering, commissioning, runtime, asset management, and documentation. It also contains support mechanisms to integrate existing profile descriptions in the device description, to allow references to existing objects, to access standard dictionaries, and to allow assignment of the device description to a device.

An EDD is independent of operating systems and supports the user by its uniform user and operation interface (only one tool, reliable operation, reduced training and documentation costs) and also the device manufacturer (no specific knowledge required; existing EDDs and libraries can be used).

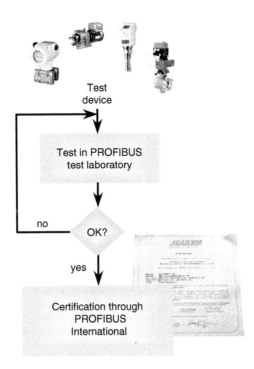

FIGURE 14.22 Certification procedure.

The EDD concept is suitable for tasks of low to middle complexity.

A DTM is software that is generated by mapping the specific functions and dialogs of a field device for parameterization, configuration, diagnostics, and maintenance, complete with user interface, in a *software component*. This component is called DTM and is integrated in the engineering tool or control system over the FDT interface. A DTM uses the routing function of an engineering system for communicating across the hierarchical levels. It works similarly as a printer driver, which the printer supplier includes in delivery, and must be installed on the PC by the user. The DTM is generated by the device manufacturer and is included in delivery of the device.

DTM generation may be performed using one of the following options:

- Specific programming in a higher programming language
- Reuse of existing components or tools through their encapsulation as DTM
- Generation from an existing device description using a compiler or interpreter
- Use of the DTM tool kit of MS Visual Basic

With DTMs it is possible to obtain direct access to all field devices for planning, diagnostics, and maintenance purposes from a central workstation. A DTM is not a stand-alone tool, but an ActiveX component with defined interfaces. The FDT/DTM concept is protocol independent and, with its mapping of device functions in software components, opens up interesting new user options. The DTM/FDT concept is very flexible, resolves interface and navigation needs, and is suitable for tasks of middle to high complexity.

14.6 Quality Assurance

In order for PROFIBUS devices of different types and manufacturers to correctly fulfill tasks in the automation process, it is essential to ensure the error-free exchange of information over the bus. The requirement for this is a standard-compliant implementation of the communication protocol and application profiles by device manufacturers (Figure 14.22). To ensure that this requirement is fulfilled, PI has established a *quality assurance procedure* whereby, on the basis of test reports, certificates are issued to devices that successfully complete the test.

Basis for the certification procedure is the standard EN 45000. PROFIBUS International has approved manufacturer-independent *test laboratories* in accordance with the specifications of this standard. Only these test laboratories are authorized to carry out device tests, which form the basis for certification.

The *test procedure*, which is the same for all test laboratories, is made up of several parts:

- The *GSD/EDD check* ensures that the device description files comply with the specification.
- The *hardware test* tests the electric characteristics of the PROFIBUS interface of the device for compliance with the specifications. This includes terminating resistors, suitability of the implemented drivers and other modules, and the quality of line level.
- The *function test* examines the bus access and transmission protocol and the functionality of the test device.
- The *conformity test* forms the main part of the test. The objective is to test conformity of the protocol implementation with the standard.
- The *interoperability test* checks the test device for interoperability with the PROFIBUS devices of other manufacturers in a multivendor plant. This checks that the functionality of the plant is maintained when the test device is added. Operation is also tested with different masters.

Once a device has successfully passed all the tests, the manufacturer can apply for a *certificate* from PROFIBUS International. Each certified device contains a certification number as a reference. The certificate is valid for 3 years but can be extended after undergoing further tests.

14.7   Implementation

For the device development or implementation of the PROFIBUS protocol, a broad spectrum of standard components and development tools (PROFIBUS ASICs, PROFIBUS stacks, monitoring and commissioning tools) as well as services are available that enable device manufacturers to realize cost-effective development. A corresponding overview is available in the product catalog of PROFIBUS International [2].

PROFIBUS interface modules are ideal for a low or medium volume of devices to be produced. These credit card-size modules implement the entire bus protocol. They are fitted on the master board of the device as an additional module.

PROFIBUS protocol chips (single chips, communication chips, protocol chips) are recommended for an individual implementation in the case of a high volume of devices.

The implementation of single-chip ASICs is ideal for *simple slaves* (IO devices). All protocol functions are already integrated on the ASIC. No microprocessor or software is required. Only the bus interface driver, the quartz, and the power electronics are required as external components.

For *intelligent slaves*, parts of the PROFIBUS protocol are implemented on a protocol chip and the remaining protocol parts implemented as software on a microcontroller. In most of the ASICs available on the market all cyclic protocol parts have been implemented, which are responsible for transmission of time-critical data.

For *complex masters*, the time-critical parts of the PROFIBUS protocol are also implemented on a protocol chip and the remaining protocol parts implemented as software on a microcontroller. Various ASICs of different suppliers are currently available for the implementation of complex master devices. They can be operated in combination with many common microprocessors.

An overview for commercially offered PROFIBUS chips and software (PROFIBUS stacks) is available at the PROFIBUS Web site [2]. For further information, please contact the suppliers directly.

Modem chips are available to realize the (low) power consumption, which is required when implementing a bus-powered field device with MBP transmission technology. Only a feed current of 10 to 15 mA over the bus cable is available for these devices, which must supply the overall device, including the bus interface and the measuring electronics. These modems take the required operating energy for the overall device from the MBP bus connection and make it available as feed voltage for the other electronic components of the device. At the same time, the digital signals of the connected protocol chip are converted into the bus signal of the MBP connection modulated to the energy supply.

14.8 Prospects

While the fieldbuses were pioneering the field of distributed automation in discrete and continuous manufacturing facilities for the past 15 years, Ethernet was gaining great success in office automation. The technology matured more and more and evolved a high degree of comfort and flexibility, such as high transmission speed, easy-to-handle cables and connectors, efficient control protocols, network devices like switches, and the tremendous success of the Internet. The fieldbus organizations now are eager to provide solutions for a steadily growing demand of the market. The solution from PROFIBUS International is PROFINET.

14.8.1 PROFINET CBA

PROFINET CBA is a new automation concept that has emerged as a result of the trend in automation technology toward modular, reusable machines (mechatronic components) and plants with distributed intelligence. With its comprehensive design (uniform model for engineering, communication, and migration architecture to other communication systems, such as PROFIBUS and OPC), PROFINET CBA fulfills all the key demands of automation technology for:

- Consistent communications from field level to corporate management level such as enterprise resource planning (ERP) and manufacturing execution systems (MES) using Ethernet
- A vendor-independent plantwide engineering model for the entire automation landscape
- Openness to other systems
- Implementation of IT standards
- Integration capability of PROFIBUS segments without changes

PROFINET CBA is available as a *specification* and as operating system-independent *source software* for Ethernet based communications [9].

14.8.2 PROFINET IO

The PROFINET component model is ideal for intelligent field devices and programmable controllers with data format interfaces that can be standardized. Simple field devices with many I/O signals do not fit into the engineering model of PROFINET CBA. Thus, PROFINET IO offers an integration methodology based on the PROFINET communication protocols such that a manufacturer of PROFIBUS DP slave devices feels comfortable to switch over. He will find the services described for PROFIBUS DP in PROFINET IO and more. The most essential feature of this integration is the use of distributed field devices with their input and output data to be processed within the application program of a PLC.

14.8.3 The PROFINET Migration Model

This model allows the integration of PROFIBUS DP segments in PROFINET using *proxies*. These assume a *proxy function* for all the devices connected to PROFIBUS. This means that when rebuilding or expanding plants, the entire spectrum of PROFIBUS devices, including products of PROFIdrive and PROFIsafe, can be implemented unchanged, thus providing users with maximum investment protection. Proxy technology also allows integration of other fieldbus systems (Figure 14.23). A second possibility is the usage of PROFINET IO devices directly connected to the host controller (PLC) via PROFINET. Intelligent field devices may be connected directly as a component. This way users have all the possibilities to migrate from the current situation to PROFINET at their convenience.

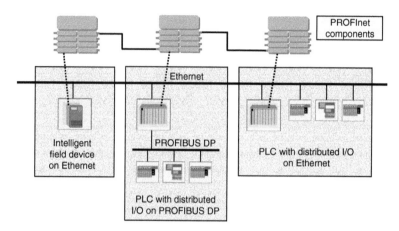

FIGURE 14.23 The migration concept of PROFINET CBA and PROFINET IO.

Abbreviations

ASIC	Application-specific integrated circuit
BIA	German Institute of Occupational Safety and Health
CBA	Component-based automation
CPU	Central processing unit
CRC	Cyclic redundancy check
DP	Decentralized peripherals
DPM1	PROFIBUS DP master class 1, usually a programmable logic controller
DPM2	PROFIBUS DP master class 2, usually a laptop or PC
DTM	Device type manager
DXB	Data exchange broadcast (slave-to-slaves communication)
EDD	Electronic device description
EMI	Electromagnetic interference
EN, prEN	European standard, preliminary European standard
FB	Function block
FDT	Field device tool
FISCO	Fieldbus intrinsically safe concept
FM	Factory Mutual Global, a commercial and industrial property insurance company with a unique focus on risk management (www.fmglobal.com)
GSD	General station description (electronically readable data sheet)
HMI	Human machine interface
HW	Hardware
IEC	International Electrotechnical Commission
I/O	Input/output
ISO/OSI	International Organization for Standardization/Open Systems Interconnection (reference model)
MS0	Cyclic master slave communication services of PROFIBUS DP
MS1/MS2	Acyclic master slave communication services of PROFIBUS DP
NAMUR	Association of users of process control technology
OPC	OLE for process control
PA	Process automation
PC	Personal computer
PDU	Protocol data unit
PLC	Programmable logic controller
PTB	Pysikalisch-Technische Bundesanstalt, national institute of natural and engineering sciences and the highest technical authority for metrology and physical safety engineering of the Federal Republic of Germany (www.ptb.de)
SW	Software
UL	Underwriters Laboratories, Inc., an independent, not-for-profit product safety testing and certification organization (www.ul.com)

References

1. IEC 61158/61784: Digital Data Communications for Measurement and Control: Fieldbus for Use in Industrial Control Systems, 2003.
2. PROFIBUS home page: www.profibus.com
3. Optical transmission technology for PROFIBUS, version 2.0, 1999, PROFIBUS Order 2.021.
4. IEC/TS 60079-27: Electrical Apparatus for Explosive Gas Atmospheres: Part 27: Fieldbus Intrinsically Safe Concept (FISCO); Parts 11, 14, and 25: Constructional and installation requirements, 2002.
5. PROFIBUS communication and proxy function blocks according to IEC 61131-3, version 1.2, July 2001, PROFIBUS Order 2.182.
6. Food and Drug Administration: 21 CFR Part 11.
7. Popp, M., The New Rapid Way to PROFIBUS DP, PROFIBUS Order 4.072.
8. PROFIBUS System Description: Technology and Application, October 2002, free download from www.profibus.com or PNO Order 4.002.
9. PROFInet System Description: Technology and Application, November 2002, free download from www.profibus.com or PNO Order 4.132.

15

The CIP Family of Fieldbus Protocols

Viktor Schiffer
Rockwell Automation

15.1 Introduction

In the past, typical fieldbus protocols (e.g., Profibus, Interbus-S, FIP (Factory Instrumentation Protocol), P-Net, AS-i (Actuator/Sensor Interface)) have been isolated implementations of certain ideas and functionalities that the inventors thought were best suited to solve a certain problem or do a certain job. This has led to quite effective fieldbuses that do their particular job quite well, but they are optimized for certain layers within the automation pyramid or are limited in their functionality (e.g., strict single master systems running a Master/Slave protocol). This typically results in barriers within the automation architecture that are difficult to penetrate and that require complex gateway devices without being able to fully bridge the gap between the various systems that can be quite different in nature.

In contrast, the CIP™* family of protocols (CIP = Common Industrial Protocol) offers a scalable solution that allows a uniform protocol to be employed from the top level of an automation architecture down to the device level without burdening the individual devices.

DeviceNet™* is the first member of this protocol family introduced in 1994. DeviceNet is a CIP implementation using the very popular Controller Area Network (CAN) data link layer. CAN in its typical form (ISO 11898 [11]) defines layers 1 and 2 of the OSI seven-layer model [14] only, while DeviceNet covers the rest. The low cost of implementation and the ease of use of the DeviceNet protocol has led to a large number of manufacturers, with many of them organized in the Open DeviceNet Vendor Association (ODVA; see http://www.odva.org).

---

*CIP™ and DeviceNet™ are trademarks of ODVA.

Layers according to ISO/OSI

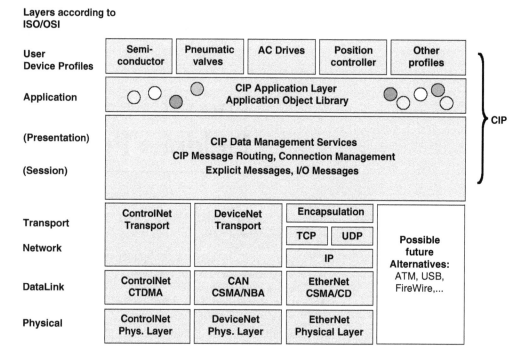

FIGURE 15.1 Relationship between CIP, its implementations, and the ISO/OSI layer model.

ControlNet™,* introduced a few years later (1997), implemented the same basic protocol on new data link layers that allow for much higher speed (5 Mbps), strict determinism, and repeatability while extending the range of the bus (several kilometers with repeaters) for more demanding applications. Vendors and users of ControlNet products are organized within ControlNet International (CI; see http://www.controlnet.org) to promote the use of these products.

In 2000, ODVA and ControlNet International introduced the newest member of the CIP family — EtherNet/IP™,** where IP stands for Industrial Protocol. In this network adaptation, CIP runs over TCP/IP and therefore can be deployed over any Transmission Control Protocol (TCP)/Internet Protocol (IP)-supported data link and physical layers, the most popular of which is IEEE 802.3 [12], commonly known as Ethernet.

The universal principles of CIP easily lend themselves to possible future implementations on new physical/data link layers, e.g., ATM, USB, or FireWire. The overall relationship between the three implementations of CIP and the ISO/OSI layer model is shown in Figure 15.1.

Two significant additions to CIP are currently being worked on: CIP Sync™ and CIP Safety™.*** CIP Sync allows synchronization of applications in distributed systems through precision real-time clocks in all devices. These real-time clocks are kept in tight synchronization by background messages between clock masters and clock slaves using the new IEEE 1588:2002 standard [24]. A more detailed description of this CIP extension is given in Section 15.5.1. CIP Safety is a protocol extension that allows the transmission of safety-relevant messages. Such messages are governed by additional timing and integrity mechanisms that are guaranteed to detect system flaws to a very high degree, as required by international standards such as IEC 61508 [15]. If anything goes wrong, the system will be brought to a safe state, typically taking the machine to a standstill. A more detailed description of this CIP extension is given in Section 15.5.2. In both cases, ordinary devices can operate with CIP Sync or CIP Safety devices side by side in the same system.

---

*ControlNet™ is a trademark of ControlNet International.

**EtherNet/IP™ is a trademark of ControlNet International under license by ODVA.

***CIP Sync™ and CIP Safety™ are trademarks of ODVA.

There is no need for strict segmentation into Standard, Sync, and Safety networks. It is even possible to have any combination of all three functions in one device.

15.2 Description of CIP

CIP is a very versatile protocol that has been designed with the automation industry in mind. However, due to its very open nature, it can be applied to many more areas. The overall CIP Specification is divided into several volumes:

- Volume 1 is the CIP Specification. It contains all general parts of the specification that apply to all the network variants.
- Volume 2 is the EtherNet/IP Specification. It contains the adaptation of CIP to the Ethernet TCP/IP and User Datagram Protocol (UDP)/IP transportation layers and all details that apply specifically to EtherNet/IP, including extensions and any modifications of the CIP Specification.
- Volume 3 is the DeviceNet Specification. It contains the adaptation of CIP to the CAN data link layer and all details that apply specifically to DeviceNet, including extensions and any modifications of the CIP Specification.
- Volume 4 is the ControlNet Specification. It contains the adaptation of CIP to the ControlNet data link layer and all details that apply specifically to ControlNet, including extensions and any modifications of the CIP Specification.
- Volume 5 will contain CIP Safety; it is planned to be published in early 2005.

The CIP Specification [4] is available from ODVA. It is beyond the scope of this handbook to fully describe each and every detail of this specification, but the key features will be presented. The specification is subdivided into several chapters and appendices that describe the following features:

- Object modeling
- Messaging protocol
- Communication objects
- General object library
- Device profiles
- Electronic Data Sheets
- Services
- Bridging and routing
- Data management

There are a few more chapters containing descriptions of further CIP elements, but they are not of significance in the context of this book.

A few terms used throughout this section should be described here to ensure they are well understood:

- Client: Within a Client/Server architecture, the client is the device that sends a request to a server. The client expects a response from the server.
- Server: Within a Client/Server architecture, the server is the device that receives a request from a client. The server is expected to give a response to the client.
- Producer: Within a Producer/Consumer architecture, the producing device places a message on the network for consumption by one or several consumers. The produced message is in general not directed to a specific consumer.
- Consumer: Within a Producer/Consumer architecture, the consumer is one of potentially several consuming devices that pick up a message placed on the network by a producing device.
- Producer/Consumer model: CIP makes use of the Producer/Consumer model as opposed to the traditional Source/Destination message addressing scheme (Figure 15.2). It is inherently multicast. Nodes on the network determine if they should consume the data in a message based on the Connection ID in the packet.

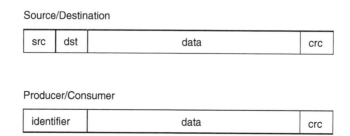

FIGURE 15.2 Source/Destination vs. Producer/Consumer model.

- Explicit Message: Explicit Messages contain addressing and service information that directs the receiving device to perform a certain service (action) on a specific part (e.g., an attribute) of a device.
- Implicit (Input/Output (I/O)) Message: Implicit Messages do not carry address or service information; the consuming node(s) already know what to do with the data based on the Connection ID that was assigned when the connection was established. They are called Implicit Messages because the meaning of the data is implied by the Connection ID.

Let us now have a look at the individual elements of CIP.

15.2.1 Object Modeling

CIP makes use of abstract object modeling to describe:

- The suite of available communication services
- The externally visible behavior of a CIP node
- A common means by which information within CIP products is accessed and exchanged

Every CIP node is modeled as a collection of objects. An object provides an abstract representation of a particular component within a product. Anything not described in object form is not visible through CIP. CIP objects are structured into classes, instances, and attributes.

A class is a set of objects that represents the same kind of system component. An object instance is the actual representation of a particular object within a class. Each instance of a class has the same attributes, but it has its own particular set of attribute values. As Figure 15.3 illustrates, multiple object instances within a particular class can reside within a CIP node. In addition to the instance attributes, an object class may also have class attributes. These are attributes that describe properties of the whole object class, e.g., how many instances of this particular object exist. Furthermore, both object instances and the class itself exhibit a certain behavior and allow certain services to be applied to the attributes, instances, or whole class. All publicly defined objects that are implemented in a device must follow at least the mandatory requirements of the CIP specification. Vendor-specific objects may also be defined with a set of instances, attributes, and services according to the requirements of the vendor. However, they need to follow certain rules described in Chapter 4 of the CIP Specification [4].

The objects and their components are addressed by a uniform addressing scheme consisting of:

- Node Identifier: An integer identification value assigned to each node on a CIP network. On DeviceNet and ControlNet, this is also called MAC ID (Media Access Control Identifier) and is nothing more than the node number of the device. On EtherNet/IP, the Node ID is the IP address.
- Class Identifier (Class ID): An integer identification value assigned to each object class accessible from the network.
- Instance Identifier (Instance ID): An integer identification value assigned to an object instance that identifies it among all instances of the same class.

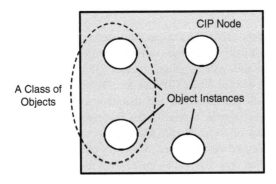

FIGURE 15.3 A class of objects.

- Attribute Identifier (Attribute ID): An integer identification value assigned to a class or instance attribute.
- Service Code: An integer identification value that denotes an action request that can be directed at a particular object instance or object class (see Section 15.2.2).

Object Class Identifiers are divided into open objects, defined in the CIP Specifications (ranging from 0x00 to 0x63 and 0x00F0 to 0x02FF), and vendor-specific objects (ranging from 0x64 to 0xC7 and 0x0300 to 0x04FF); all other Class Identifiers are reserved for future use. In some cases, e.g., within the Assembly Object class, Instance Identifiers are divided into open instances, defined in the CIP Specifications (ranging from 0x00 to 0x63 and 0x0100 to 0x02FF), and vendor-specific instances (ranging from 0x64 to 0xC7 and 0x0300 to 0x04FF); all other Instance Identifiers are reserved for future use. Attribute Identifiers are divided into open attributes, defined in the CIP Specifications (ranging from 0x00 to 0x63), and vendor-specific attributes (ranging from 0x64 to 0xC7); the other Attribute Identifiers are reserved for future use. Vendor-specific objects can be created with a lot of freedom, but they still have to adhere to certain rules specified for CIP, e.g., they can use whatever Instance and Attribute IDs they wish, but their class attributes must follow the CIP Specification.

Figure 15.4 shows an example of this object addressing scheme. More details on object modeling can be found in Chapters 1 and 4 of the CIP Specification [4].

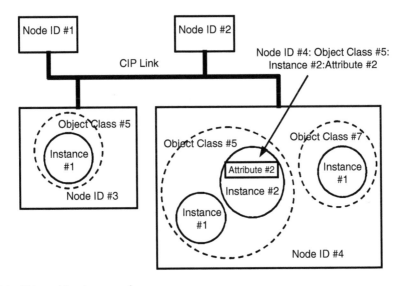

FIGURE 15.4 Object addressing example.

15.2.2   Services

Service Codes are used to define the action that is requested to take place when an object or parts of an object are addressed through Explicit Messages using the addressing scheme described in Section 15.2.1. Apart from the simple read and write functions, a set of CIP Common Services (totaling 22, currently described in [4]) have been defined. These CIP Common Services are common in nature, which means that they can be used in all CIP networks and that they are useful for a large variety of objects. Furthermore, there are object-specific Service Codes that may have a different meaning for the same code, depending on the class of object. Finally, there is a possibility to define vendor-specific services according to the requirements of the developer. While this gives a lot of flexibility, the disadvantage of vendor-specific services is that they may not be understood universally.

Complete details of the CIP Service Codes can be found in Appendix A of the CIP common Specification [4].

15.2.3   Messaging Protocol

CIP is a connection-based protocol. A CIP Connection provides a path between multiple application objects. When a connection is established, the transmissions associated with that connection are assigned a Connection ID (CID) (Figure 15.5). If the connection involves a bidirectional exchange, then two Connection ID values are assigned.

The definition and format of the Connection ID is network dependent. For example, the Connection ID for CIP Connections over DeviceNet is based on the CAN Identifier field.

Since most messaging on a CIP network is done through connections, a process has been defined to establish such connections between devices that are not connected yet. This is done through the Unconnected Message Manager (UCMM) function, which is responsible for the processing of Unconnected Explicit Requests and Responses.

The general method to establish a CIP Connection is by sending a UCMM Forward_Open Service Request Message. While this is the method used on ControlNet and EtherNet/IP (all devices that allow Connected Messaging support it), it is rarely used on DeviceNet so far. For DeviceNet, the simplified methods described in Sections 15.3.1.11 and 15.3.1.12 are typically used. DeviceNet Safety™* (see Section 15.5.2), on the other hand, fully utilizes this service.

A Forward_Open request contains all information required to create a connection between the originator and the target device and, if requested, a second connection between the target and the originator. In particular, the Forward_Open request contains information on the following:

- Time-out information for this connection
- Network Connection ID for the connection from the originator to the target
- Network Connection ID for the connection from the target to the originator
- Information on the identity of the originator (Vendor ID and Serial Number)
- (Maximum) data sizes of the messages on this connection

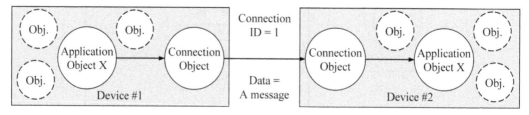

FIGURE 15.5 Connections and Connection IDs.

---

*DeviceNet Safety™ is a trademark of ODVA.

- Trigger mechanisms, e.g., Cyclic, Change of State (COS)
- Connection Path for the application object data in the node

The Connection Path may also contain a Routing Segment that allows connections to exist across multiple CIP networks. The Forward_Open request may also contain an electronic key of the target device (Vendor ID, Device Type, Product Code, Revision), as well as configuration information that will be forwarded to the Configuration Assembly of the target device.

Some networks, like ControlNet and EtherNet/IP, may also make extensive use of Unconnected Explicit Messaging, while DeviceNet uses Unconnected Messaging only to establish connections.

All connections in a CIP network can be divided into I/O Connections and Explicit Messaging Connections:

- I/O Connections provide dedicated, special-purpose communication paths between a producing application and one or more consuming applications. Application-specific I/O data move through these ports and are often referred to as Implicit Messaging. These messages are typically multicast.
- Explicit Messaging Connections provide generic, multipurpose communication paths between two devices. These connections are often referred to as just Messaging Connections. Explicit Messages provide the typical request/response-oriented network communications. These messages are typically point-to-point.

The actual data transmitted in CIP I/O Messages are the I/O data in an appropriate format — it may be prepended by a Sequence Count value. This Sequence Count value can be used to distinguish old data from new, e.g., if a message has been re-sent as a heartbeat in a COS Connection. The two states Run and Idle can be indicated with an I/O Message either by prepending a Run/Idle header, used for ControlNet and EtherNet/IP, or by sending I/O data (Run) or no I/O data (Idle), mainly used for DeviceNet. Run is the normal operative state of a device; the reaction to receiving an Idle event is vendor-specific and application-specific. Typically, this means bringing all outputs of the device to an Idle state, and that typically means "off," i.e., de-energized.

Explicit Messaging requests, on the other hand, contain a Service Code with path information to the desired object (attribute) within the target device followed by data (if any). The associated responses repeat the Service Code followed by status fields followed by data (if any). DeviceNet uses a condensed format for Explicit Messages, while ControlNet and EtherNet/IP use the full format.

More details of the messaging protocol can be found in Chapter 2 of the CIP Specification [4].

15.2.4 Communication Objects

The CIP communication objects manage and provide the runtime exchange of messages. While these objects follow the overall principles and guidelines for CIP objects, the communication objects are unique in a way since they are the focal point for all CIP communication. It therefore makes sense to have a look at them in more detail.

Every instance of a communication object contains a link producer part or a link consumer part, or both. I/O Connections may be either producing or consuming, or producing and consuming, while Explicit Messaging Connections are always producing and consuming.

Figure 15.6 and Figure 15.7 show the typical connection arrangement for CIP I/O Messaging and CIP Explicit Messaging. The attribute values in the Connection Objects define a set of attributes that describe vital parameters of this connection. Note that Explicit Messages are always directed to the Message Router Object.

First of all, they state what kind of connection this is. They specify whether this is an I/O Connection or an Explicit Messaging Connection, but also the maximum size of the data to be exchanged across this connection, and the source and sink of this data. Note that Explicit Messages are always directed to the Message Router Object.

Further attributes define the state of this connection and what kind of behavior this connection is to show. Of particular importance is how messages are triggered (from the application, through Change of State or Change of Data, through Cyclic events or network events) and the timing of the connections

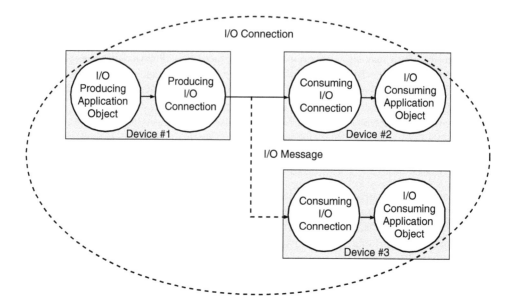

FIGURE 15.6 CIP I/O Multicast Connection.

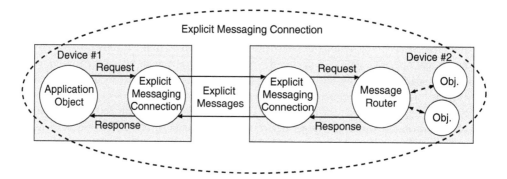

FIGURE 15.7 CIP Explicit Messaging Connection.

(time-out associated with this connection and predefined action if a time-out occurs). CIP allows multiple connections to coexist in a device, although simple devices, e.g., simple DeviceNet slaves, will typically have only one or two connections alive at any given point in time.

Complete details of the communication objects can be found in Chapter 3 of the CIP Specification [4].

15.2.5 Object Library

The CIP family of protocols contains a very large collection of commonly defined objects (currently 48 object classes). The overall set of object classes can be subdivided into three types:

- General-use objects
- Application-specific objects
- Network-specific objects

Apart from the objects that are network-specific, all other objects are used in all three CIP network types.

Figure 15.8 shows the general-use objects, Figure 15.9 shows a group of application-specific objects, and Figure 15.10 shows a group of network-specific objects. New objects are added on an ongoing basis.

The general-use objects can be found in many different devices, while the application-specific objects are typically only found in devices hosting such applications.

- Identity Object, see Section 15.2.5.1
- Message Router Object
- Assembly Object, see Section 15.2.5.3
- Connection Object, see Section 15.2.4
- Connection Manager Object, see Section 15.2.4
- Register Object

- Parameter Object, see Section 15.2.5.2
- Parameter Group Object
- Acknowledge Handler Object
- Connection Configuration Object
- Port Object
- Selection Object
- File Object

FIGURE 15.8 General-use objects.

- Discrete Input Point Object
- Discrete Output Point Object
- Analog Input Point Object
- Analog Output Point Object
- Presence Sensing Object
- Group Object
- Discrete Input Group Object
- Discrete Output Group Object
- Discrete Group Object
- Analog Input Group Object
- Analog Output Group Object
- Analog Group Object
- Position Sensor Object
- Position Controller Supervisor Object
- Position Controller Object

- Sequencer Object
- Command Block Object
- Motor Data Object
- Control Supervisor Object
- AD/DC Drive Object
- Overload Object
- Softstart Object
- S-Device Supervisor Object
- S-Analog Sensor Object
- S-Analog Actor Object
- S-Single Stage Controller Object
- S-Gas Calibration Object
- Trip Point Object
- S-Partial Pressure Object

FIGURE 15.9 Application-specific objects.

- DeviceNet Object, see Section 15.3.1.4.1
- ControlNet Object, see Section 15.3.2.4.1
- ControlNet Keeper Object, see Section 15.3.2.4.2
- ControlNet Scheduling Object, see Section 15.3.2.4.3
- TCP/IP Interface Object, see Section 15.3.3.5.1
- Ethernet Link Object, see Section 15.3.3.5.2

FIGURE 15.10 Network-specific objects.

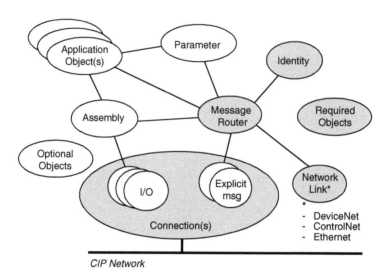

FIGURE 15.11   Typical device object model.

This looks like a large number of object types, but typical devices only implement a subset of these objects. Figure 15.11 shows the object model of such a typical device.

The objects required in a typical device are:

- Either a Connection Object or a Connection Manager Object
- An Identity Object
- One or several network link-related objects (depends on network)
- A Message Router Object (at least its function)

Further objects are added according to the functionality of the device. This allows very good scalability of implementations so that small devices such as a proximity sensor on DeviceNet are not burdened with unnecessary overhead. Developers typically use publicly defined objects (see above list), but can also create their own objects in the vendor-specific areas, e.g., Class IDs 100 to 199. However, it is strongly encouraged to work with the Special Interest Groups (SIGs) of ODVA and ControlNet International to create common definitions for further objects instead of inventing private ones.

Out of the general-use objects, several will be described in more detail below.

15.2.5.1   Identity Object (Class Code 0x01)

This object is described in more detail for two reasons: (1) being a relatively simple object, it can easily be used to show the general principles, and furthermore, (2) every device must have an Identity Object. Therefore, it is of general interest in this context.

The vast majority of devices only support one instance of the Identity Object. Thus, there are typically no requirements for any class attributes that would describe further class details, e.g., how many instances exist in the device; only instance attributes are required in most cases. There are mandatory attributes (Figure 15.12) and optional attributes (Figure 15.13).

- The Vendor ID attribute allows an identification of the vendor of every device. This UINT (Unsigned Integer) value (for Data Type descriptions, see Section 15.2.9) is assigned to a specific vendor by ODVA or ControlNet International. If a vendor intends to build products for more than one CIP network, he will get the same Vendor ID for all networks.
- The Device Type specifies which profile has been used for this device. It must be one of the Device Types described in Chapter 6 of the CIP Specification [4] or a vendor-specific type (see Section 15.2.6).
- The Product Code is a UINT number defined by the vendor of the device. This is used to distinguish multiple products of the same Device Type from the same vendor.

- Vendor ID
- Device Type
- Product Code
- Revision
- Status
- Serial Number
- Product Name

FIGURE 15.12 Mandatory attributes.

- State
- Configuration Consistency Value
- Heartbeat Interval
- Languages Supported

FIGURE 15.13 Optional attributes.

- The Revision is split into two USINT (Unsigned Short Integer) values specifying a Major Revision and a Minor Revision. Any change of the device that results in a modified behavior of the device on the network must be reflected in a change of at least the Minor Revision. Any change in the device that needs a revised Electronic Data Sheet (EDS; see Section 15.2.7) must be reflected in a change of the Major Revision. Vendor ID, Device Type, Product Code, and Major Revision allow an unambiguous identification of an EDS for this device.
- The Status attribute provides information on the status of the device, e.g., whether it is owned (controlled by another device), whether it is configured (to something different from the out-of-the-box default), and whether any major or minor faults have occurred.
- The Serial Number is used to uniquely identify individual devices in conjunction with the Vendor ID, i.e., no two CIP devices of a vendor may carry the same Serial Number. The 32 bits of the Serial Number allow ample space for a subdivision into number ranges that could be used by different divisions of larger companies.
- The Product Name attribute allows the vendor to give a meaningful ASCII name string (up to 32 characters) to the device.
- The State attribute describes the state of a device in a single UINT value; it is thus less detailed than the Status attribute.
- The Configuration Consistency Value allows a distinction between a configured and an unconfigured device or between different configurations in a device. This helps avoid unnecessary configuration downloads.
- The Heartbeat Interval allows enabling of the Device Heartbeat Message and setting the maximum time between two heartbeats to 1 to 255 s.

The services supported by the class and instance attributes are either Get_Attribute_Single (typically implemented in DeviceNet devices) or Get_Attributes_All (typically implemented in ControlNet and EtherNet/IP devices). None of the attributes is settable, except for the Heartbeat Interval (if implemented). The only other service that is typically supported by the Identity Object is the reset service.

The behavior of the Identity Object is described through a state transition diagram. This and further details of the Identity Object can be found in Chapter 5 of the CIP Specification [4].

15.2.5.2 Parameter Object (Class Code 0x0F)

This object is described in some detail since its concept is referred to in Section 15.2.7, "Configuration and Electronic Data Sheets." This object, when used, comes in two "flavors": a complete object and an abbreviated version (Parameter Object Stub). This abbreviated version is mainly used by DeviceNet

Parameter Value	This is the actual parameter.
Link Path Size	These two attributes contain information on what application object/instance/
Link Path	attribute the parameter value is retrieved from.
Descriptor	This describes parameter properties, e.g., read-only, monitor parameter, etc.
Data Type	This must be one of the Data Types described in Chapter C-6.1 of the CIP Specification; see Section 15.2.9.
Data Size	Data size in bytes.

FIGURE 15.14 Parameter Object Stub attributes.

devices that only have small amounts of memory available. The Object Stub in conjunction with the Electronic Data Sheet has more or less the same functionality as the full object (see Section 15.2.7).

The purpose of this object is to provide a general means to allow access to many attributes of the various objects in the device without a simple tool (such as a handheld terminal) having to know anything about the specific objects in the device.

The class attributes of the Parameter Object contain information on how many instances exist in this device and a Class Descriptor indicating, among other properties, whether a full or stub version is supported. Furthermore, they tell whether a Configuration Assembly is used and what language is used in the Parameter Object.

Of the instance attributes, the first six are those required for the Object Stub. These are listed in Figure 15.15. These six attributes already allow access, interpretation, and modification of the parameter value, but the remaining attributes make life a lot better:

- The next three attributes provide ASCII strings with the name of the parameter, its engineering units, and an associated help text.
- Another three attributes contain the minimum, maximum, and default values of the parameter.
- The next four attributes that follow allow scaling of the parameter value so that the parameter can be displayed in a more meaningful way, e.g., raw value in multiples of 10 mA, scaled value displayed in amps.
- Another four attributes follow that can link the scaling values to other parameters. This feature allows variable scaling of parameters, e.g., percentage scaling to a full range value that is set by another parameter.
- Attribute 21 defines how many decimal places are to be displayed if the parameter value is scaled.
- Finally, the last three attributes are an international language version of the parameter name, its engineering units, and the associated help text.

15.2.5.3 Assembly Object (Class Code 0x04)

Using the Assembly Object gives the option of mapping data from attributes of different instances of various classes into one single attribute (attribute 3) of an instance of the Assembly Object. This mapping is generally used for I/O Messages to maximize the efficiency of the control data exchange on the network. Due to the Assembly mapping, the I/O data are available in one block; thus, there are fewer Connection Object instances and fewer transmissions on the network. The process data are normally combined from different application objects. An Assembly Object can also be used to configure a device with a single data block, rather than having to set individual parameters.

CIP makes a distinction between Input and Output Assemblies. Input and output in this context are viewed from the network. An Input Assembly reads data from the application and produces it on the network. An Output Assembly consumes data from the network and writes the data to the application. This data mapping is very flexible; even mapping of individual bits is permitted. Assemblies can also be

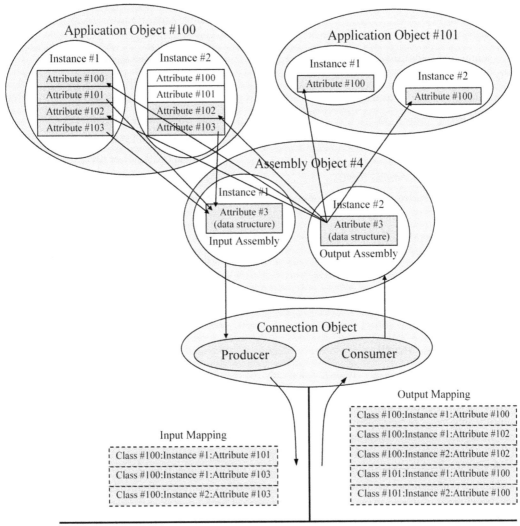

FIGURE 15.15 Example of an Assembly mapping.

used to transmit a complete set of configurable parameters instead of accessing them individually. These Assemblies are called Configuration Assemblies.

Figure 15.15 shows an example of an Assembly mapping. The data from application objects 100 and 101 are mapped in two instances of the Assembly Object. Instance 1 is set up as an Input Assembly for the input data and instance 2 as an Output Assembly for output data. The data block is always accessed via attribute 3 of the relevant Assembly instance. Attributes 1 and 2 contain mapping information.

The I/O Assembly mapping is specified for certain Device Profiles (e.g., Motor Starters) by the ODVA. Device developers can then choose which Assemblies they support in their products. If none of the publicly defined Assemblies fully represent the functionality of the product, a device vendor may implement additional vendor-specific Assemblies (Instance IDs 100 to 199).

CIP defines static and dynamic Assembly Objects. Whereas mapping for static Assemblies is permanently programmed in the device (ROM), it can be modified and extended for dynamic mapping (RAM). Most simple CIP devices support only static Assembly Objects. Dynamic Assembly Objects tend to be used in more complex devices.

• Generic Device (0x00)	• Motor Starter (0x16)
• AC Drives (0x02)	• Soft Starter (0x17)
• Motor Overload (0x03)	• Human Machine Interface (0x18)
• Limit Switch (0x04)	• Mass Flow Controller (0x1A)
• Inductive Proximity Switch (0x05)	• Pneumatic Valve(s) (0x1B)
• Photoelectric Sensor (0x06)	• Vacuum/Pressure Gauge (0x1C)
• General Purpose Discrete I/O (0x07)	• Process Control Valve (0x1D)
• Resolver (0x09)	• Residual Gas Analyzer (0x1E)
• Communications Adapter (0x0C)	• DC Power Generator (0x1F)
• ControlNet Programmable Logic Controller (0x0E)	• RF Power Generator (0x20)
• Position Controller (0x10)	• Turbomolecular Vacuum Pump (0x21)
• DC Drives (0x13)	• ControlNet Physical Layer (0x32)
• Contactor (0x15)	(this is not a "normal" profile; it does not contain any objects)

FIGURE 15.16 CIP Device Types.

15.2.6 Device Profiles

It would be possible to design products using only the definitions of communication links and objects, but this could easily result in similar products having quite different data structures and behavior. To overcome this situation and to make the application of CIP devices much easier, devices of similar functionality have been grouped into Device Types with associated profiles. Such a CIP profile contains the full description of the object structure and behavior. Figure 15.16 shows the Device Types and associated profiles that have been defined so far in the CIP Specification [4] (profile numbers are in parentheses).

Device developers must use a profile. Any device that does not fall into the scope of one of the specialized profiles must use the Generic Device profile or a vendor-specific profile. What profile is used and which parts of it are implemented must be described in the user documentation of the device.

Every profile consists of a set of objects, some required, some optional, and a behavior associated with that particular type of device. Most profiles also define one or several I/O data formats (Assemblies) that define the meaning of the individual bits and bytes of the I/O data. In addition to the publicly defined object set and I/O data Assemblies, vendors can add objects and Assemblies of their own if they have devices that have additional functionality. If that is still not appropriate, vendors can create profiles of their own within the vendor-specific profile range. They are then free to define whatever behavior and objects are required for their device as long as they stick to some general rules for profiles. Whenever additional functionality is used by multiple vendors, ODVA and ControlNet International encourage coordinating these new features through discussion in the Special Interest Groups (SIGs). They can then create new profiles and additions to the existing profiles for everybody's use and the benefit of the device users.

All open (ODVA/CI defined) profiles carry numbers in the 0x00 through 0x63 or 0x0100 through 0x02FF ranges, while vendor-specific profiles carry numbers in the 0x64 through 0xC7 or 0x0300 through 0x02FF ranges. All other profile numbers are reserved by CIP.

Complete details of the CIP profiles can be found in Chapter 6 of the CIP Specification [4].

15.2.7 Configuration and Electronic Data Sheets

CIP has made provisions for several options to configure devices:

- A printed data sheet
- Parameter Objects and Parameter Object Stubs
- An Electronic Data Sheet (EDS)
- A combination of an EDS and Parameter Object Stubs
- A Configuration Assembly combined with any of the above methods

When using configuration information collected on a printed data sheet, configuration tools can only provide prompts for service, class, instance, and attribute data and relay this information to a device. While this procedure can do the job, it is the least desirable solution since it does not determine the context, content, or format of the data.

Parameter Objects, on the other hand, provide a full description of all configurable data of a device. This allows a configuration tool to gain access to all parameters and maintain a user-friendly interface since the device itself provides all the necessary information. However, this method imposes a burden on a device with full parameter information, which may be excessive for a small device, e.g., a simple DeviceNet slave. Therefore, an abbreviated version of the Parameter Object, called Parameter Object Stub, may be used (see Section 15.2.5.2). This still allows access to the parameter data, but it does not describe any meaning of this data. Parameter Stubs in conjunction with a printed data sheet are usable, but certainly not optimal. On the other hand, an EDS supplies all the information that a full Parameter Object contains in addition to I/O Connection information. The EDS thus provides the full functionality and ease of use of the Parameter Object without imposing an excessive burden on the individual devices. Another value of the EDS is that it provides a means for tools to do offline configuration and download the configuration data to the device at a later point in time.

An EDS is a simple ASCII text file that can be generated on any ASCII editor. The CIP Specification lays down a set of rules for the overall design and syntax of an EDS. The main purpose of the EDS is to give information on several aspects of the device's capabilities, the most important ones being the I/O Connections it supports and what parameters for display or configuration exist within the device. It is highly recommended that all supported I/O Connections are described within the EDS; this makes the application of a device much easier. When it comes to parameters, it is up to the developer to decide which items to make accessible to the user.

Let us look at some details of the EDS: EDSs are structured into sections. Every section starts with a section name in square brackets []. The first two sections are mandatory for all EDSs.

- [File]: Describes the contents and revision of the file
- [Device]: Is equivalent to the Identity Object information and is used to match an EDS to a device
- [Device Classification]: Describes what network the device can be connected to, optional for DeviceNet, required for ControlNet and EtherNet/IP
- [IO_Info]: Describes connection methods and I/O sizes, DeviceNet only
- [Variant_IO_Info]: Describes multiple I/O_Info data sets, DeviceNet only
- [ParamClass]: Describes class-level attributes of the Parameter Objects
- [Params]: Identifies all configuration parameters in the device; follows the Parameter Object definition, further details below
- [EnumPar]: Enumeration list of parameter choices to present to the user; old method specified for DeviceNet only
- [Assembly]: Describes the structure of data items
- [Groups]: Identifies all parameter groups in the device and lists group name and Parameter Object instance numbers
- [Connection Manager]: Describes connections supported by the device, typically used in ControlNet and EtherNet/IP
- [Port]: Describes the various network ports a device may have
- [Modular]: Describes modular structures inside a device
- [Capacity]: Brand new EDS section to specify communication capacity of EtherNet/IP and ControlNet devices

With these sections, a very detailed description of a device can be made. Only a few of these details are described here, and further reading is available in [25] and [26].

A tool with a collection of EDSs will first use the device section to try to match an EDS with each device it finds on a network. Once this is done and a particular device is chosen, the tool can then display device properties and parameters and allow their modification (if necessary). A tool may also display what I/O Connections a device may allow and which of these are already in use. EDS-based tools are mainly used for slave or adapter devices; scanner devices are typically too complex to be configured through EDSs. For those devices, the EDS is mainly used to identify the device and then guide the tool to call a matching configuration applet.

A particular strength of the EDS approach lies in the methodology of parameter configuration. A configuration tool typically takes all the information supplied by the Parameter Object and an EDS and displays it in a user-friendly manner. This enables the user to configure a device in many cases without the need of a detailed manual; the tool presentation of the parameter information together with help texts allows one to make the right decisions for a complete device configuration, provided, of course, that the developer has supplied all information required.

A complete description of what can be done with EDSs goes well beyond the scope of this handbook. For further details, consult [25, 26] and Chapter 7 of the CIP Specification [4].

15.2.8 Bridging and Routing

CIP has defined mechanisms that allow the transmission of messages across multiple networks, provided the bridging devices (routers) between the various networks are equipped with a suitable set of capabilities (objects and support of services). Once this is the case, the message will be forwarded from router to router until it has reached its destination node.

Here is how it works. For Unconnected Explicit Messaging, the actual Explicit Message to be executed on the target device is wrapped up using another type of Explicit Message service, the so-called Unconnected_Send Message. This Unconnected_Send Message (Service Code 0x52 of the Connection Manager Object) contains complete information on the transport mechanism, in particular time-outs (they may be different while the message is still en route) and path information. The first router device that receives an Unconnected_Send Message will take its contents and forward them to the next network as specified within the path section of the message. Before the message is actually sent, the used part of the path is removed, but remembered by the intermediate router device for the return of any response. This process is executed for every hop until the final destination network is reached. The number of hops is theoretically limited by the message length.

Once the Unconnected_Send Message has arrived at the target network, the inner Explicit Message is then sent to the target device, which executes the requested service and generates a response. This response is then routed back through all the routers it has gone through during its forward journey until it has finally reached the originating node. It is important to note in this context that the transport mechanism may have been successful in forwarding the message and returning the response, but the response could still contain an indication that the desired service could not be performed successfully in the target network/device. Through this mechanism, the router devices do not need to know anything about the message paths ahead of time. Thus, no programming of any of the router devices is required. This is often referred to as seamless routing.

When a connection (I/O or Explicit) is set up using the Forward_Open service (see Section 15.3.2.10), it may go to a target device on another network. To enable the appropriate setup process, the Forward_Open Message may contain a field with path information describing a route to the target device. This is very similar to the Unconnected_Send service described above. This routing information is then used to create routed connections within the routing devices between the originator and the target of the message. Once set up, these connections automatically forward any incoming messages for this connection to the outgoing port en route to the target device. Again, this is repeated until the message has reached its target node. As with routed Unconnected Explicit Messages, the number of hops is generally limited

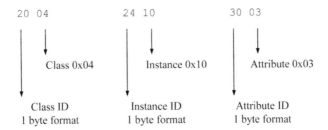

FIGURE 15.17 Logical Segment encoding example.

only by the capability of the devices involved in this process. In contrast to routed Unconnected Messages, routed Connected Messages do not carry path information. Since Connected Messages always use the same path for any given connection, the path information that was given to the routing devices during connection setup is held there as long as the connection exists. Again, the routing devices do not have to be preprogrammed; they are self-configured during the connection establishment process.

15.2.9 Data Management

The Data Management part of the CIP Specification describes addressing models for CIP entities and the data structure of the entities themselves.

The entity addressing is done by so-called Segments, a method that allows very flexible usage so that many different types of addressing methods can be accommodated. The first byte of a CIP Segment allows a distinction between a Segment Address (0x00 to 0x9F) and a Data Type description (0xA0 to 0xDF). Two uses of this addressing scheme (Logical Segments and Data Types) are looked at in a little more detail here; all of them are described in Appendix C of the CIP Specification [4].

15.2.9.1 Logical Segments

Logical Segments (first byte = 0x20 to 0x3F) are addressing segments that can be used to address objects and their attributes within a device. They are typically structured into [Class ID] [Instance ID] [Attribute ID, if required].

Each element of this structure allows various formats (1, 2, and 4 bytes). Figure 15.17 shows a typical example of this addressing method.

This type of addressing is commonly used to point to assemblies, parameters, or any other addressable attribute within a device. It is extensively used in EDSs, but also within Unconnected Messages, to name just a few application areas. A complete list of all Segment types and their encoding can be found in Appendix C of the CIP Specification [4].

15.2.9.2 Data Types

Data Types (first byte = 0xA0 to 0xDF) can be either structured (first byte = 0xA0 to 0xA3) or elementary (first and only byte = 0xC1 to 0xDE) Data Types. Structured Data Types can be arrays of elementary Data Types or any assembly of arrays or elementary Data Types. Of particular importance in the context of this handbook are the elementary Data Types. They are used within EDSs to specify the Data Types of parameters and other entities.

Here is a list of commonly used Data Types:

- 1 bit (encoded into 1 byte):
  - Boolean, BOOL, Type Code 0xC1
- 1 byte:
  - Bit string, 8 bits, BYTE, Type Code 0xD1
  - Unsigned 8-bit integer, USINT, Type Code 0xC6
  - Signed 8-bit integer, SINT, Type Code 0xC2
- 2 bytes:
  - Bit string, 16 bits, WORD, Type Code 0xD2

- Unsigned 16-bit integer, UINT, Type Code 0xC7
- Signed 16-bit integer, INT, Type Code 0xC3
- 4 bytes:
  - Bit string, 32 bits, DWORD, Type Code 0xD3
  - Unsigned 32-bit integer, UDINT, Type Code 0xC8
  - Signed 32-bit integer, DINT, Type Code 0xC4

The Data Types in CIP follow the requirements of IEC 61131-3 [9]. A complete list of all Data Types and their encodings can be found in Appendix C of the CIP Specification [4].

15.2.9.3 Maintenance and Further Development of the Specifications

Both ODVA and ControlNet International have a set of working groups that have the task of maintaining the specifications and creating protocol extensions, e.g., new profiles or functional enhancements such as CIP Sync and CIP Safety. These groups are called Special Interest Groups (SIGs) for DeviceNet and ControlNet and Joint Special Interest Groups (JSIGs) for EtherNet/IP. JSIGs are called "joint" since it is a combination of ODVA and ControlNet International members that do the work, since the EtherNet/IP technology is jointly administered by both groups.

The results of these SIGs are written up as DSEs (DeviceNet Specification Enhancements), CSEs (ControlNet Specification Enhancements), ESEs (EtherNet/IP Specification Enhancements), or CIPSEs (CIP Specification Enhancements), presented to the Technical Review Board (TRB) for approval and then incorporated into the specifications. Only ODVA or ControlNet International members can work within the SIGs, and those participants have the advantage of advance knowledge of technical changes. Participation in one or several SIGs is therefore highly recommended.

15.3 Network Adaptations of CIP

Up to now there are three public derivatives of CIP. These three derivatives are based on quite different data link layers and transport mechanisms, but they maintain the principles of CIP.

15.3.1 DeviceNet

DeviceNet was the first public implementation of CIP. As already mentioned in Section 15.2, DeviceNet is based on the Controller Area Network (CAN). The adaptations of CIP are done to accommodate certain limitations of the CAN protocol (up to 8 bytes payload only) and to allow for a simple device with only minimal processing power; for a more detailed description of the CAN protocol and some of its applications, see [10]. DeviceNet uses a subset of the CAN protocol (11-bit identifier only, no remote frames).

Figure 15.18 shows the relationship between CIP, DeviceNet, and the ISO/OSI layer model.

15.3.1.1 Physical Layer and Relationship to CAN

The physical layer of DeviceNet is an extension of ISO 11898 [11]. This extension defines the following additional details:

- Improved transceiver characteristics that allow the support of up to 64 nodes per network
- Additional circuitry for overvoltage and miswiring protection
- Several types of cables for a variety of applications
- Several types of connectors for open (IP20) and enclosed (IP65/67) devices

These extensions result in a communication system with the following physical layer characteristics:

- Trunkline/dropline configuration
- Support for up to 64 nodes
- Node removal without severing the network
- Simultaneous support for both network-powered (sensors) and separately powered (actuators) devices

Layers according to
ISO/OSI

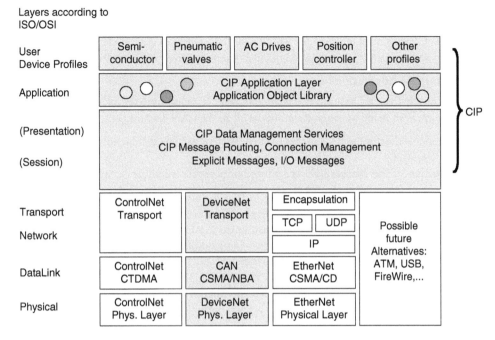

FIGURE 15.18 Relationship between CIP and DeviceNet.

- Use of sealed or open-style connectors
- Protection from wiring errors
- Selectable data rates of 125, 250, and 500 kBaud
- Adjustable power configuration to meet individual application needs
- High current capability (up to 16 amps per supply)
- Operation with off-the-shelf power supplies
- Power taps that allow the connection of several power supplies from multiple vendors that comply with DeviceNet standards
- Built-in overload protection
- Power available along the bus: both signal and power lines contained in the cable

The cables described in the DeviceNet Specification have been specially designed to meet minimum propagation speed requirements to make sure they can be used up to the maximum system length. Using the specified cables, in conjunction with suitable transceiver circuits, results in overall systems, as specified in Figure 15.19.

ODVA has issued a guideline [7] that gives complete details on how to build the physical layer of a DeviceNet network.

Developers of DeviceNet devices have the choice of creating DeviceNet circuits with or without physical layer isolation; both versions are fully specified. Furthermore, a device may take some or all of its power

	Trunk Distance			Drop Length	
Data Rate	Thick Cable	Thin Cable	Flat Cable	Maximum	Cumulative
125 kBaud	500 meters		420 meters		156 meters
250 kBaud	250 meters	100 meters	200 meters	6 meters	78 meters
500 kBaud	100 meters		75 meters		39 meters

FIGURE 15.19 Data rate vs. trunk and drop length.

from the bus, thus avoiding extra power lines for devices that can live on the power supplied through the DeviceNet cable.

All DeviceNet devices must be equipped with one of the connectors described in the DeviceNet Specification. Hard wiring of a device is allowed, provided the node is removable without severing the trunk.

15.3.1.2 Protocol Adaptations

On the protocol side, there are basically two adaptations of CIP (apart from the addition of the DeviceNet Object) that have been made to better accommodate it to the CAN data frame:

- Limitation to short messages (8 bytes or less) where possible; introduction of fragmentation for longer messages.
- Introduction of the Master/Slave communication profile minimizes connection establishment management (see Section 15.3.1.12).

These two features have been created to allow the use of simple and thus inexpensive microcontrollers. This is particularly important for small, cost-sensitive devices like photo-eyes or proximity sensors. As a result of this specialization, the DeviceNet protocol in its simplest form has been implemented in 8-bit microprocessors with as little as 4 kbytes of code memory and 175 bytes of RAM.

The fragmentation of messages comes in two varieties. For I/O Messages typically sent with a fixed length, the use of fragmentation is defined through the maximum length of data to be transmitted through a connection. Any connection that has more than 8 bytes to transmit always uses the fragmentation protocol, even if the actual data to be transmitted are 8 bytes or less, e.g., an Idle Message. For Explicit Messaging, the use of the fragmentation protocol is indicated with every message, since the actual message will vary in length. The actual fragmentation protocol is contained in one extra byte within the message that indicates whether the fragment is a start, middle, or end fragment. A modulo 64 rolling fragment counter allows very long fragmented messages, in theory limited only by the maximum Produced or Consumed Connection Sizes (65,535 bytes). In reality, it is the capability of the devices that limits the message sizes.

15.3.1.3 Indicators and Switches

DeviceNet devices may be built with indicators or without, but it is recommended to incorporate some of the indicators described in the specification. These indicators allow the user to determine the state of the device and its network connection. Devices may have additional indicators with a behavior not described in the specification. However, any indicators that carry names of those described in the specification must also follow their specified behavior.

Devices may be built with or without switches or other directly accessible means for configuration. If switches for MAC ID and baud rate exist, then certain rules apply regarding how these values have to be used at power-up and during the operation of the device.

15.3.1.4 Additional Objects

The DeviceNet Specification defines one additional object, the DeviceNet Object.

15.3.1.4.1 DeviceNet Object (Class Code 0x03)

A DeviceNet Object is required for every DeviceNet port of the device. The instance attributes of this object contain information on how this device uses the DeviceNet port. In particular, there is information about the MAC ID of the device and the (expected) baud rate of the DeviceNet network this device is attached to. Both attributes are always expected to be nonvolatile, i.e., after a power interruption, the device is expected to try to go online again with the same values that were stored in these attributes before the power interruption. Devices that set these values through switches typically override any stored values at power-up.

15.3.1.5 Network Access

DeviceNet uses the network access mechanisms described in the CAN specification, i.e., bitwise arbitration through the CAN Identifier for every frame to be sent. This requires a system design that does not allow multiple uses of any of these identifiers. Since the node number of every device is coded into the CAN Identifier (see Section 15.3.1.10), it is generally sufficient to make sure that none of the node numbers exist more than once on any given network. This is guaranteed through the Network Access algorithm (see Section 15.3.1.6).

15.3.1.6 Going Online

Any device that wants to communicate on DeviceNet must go through a Network Access algorithm before any communication is allowed. The main purpose of this process is to avoid duplicate Node IDs on the same network. Every device that is ready to go online sends a Duplicate MAC ID Check Message containing its Port Number, Vendor ID, and Serial Number. If another device is already online with this MAC ID or is in the process of going online, it responds with a Duplicate MAC ID Response Message that directs the checking device to go offline and not communicate any further.

If two or more devices with the same MAC ID should happen to try to go online at exactly the same time, all of them will win arbitration at the same time (same CAN ID) and will proceed with their messages. However, since they will exhibit different values in the data field of the message, all devices on the link will flag Cyclic Redundancy Check (CRC) errors and thus cause a repetition of the message. This may eventually result in a Bus-Off condition for these devices, but a situation with duplicate Node ID is safely avoided.

15.3.1.7 Offline Connection Set

The Offline Connection Set is a set of messages that have been created to communicate with devices that have failed to go online (see Section 15.3.1.6), e.g., to allow setting a new MAC ID. Full details of these messages can be found in [5] or [10].

15.3.1.8 Explicit Messaging

All Explicit Messaging in DeviceNet is done via connections and the associated Connection Object instances. However, these objects must first be set up in the device. This can be done by using the Predefined Master/Slave Connection Set to activate a static Connection Object already available in the device or by using the Unconnected Message Manager (UCMM) port of a device, through which this kind of Connection Object can be dynamically set up. The only messages that can be sent to the UCMM are Open or Close requests that set up or tear down a Messaging Connection, while the only messages that can be sent to the Master/Slave equivalent are an Allocate or Release request (see also Section 15.3.1.12). Explicit Messages always pass via the Message Router Object to the individual objects (refer to Figure 15.11).

As mentioned in Section 15.2.3, Explicit Messages on DeviceNet have a very compact structure to make them fit into the 8-byte frame in most cases. Figure 15.20 shows a typical example of a request message.

Byte offset	Bit number								
	7	6	5	4	3	2	1	0	
0	Frag [0]	XID	MAC ID						Message header
1	R/R [0]	Service Code							Message body
2	Class ID								
3	Instance ID								
4 ... 7	Service data ... (optional)								

FIGURE 15.20 Format of nonfragmented Explicit Request Message using the 8/8 message body format (1 byte for Class ID, 1 byte for Instance ID).

Byte offset	Bit number								
	7	6	5	4	3	2	1	0	
0	Frag [0]	XID	MAC ID						Message header
1	R/R [1]	Service Code							
2 ... 7	Service data ... (optional)								Message body

FIGURE 15.21 Format of a nonfragmented 8/8 Explicit Response Message.

The consumer of this Explicit Message responds in the format shown in Figure 15.21. The consumer sets the R/R (Request/Response) bit and repeats the Service Code of the request message. If data are transferred with the response, this is entered in the service data field.

Most messages will use the 8/8 format shown in Figure 15.20, since they only need to address Class and Instance IDs up to 255. If there is a need to address any Class/Instance combinations above 255, then this is negotiated between the two communication partners during the setup of the connection.

Should an error occur, the receiver responds with the Error Response Message. The Service Code for this message is 0x14; 2 bytes of error code are returned in the service data field. Further details of the message encoding, including the use of fragmentation, can be found in [5, 10].

15.3.1.9 I/O Messaging

I/O Messages have a very compact structure; only the naked data are transmitted without the Run/Idle header and Sequence Count value used in ControlNet and EtherNet/IP. For messages up to 8 bytes long, the full CAN data field is used for I/O data. I/O Messages that are longer use 1 byte of the CAN data field for the fragmentation protocol (Figure 15.22 and Figure 15.23). I/O Messages without data (i.e., with zero length data) indicate the Idle state of the producing application. Any producing device can do this — master, slave, or peer.

Byte offset	Bit number							
	7	6	5	4	3	2	1	0
0 ... 7	Process data (0 – 8 bytes)							

FIGURE 15.22 Format of a nonfragmented I/O Message, 0 to 8 bytes.

Byte offset	Bit number							
	7	6	5	4	3	2	1	0
0	Fragmentation protocol							
1 ... 7	Process data (0 – 7 bytes)							

FIGURE 15.23 Format of the fragmented I/O Message.

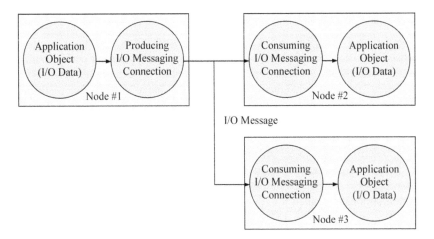

FIGURE 15.24 I/O Messaging Connections.

As already mentioned, I/O Messages are used to exchange high-priority application and process data via the network, and this communication is based on the Producer/Consumer model. The associated I/O data are always transferred from one producing application object to one or more consuming application objects. This is undertaken using I/O Messages via I/O Messaging Connection Objects (Figure 15.24 shows two consuming applications) that must have been previously set up in the device.

This can be done in one of two ways by using:

- The Predefined Master/Slave Connection Set to activate a static I/O Connection Object already available in the device
- An Explicit Messaging Connection Object already available in the device to dynamically create and set up an appropriate I/O Connection Object

I/O Messages usually pass directly to the data of the assigned application object. The Assembly Object is the most common application object used with I/O Connections. Refer to Figure 15.11.

15.3.1.10 Using the CAN Identifier

DeviceNet is based on the standard CAN protocol and therefore uses an 11-bit message identifier. A distinction can therefore be made between $2^{11} = 2048$ messages. Six bits is sufficient to identify a device because a DeviceNet network is limited to a maximum of 64 participants. This 6-bit Device Identifier (node address) is also called MAC ID.

The overall CAN Identifier range is divided into four Message Groups of varying sizes (Figure 15.25).

In DeviceNet, the CAN Identifier is the Connection ID. This is composed of the ID of the Message Group, the Message ID within this group, and the MAC ID of the device. The source or destination address is possible as the MAC ID. The definition depends on the Message Group and the Message ID. The significance of the message within the system is defined by the Message Group and Message ID.

Connection ID = CAN Identifier (bits 10:0)											Used for
10	9	8	7	6	5	4	3	2	1	0	
0	Message ID				Source MAC ID						Message Group 1
1	0	MAC ID						Message ID			Message Group 2
1	1	Message ID			Source MAC ID						Message Group 3
1	1	1	1	1	Message ID						Message Group 4
1	1	1	1	1	1	1	x	x	x	x	Invalid CAN Identifiers

FIGURE 15.25 Definition of the Message Groups.

The four Message Groups are used as follows:

- Message Group 1: Assigned 1024 CAN Identifiers (0x0000 to 0x03FF), 50% of all identifiers available. Up to 16 different Message IDs are available to the user per device (node) within this group. The priority of a message from this group is primarily determined by the Message ID (the significance of the message) and only after that by the source MAC ID (the producing device). If two devices transmit at the same time, then the device with a lower Message ID will always win the arbitration. However, if two devices transmit the same Message ID at the same time on the CAN bus, then the device with the lower node number will win. A 16-stage priority system can be set up relatively easily in this manner. The messages of Group 1 are therefore well suited for the exchange of high-priority process data.

- Message Group 2: Assigned 512 identifiers (0x0400 to 0x05FF). Most of the Message IDs of this group are optionally defined for what is commonly referred to as the Predefined Master/Slave Connection Set (see Section 15.3.1.12). One Message ID is defined for network management (Section 15.3.1.6). The priority here is primarily determined by the device address, and only after that by the Message ID. If you consider the bit positions in detail, you will see that a CAN controller with an 8-bit mask is able to filter out its Group 2 Messages based on MAC ID.

- Message Group 3: Has 448 CAN Identifiers (0x0600 to 0x07BF) and a structure similar to that of Message Group 1. Unlike this group, however, low-priority process data are mainly exchanged. In addition to this, the main use of this group is also the setting up of dynamic Explicit Connections. Seven Message IDs are possible per device; two of these are reserved for what is commonly referred to as the UCMM port (Section 15.3.1.11).

- Message Group 4: Has 48 CAN Identifiers (0x07C0 to 0x07EF) and does not include any device addresses, but only Message IDs. The messages of this group are only used for network management. Four Message IDs are currently assigned for services of the Offline Connection Set.

The other 16 CAN Identifiers (0x07F0 to 0x07FF) are invalid CAN IDs and thus not permitted in DeviceNet systems.

This type of CAN Identifier issuing system means that unused Connection IDs (CAN Identifiers) cannot be used by other devices. Each device has exactly 16 Message IDs in Group 1, 8 Message IDs in Group 2, and 7 Message IDs in Group 3. One advantage of this system is that the CAN Identifiers used in the network can always be clearly assigned to a device. Devices are responsible for managing their own identifiers. This simplifies not only the design but also troubleshooting and diagnosis in DeviceNet systems. A central tool that keeps a record of all assignments on the network is not needed.

15.3.1.11 Connection Establishment

As is described in Sections 15.3.1.8 and 15.3.1.9, messages in DeviceNet are always exchanged in a connection-based manner. Communication objects must be set up for this purpose. These are not initially available when a device is switched on; they first have to be created. The only port by which a DeviceNet device can be addressed when first switched on is the Unconnected Message Manager port (UCMM port) or the Group 2 Only Unconnected Explicit Message port of the Predefined Master/Slave Connection Set. Picture these ports like doors to the device. Only one particular key will fit in each lock. The appropriate key to this lock is the Connection ID, i.e., the CAN Identifier of the selected port. Other doors in the device can only be opened once the appropriate key is available and other Connection Objects are set up.

The setting up of a link via the UCMM port represents a general procedure to be strived for with all DeviceNet devices. Devices that in addition to having the Predefined Master/Slave Connection Set are also UCMM capable are called Group 2 Servers. A Group 2 Server can be addressed by one or more connections from one or more clients.

Since UCMM capable devices need a good amount of processing power to service multiple communication requests, a simplified communication establishment and I/O data exchange method has been created for low-end devices. This is called the Predefined Master/Slave Connection Set (see Section 15.3.1.12). This covers as many as five predefined connections that can be activated (assigned) when accessing the device. The Predefined Master/Slave Connection Set represents a subset of the general connection establishment method. It is limited to pure Master/Slave relations. Slave devices that are not UCMM capable, and only support this subset, are called Group 2 Only Servers in DeviceNet speak. Only the master that allocates it can address a Group 2 Only Server. All messages received by this device are defined in Message Group 2.

More details of the connection establishment using UCMM and the Master/Slave Connection Set can be found in [5] and [10].

15.3.1.12 Predefined Master/Slave Connection Set

Establishing a connection via the UCMM port requires a relatively large number of individual steps that have to be conducted to allow for data exchange via DeviceNet. The devices must provide resources to administer the dynamic connections. Because every device can set up a connection with every other device, and the source MAC ID of the devices is contained in the Connection ID, the CAN Identifier (Connection ID) may have to be filtered via software. This depends on how many connections a device supports, and the type and number of screeners (hardware CAN ID filters) of the CAN chip used in the device's implementation.

While this approach provides for taking full advantage of the multicast, peer-to-peer, and Producer/Consumer capabilities of CAN, a simpler method that needs fewer CPU resources is needed for low-end devices. To that end, the Predefined Master/Slave Connection Set was defined. The Group 2 Only Unconnected Explicit Message port of the Predefined Master/Slave Connection Set therefore provides an interface for a set of five preconfigured connection types in a node.

The basis of this model is a 1:n communication structure consisting of one control device and decentralized I/O devices. The central instance of such a system is known as the Master, and the decentralized devices are known as Slaves. Multiple masters are allowed on the network, but a slave can only be allocated to one master at any point in time.

The predefined Connection Objects occupy instances 1 to 5 in the Connection Object (Class ID 0x05; see Section 15.2.4):

- Explicit Messaging Connection:
  - Group 2 Explicit Request/Response Message (Instance ID 1)
- I/O Messaging Connections:
  - Polled I/O Connection (Instance ID 2)
  - Bit-Strobe I/O Connection (Instance ID 3)
  - Change of State or Cyclic I/O Connection (Instance ID 4)
  - Multicast Polling I/O Connection (Instance ID 5)

The messages to the slave are defined in Message Group 2, and some of the responses from the slave are contained in Message Group 1. The distribution of Connection IDs for the Predefined Master/Slave Connection Set is defined as shown in Figure 15.26.

Because the CAN ID of most of the messages the master produces contains the destination MAC ID of the slave, it is imperative that only one master talks to any given slave. Therefore, before a master can use this Predefined Connection Set, she must first allocate it with the device. The DeviceNet Object manages this important function in the slave device. It allows only one master to allocate its Predefined Connection Set, thereby preventing duplicate CAN IDs from appearing on the wire.

The two services used are called Allocate_Master/Slave_Connection_Set (Service Code 0x4B) and Release_Group_2_Identifier_Set (Service Code 0x4C). These two services always access Instance 1 of the DeviceNet object (Class ID 0x03) (Figure 15.27).

Figure 15.27 shows the Allocate Message with 8-bit Class ID and 8-bit Instance ID, a format that is always used when it is sent as a Group 2 Only Unconnected Message. It may also be sent across an existing connection and in a different format if a format other than 8/8 was agreed upon during the connection establishment.

The Allocation Choice Byte is used to set which predefined connections are to be allocated (Figure 15.28).

Connection ID = CAN Identifier (bits 10:0)											Used for
10	9	8	7	6	5	4	3	2	1	0	
0	Group 1 Message ID				Source MAC ID						Group 1 Messages
0	1	1	0	0	Source MAC ID						Slave's I/O Multicast Poll Response
0	1	1	0	1	Source MAC ID						Slave's I/O Change of State or Cyclic Message
0	1	1	1	0	Source MAC ID						Slave's I/O Bit-Strobe Response Message
0	1	1	1	1	Source MAC ID						Slave's I/O Poll Response or COS/Cyclic Ack Message
1	0	MAC ID					Group 2 Message ID				Group 2 Messages
1	0	Source MAC ID					0	0	0		Master's I/O Bit-Strobe Command Message
1	0	Source MAC ID					0	0	1		Master's I/O Multicast Poll Group ID
1	0	Destination MAC ID					0	1	0		Master's Change of State or Cyclic Acknowledge Message
1	0	Source MAC ID					0	1	1		Slave's Explicit/Unconnected Response Messages
1	0	Destination MAC ID					1	0	0		Master's Explicit Request Messages
1	0	Destination MAC ID					1	0	1		Master's I/O Poll Command/COS/Cyclic Message
1	0	Destination MAC ID					1	1	0		Group 2 Only Unconnected Explicit Request Messages
1	0	Destination MAC ID					1	1	1		Duplicate MAC ID Check Messages

FIGURE 15.26 Connection IDs of the Predefined Master/Slave Connection Set.

Byte offset	Bit number								
	7	6	5	4	3	2	1	0	
0	Frag [0]	XID	MAC ID						Message header
1	R/R [0]	Service Code [0x4B]							Message body
2...5		Class ID [0x03]							
		Instance ID [0x01]							
		Allocation Choice							
	0	0	Allocator's MAC ID						

FIGURE 15.27 Allocate_Master/Slave_Connect_Set Request Message.

Bit number							
7	6	5	4	3	2	1	0
Reserved	Ack Suppression	Cyclic	Change of State	Multicast Polling	Bit-Strobe	Polled	Explicit Message

FIGURE 15.28 Format of the Allocation Choice Byte.

The associated connections are activated by setting the appropriate bits. Change of State and Cyclic Connections are mutually exclusive choices. The Change of State/Cyclic Connection may be configured as not acknowledged using acknowledge suppression. The individual connection types are described in more detail below.

The allocator's MAC ID contains the address of the node (master) that wants to assign the Predefined Master/Slave Connection Set. Byte 0 of this message differs from the allocator's MAC ID if this service has been passed on to a Group 2 Only Server via a Group 2 Only Client (what is commonly referred to as a proxy function).

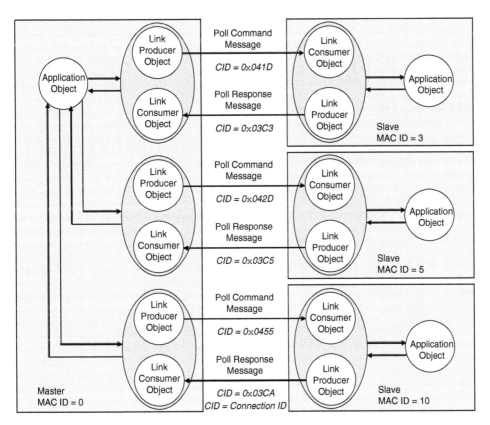

FIGURE 15.29 Polled I/O Connections.

The slave, if not already claimed, responds with a Success Message. The connections are now in configuring status. Setting the Expected_Packet_Rate (EPR) (Set_Attribute_Single service to attribute 9 in the appropriate Connection Object, value in ms) starts the connection's time-monitoring function. The connection then changes into established state and I/O Messages begin transferring via this connection.

The allocated connections can be released individually or collectively through the Release_Group_2_Identifier_Set service (Service Code 0x4C), using the same format as in Figure 15.27, except that the last byte (allocator's MAC ID) is omitted.

The following is an explanation of the four I/O Connection types in the Predefined Master/Slave Connection Set.

15.3.1.12.1 Polled I/O Connection

A Polled I/O Connection is used to implement a classic Master/Slave relationship between a control unit and a device. In this setup, a master can transfer data to a slave using the poll request and receive data from the slave using the poll response. Figure 15.29 shows the exchange of data between one master and three slaves in the Polled I/O mode.

In a message between a master and a slave using the Polled I/O Connection, the amount of data transferred via this message can be any length. If the length exceeds 8 bytes, the fragmentation protocol is automatically used. A Polled I/O Connection is always a point-to-point connection between a master and a slave. The slave consumes the Poll Message and sends back an appropriate response, normally its input data.

The Polled Connection is subject to a time-monitoring function (that can be adjusted) in the device. A Poll Command must have been received within this time ($4 \times$ EPR); otherwise, the connection changes over into time-out mode. When a connection times out, the node may optionally go to a preconfigured fault state as set up by the user. A master usually polls all the slaves in a round-robin manner.

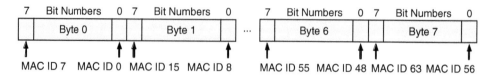

FIGURE 15.30 Data format of the Bit-Strobe I/O Connection.

A slave's response time to a poll command is not defined in the DeviceNet Specification. This provides much flexibility for slave devices to be designed appropriate to their primary applications, but it may also exclude the device from use in higher-speed applications.

15.3.1.12.2 Bit Strobe I/O Connection
The master's transmission on this I/O Connection is also known as a Bit-Strobe Command. Using this command, a master multicasts one message to reach all its slaves allocated for the Bit-Strobe Connection. The frame sent by the master using a Bit-Strobe Command is always 8 or 0 bytes (if Idle). From these 8 bytes, each slave is assigned one bit (Figure 15.30). Each slave can send back as many as 8 data bytes in its response.

A Bit-Strobe I/O Connection represents a multicast connection between one master and any number of strobe-allocated slaves (Figure 15.31). Since all devices in a network receive the Bit-Strobe Command at the same time, they can be synchronized by this command. When the Bit-Strobe Command is received, the slave may consume its associated bit and then send a response of up to 8 bytes.

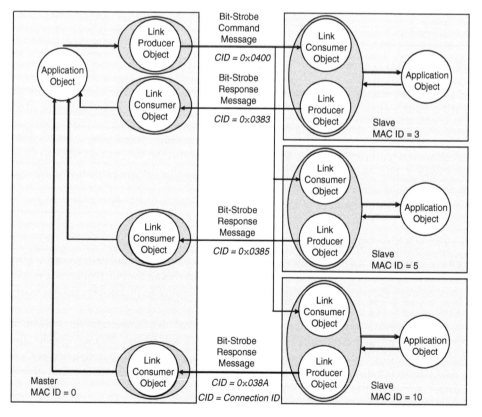

FIGURE 15.31 Bit-Strobe I/O Connections.

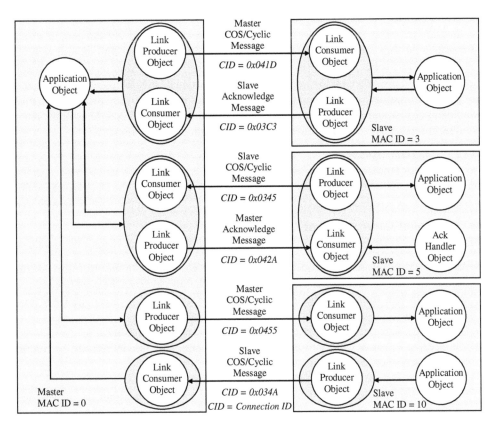

FIGURE 15.32 COS/Cyclic I/O Connections.

Since this command uses the source MAC ID in the Connection ID (Figure 15.26), devices that support the Bit-Strobe I/O Connection and have a CAN chip with screening limited to only 8 bits of the CAN ID (11 bits) must perform software screening of the CAN Identifier.

15.3.1.12.3 *Change of State/Cyclic I/O Connection*
The COS/Cyclic I/O Connection differs from the other types of I/O Connections in that both endpoints produce their data independently. This can be done in a change-of-state or cyclic manner. In the former case, the COS I/O Connection recognizes that the application object data that the Produced_Connection_ Path is indicating have changed. In the latter case, a timer of the Cyclic I/O Connection expires and therefore triggers the message transfer of the latest data from the application object.

A COS/Cyclic I/O Connection can be set up as acknowledged or unacknowledged. When acknowledged, the consuming side of the connection must set up a defined path to the Acknowledge Handler Object to ensure that the retries, if needed, are properly managed.

Figure 15.32 shows the various COS/Cyclic I/O Connection possibilities.

A COS/Cyclic I/O Connection can also originate from a master. This connection then seems like a Polled I/O Connection to the slave. This can be seen in Figure 15.26 since the same Connection ID is issued for the master's Polled I/O Message as is issued for the master's COS/Cyclic I/O Message.

COS Connections have two additional behaviors. The Expected Packet Rate (EPR) is used as a default production trigger so that if the data have not changed after the EPR timer has expired, the data will be re-sent as a heartbeat. This is so the consuming node can know the difference between a dead node and one whose data has not changed. COS Connections also have a Production Inhibit timer to prevent a chattering node from using too much bandwidth.

15.3.1.12.4 *Multicast Polled I/O Connection*

This connection is similar to the regular I/O poll except that all of the slaves belonging to a multicast group consume the same output data from the master. Each slave responds with its own reply data. A unique aspect of this connection is that the master picks the CAN ID from one of the slaves in the multicast group and must then set to that same value the consumed CAN ID in each of the other slaves. If during runtime that slave's connection times out, the master must either stop producing its multicast poll command or pick another slave in the group and reset the command CAN ID in all the remaining slaves in the group to that value before sending another Multicast Poll Command.

15.3.1.12.5 *I/O Data Sharing*

Due to the inherent broadcast nature of all CAN frames, applications can be set up to listen to the data produced by other applications. Such a listen-only mode is not described in the DeviceNet Specification, but some vendors have created products that do exactly that, e.g., Shared Inputs in Allen-Bradley scanners.

15.3.1.12.6 *Typical Master/Slave Start Sequence*

A typical start-up of a DeviceNet network with a scanner and a set of slaves is executed as follows:

- All devices run their self-test sequence and then try to go online with the algorithm described in Section 15.3.1.6. Any device that uses an autobaud mechanism to detect the baud rate of a network will have to wait with its Duplicate Node ID Message until it has seen enough CAN frames to detect the correct baud rate.
- Once online, slave devices will do nothing until their master allocates them.
- Once online, a master will try to allocate each slave configured into its scan list by running the following sequence of messages:
  - Try to open a connection to the slave using a UCMM Open Message.
  - If successful, the master can then use this connection for further communication with the slave.
  - If not successful, the master will try again after a minimum wait time of 1 s.
  - If unsuccessful again, the master will try to allocate the slave using the Group 2 Only Uncon-nected Explicit Request Message (at least for Explicit Messaging).
  - If successful, the master can then use this connection for further communication with the slave.
  - If not successful, the master will try again after a minimum wait time of 1 s.
  - If unsuccessful again, the master will start all over again with the UCMM Message. This process will carry on indefinitely or until the master has allocated the slave.
- Once the master has allocated the slave, it may carry out some verification to see whether it is safe to start I/O Messaging with the slave. The master may also apply some further configuration to the connections it has established, e.g., setting the Explicit Messaging Connection to "Deferred Delete."
- Setting the EPR value(s) brings the I/O Connection(s) to the Established State so that I/O Mes-saging can commence.

15.3.1.12.7 *Master/Slave Summary*

The task of supporting the Predefined Master/Slave Connection Set represents a solution that can be easily implemented for the device manufacturer. Simple BasicCAN controllers may be used; software screening of the CAN Identifier is generally not necessary, enabling the use of low-cost 8-bit controllers. This may represent an advantage as far as the devices are concerned but entails disadvantages for the system design.

Group 2 Only (i.e., UCMM incapable) devices permit only one Explicit Connection between client (master) and server (slave), whereas UCMM capable devices can maintain Explicit Messaging Connec-tions with more than one client at the same time.

If a device wants to communicate with one of the allocated slaves that do not support UCMM, the master recognizes this situation and sets up a communication link with the requestor instead. Any communication between the requestor is then automatically routed via the master. This is called the proxy function. Since this puts an additional burden on the master and on network bandwidth, it is recommended that slave devices support UCMM.

Although not explicitly defined in the DeviceNet Specification, DeviceNet masters can, under certain conditions, automatically configure their scan lists or the devices contained in their scan lists. This functionality simply makes use of the messaging capabilities of masters and slaves that allow the master to read from a slave whatever information is required to start an I/O communication and to download any configurable parameter that has been communicated to the master via EDS. This functionality facilitates the replacement of even complex slave devices without the need for a tool, reducing downtime of a system dramatically.

15.3.1.13 Device Profiles

DeviceNet uses the full set of profiles described in Chapter 6 of the CIP Specification [4].

15.3.1.14 Configuration

EDSs of DeviceNet devices can make full use of all EDS features, but they do not necessarily contain all sections. Typical DeviceNet devices contain (apart from the mandatory sections) at least an IO_Info section. This section specifies which types of Master/Slave connections are supported and which one(s) should be enabled as default. It also declares which I/O Connections may be used in parallel. Chapter 7 of the DeviceNet Specification [5] gives full details of this section of a DeviceNet EDS.

A full description of what can be done in DeviceNet EDSs would go well beyond the scope of this handbook, so References [25] and [26] are recommended for further reading.

15.3.1.15 Conformance Test

At an early stage, the ODVA defined test and approval procedures for DeviceNet devices and systems. Manufacturers are given the opportunity to have their devices checked for conformance with the DeviceNet Specification in one of several independent DeviceNet conformance test centers. Only then do two key characteristics of all DeviceNet devices become possible: *interoperability* and *interchangeability*.

Interoperability means that DeviceNet devices from all manufacturers can be configured to operate with each other on the network. Interchangeability goes one step farther by providing the means for devices of the same type (i.e., they comply with the same Device Profile) to be logical replacements for each other, regardless of the manufacturer.

The conformance test checks both of these characteristics. This test is divided into three parts:

- A software test to verify the function of the DeviceNet protocol. Depending on the complexity of the device, as many as several thousand messages are transmitted to the device under test (DUT). To allow a test that is closely adapted to the characteristics of the DUT, a formal description of all relevant features of the DUT must be provided by the manufacturer.
- A hardware test to check conformance with the characteristics of the physical layer. This test checks all requirements of the specification like miswiring protection, overvoltage withstand, grounding, CAN transceiver, etc. The test may be destructive for noncompliant devices.
- A system interoperability test that verifies that the device can function in a network with more than 60 nodes and a variety of scanners from various manufacturers.

The software test is available from ODVA. It is a Windows-based tool, running on various PC CAN interface cards from a number of suppliers. It is recommended that device developers run this test in their own lab before taking devices to the official ODVA test. The hardware test and the system interoperability test involve more complex test setups that are typically not available to device developers.

When a device passes the test, it is said to be DeviceNet CONFORMANCE TESTED®.* Many DeviceNet users now demand this seal. A device that has not been tested accordingly has a significant market disadvantage. Devices that have passed conformance testing are published on the ODVA Web site.

---

*DeviceNet CONFORMANCE TESTED® is a registered certification mark of ODVA.

15.3.1.16 Tools

Tools for DeviceNet networks can be divided into three groups:

- Physical layer tools: Tools (hardware and/or software) that verify the integrity and conformance of the physical layer or monitor the quality of the data transmission.
- Configuration tools: Software tools that are capable of communicating with individual devices for data monitoring and configuration purposes. They can range from very basic software operating from handheld devices to powerful PC-based software packages to configure complete networks. Most configuration tools are EDS-based; however, more complex devices like scanners tend to have their own configuration applets that are only partially based on EDSs. Some of these tools support multiple access paths to the network, e.g., via Ethernet and suitable bridging devices, and thus allow remote access. High-level tools also actively query the devices on the network to identify them and monitor their health.
- Monitoring tools: Typically PC-based software packages that can capture and display the CAN frames on the network. A raw CAN frame display may be good enough for some experts, but it is recommended that a tool that allows both raw CAN display and DeviceNet interpretation of the frames be used.

For a typical installation, a configuration tool is all that is needed. However, to ensure the network is operating reliably, a check with a physical layer tool is highly recommended. Experience shows that the overwhelming majority of DeviceNet network problems are caused by inappropriate physical layer installation. Protocol monitoring tools are mainly used to investigate interoperability problems and to assist during the development process.

Turn to the DeviceNet product catalog on the ODVA Web site to access a list of vendors that provide tools for DeviceNet.

15.3.1.17 Advice for Developers

Before any development of a DeviceNet product is started, the following issues should be considered in detail:

- What functionality does the product require today and in future applications?
  - Slave functionality
  - Master functionality
  - Peer-to-peer messaging
  - Combination of the above
- What are the physical layer requirements? Is IP 65/67 required or is IP 20 good enough?
- What type of hardware should be chosen for this product?
- What kind of firmware should be used for this product? Will a commercially available communication stack be used?
- Will the development of hardware and software be done internally or will it be designed by an outside company?
- What kind of configuration software should be used for this product? Will a commercially available software package be used; i.e., is an EDS adequate to describe the device or is custom software needed?
- What are the configuration requirements?
- Will the product be tested for conformance and interoperability (highly recommended)?
- What design and verification tools should be used?
- What is an absolute must before the products can be placed on the market (own the specification, have a Vendor ID)?

A full discussion of these issues goes well beyond the scope of this book; see Reference [27] for further reading.

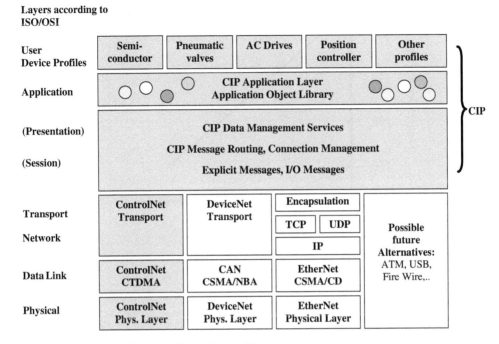

FIGURE 15.33 Relationship between CIP and ControlNet.

15.3.1.18 DeviceNet Overall Summary

Since its introduction in 1994, DeviceNet has been used successfully in millions of nodes in many different applications. It is a *de facto* standard in many countries, and this situation is reflected in several national and international standards [16–18]. Due to its universal communication characteristics, it is one of the most versatile fieldbuses for low-end devices. While optimized for devices with small amounts of I/O, it can easily accommodate larger devices as well. Powerful EDS-based configuration tools allow easy commissioning and configuration of even complex devices without the need to consult manuals.

While most applications are of the Master/Slave type, peer-to-peer communication is used in a rising number of applications, greatly simplifying the design, operation, and maintenance of these networks. With the introduction of CIP Safety on DeviceNet, many machine-level applications that today need a set of dedicated networks will soon be accommodated in only one DeviceNet network.

Finally, its use of CIP and object structure allows the blending of DeviceNet networks into an overall CIP network structure that permits seamless communication, just as if it was only one network.

15.3.2 ControlNet

ControlNet is based on a physical layer and a bus access mechanism that was specifically developed for this network to provide absolute determinism. All other features are based on CIP.

Figure 15.33 shows the relationship between CIP, ControlNet, and the ISO/OSI layer model.

15.3.2.1 Physical Layer and Frame Structure

The physical layer of ControlNet has specifically been designed for this network; it does not reuse any existing open technology. The basis of the physical layer is a 75-Ω coax trunk line cable (typically of the RG6 type) terminated at both ends. To reduce impedance mismatch, all ControlNet devices are connected to the network through special taps that consist of a coupling network and a specific length of dropline (1 m). There is no minimum distance requirement between any two of these taps, but since every tap introduces some signal attenuation, each tap reduces the maximum length of the trunkline by 16.3 m.

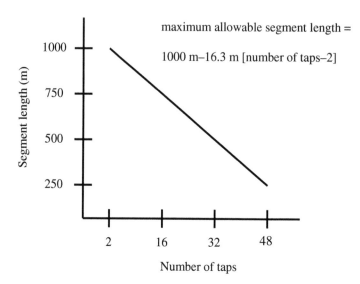

FIGURE 15.34 Coax medium topology limits.

This results in a full-length trunkline of 1000 m with only two taps at the ends, while a fully populated physical network with 48 taps allows a trunkline length of 250 m (Figure 15.34).

This physical layer limitation was taken into account from the very beginning by including repeaters into the design that can increase the network size without lowering the speed. Therefore, if a network is to be built with a higher number of nodes (up to 99 nodes are possible) or with a topology that goes beyond the single trunkline limitations, repeaters can be used to create any tree topology or even a ring topology using a special type of repeater. There are also repeaters for fiber-optic media that can be used either to increase the system size even further or to allow very good isolation of network segments in harsh EMC (Electromagnetic Compatibility) environments or for high-voltage applications.

The number of repeaters in series between any two nodes used to be limited to five until recently. Better repeater technology now allows up to 20 repeaters in series. However, whatever media technology is used, the overall length of a ControlNet system (distance between any two nodes on the network) is limited. This fundamental limit is due to propagation delay. With currently available media, this translates into approximately 20 km.

To better accommodate industry requirements, ControlNet supports redundant media, allowing bumpless transfer from primary to secondary media or vice versa if one of them should fail or deteriorate. Developers are encouraged to support this redundant media feature in their design. For cost-sensitive applications, less expensive device variants may then be created by populating one channel only.

Another feature often used in the process industry is the capability to run ControlNet systems into areas with an explosion hazard. The system is fully approved to meet worldwide standards for intrinsic safety (explosion protection).

The connectors used for copper media are of the BNC type; TNC type connectors have been introduced recently for applications that require IP 67 protection. Devices may also implement a Network Access Port (NAP). This feature takes advantage of the repeater function of the ControlNet application-specific integrated circuits (ASICs). It uses an additional connector (RJ45) with RS 422-based signals that provides easy access to any node on the network for configuration devices.

The signal transmitted on the copper media is a 5 Mbit/s Manchester encoded signal with an amplitude of up to 9.5 V (pk-pk) at the transmitter that can be attenuated down to 510 mV (pk-pk) at the receiving end. The transmitting and receiving circuits, coupled to the cable through transformers, are described in full detail in the ControlNet Specification [3].

15.3.2.2 Protocol Adaptation

ControlNet can use all features of CIP. The ControlNet frame is big enough that fragmentation is rarely required. Since ControlNet is not expected to be used in very simple devices, there is no scaling.

15.3.2.3 Indicators and Switches

ControlNet devices must be built with device status and network status indicators, as described in the specification. Devices may have additional indicators, which must not carry any of the names of those described in the specification.

Devices may be built with or without switches or other directly accessible means for configuration. If switches for the MAC ID exist, then certain rules apply about how these values have to be used at power-up and during the operation of the device.

15.3.2.4 Additional Objects

The ControlNet Specification defines three additional objects, the ControlNet Object (Class Code 0xF0), the Keeper Object (Class Code 0xF1), and the Scheduling Object (Class Code 0xF2).

15.3.2.4.1 ControlNet Object

The ControlNet Object contains a host of information on the state of the ControlNet link of the device, among them diagnostic counters, data link and timing parameters, and the MAC ID. A ControlNet Object is required for every physical layer attachment of the device. A redundant channel pair counts as one attachment.

15.3.2.4.2 Keeper Object

The Keeper Object (not required for every device) holds (for link scheduling software) a copy of the Connection Originator schedule data for all Connection Originator devices using a network. Every ControlNet network with scheduled I/O traffic must have at least one device with a Keeper Object (typically a Programmable Logic Controller (PLC) or another Connection Originator). If there are multiple Keeper Objects on a link, they perform negotiations to determine which Keeper is the Master Keeper and which Keeper(s) performs Backup Keeper responsibilities. The Master Keeper is the Keeper actively distributing attributes to the nodes on the network. A Backup Keeper is one that monitors Keeper-related network activity and can transition into the role of Master Keeper should the original Master Keeper become inoperable.

15.3.2.4.3 Scheduling Object

The Scheduling Object is required in every device that can originate an I/O Messaging Connection. Whenever a link scheduling tool accesses a Connection Originator on a ControlNet link, an instance of the Scheduling Object is created and a set of object-specific services is used to interface with this object. Once the instance is created, the link scheduling tool can then read and write connection data for all connections to originate from this device. After having read all connection data from all Connection Originators, the link scheduling tool can calculate an overall schedule for the ControlNet link and write this data back to all Connection Originators. The scheduling session is ended by deleting the instance of the Scheduling Object.

15.3.2.5 Network Access

The bus access mechanism of ControlNet allows full determinism and repeatability while still maintaining sufficient flexibility for various I/O Message triggers and Explicit Messaging.

This bus access mechanism is called Concurrent Time Domain Multiple Access (CTDMA); it is illustrated in Figure 15.35.

The time axis is divided into equal intervals called Network Update Time (NUT). Within each NUT there is a subdivision into a Scheduled Service Time, an Unscheduled Service Time, and the Guardband.

Figure 15.36 shows the function of the Scheduled Service. Every node up to and including the SMAX node (maximum node number participating in the Scheduled Service) has a chance to send a message

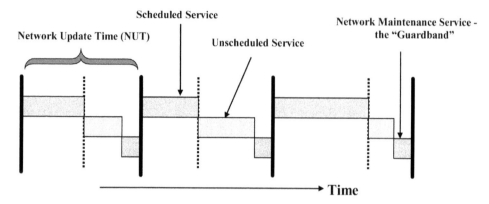

FIGURE 15.35 Media access through CTDMA.

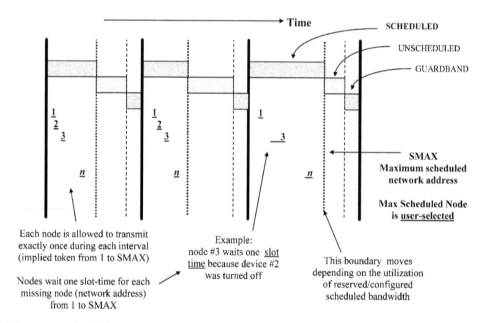

FIGURE 15.36 Scheduled Service.

within the Scheduled Service. If a particular node has no data to send, it will nevertheless send a short frame to indicate that it is still alive. If a node fails to send its frame, the next-higher node number will step in after a very short, predetermined waiting time. This makes sure that the failure of a node will not lead to an interruption of the NUT cycle.

Figure 15.37 shows the function of the Unscheduled Service. Since this service is designed for non-time-critical messages, only one node is guaranteed to get access to the bus during the Unscheduled Service Time. If there is time left, then the other nodes (with higher node numbers) will also get a chance to send. As with the Scheduled Service Time, if a node fails to send when it is its turn, the next node will step in. The node number that is allowed to send first within the Unscheduled Service Time is increased by 1 in each NUT. This guarantees an equal chance to all nodes. All node sequencing in this interval wraps; UMAX is followed by the lowest node number (typically 1) on the network.

Those two service intervals combined with the Guardband guarantee determinism and repeatability while still maintaining sufficient freedom to allow for unscheduled message transmissions, e.g., for parameterization.

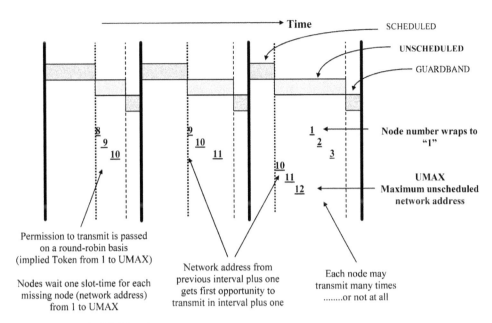

FIGURE 15.37 Unscheduled Service.

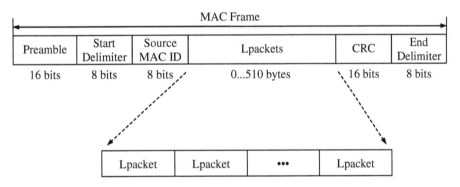

FIGURE 15.38 MAC frame format.

15.3.2.6 Frame Description

Every frame transmitted on ControlNet has the format of the MAC frame (Figure 15.38).

Within every MAC frame, there is a field of up to 510 bytes that is available for the transmission of data or messages. This field may be populated with one or several Lpackets (link packets). These Lpackets carry the individual messages (I/O or Explicit) of CIP. There are also some specialized Lpackets used for network management. Since every node always listens to all MAC frames, they have no problem consuming any of the Lpackets in the frame that might be unicast, multicast, or broadcast in nature. This feature allows fine-tuned multicasting of small amounts of data to different sets of consumers without too much overhead.

There are two types of Lpacket formats: fixed tag and generic tag. The fixed tag Lpackets are used for Unconnected Messaging and network administration, while the generic tag Lpackets are used for all Connected Messaging (I/O and Explicit).

Figure 15.39 shows the format of a fixed tag Lpacket. By including the destination MAC ID, this format reflects the fact that these Lpackets are always directed from the requesting device (sending the MAC frame) to the target device (the destination MAC ID). The service byte within the fixed tag Lpacket does

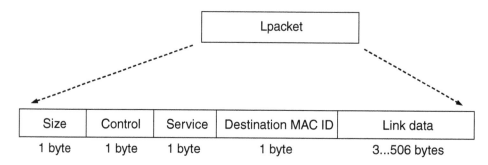

FIGURE 15.39   Fixed tag Lpacket format.

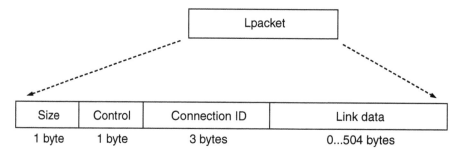

FIGURE 15.40   Generic tag Lpacket format.

not represent the service of an Explicit Message, but a more general service type since the fixed tag Lpacket format can be used for a variety of actions such as network administration.

Figure 15.40 shows the format of a generic tag Lpacket. The size byte specifies the number of words within the Lpacket; the control byte gives information on what type of Lpacket this is. The 3-byte Connection Identifier specifies which connection this Lpacket belongs to. These three bytes are the three lower bytes of the 4-byte Connection ID specified in the Forward_Open Message. The ControlNet Specification gives full details on how to assemble the three lower bytes of the Connection ID; the uppermost byte is always zero. For a device that receives the MAC frame, the Connection ID is the indication whether to ignore the Lpacket (the device is not part of the connection), to consume the data and forward it to the application (the device is an endpoint of this connection), or to forward the data to another network (the device acts as a bridge in a bridged connection).

15.3.2.7   Network Start-Up

After power-up, every ControlNet device goes through a process of getting access to the ControlNet communication link and learning the current NUT and other timing requirements. This is a fairly complex process typically handled by the commercially available ControlNet ASICs. It would go beyond the scope of this handbook to describe all the details here.

15.3.2.8   Explicit Messaging

Unlike DeviceNet, Explicit Messages on ControlNet can be sent connected or unconnected; both are typically transmitted within the unscheduled part of the NUT. Connected Explicit Messaging requires setting up a connection first (see Section 15.3.2.10). This, of course, means that all resources required for the management of the connection must stay reserved for this purpose as long as the connection exists. To avoid tying up these resources, most Explicit Messages can also be sent unconnected. Every part of an Explicit Message (request, response, acknowledgments) is wrapped into an Lpacket using the fixed tag Lpacket format for Unconnected Messaging (Figure 15.39) and the generic tag Lpacket format for Connected Messaging (Figure 15.40). The service/class/instance/attribute fields (see Section 15.2.3) of the Explicit Message are contained in the link data field.

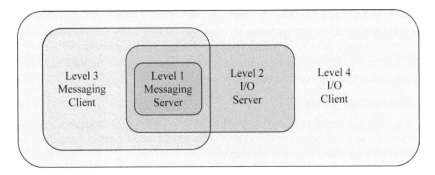

FIGURE 15.41 Device levels.

15.3.2.9 I/O Messaging

ControlNet I/O Messaging, like any other CIP I/O Messaging, is done across connections and it always takes place in the scheduled part of NUT. Only one MAC frame may be transmitted by any device within its time slot, but this MAC frame may contain multiple Lpackets so that data can be sent to multiple nodes in one NUT. The individual Lpackets may be consumed by one node only or by multiple nodes if they are set up to consume the same data.

I/O Messages use the generic tag Lpacket format (Figure 15.40). The link data field contains the I/O data prepended with a 16-bit sequence count number for the packet. I/O data transmission without the Sequence Count Number is possible in principle, but is not used today. Run/Idle can be indicated within a prepended Run/Idle header or by sending the data packet (Run) or no data packet (Idle). Which of the two methods that is used is indicated in the connection parameters in the Connection Manager section of the EDS. However, only the Run/Idle header method has been in use for ControlNet up to now.

15.3.2.10 Connection Establishment

All connections on ControlNet are established using a UCMM Forward_Open message (see Section 15.2.3); therefore, all devices must support the UCMM function.

15.3.2.11 Device Levels

While not official categories, it is useful to distinguish among several levels of devices (Figure 15.41); you only have to implement the functionality you need. The minimal device function (level 1) is that of a Messaging Server. It is used for Explicit Messaging applications only and acts as a target for Connected and Unconnected Explicit Messages, e.g., for program upload/download, data collection, status monitoring, etc. The next class of device (level 2) is an I/O Server. It adds I/O Messaging support to a level 1 device and acts as a target for both Explicit and I/O Messages, e.g., simple I/O devices, pneumatic valves, AC drives, etc. These devices are also called adapters. Another class of devices is a Messaging Client (level 3). It adds client support to level 1 Explicit Messaging applications and acts as a target and an originator for messaging applications, e.g., computer interface cards or Human-Machine Interface (HMI) devices. Finally, the most powerful class of device is a scanner (level 4). It adds I/O Message origination support to levels 1, 2, and 3 and acts as a target and an originator for Explicit and I/O Messages, e.g., PLCs, I/O scanners, etc.

15.3.2.12 Device Profiles

ControlNet uses the full set of profiles described in Chapter 6 of the CIP Specification [4].

15.3.2.13 Configuration

ControlNet devices typically come with EDSs, as described in Section 15.2.7. For EDS-based configuration tools, the EDS should contain a Connection Manager section to describe the details of the connections that can be made into the device. This section basically is a mirror of what is contained in the

Forward_Open message that a Connection Originator would send to the device. Multiple connections can be specified within an EDS, and then one or more can be chosen by the configuration tool.

An EDS may also contain individual parameters or a Configuration Assembly with a complete description of all parameters within this Assembly. In many applications, the Configuration Assembly is transmitted as an attachment to the Forward_Open Message.

15.3.2.14   Conformance Test

ControlNet International has defined a conformance test for ControlNet devices. Currently, this test is a protocol conformance test only, since it is expected that most implementations use the commercially available components for transformers and drivers.

As many as several thousand messages are transmitted to the DUT, depending on the complexity of the device. To allow a test that is closely adapted to the characteristics of the DUT, a formal description of all relevant features of the DUT must be provided by the manufacturer.

The software test is available from ControlNet International. It is a Windows-based tool, running on a PC interface card through a NAP connection (see Section 15.3.2.1). It is recommended that device developers run this test in their own labs before taking devices to the official ControlNet International test.

When a device passes the test, it is said to be ControlNet CONFORMANCE TESTED™.* Many ControlNet users now demand this seal. A device that has not been tested accordingly has a significant market disadvantage. Devices that have passed conformance testing are published on the ControlNet International Web site.

15.3.2.15   Tools

Tools for ControlNet networks can be divided into three groups:

- Physical layer tools: Tools (hardware and software) that verify the integrity and conformance of the physical layer or monitor the quality of the data transmission.
- Configuration tools: Software tools that are capable of communicating with individual devices for data monitoring and configuration purposes. Most configuration tools are EDS-based; however, more complex devices like scanners tend to have their own configuration applets that are only partially based on EDSs. Some of these tools support multiple access paths to the network, e.g., via Ethernet and suitable bridging devices, and thus allow remote access. High-level tools also actively query the devices on the network to identify them and monitor their health. Configuration tools may also be integrated into other packages like PLC programming software.
- Monitoring tools: Typically PC-based software packages that can capture and display the ControlNet frames on the network. A raw ControlNet frame display may be good enough for some experts, but it is recommended that a tool that allows both raw ControlNet frame display and interpreted frames be used.

For a typical installation, a configuration tool is all that is needed. However, to ensure that the network is operating reliably, a check with a physical layer tool is highly recommended. Experience shows that the overwhelming majority of ControlNet network problems are caused by inappropriate physical layer installation. Protocol monitoring tools are mainly used to investigate interoperability problems and to assist during the development process.

Turn to the ControlNet product catalog on the ControlNet International Web site to access a list of vendors that provide tools for ControlNet.

15.3.2.16   Advice for Developers

Before any development of a ControlNet product is started, the following issues should be considered in detail:

---

*ControlNet CONFORMANCE TESTED™ is a certification mark of ControlNet International.

- What functionality (device class; see Section 15.3.2.11) does the product require today and in future applications?
  - Messaging server only
  - Adapter functionality
  - Messaging client
  - Scanner functionality
- What are the physical layer requirements? Is IP 65/67 required or is IP 20 good enough?
- Will the development be based on commercially available hardware components and software packages (recommended) or designed from scratch (possible but costly)?
- What are the configuration requirements?
- Will the product be tested for conformance (highly recommended)?
- What design and verification tools should be used?
- What is an absolute must before products can be placed on the market (own the specification, have a Vendor ID)?

ControlNet chip sets and associated software packages are available from Rockwell Automation and through ControlNet International. Turn to the ControlNet International Web site for a list of companies that can support ControlNet developments.

15.3.2.17 ControlNet Overall Summary

Since its introduction in 1997, ControlNet has been used successfully in hundreds of thousands of nodes in many different applications. It is the network of choice for many high-speed I/O and PLC interlocking applications. Like DeviceNet, ControlNet has been turned into an international standard [19]. Due to its universal communication characteristics, it is one of the most powerful controller-level fieldbuses.

The specific strength of ControlNet is its full determinism and repeatability, which make it ideally suited for many high-speed applications while maintaining full Explicit Messaging capabilities without compromising its real-time behavior.

Finally, its use of CIP and object structure allows the blending of ControlNet networks into an overall CIP network structure that permits seamless communication, just as if it was only one network.

15.3.3 EtherNet/IP

EtherNet/IP is the newest member of the CIP family; it is a technology supported by both ODVA and ControlNet International. EtherNet/IP has evolved from ControlNet and is therefore very similar to ControlNet in the way the CIP Specification is applied. Due to the length of the Ethernet frames and the typical multimaster structure of Ethernet networks, there are no particular limitations in the EtherNet/IP implementation of CIP. Basically all that is required is a mechanism to encode CIP Messages into Ethernet frames.

Figure 15.42 shows that there is an encapsulation mechanism (see Section 15.3.3.6) that specifies how I/O and Explicit Messages are wrapped into Ethernet frames. The well-known TCP/IP protocol is used for the encapsulation of Explicit Messages, while UDP/IP is used for the encapsulation of I/O Messages. The use of the very popular TCP/IP and UDP/IP stacks for encapsulation means that many applications will not require extra middleware for this purpose, since these stacks are already in use in many applications anyway.

Even with the use of certain infrastructure devices (see Section 15.3.3.16) it is difficult to make today's Ethernet fully deterministic. Therefore, many CIP users may prefer ControlNet for applications that require full determinism and repeatability. However, future extensions to CIP such as CIP Sync (see Section 15.5.1) will allow EtherNet/IP to be used in highly synchronous and deterministic applications like coordinated drives.

15.3.3.1 Physical Layer Adaptation

Since EtherNet/IP is taking the Ethernet protocol to the factory floor, there are some restrictions and further requirements on the physical layer [12] that is to carry EtherNet/IP in a typical factory automation

**Layers according to
ISO/OSI**

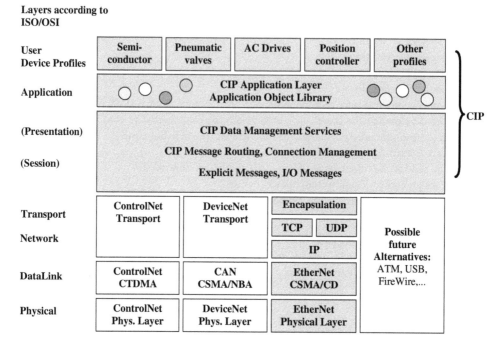

FIGURE 15.42   Relationship between CIP and EtherNet/IP.

environment. The actual signaling is left unchanged, but there are some additional specifications on connectors and cabling. For IP 20 applications, the well-known RJ45 connector is used, but for applications that require a higher degree of protection, suitable connectors have been specified. The EtherNet/IP specification lists a sealed connector based on the RJ45 type. A second connector (D-coded M12) is a recent addition for devices that require a more compact connector. This connector has also been specified by a number of other organizations, so it is expected that it will become the *de facto* standard for field devices.

Cat 5E or Cat 6 shielded or unshielded cables are recommended for EtherNet/IP. The use of shielded cables is specifically recommended in application where adjacent material, such as metal cable ducts, may have substantial influence on the characteristics of the cable. Copper media may only be used for distances up to 100 m. Fiber-optic media are recommended for longer distances. Fiber-optic media may also be advisable for applications with very high electromagnetic disturbances or high-voltage potential differences between devices.

15.3.3.2   Frame Structure

EtherNet/IP uses standard Ethernet TCP/IP and UDP/IP frames as defined by international standards [12]. Therefore, no further frame details are described here.

15.3.3.3   Protocol Adaptation

EtherNet/IP can use all features of CIP. The Ethernet frame is big enough that fragmentation is rarely required. Since EtherNet/IP is not expected to be used in very simple devices, no further scaling than that described in Section 15.3.3.10 is required.

15.3.3.4   Indicators and Switches

EtherNet/IP devices that need to conform to the industrial performance level must have the set of indicators described in Chapter 9 of the EtherNet/IP Specification [6]. Devices may have additional indicators, which must not carry any of the names of those described in the specification.

Devices may be built with or without switches or other directly accessible means for configuration.

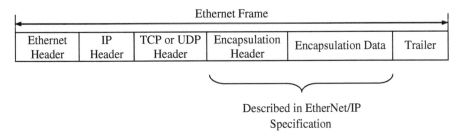

FIGURE 15.43 Relationship between CIP and Ethernet frames.

15.3.3.5 Additional Objects

The EtherNet/IP Specification defines two additional objects: the TCP/IP Object (Class Code 0xF5) and the Ethernet Link Object (Class Code 0xF6).

15.3.3.5.1 TCP/IP Object

The TCP/IP interface object provides a mechanism to configure a device's TCP/IP network interface. Examples of configurable items include the device's IP address, network mask, and gateway address.

15.3.3.5.2 Ethernet Link Object

The Ethernet link object maintains link-specific counters and status information for an Ethernet 802.3 communications interface. Each device has exactly one instance of the Ethernet link object for each Ethernet 802.3 communications interface. A request to access instance 1 of the Ethernet link object always refers to the instance associated with the communications interface over which the request was received.

15.3.3.6 EtherNet/IP Encapsulation

EtherNet/IP is completely based on existing TCP/IP and UPD/IP technologies and uses these principles without any modification. TCP/IP is mainly used for the transmission of Explicit Messages, while UDP/IP is used mainly for I/O Messaging.

The encapsulation protocol defines a reserved TCP port number that is supported by all EtherNet/IP devices. All EtherNet/IP devices accept at least two TCP connections on TCP port number 0xAF12.

The encapsulation protocol also defines a reserved UDP port number that is supported by all EtherNet/IP devices. All devices accept UDP packets on UDP port number 0xAF12. However, most UDP port assignments in EtherNet/IP are determined by (TCP) Explicit Messages; most EtherNet/IP UDP messages do not, in fact, use port 0xAF12. Since UDP, unlike TCP, does not have an ability to reorder packets, whenever UDP is used to send an encapsulated message, the entire message is sent in a single UDP packet and only one encapsulated message is present in any UDP packet.

15.3.3.6.1 General Use of the Ethernet Frame

Since EtherNet/IP is completely based on Ethernet with TCP/IP and UDP/IP, all CIP-related messages sent on an EtherNet/IP network are based on Ethernet frames with an IP header (Figure 15.43).

The Ethernet, IP, and TCP or UDP headers are described through international standards (see Section 15.3.3.2); therefore, details of these headers are only mentioned in the EtherNet/IP Specification when necessary to understand how they are used.

The encapsulation header is a description of the meaning of the encapsulation data. Most encapsulation data use the so-called Common Packet Format. I/O Messages sent in UDP frames do not carry an encapsulation header, but they still follow the Common Packet Format.

15.3.3.6.2 Encapsulation Header and Encapsulation Commands

The overall encapsulation packet has the structure described in Figure 15.44.

While the description of some of the encapsulation header details would go beyond the scope of this handbook, the command field needs some more attention here. However, only those commands that are

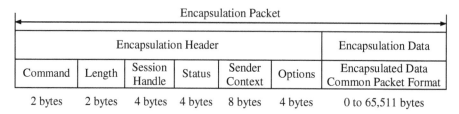

FIGURE 15.44 Structure of the encapsulation packet.

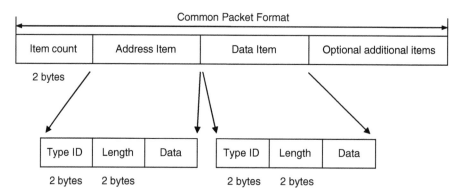

FIGURE 15.45 Common Packet Format.

needed to understand EtherNet/IP are described here, and their descriptions only list the main features. The encapsulated data as such follows the Common Packet Format (see Section 15.3.3.6.2.4).

15.3.3.6.2.1 ListIdentity Command — The ListIdentity command is a broadcast UDP message that tells all EtherNet/IP devices to return a data set with identity information. This command is typically used by software tools to browse a network.

15.3.3.6.2.2 RegisterSession/UnRegisterSession Commands — These two commands are used to register and unregister a CIP Session between two devices. Once such a session is established, it can be used to exchange further messages. Multiple sessions may exist between two devices, but this is not common.

The device requesting the session creates a sender context value; the device receiving the session request creates a session handle. Both values are used to identify messages between the two devices.

15.3.3.6.2.3 SendRRData/SendUnitData Commands — The SendRRData command is used for Unconnected Messaging; the SendUnitData command is used for Connected Explicit Messaging.

15.3.3.6.2.4 Common Packet Format — The Common Packet Format is a construct that allows packing of multiple items into one encapsulation frame (Figure 15.45). However, in most cases, only one Address Item and one Data Item are represented.

All encapsulated messages are then assembled using at least these two items within the Common Packet Format. Full details of this encapsulation can be found in Chapter 2 of the EtherNet/IP Specification [6].

15.3.3.7 IP Address Assignment

Since the initial development of TCP/IP, numerous methods for configuring a device's IP address have evolved. Not all of these methods are suitable for industrial control devices. In the office environment, for example, it is common for a PC to obtain its IP address via the Dynamic Host Configuration Protocol (DHCP), potentially getting a different address each time the PC reboots. This is acceptable because the PC is typically a client device that only makes requests, so there is no impact if its IP address changes.

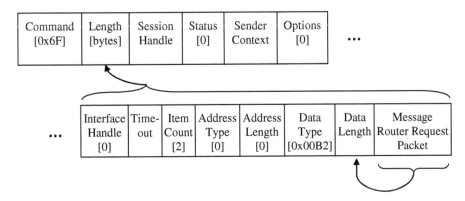

FIGURE 15.46 UCMM request encapsulation.

However, for an industrial control device that is a target of communication requests, the IP address cannot change at each power-up. If you're talking to a particular PLC, you want that PLC to be at the same address the next time it powers up.

To further complicate matters, the only interface common to all EtherNet/IP devices is an Ethernet communications port. Some devices may also have a serial port, user interface display, hardware switches, or other interfaces, but these are not universally shared across all devices. Since Ethernet is the common interface, the initial IP address must at least be configurable over Ethernet.

The EtherNet/IP Specification, via the TCP/IP Interface Object, defines a number of ways to configure a device's IP address. A device may obtain its IP address via Bootstrap Protocol (BOOTP), DHCP, or an explicit Set_Attribute (single or set-all) service. None of these methods are mandated however. As a result, vendors could choose different methods for configuring IP addresses.

From the user's perspective, it is desirable for vendors to support some common mechanism(s) for IP address configuration. Therefore, ODVA, Profibus User Organization (PNO), and Modbus/IDA (Interface for Distributed Automation) are currently working on mandating a set of common methods to assign an IP address across the Ethernet link. The current ODVA recommendations on this subject can be downloaded from the ODVA Web site [8].

15.3.3.8 Use of the Encapsulation Data

15.3.3.8.1 Explicit Messaging

Unlike DeviceNet, Explicit Messages on EtherNet/IP can be sent connected or unconnected. Connected Explicit Messaging requires setting up a connection first (see Section 15.3.3.9). This, of course, means that all resources required for the management of the connection must stay reserved for this purpose as long as the connection exists. To avoid tying up these resources, most Explicit Messages can also be sent unconnected. Explicit Messages on EtherNet/IP are sent with a TCP/IP header and use encapsulation with the SendRRData Command (unconnected) and the SendUnitData Command (connected). As an example, the full encapsulation of a UCMM request is shown in Figure 15.46.

The Message Router Request Packet, containing the message as such, follows the general format of Explicit Messages defined in Chapter 2 of the CIP Specification [4].

15.3.3.8.2 I/O Messaging

I/O Messages on EtherNet/IP are sent with a UDP/IP header. No encapsulation header is required, but the message still follows the Common Packet Format (e.g., Figure 15.47).

The data field contains the I/O data prepended with a 16-bit Sequence Count Number for the packet. I/O data transmission without the Sequence Count Number is possible in principle, but is not used today. Run/Idle can be indicated within a Run/Idle header or by sending the data packet (Run) or no data packet (Idle). Which of the two methods is used is indicated in the connection parameters of the

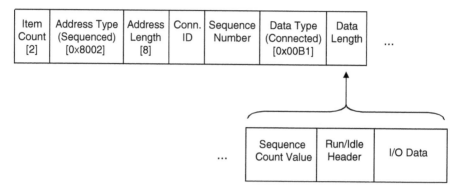

FIGURE 15.47 I/O Message encapsulation.

Connection Manager section of the EDS. However, the Run/Idle header method is recommended for use in EtherNet/IP, and this is what is shown in Figure 15.47.

I/O Messages from the originator to the target are typically sent as UDP unicast frames, while those sent from the target to the originator are typically sent as UDP multicast frames. This allows other EtherNet/IP devices to listen to this input data. To avoid these UDP multicast frames propagating all over the network, it is highly recommended that switches that support Internet Group Management Protocol (IGMP) Snooping be used. IGMP (see [41]) is a protocol that allows the automatic creation of multicast groups. Using this functionality, the switch will automatically create and maintain a multicast group consisting of the devices that need to consume these multicast messages. Once the multicast groups have been established, the switch will direct such messages only to those devices that have subscribed to the multicast group of that message.

15.3.3.9 Connection Establishment

All connections on EtherNet/IP are established using a UCMM Forward_Open Message (see Section 15.2.3); therefore, all devices must support the UCMM function.

15.3.3.10 Device Levels (Clients, Servers)

While not official categories, it is useful to distinguish among several levels of devices (Figure 15.48); one only has to implement the functionality needed. The minimal device function (level 1) is that of a Messaging Server. It is used for Explicit Messaging applications only and acts as a target for Connected and Unconnected Explicit Messages, e.g., for program upload/download, data collection, status monitoring, etc. The next class of device (level 2) is an I/O Server. It adds I/O Messaging support to a level 1 device and acts as a target for both Explicit and I/O Messages, e.g., simple I/O devices, pneumatic valves, AC drives, etc. These devices are also called adapters. Another class of device is a Messaging Client (level 3). It adds client support to level 1 Explicit Messaging applications and acts as a target and an originator for messaging applications, e.g., computer interface cards or HMI devices. Finally, the most powerful class of device is a scanner (level 4). It adds I/O Message origination support to levels 1, 2, and 3 and acts as a target and an originator for Explicit and I/O Messages, e.g., PLCs, I/O scanners, etc.

15.3.3.11 Device Profiles

EtherNet/IP uses the full set of profiles described in Chapter 6 of the CIP Specification [4].

15.3.3.12 Configuration

EtherNet/IP devices typically come with EDSs, as described in Section 15.2.7. For EDS-based configuration tools, the EDS should contain a Connection Manager section to describe the details of the connections that can be made into the device. This section basically is a mirror of what is contained in the Forward_Open message that a Connection Originator would send to the device. Multiple connections can be specified within an EDS that can then be chosen by the configuration tool.

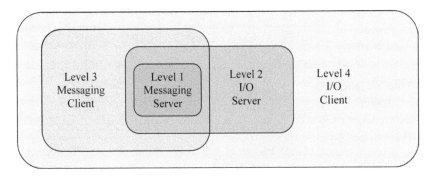

FIGURE 15.48 Device levels.

An EDS may also contain individual parameters or a Configuration Assembly with a complete description of all parameters within this Assembly. In many applications, the Configuration Assembly is transmitted as an attachment to the Forward_Open Message.

15.3.3.13 Conformance Test

Conformance testing is mandatory for all EtherNet/IP devices. Currently, this test is a protocol conformance test only since it is expected that most implementations use commercially available components for media access and physical attachments.

Depending on the complexity of the device, as many as several thousand messages are transmitted to the DUT. To allow a test that is closely adapted to the characteristics of the DUT, a formal description of all relevant features of the DUT must be provided by the manufacturer.

The software test is available from ODVA. It is a Windows-based tool, running on a PC with a standard Ethernet card. It is recommended that device developers run this test in their own labs before taking devices to the official ODVA test.

When a device passes the test, it is said to be EtherNet/IP CONFORMANCE TESTED™.* Devices that have passed conformance testing are published on the ODVA Web site.

15.3.3.14 Requirements for TCP/IP Support

In addition to the various requirements set forth in the EtherNet/IP Specification, all EtherNet/IP hosts are required to have a minimally functional TCP/IP suite and transport mechanism. The minimum host requirements for EtherNet/IP hosts shall be those covered in RFC 1122 [36], RFC 1123 [37], and RFC 1127 [38] and the subsequent documents that may supersede them. Whenever a feature or protocol is implemented by an EtherNet/IP host, that feature shall be implemented in accordance with the appropriate RFC documents, regardless of whether the feature or protocol is considered required or optional by this specification. The Internet and RFCs are dynamic. There will be changes to the RFCs and to the requirements included in this section as the Internet and this specification evolve, and these changes will not always provide for backward compatibility.

All EtherNet/IP devices shall at a minimum support:

- Internet Protocol (IP version 4) (RFC 791 [29])
- User Datagram Protocol (UDP) (RFC 768 [28])
- Transmission Control Protocol (TCP) (RFC 793 [31])
- Address Resolution Protocol (ARP) (RFC 826 [32])
- Internet Control Messaging Protocol (ICMP) (RFC 792 [30])
- Internet Group Management Protocol (IGMP) (RFC 1112 [35] and RFC 2236 [41])
- IEEE 802.3 (Ethernet) as defined in RFC 894 [33]

---

*EtherNet/IP CONFORMANCE TESTED™ is a certification mark of ODVA.

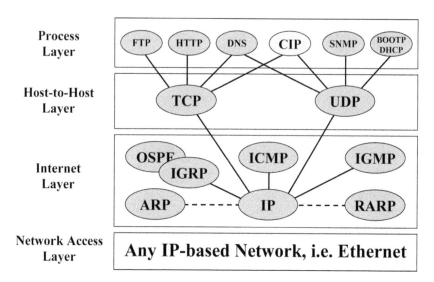

FIGURE 15.49 Relationship of CIP to other typical Ethernet protocols.

Although the encapsulation protocol is suitable for use on other networks besides Ethernet that support TCP/IP, and products may be implemented on these other networks, conformance testing of EtherNet/IP products is limited to those products on Ethernet. Other suitable networks include:

- Point-to-Point Protocol (PPP) (RFC 1171 [39])
- ARCNET (RFC 1201 [40])
- Fiber Distributed Data Interface (FDDI) (RFC 1103 [34])

15.3.3.15 Coexistence of EtherNet/IP and Other Ethernet-Based Protocols

EtherNet/IP devices are encouraged but not required to support other Internet protocols and applications not specified in the EtherNet/IP Specification. For example, they may support Hypertext Transfer Protocol (HTTP), Telnet, File Transfer Protocol (FTP), etc. The EtherNet/IP Specification makes no requirements with regard to these protocols and applications.

Figure 15.49 shows the relationship between CIP and other typical Ethernet-based protocol stacks. Since EtherNet/IP, like many other popular protocols, is based on TCP/IP and UDP/IP, coexistence with many other services and protocols is no problem at all and CIP blends nicely into the set of already existing functions. This means that anybody already using some or all of these popular Ethernet services can add CIP without too much of a burden; the existing services like HTTP or FTP may remain as before, and CIP will become another service on the process layer.

15.3.3.16 Ethernet Infrastructure

To successfully apply EtherNet/IP to the automation world, the issue of determinism has to be considered. The inherent principle of the Ethernet bus access mechanism whereby collisions are detected and nodes back off and try again after a while cannot guarantee determinism. While Ethernet in its present form cannot be made strictly deterministic, there are ways to improve this situation.

First, the hubs typically used in many office environments have to be replaced by the more intelligent switches that will forward only those Ethernet frames that are intended for nodes connected to this switch. With the use of full-duplex switch technology, collisions are completely avoided; instead of colliding, multiple messages sent to the same node at the same time are queued up inside the switch and are then delivered one after another.

As already mentioned in Section 15.3.3.8.2, it is highly recommended that switches that support IGMP Snooping be used.

If EtherNet/IP networks are to be connected to a general company network, then this should always be done through a router. The router keeps the UDP multicast messages from propagating into the company network and makes sure that broadcast or multicast office traffic does not congest the control network. Even though the router separates the two worlds, it can be set up to allow the TCP/IP-based Explicit Messages to pass through so that a configuration tool sitting in a PC in the office environment may very well be capable of monitoring and configuring devices on the control network.

15.3.3.17 Tools

Tools for EtherNet/IP networks can be divided into four groups:

- Physical layer tools: Tools (hardware and software) that verify the integrity and conformance of the physical layer or monitor the quality of the data transmission.
- Commissioning tools: All EtherNet/IP devices need an IP address. In some cases, the setting of this address can only be achieved through the Ethernet link (see Section 15.3.3.7). In these cases, a BOOTP/DHCP server tool is required such as the free BOOTP/DHCP routine downloadable from the Rockwell Automation Web site.
- Configuration tools: Software tools that are capable of communicating with individual devices for data monitoring and configuration purposes. Most configuration tools are EDS based; however, more complex devices like scanners tend to have their own configuration applets that are only partially based on EDSs. Some of these tools support multiple access paths to the network, e.g., via suitable bridging devices. High-level tools also actively query the devices on the network to identify them and monitor their health. Configuration tools may also be integrated into other packages like PLC programming software.
- Monitoring tools: Typically PC-based software packages (called sniffers) that can capture and display the Ethernet frames on the network. A raw Ethernet frame display may be good enough for some top experts, but it is recommended that a tool that allows both raw Ethernet frame display and multiple levels of frame interpretation (IP, TCP/UDP, EtherNet/IP header interpretation) be used. Due to the popularity of Ethernet, a large number of sniffers are available, but not all of them support EtherNet/IP decoding.

For a typical installation, a commissioning tool and a configuration tool are all that is needed.

Protocol monitoring tools are mainly used to investigate interoperability problems and to assist during the development process.

Turn to the EtherNet/IP product catalog on the ODVA Web site to access a list of vendors that provide tools for EtherNet/IP.

15.3.3.18 Advice for Developers

Before any development of an EtherNet/IP product is started, the following issues should be considered in detail:

- What functionality (device class; see Section 15.3.3.10) does the product require today and in future applications?
  - Messaging server only
  - Adapter functionality
  - Messaging client
  - Scanner functionality
- What are the physical layer requirements? Is IP 65/67 required or is IP 20 good enough?
- Will the development be based on commercially available hardware components and software packages (recommended) or designed from scratch (possible but costly)?
- What are the configuration requirements?
- What design and verification tools should be used?

- What is an absolute must before products can be placed on the market (own the specification, have a Vendor ID, have the product conformance tested)?

Ethernet chip sets and associated base software packages are available from a large number of vendors on the market. For support of the EtherNet/IP part of the development, turn to the ODVA Web site for a list of companies that can support EtherNet/IP developments.

15.3.3.19 EtherNet/IP Overall Summary

Since its introduction in 2000, EtherNet/IP has shown remarkable growth in many applications that used to be done with traditional fieldbuses. This success is largely attributed to the fact that this TCP/UDP/IP-based Ethernet system has introduced real-time behavior into the Ethernet domain without giving up any of its highly appreciated features such as company-wide access with standard and specialized tools through corporate networks.

The specific strength of EtherNet/IP is the fact that it does not require a modified or highly segregated network; standard switches and routers as known in the office world can be used without modification. At the same time, this means that all existing transport-level or TCP/UDP/IP-level protocols can continue to be used without any need for special bridging devices. The substantially improved real-time behavior of CIP Sync and the introduction of CIP Safety will soon allow EtherNet/IP to be used in applications that today need a set of several dedicated fieldbuses.

Finally, its use of CIP and object structure allows the blending of EtherNet/IP networks into an overall CIP network structure that allows seamless communication, just as if it was only one network.

15.4   Benefits of the CIP Family

The benefits of the CIP family can be subdivided into two groups:

- Benefits for the manufacturer of devices
- Benefits for the user of devices and systems

15.4.1   Benefits for the Manufacturer of Devices

Major benefits for manufacturers come from the fact that existing knowledge can be reused from one protocol to another. This results in lower training costs for development, sales, and support personnel. Reduced development costs can be achieved since certain parts (e.g., parameters, profiles) of the embedded firmware can be reused from one network to another since they are identical. As long as these parts are written in a high-level language, the adaptation is simply a matter of running the right compiler for the new system.

Another very important advantage for manufacturers is the easy routing of messages from one system to another. Any routing device can be designed very easily, since there is no need to invent a translation from one system to another; both systems already speak the same language. Manufacturers also benefit from dealing with the same organizations for support and conformance testing.

15.4.2   Benefits for the Users of Devices and Systems

Major benefits for users come from the fact that existing knowledge can be reused from one protocol to another, e.g., Device Profiles and the behavior of devices are identical from one system to another. This results in lower training costs. Technical personnel and users do not have to make very large changes to adapt an application from one type of CIP network to another. The system integrator can choose the CIP network that is best suited to his application without having to sacrifice functionality.

A further, very important benefit comes from the ease of bridging and routing within the CIP family. Moving information between noncompatible fieldbuses is always difficult and cumbersome, since it is almost impossible to translate functionality from one fieldbus to another. This is where the full benefits of CIP can be reaped. Forwarding of data and messages from top to bottom and back again is very easy to implement and uses very little system overhead. There is no need to translate from one data structure to another — they are the same. Services and status codes share the same benefit: these, too, are identical over all CIP networks. Finally, creating a message that runs through multiple hops of CIP networks is simply a matter of inserting the full path from the originating to the target device. Not a single line of code or any other configuration is required in the routing devices. This results in fast and efficient services that are easy to create and maintain. Even though these networks may be used in different parts of the application, messaging from one end to another really works as if there is only one network.

Finally, the very efficient Producer/Consumer mechanisms used in all CIP networks result in very fast and efficient use of the transmission bandwidth, with the result that system performance is often much higher than that with other fieldbuses running at higher raw baud rates. Only the data that are really important will be transmitted, instead of repeating old data again and again.

Planned and future protocol extension will always be integrated in a manner that allows coexistence of normal devices with enhanced devices like those supporting CIP Sync and CIP Safety. Therefore, no strict segmentation into Standard, CIP Sync, and CIP Safety networks is required unless there is a compelling reason, e.g., unacceptably high response time due to high bus load.

15.5 Protocol Extensions under Development

15.5.1 CIP Sync

15.5.1.1 General Considerations

CIP networks as described in References [3–6] have a real-time behavior that is appropriate for many applications, but there are a growing number of applications that require much tighter control of certain real-time parameters. Let us have a look at some of them:

- Real-time: This term is being used in a large number of differing meanings in various contexts. For further use in this section, the following definition is used:

  A system exhibits real-time behavior when it can react to an external stimulus within a predetermined time. How short or how long this time is depends on the application. Demanding industrial control applications require reactions in the millisecond range, while some process control applications can often live with reaction times of several seconds or more.

- Determinism: A deterministic system allows worst-case determination (not a prediction or a probability) of when a certain action takes place. Industrial communication systems may offer determinism to a higher or lesser degree depending on how they are implemented and used. Networks featuring message transmission at a predetermined point in time, such as ControlNet, SERCOS interface, and Interbus-S, are often said to offer absolute determinism. On the other hand, networks such as Ethernet may become undeterministic under certain load conditions, specifically when deployed in half-duplex mode with hubs. However, when Ethernet is deployed with full-duplex, high-speed switches, it operates in a highly deterministic manner (see Section 15.3.3.16).

- Reaction time: In an industrial control system, the overall system reaction time is what determines the real-time behavior. The communication system is only one of several contributing factors to the overall reaction time. In general, it is the time from an input stimulus to a related output action.

- Jitter: The term *jitter* is used to define the time deviation of a certain event from its average occurrence. Some communication systems rely on a very small message jitter, while most applications only require that a certain jitter is not exceeded for actions at the borders of the system, such as input sampling jitter and output action jitter.

- Synchronicity: Distributed systems often require certain actions to take place in a coordinated fashion; i.e., these actions must take place at a predetermined moment in time independent of where the action is to take place. A typical application is coordinated motion or electronic gearing. Some of these applications require a synchronicity in the microsecond range.

- Data throughput: This is the capability of a system to process a certain amount of data within a certain time span. For communication systems, protocol efficiency, the communication model (e.g., Producer/Consumer), and endpoint processing power are most important, while the wire speed only sets the limit of how much raw data can be transmitted across the physical media.

CIP Sync is a CIP-based communication principle that enables synchronous low-jitter system reactions without the need for low-jitter data transmission. This is of great importance in systems that do not provide absolute deterministic data transmission or where it is desirable for a variety of higher-layer protocols to run in parallel to the application system protocol. The latter situation is characteristic for Ethernet. Most users of TCP/IP-based Ethernet want to keep using it as before without the need to resort to a highly segregated network segment to run the real-time protocol. The CIP Sync communication principle meets these requirements.

15.5.1.2   Using IEEE 1588 Clock Synchronization

The recently published IEEE standard 1588 — Standard for a Precision Clock Synchronization Protocol for Networked Measurement and Control Systems [24] — lays the foundation for a precise synchronization of real-time clock in a distributed system. An IEEE 1588 system consists of a Time Master that distributes its system time to Time Slaves in a tree-like structure. The Time Master may be synchronized with another real-time clock higher up in the hierarchy, while the Time Slaves may be Time Masters for other devices below them. A Time Slave that is Time Master to another set of devices (typically in another part of the system) is also called a Boundary Clock. The time distribution is done by multicasting a message with the actual time of the master clock. This message originates in a relatively high layer of the communication stack, and, therefore, the actual transmission takes place at a slightly later point in time. Also, there will be a variation of the stack processing time from one message to another. To compensate this delay and its jitter, the actual transmission time can be captured in a lower layer of the communication stack, such as noting the "transmit complete" feedback from the communication chip. This update time capture is then distributed in a follow-up message. The average transmission delay is also determined so that the time offset between master and slave clock can also be compensated. This protocol has been fully defined for Ethernet UDP/IP systems, and the protocol details for further industrial communication systems are to follow. The clock synchronization accuracy that can be achieved with this system largely depends on the precision time capture of the master clock broadcast message. Hardware-assisted time capture systems can reach a synchronization accuracy of 250 ns or less. It is expected that Ethernet chip manufacturers will offer integrated IEEE 1588 hardware support in the very near future.

15.5.1.3   Additional Object

CIP Sync will require the addition of a new time synchronization object. This object manages the real-time clock inside a CIP Sync device and provides access to the IEEE 1588 timing information. Figure 15.50 shows the relationship of the additional object required for CIP Sync.

15.5.1.4   CIP Sync Communication Principles

Real-time clocks coordinated through the IEEE 1588 protocol on their own do not constitute a real-time system yet. Additional details to show how time stamping is used for input sampling and for the coordination of output actions will be added. Some Device Profiles will be extended as well to incorporate time information in their I/O Assemblies. Details of this activity are under discussion in the ODVA Distributed Motion Control JSIG.

15.5.1.5   Message Prioritization

Combining these three elements (Sections 15.5.1.2, 15.5.1.3, and 15.5.1.4) with collision-free infrastructure (see Section 15.3.3.16) is sufficient to build a real-time system. However, it is necessary to consider all traffic within the system and arrange all application messages containing time-critical data in such a

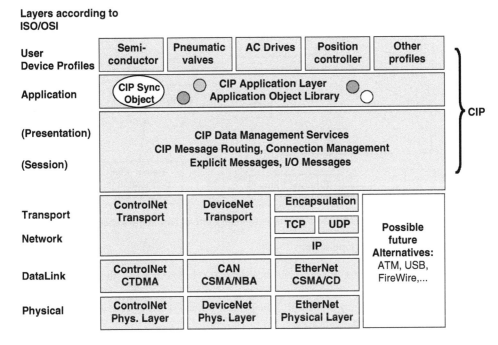

FIGURE 15.50 CIP extensions required for CIP Sync.

way that they are guaranteed to arrive at all consumers in time. When other Ethernet protocols, such as HTTP or FTP with possibly very long frames, need to coexist in the same system, the situation may need careful configuration. Ethernet frames with up to 1500 bytes of payload (approximately 122 μs long in a 100 Mbit/second system) can easily congest the system and delay important messages by an undetermined amount of time, possibly too long for correct functioning of the system.

This is where message prioritization becomes an important element. Of the many prioritization schemes in use or proposed for Ethernet today, EtherNet/IP uses message prioritization according to IEEE 802.3:2002 [13]. This is a scheme supported by many switches available today. It allows preferential treatment of Ethernet frames in such a way that those frames with the highest priority will jump the message queues in a switch and will get transmitted first. Messages with high priority will get transmitted, while those with lower priority typically have to wait. Suitable priority assignment for all time-critical messages then guarantees their preferential treatment. Standard EtherNet/IP and other Ethernet messages will get low or no priority and thus have to wait until all higher-priority messages have passed.

Once this prioritization scheme is implemented, one full-length frame can be tolerated within every communication cycle consisting of a set of prioritized input (port A through port E) and output (port F) messages. Figure 15.51 illustrates this process.

15.5.1.6 Applications of CIP Sync

Typical applications for CIP Sync are time-stamping sensor inputs, distributed time-triggered outputs, and distributed motion such as electronic gearing or camming applications. For example, in motion applications, the sensors sample their actual position at a predetermined time, i.e., in a highly synchronized way, and transmit them to the application master that coordinates the motion. The application master then calculates the new reference values and sends them to the motion drives. Using CIP Sync, it is no longer necessary to have extremely low jitter in the communication system; it is sufficient to transmit all time-critical messages in time, and their exact arrival time becomes irrelevant. The assignment of suitable priorities to CIP Sync communication guarantees that all time-critical messages always get the bandwidth they need and all other traffic is automatically limited to the remaining bandwidth.

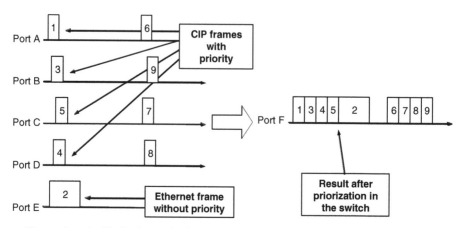

The numbers inside the frames indicate their relative arrival time at the switch port

FIGURE 15.51 Ethernet frame prioritization.

As a result of these measures, CIP Sync devices can coexist side by side with other EtherNet/IP devices without any need for network segmentation or special hardware. Even non-EtherNet/IP devices — provided they do not override any of the CIP Sync prioritizations — can be connected without any loss of performance in the CIP Sync application.

15.5.1.7 Expected Performance of CIP Sync Systems

As already mentioned, CIP Sync systems can be built to maintain a synchronization accuracy of better than 250 ns, in many cases without the use of Boundary Clocks. The communication cycle and thus the reaction delay to unexpected events is largely governed by the number of CIP Sync devices in a system. Allowing some bandwidth (approximately 40%) for non-CIP Sync messages, as described in Section 15.5.1.5, the theoretical limit (close to 100% wire load) for the communication cycle of a CIP Sync system based on a 100 Mbit/s Ethernet link is around 500 μs for 30 coordinated motion axes, with 32 bytes of data each.

15.5.1.8 CIP Sync Summary

CIP Sync based on EtherNet/IP is a natural extension of the EtherNet/IP system into the very fast real-time domain. In contrast to many other proposed or existing real-time extensions, it does not require any strict network segmentation between high-performance real-time sections and other parts of the communication system. This results in truly open systems that can tolerate the vast majority of parallel TCP/IP-based protocols found in today's industrial communication architecture without compromising performance.

In a first phase, the CIP Sync principles will be applied to EtherNet/IP, while an extension to the other CIP implementations will follow at a later time.

15.5.2 CIP Safety

CIP Safety, like other safety protocols based on industry standard networks, adds additional services to transport data with high integrity. Unlike other networks, the user of CIP Safety does not have to change his approach when going from one network or media to another. CIP Safety presents a scalable, network-independent approach to safety network design, where the safety services are described in a well-defined layer. This approach also enables the routing of safety data, allowing the user to create end-to-end safety chains across multiple links without being forced to difficult-to-manage gateways.

15.5.2.1 General Considerations

In the past and still today, hardwired safety systems employed safety relays that are interconnected to provide a safety function. Hardwired systems are difficult to develop and maintain for all but the most trivial applications. Furthermore, these systems place significant restrictions in the distance between devices.

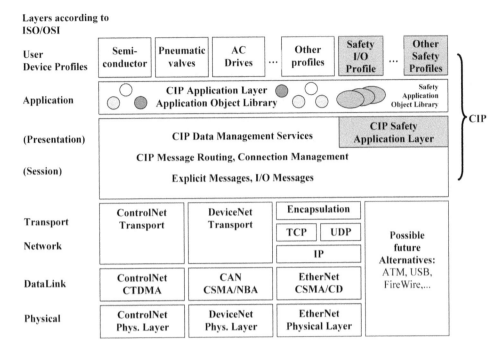

FIGURE 15.52 CIP communications layers, including safety.

Because of these issues, as well as distance and cost considerations, it is desirable to allow safety services to be implemented on standard communication networks. The key to the development of safety networks was not to create a network that could not fail, but to create a system where failures in the network would cause safety devices to go to a known state. If the user knew which state the system would go to, he could make his application safe. But this meant that significantly more checking and redundant coding information would be required.

To determine the additional safety requirements, an existing railway standard [20] was used and later extended by the German Safety Bus committee [21]. This committee provided design guidelines to safety network developers to allow their networks and safety devices to be certified according to IEC 61508 [15].

Using these results, the Common Industrial Protocol, which allows network-independent routing of standard data, was extended to allow high-integrity safety services. The result is a scalable, routable, network-independent safety layer, thus removing the requirement for dedicated safety gateways. Since all safety devices execute the same protocol, independent of which media they reside on, the user approach is consistent and independent of media or network used.

CIP Safety is an extension to standard CIP that has been approved by TÜV Rheinland for use in IEC 61508 SIL 3 and EN 954-1 category 4 applications. It extends the model by adding CIP Safety application layer functionality, as shown in Figure 15.52. The additions include several safety-related objects and Safety Device Profiles.

Because the safety application layer extensions do not rely on the integrity (see Section 15.5.2.3) of the underlying standard CIP as described in Section 15.2 and data link layers as described in Sections 15.3.1, 15.3.2, and 15.3.3, single-channel (nonredundant) hardware can be used for the data link communication interface. This same partitioning of functionality allows standard routers to be used to route safety data, as shown in Figure 15.53. The routing of safety messages is possible, because the end device is responsible for ensuring the integrity of the data. If an error occurs in the transmission of data or in the intermediate router, the end device will detect the failure and take an appropriate action.

This routing capability allows the creation of DeviceNet Safety cells with quick reaction times to be interconnected with other cells via a backbone network such as EtherNet/IP for interlocking, as shown in Figure 15.54. Only the safety data that are needed are routed to the required cell, which reduces the

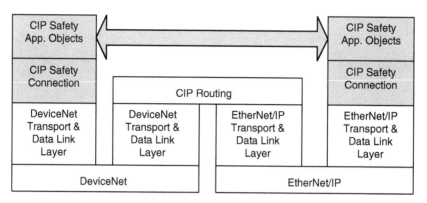

FIGURE 15.53 Routing of safety data.

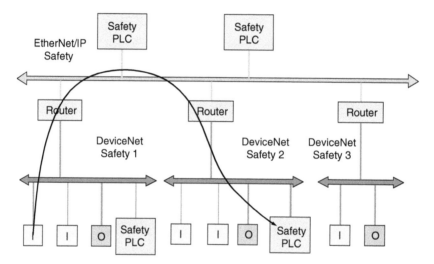

FIGURE 15.54 Network routing.

individual bandwidth requirements. The combination of fast responding local safety cells and the intercell routing of safety data allows users to create large safety applications with fast response times. Another benefit is the ability to multicast safety messages across multiple networks.

15.5.2.2 Implementation of Safety

As indicated in Figure 15.52, all CIP Safety devices also have an underlying standard CIP functionality. The extension to the CIP Safety application layer is specified using a Safety Validator Object. This object is responsible for managing the CIP Safety Connections (standard CIP Connections are managed through communication objects) and serves as the interface between the safety application objects and the link layer connections, as shown in Figure 15.55. The Safety Validator ensures the integrity of the safety data transfers by applying the integrity-ensuring measures described in Section 15.5.2.3.

- The producing safety application uses an instance of a Client Validator to produce safety data and ensure time coordination.
- The client uses a link data producer to transmit the data and a link consumer to receive time coordination messages.
- The consuming safety application uses a Server Validator to receive and check data.
- The server uses a link consumer to receive data and a link producer to transmit time coordination messages.

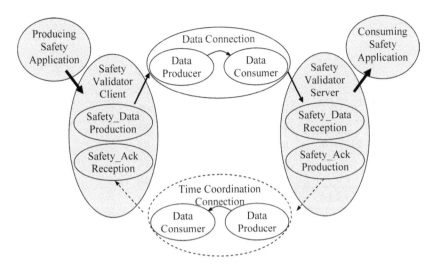

FIGURE 15.55 Relationship of Safety Validators.

The link producers and consumers have no knowledge of the safety packet and fulfill no safety function. The responsibility for high-integrity transfer and checking of safety data lies within the Safety Validators.

15.5.2.3 Ensuring Integrity

CIP Safety does not prevent communication errors from occurring, but it ensures transmission integrity by detecting errors and allowing devices to take appropriate actions. The Safety Validator is responsible for detecting these communication errors. The nine communication errors that must be detected are shown in Figure 15.56 along with the five measures CIP Safety used to detect these errors, based on Reference [21].

Communication Errors	Measures to detect communication errors				
	Time Expectation via time stamp	ID for send and receive	Safety CRC	Redundancy with Cross Checking	Diverse measure
Message Repetition	X		X*		
Message Loss	X		X*		
Message Insertion	X	X	X*		
Incorrect Sequence	X		X*		
Message Corrupt			X	X	
Message Delay	X				
Coupling of safety and safety data		X			
Coupling of safety and standard data	X	X	X	X	X
Increased age of data in bridge	X				

* The Safety CRC provides additional protection for communication errors in fragmented messages.

FIGURE 15.56 Error detection measures.

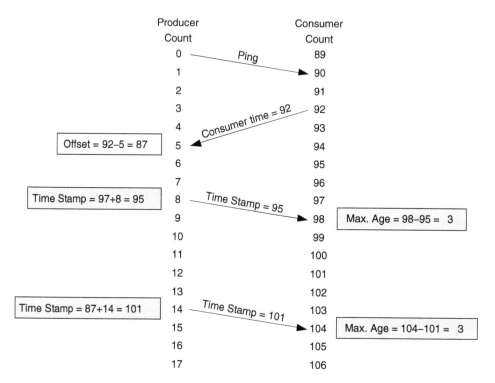

FIGURE 15.57 Time stamp.

15.5.2.3.1 *Time Expectation via a Time Stamp*

All CIP Safety data are produced with a time stamp, which allows Safety Consumers to determine the age of the produced data. This detection measure is superior to the more conventional reception timers. Reception timers can tell how much time has elapsed since a message was last received, but they do not convey any information about the actual age of the data. A time stamp allows transmission, media access/ arbitration, queuing, retry, and routing delays to be detected.

Time is coordinated between producers and consumers using ping requests and ping responses, as shown in Figure 15.57. After a connection is established, the producer will produce a ping request, which causes the consumer to respond with its consumer time. The producer will note the time difference between the ping production and the ping response and store this as an offset value. The producer will add this offset value to its producer time for all subsequent data transmissions. This value is transmitted as the time stamp. When the consumer receives a data message it subtracts its internal clock from the time stamp to determine the data age. If the data age is less than the maximum age allowed, the data are applied; otherwise, the connection goes to the safety state. The device application is notified so that the connection safety state can be appropriately reflected.

The ping request-and-response sequence is repeated periodically to correct for any drift in producer or consumer time base drift.

15.5.2.3.2 *Production Identifier*

A Production Identifier (PID) is encoded in each data production of a Safety Connection to ensure that each received message arrives at the correct consumer. The PID is derived from an electronic key, the device Serial Number, and the CIP Connection Serial Number. Any safety device inadvertently receiving a message with the incorrect PID will go to a safety state. Any safety device that does not receive a message within the expected time interval with the correct PID will also go to a safety state. This measure ensures that messages are routed correctly in multilink applications.

15.5.2.3.3 *Safety CRC*

All safety transfers on CIP Safety use Safety CRCs to ensure the integrity of the transfer of information. The Safety CRCs serve as the primary measure to detect possible corruption of transmitted data. They provide detection up to a Hamming distance of 4 for each data transfer section, though the overall Hamming distance coverage is greater for the complete transfer due to the redundancy of the protocol. The Safety CRCs are generated in the Safety Producers and checked in the Safety Consumers. Intermediate routing devices do not examine the Safety CRCs. Thus, by employing end-to-end Safety CRCs, the individual data link CRCs are not part of the safety function. This eliminates certification requirements for intermediate devices and helps to ensure that the safety protocol is independent of the network technology. The Safety CRC also provides a strong protection mechanism that allows underlying data link errors such as bit stuffing or fragmentation errors to be detected.

The individual link CRCs are not relied on for safety, but they are still enabled. This provides an additional level of protection and noise immunity, by allowing data retransmission for transient errors at the local link.

15.5.2.3.4 *Redundancy and Cross-Check*

Data and CRC redundancy with cross-checking provides an additional measure of protection by detecting possible corruption of transmitted data. They effectively increase the Hamming distance of the protocol. These measures allow long safety data packets, up to 250 bytes, to be sent with high integrity. For short packets of 2 bytes or less, data redundancy is not required; however, redundant CRCs are cross-checked to ensure integrity.

15.5.2.3.5 *Diverse Measures for Safety and Standard*

The CIP Safety protocol is present only in safety devices; this prevents standard devices from masquerading as a safety device.

15.5.2.4 Safety Connections

CIP Safety provides two types of Safety Connections:

- Unicast
- Multicast

A unicast connection, as shown in Figure 15.58, allows a Safety Validator client to be connected to a Safety Validator server using two link layer connections.

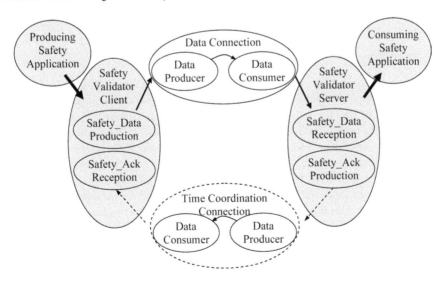

FIGURE 15.58 Unicast connection.

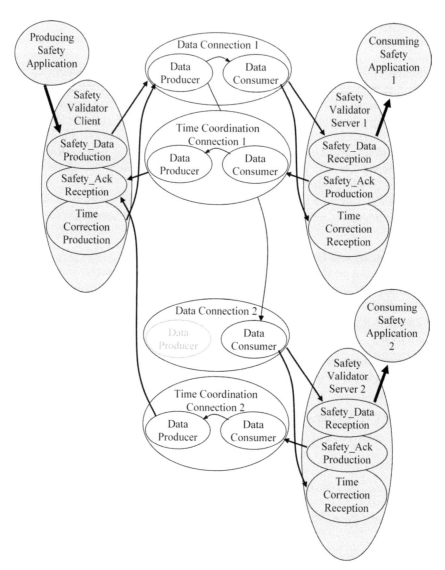

FIGURE 15.59 Multicast connection.

A multicast connection, as shown in Figure 15.59, allows up to 15 Safety Validator servers to consume safety data from a Safety Validator client. When the first Safety Validator server establishes a connection with a Safety Validator client, a pair of link layer connections is established: one for data-and-time correction and one for time coordination. Each new Safety Validator server will use the existing data-and-time correction connection and establish a new time coordination connection with the Safety Validator client.

To optimize the throughput on DeviceNet, three data link connections are used for each multicast connection, as shown in Figure 15.60. The data-and-time correction messages are sent on separate connections. This allows short messages to be transmitted on DeviceNet within a single CAN frame and reduces the overall bandwidth, since the time correction and time coordination messages are sent at a much slower periodic interval.

When multicast messages are routed off link, the router combines the data-and-time correction messages from DeviceNet and separates them when messages reach DeviceNet. Since the safety message contents are unchanged, the router provides no safety function.

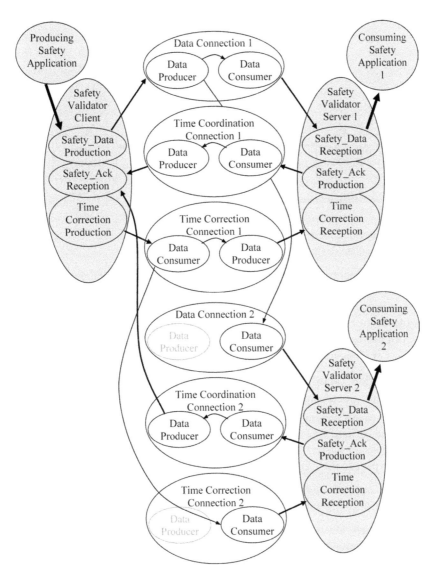

FIGURE 15.60 Multicast connection on DeviceNet.

15.5.2.5 Message Packet Sections

CIP Safety has four message sections:

- Data section
- Time-stamp section
- Time correction section
- Time coordination section

The description of these formats would go beyond the scope of this handbook. Reference [23] provides further details.

15.5.2.6 Configuration

Before safety devices can be used in a safety system, they must first be configured and connections must be established. The process of configuration requires configuration data from a configuration tool to be placed in a safety device. There are two possible sequences for configuration:

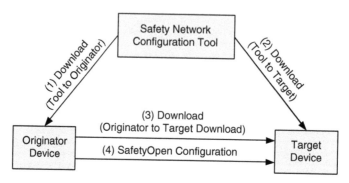

FIGURE 15.61 Configuration transfers.

- Configuration tool directly to device
- Via an intermediate device

In the configuration tool-to-device case, as shown in Figure 15.61, the configuration tool writes directly to the device to be configured (1 and 2).

In the case of intermediate device configuration, the tool first writes to an originator (1) and the originator writes to the target using an originator-to-target download (3) or a Safety_Open service (4). The Safety_Open service (4) is unique in that it allows a Safety Connection to be established at the same time that a device is configured.

15.5.2.7 Connection Establishment

CIP provides a connection establishment mechanism, using a Forward_Open service that allows producer-to-consumer connections to be established locally or across multiple links via intermediate routers. An extension of the Forward_Open, called the Safety_Open service, has been created to allow the same multilink connections for safety.

There are two types of Safety_Open requests:

- Type 1: With configuration
- Type 2: Without configuration

With the Type 1 Safety_Open request, configuration and connections are established at the same time. This allows rapid configuration of devices with simple and relatively small configuration data.

With the Type 2 Safety_Open request, the safety device must first be configured, and the Safety_Open request then establishes a Safety Connection. This separation of configuration and connection establishment allows the configuration of devices with large and complex configuration data.

In both cases, the Safety_Open request establishes all underlying link layer connections — across the local link as well as any intermediate links and routers.

15.5.2.8 Configuration Implementation

CIP Safety provides the following protection measures to ensure the integrity of configuration:

- Safety Network Number
- Password protection
- Configuration ownership
- Configuration locking

15.5.2.8.1 Safety Network Number

The Safety Network Number provides a unique network identifier for each network in the safety system. The Safety Network Number combined with the local device address allows any device in the safety system to be uniquely addressed.

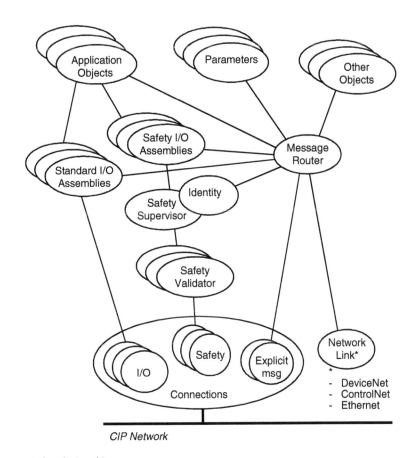

CIP Network

FIGURE 15.62 Safety device objects.

15.5.2.8.2 Password Protection
All safety devices support the use of an optional password. The password mechanism provides an additional protection measure, prohibiting the reconfiguration of a device without the correct password.

15.5.2.8.3 Configuration Ownership
The owner of a CIP Safety device can be specified and enforced. Each safety device can specify that its configuration is configured by a selected originator or that the configuration is only configured by a configuration tool.

15.5.2.8.4 Configuration Locking
Configuration Locking provides the user with a mechanism to ensure that all devices have been verified and tested prior to being used in a safety application.

15.5.2.9 Safety Devices
The relationship of the objects within a safety device is shown in Figure 15.62. Note that CIP Safety extends the CIP object model, with the addition of Safety I/O Assemblies and Safety Validator and Safety Supervisor Objects.

15.5.2.10 Safety Supervisor
The Safety Supervisor Object provides a common configuration interface for safety devices. The Safety Supervisor Object centralizes and coordinates application object state behavior and related status information, exception status indications (alarms and warnings), and defines a behavior model that is assumed by objects identified as belonging to safety devices.

15.5.2.11 CIP Safety Summary

The concept presented here demonstrates a scalable, routable, network-independent safety protocol based on extensions to the CIP architecture. This concept can be used in solutions ranging from device-level networks such as DeviceNet to higher-level networks such as EtherNet/IP. By designing network independence into CIP Safety, multilink routing of Safety Connections can be supported. Functions such as multilink routing and multicast messaging provide a strong foundation that enables users to create the fast-responding local cells and interconnect remote cells that are required for today's safety applications. The design also enables expansion to future network technologies as they become available.

15.6 Conclusion

The CIP family of protocols is a very versatile set of fieldbus protocols that are scalable to allow their use in many applications and many levels of the automation architecture. Due to the universal applicability of the underlying protocol, it is very easy to switch from one system to another. The Producer/Consumer principle, together with the open-object architecture used in the CIP family, allows very efficient use of the communication bandwidth and ensure that these modern systems can be used for many years to come.

References*

1. CIP Common Specification, Release 1.0, © 2000, 2001 by ControlNet International and Open DeviceNet Vendor Association.
2. DeviceNet Specification, Release 2.0, including Errata 5, March 31, 2002, © 1995–2002 by Open DeviceNet Vendor Association.
3. ControlNet Specification, Release 2.0, including Errata 2, December 31, 1999, © 1998, 1999 by ControlNet International.
4. CIP Common Specification, Edition 2.0, © 2001–2004 by ODVA and ControlNet International.
5. DeviceNet Adaptation of CIP Specification, Edition 1.0, December 15, 2003, © 1994–2004 by Open DeviceNet Vendor Association.
6. EtherNet/IP Specification, Release 1.0, June 5, 2001, © 2000, 2001 by ControlNet International and Open DeviceNet Vendor Association.
7. *Planning and Installation Manual*, DeviceNet Cable System, Publication PUB00027R1, downloadable from ODVA Web site (http://www.odva.org/).
8. *Recommended IP Addressing Methods for EtherNet/IP Devices*, Publication PUB00028R0, downloadable from ODVA Web site (http://www.odva.org/).
9. IEC 61131-3:1993, Programmable Controllers: Part 3: Programming Languages.
10. *Controller Area Network: Basics, Protocols, Chips and Application*, IXXAT Automation, 2001.
11. ISO 11898:1993, Road Vehicles: Interchange of Digital Information: Controller Area Network (CAN) for High-Speed Communication.
12. IEEE 802.3:2000, ISO/IEC 8802-3:2000, Information Technology: Local and Metropolitan Area Networks: Part 3: Carrier Sense Multiple Access with Collision Detection (CSMA/CD) Access Method and Physical Layer Specification.
13. IEEE 802.3:2002, Information Technology: Telecommunication and Information Exchange between Systems: LAN/MAN: Specific Requirements: Part 3: Carrier Sense Multiple Access with Collision Detection (CSMA/CD) Access Method and Physical Layer Specifications.
14. ISO/IEC 7498-1:1994, Information Technology: Open Systems Interconnection: Basic Reference Model.
15. IEC 61508, Functional Safety of Electrical/Electronic/Programmable Electronic Safety-Related Systems, 1998.

---

*All RFCs are downloadable from http://www.faqs.org/rfcs/

16. IEC 62026-3, Low-Voltage Switchgear and Controlgear: Controller-Device Interfaces (CDIs): Part 3: DeviceNet, 2000.
17. EN 50325-2, Industrial Communications Subsystem Based on ISO 11898 (CAN) for Controller-Device Interfaces: Part 2: DeviceNet, 2000.
18. GB/T 18858 (Chinese national standard), Low-Voltage Switchgear and Controlgear Controller-Device Interface, 2003.
19. IEC 61158, Digital Data Communications for Measurement and Control: Fieldbus for Use in Industrial Control Systems, 2000.
20. EN 50159-1:2001, Railway Applications, Communication, Signaling and Processing Systems.
21. *Draft Proposal Test and Certification Guideline, Safety Bus Systems*, BG Fachausschuβ Elektrotechnik, May 28, 2000.
22. IEC 61508, Functional Safety of Electrical/Electronic/Programmable Electronic Safety-Related Systems.
23. Vasko, David A. and Suresh R. Nair, 2003, CIP Safety: Safety Networking for the Future, paper presented at *Proceedings of the 9th International CAN Conference*.
24. IEEE 1588:2002, Standard for a Precision Clock Synchronization Protocol for Networked Measurement and Control Systems.
25. Schiffer, Viktor, 2003, Modular EDSs and Other EDS Enhancements for DeviceNet, paper presented at *Proceedings of the 9th International CAN Conference*.
26. Schiffer, Viktor, 2003, Device Configuration Using Electronic Data Sheets, *ODVA Conference and 9th Annual Meeting*, downloadable from ODVA Web site.
27. Schiffer, Viktor and Ray Romito, DeviceNet Development Considerations, downloadable from ODVA Web site, 2000.
28. RFC 768, User Datagram Protocol, 1980.
29. RFC 791, Internet Protocol, 1981.
30. RFC 792, Internet Control Message Protocol, 1981.
31. RFC 793, Transmission Control Protocol, 1981.
32. RFC 826, Ethernet Address Resolution Protocol, or Converting Network Protocol Addresses to 48.bit Ethernet Address for Transmission on Ethernet Hardware, 1982.
33. RFC 894, Standard for the Transmission of IP Datagrams over Ethernet Networks, 1984.
34. RFC 1103, Proposed Standard for the Transmission of IP Datagrams over FDDI Networks, 1989.
35. RFC 1112, Host Extensions for IP Multicasting, 1989.
36. RFC 1122, Requirements for Internet Hosts: Communication Layers, 1989.
37. RFC 1123, Requirements for Internet Hosts: Application and Support, 1989.
38. RFC 1127, Perspective on the Host Requirements RFCs, 1989.
39. RFC 1171, Point-to-Point Protocol for the Transmission of Multi-Protocol Datagrams over Point-to-Point Links, 1990.
40. RFC 1201, Transmitting IP Traffic over ARCNET Networks, 1991.
41. RFC 2236, Internet Group Management Protocol, Version 2, 1997.

16

Configuration and Management of Fieldbus Systems

Stefan Pitzek
Vienna University of Technology

Wilfried Elmenreich
Vienna University of Technology

16.1 Introduction

Fieldbus systems are often evaluated by their technical merits, like performance, efficiency, and suitability for a particular application. Being designed to perform control applications, most industrial communication networks are well capable of performing their respective application tasks. Besides these ostensible criteria, however, there are some other capabilities a fieldbus system must provide, which in some cases might actually have a greater influence on the usability of a particular system than the technical ability to fulfill the given control requirements. These capabilities deal with the configuration and management, i.e., the setup, configuration, monitoring, and maintenance of the fieldbus system.

Powell [32] describes the problematic situation in the past: "Fifteen years ago, a typical process automation plant consisted of various field devices from half a dozen of vendors. Each device had its own setup program with different syntax for the same semantics. The data from the devices often differed in the data formats and the routines to interface each device."

Since that time, a lot of concepts and methods have been devised in order to support these configuration and management tasks. Many of the concepts have been implemented in fieldbus technologies such as HART (highway addressable remote transducer), Profibus, Foundation Fieldbus, LON (local operating network), etc. It is the objective of this chapter to give an introduction to the state-of-the-art concepts and methods for the configuration and management of fieldbus systems.

The remainder of the chapter is organized as follows. Section 16.2 gives definitions of the concepts and terms in the context of configuration and management of fieldbus systems. Section 16.3 investigates the requirements for configuration and management tasks. Section 16.4 analyzes the necessary interfaces of a field device and proposes a meaningful distinction of interface types. Section 16.5 discusses profiles and other representation mechanisms for system properties in several fieldbus systems. Section 16.6 gives an overview of application development methods and their implications for configuration and management of fieldbus networks. Section 16.7 examines the initial setup of a system in terms of hardware and software configuration. Section 16.8 deals with approaches for the management of fieldbus systems, like application download, diagnosis, and calibration of devices. Section 16.9 presents maintenance methods for reconfiguration, repair, and reintegration of fieldbus devices.

16.2   Concepts and Terms

The purpose of this section is to introduce and define some important concepts and terms that are used throughout this chapter.

16.2.1   Configuration vs. Management

The term *configuration* is used for a wide range of actions. Part of the configuration deals with setting up the hardware infrastructure of a fieldbus network and its nodes, i.e., physically connecting nodes (cabling) and configuring (e.g., by using switches, jumpers) nodes in a network. On the other hand, configuration also involves setting up the network on the logical (i.e., software) level. Depending on the network topology and underlying communication paradigm (and other design decisions), this leads to very different approaches to how configuration mechanisms are implemented.

In contrast, *management* deals with handling an already built system and includes maintenance, diagnosis, monitoring, and debugging. As with configuration, different fieldbus systems can greatly differ in their support and capabilities for these areas.

Often configuration and management are difficult to separate since procedures such as plug-and-play (see Section 16.2.3) involve configuration as well as management tasks.

16.2.2   Smart Devices

The term *smart* or *intelligent* device was first used in this context by Ko and Fung [21], meaning a sensor or actuator device that is equipped with a network interface in order to support an easy integration into a distributed control application.

In the context of fieldbus systems, a *smart* device supports its configuration and management by providing its data via a well-defined network interface [23] or offering a self-description of its features. The description usually comes in a machine-readable form (e.g., as an Extensible Markup Language (XML) description) that resides either locally at the fieldbus device (e.g., IEEE 1451.2 [17]) or at a higher network level being referenced by a series number (e.g., OMG Smart Transducer Interface [26]).

16.2.3   Plug-and-Play vs. Plug-and-Participate

Plug-and-play describes a feature for the automatic integration of a newly connected device into a system without user intervention. While this feature works well for personal computers within an office environment, it is quite difficult to achieve this behavior for automation systems, since without user intervention the system would not be able to guess what sensor data should be used and what actuator should be instrumented by a given device. Therefore, in the automation domain the more correct term *plug-and-participate* should be used, describing the initial configuration and integration of a new device that can be automated. For example, after connecting a new sensor to a network, it could be automatically detected, given a local name, and assigned to a communication slot. The task of a human system integrator is then reduced to decide on the further processing and usage of the sensor data.

16.2.4 State

Zadeh states that the "notion of state of a system at any given time is the information needed to determine the behavior of the system from that time on" [40, p. 3]. In real-time computer systems, we distinguish between the *initialization state* (i-state) and the *history state* (h-state) [22].

The i-state encompasses the static data structure of the computer system, i.e., data that are usually located in the static (read-only) memory of the system. The i-state does not change during the execution of a given application, e.g., calibration data of a fieldbus node. The h-state is the "dynamic data structure ... that undergoes change as the computation progresses" [22, p. 91]. An example for an h-state is the cached results of a sequence of measurements that are used to calculate the current state of a process variable.

The size of the h-state at a given level of abstraction may vary during execution. A good system design will aim at having a *ground state*, i.e., when the size of the h-state becomes zero. In a distributed system, this usually requires that no task is active and no messages are in transit.

16.3 Requirements on Configuration and Management

The requirements on a configuration and management framework are driven by several factors. We have identified the following points:

- *(Semi)automatic configuration*: The requirement for a plug-and-play-like configuration can be justified by three arguments:
    1. An automatic or semiautomatic configuration saves time and therefore leads to better maintainability and lower costs.
    2. The necessary qualification of the person who sets up the system may be lower if the overall system is easier to configure.
    3. The number of configuration faults will decrease, since monotone and error-prone tasks like looking up configuration parameters in heavy manuals are done by the computer. In most cases, a fully automatic configuration will only be possible if the functionality of the system is reduced to a manageable subset. For more complex applications, consulting the human mind is unavoidable. Thus, we distinguish two cases:
        (i) The automatic setup of simple subsystems. This use case mostly deals with systems that require an *automatic* and *autonomous* (i.e., without human intervention) reconfiguration of network and communication participants in order to adapt to different operating environments. Usually, such systems either use very sophisticated (and often costly) negotiation protocols or work only on closely bounded and well-known application domains.
        (ii) Computer-supported configuration of large distributed systems. This case is the usual approach.
- *Comprehensible interfaces*: In order to minimize errors, all interfaces will be made as comprehensible as possible. This includes the uniform representation of data provided by the interfaces and the capability of selectively restricting an interface to the data required by its user. The comprehensibility of an interface can be expressed by the *mental load* that it puts on to the user. Different users need different specialized interfaces, each with a minimum of mental load. For example, an application developer mostly has a service-centered view of the system. Physical network details and other properties not relevant for the application should be hidden from the developer [27].
- *Uniform data structures*: The configuration and management of fieldbus systems require representations of system properties that are usable by software tools. In order to avoid a situation where each application deals with the required information in its own way, these representations should be generic, highly structured, and exactly specified.

- *Low overhead on embedded system*: Fieldbus systems employ embedded hardware for reasons of cost, size, power consumption, and mechanical robustness. Such embedded hardware usually provides far less memory and processing power than average desktop systems. Currently, typical microcontrollers provide about several hundred bytes of RAM and few kilobytes of Flash ROM. Clocked by an internal oscillator, these microcontrollers provide about 0.5 up to 16 MIPS of processing power. Therefore, the designers of configuration and management tools must take care that there is as little overhead on the embedded system nodes as possible (e.g., static data required for management should be stored in a central repository outside the network).
- *Use of standard software/hardware*: Computers running standard Windows or Linux operating systems do not provide guaranteed response times for programs, and most hardware interfaces are controlled by the operating system. Since this might violate the special timing requirements of a fieldbus protocol, it is often not possible to directly connect a configuration host computer to the fieldbus network using the fieldbus protocol itself. Instead, a configuration tool must use some other means of communication, such as standard communication protocols or interfaces like Transmission Control Protocol (TCP)/Internet Protocol (IP), RS232, universal serial bus (USB), or standard middleware like CORBA (Common Object Request Broker Architecture). Since fieldbus nodes might not be powerful enough to implement these mechanisms, communication will often be performed using dedicated gateway nodes. In order to reduce the complexity of the involved conversion and transformation steps, the interface to and from the fieldbus node must be comprehensible, structurally simple, and easy to access.

16.4 Interface Separation

If different user groups access a system for different purposes, they should be provided with only interfaces to the information relevant for their respective purposes [33].

Interfaces for different purposes may differ by the accessible information and in the temporal behavior of the access across the interface.

Kopetz et al. [23] have identified three interfaces to transducer nodes of a fieldbus:

1. The *configuration and planning* (CP) interface allows the integration and setup of newly connected nodes. It is used to generate the "glue" in the network that enables the components of the network to interact in the intended way. Usually, the CP interface is not time critical.
2. The *diagnostic and management* (DM) interface is used for parametriation and calibration of devices and to collect diagnostic information to support maintenance activities. For example, a remote maintenance console can request diagnostic information from a certain sensor. The DM interface is usually not time critical.
3. The *real-time service* (RS) interface is used to communicate the application data, e.g., sensor measurements or set values for an actuator. This interface usually has to fulfill timing constraints such as a bounded latency and a small communication jitter. The RS interface has to be configured by means of the CP (e.g., communication schedules) or DM (e.g., calibration data or level monitors) interface.

The TTP/A (time-triggered protocol for SAE class A applications) fieldbus system [24] uses time-triggered scheduling that provides a deterministic communication scheme for the RS interface. A specified part of the bandwidth is reserved for arbitrary CP and DM activities. Therefore, it is possible to perform configuration and planning tasks while the system is in operation without a probe effect on the real-time service [15].

16.4.1 The Interface File System Approach

The concept of the Interface File System (IFS) was introduced by Kopetz et al. [23]. The IFS provides a unique addressing scheme to all relevant data of the nodes in a distributed system. Thus, the IFS maps real-time data, all kinds of configuration data, self-describing information, and internal state reports for diagnosis purposes.

The IFS is organized hierarchically as follows. The *cluster name* addresses a particular fieldbus network. Within the cluster, a specific node is addressed by the *node name*. The IFS of a node is structured into *files* and *records*. Each record is a unit of four *bytes* of data.

The IFS is a generic approach that has been implemented with the TTP/A protocol [24] as a case study for the OMG Smart Transducer Interface. The IFS approach supports well the integration and management of heterogeneous fieldbus networks. The IFS provides the following benefits:

- It establishes a well-defined interface between network communication and local application. The local application uses API (Application Programming Interface) functions to read and write data from or into the IFS. The communication interface accesses the IFS to exchange data across the network.
- The IFS hides network communication from the node application and provides location transparency for a message, since a task does not have to discriminate between data that is locally provided and data that is communicated via the network.
- Since the configuration and management data are also mapped into the IFS, configuration and management tools can directly use the CORBA STI (smart transducer interface) for accessing this information from outside the network. Figure 16.1 depicts an architecture with configuration and management tools that access the IFS of a fieldbus network from the Internet.

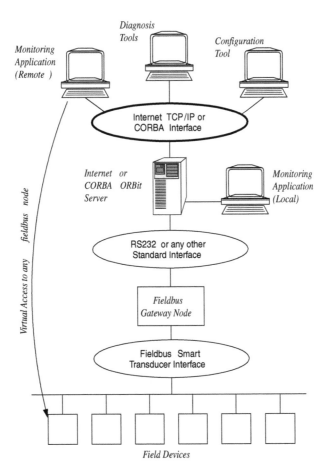

FIGURE 16.1 Architecture for remote configuration and monitoring.

The IFS maps real-time service data, configuration data, and management data all in the same way. In fact, the management interface can be used to define the real-time service data set dynamically (e.g., to select between a smoothed value or a dynamic value as the result from a sensor). While it is required to provide real-time guarantees for communication of real-time data, the access to configuration and management data is not time critical. This enables the employment of Web-based tools for remote maintenance.

Tools that interface with the IFS have been implemented using CORBA as middleware. CORBA is an object model managed by the Object Management Group (OMG) that provides transparent communication among remote objects. Objects can be implemented in different programming languages and can run on different platforms. The standardized CORBA protocol IIOP (Internet Inter-ORB Protocol) can be routed over TCP/IP, thus supporting worldwide access to and communication between CORBA objects across the Internet.

Alternatively, it is possible to use Web Services as the management interface to a fieldbus network. A case study that implements Web Services on top of the IFS of a fieldbus is described in [36].

16.5 Profiles, Data Sheets, and Descriptions

In order to build and configure systems, users require information on different properties of the parts of the targeted system. Such information comes in the form of hardware manuals or data sheets. Since this information is intended for human consumption, representation and content are typically less formal than would be required for computer processing of this information. For that reason, dedicated computer-readable representations of fieldbus system properties are required, which play a similar role as information sources for a computer-based support framework during configuration and management of a system. These representations allow for establishing common rule sets for developing and configuring applications and for accessing devices and system properties (for configuration as well as management functions). In the following, we examine several representation mechanisms.

16.5.1 Profiles

Profiles are a widely used mechanism to create interoperability in fieldbus systems. We distinguish several types of profiles, i.e., application, functional, or device profiles. Heery and Patel [16] propose a very general and short profile definition that we adopt for our discussion: "Profiles are schemas, which consist of data elements drawn from one or more name spaces,[*] combined together by implementors, and optimized for a particular local application."

In many cases, a profile is the result of the joint effort of a group of device vendors in a particular area of application. Usually, a task group is founded that tries to identify reoccurring functions, usage patterns, and properties in their domain and then creates strictly formalized specifications according to these identified parts, resulting in so-called *profiles*.

More specific, for each device type, a profile exactly defines what kind of communication objects, variables, and parameters have to be implemented so that a device conforms to the profile. Profiles usually distinguish several types of variables and parameters (e.g., process parameters, maintenance parameters, user-defined parameters) and provide a hierarchical conformance model that allows for the definition of user-defined extensions of a profile. A device profile need not necessarily correspond to a particular physical device; for example, a physical node could consist of multiple virtual devices (e.g., multipurpose input/output (I/O) controller), or a virtual device could be distributed over several physical devices.

Protocols supporting device, functional, and application profiles are CANopen [7], Profibus, and LON [25] (LonMark functional profiles). Figure 16.2 depicts, as an example, the visual specification of a LonMark[**] functional profile for an analog input object. The profile defines a set of network variables (in this example only the mandatory ones are shown) and local configuration parameters

---

[*]That is, sources.

[**]http://www.lonmark.org.

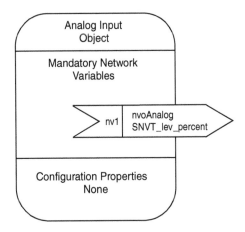

FIGURE 16.2 Functional profile for an analog input.

(none in this example). The arrow specifies that this profile outputs a digital representation of an analog value, whereas the structure of this output is defined with the standardized (in LON) network variable type SNVT_lev_percent (−163.84 to 163.84% of full scale). In addition, a profile also specifies other important properties, such as timing information, valid range, update rate, power-up state, error condition, and behavior (usually as a state diagram).

While the approach for creating profiles is comparable for different protocols, the profiles are not always interchangeable between the various fieldbuses, although advancements (at least for process control-related fieldbuses) have been made within IEC 61158 [19]. Block- and class-based concepts, such as function blocks as they are defined for the Foundation Fieldbus or Profibus DP, or component classes in IEEE 1451.1 [18], can be considered implementations of the functional profile concept.

16.5.2 Electronic Data Sheets

Classical data sheets usually provide a detailed description of mostly physical key properties of a device such as available pins, electrical properties of pins, available amount and layout of memory, processing power, etc. Electronic data sheets play a conceptually similar role, but usually with a different focus, since they often try to abstract from details of physical properties of the underlying system and describe properties of a higher-level system model (e.g., Institute of Electrical and Electronics Engineers (IEEE) digital transducer interface [17] or Interface File System [26]). Such *electronic data sheets* follow strict and formalized specification rules in order to allow computer-supported processing of the represented information.

A generic electronic data sheet format was developed as part of the smart transducer-related IEEE 1451 standards family. IEEE 1451.2 [17] specifies the transducer electronic data sheet (TEDS) and a digital interface to access that data sheet and to read sensors or set actuators.

Figure 16.3 depicts the TEDS in context of the system architecture as defined in IEEE 1451:

- *Smart Transducer Interface Module* (STIM): A STIM contains from 1 to 255 transducers of various predefined types together with their descriptions in the form of the corresponding TEDSs.
- *Network-capable application processor* (NCAP): The NCAP is the interface to the overall network. By providing an appropriate NCAP, the transducer interface is independent of the physical fieldbus protocol.
- *Transducer-independent interface* (TII): The TII is the interface between the STIM and the NCAP. It is specified as an abstract model of a transducer instrumented over 10 digital communication lines.

TEDSs describe node-specific properties, such as the structure and temporal properties of devices and transducer data. Since the transducer interface in IEEE 1451 is line based, the basic communication

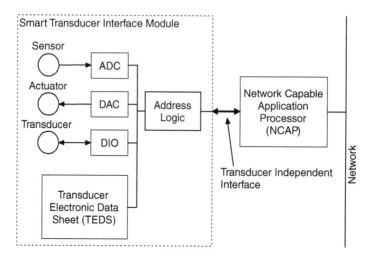

FIGURE 16.3 Smart Transducer Interface Module connected to the NCAP.

primitive is a *channel*. A channel represents a single flow path for digital data or an analog signal. One STIM may contain multiple channels and has an associated *meta-TEDS* that describes properties of the STIM, such as device identification information, number of implemented channels, command response time, or worst-case timing information. Each channel has an associated *channel TEDS* that describes channel-related information such as data structure, transducer, data conversion, timing, etc.

IEEE 1451 aims at self-contained nodes. Thus, TEDSs are stored in a memory directly located at the nodes. This requires considerable memory resources, so the representation of the configuration information for such a system must be very compact. IEEE 1451 achieves this goal by providing a large set of predefined transducer types and modes based on enumerated information, where identifiers are associated with more detailed prespecified descriptions (similar to error codes). An instance of a transducer description can be derived from the predefined types, and thus the memory requirements for the transducer description are kept low.

The *smart transducer descriptions* (STDs), as defined in [28], take a comparable role for describing properties of devices that follow the CORBA Smart Transducer Interfaces standard (the descriptions themselves are currently not part of the standard), although there are some notable differences between both approaches.

Unlike the commonly used enumeration-based description of properties, the STD and related formats use XML [39] as the primary representation mechanism for all relevant system aspects. Together with related standards, such as XML Schema or XSLT, XML provides advanced structuring, description, representation, and transformation capabilities. It is becoming the *de facto* standard for data representation and has extensive support throughout the industry. Some examples of XML used in applications in the fieldbus domain can be found in [6, 9, 37].

As the name implies, the smart transducer descriptions describe the properties of nodes in the smart transducer network. The STD format is used for describing both static properties of a device family (comparable to classic data sheets) and devices that are configured as part of a particular application (e.g., the local STD also contains the local node address). The properties described in STDs can be divided into the following categories:

- *Microcontroller information*: This block holds information on the microcontroller and clock of the smart transducer (e.g., controller vendor, clock frequency, clock drift).
- *Node information*: This block describes properties that are specific to a particular node and mostly consist of identification information, such as vendor name, device name/version, and node identifiers (serial number, local name).

```
<SmartTransducerDescription xmlns="...">
    ...
    <ProtocolBlock>
        ...
      <IFSFile>
        <rodl:fileName>16</rodl:fileName>
        <FileLength>12</FileLength>
        <FileStorage>data</FileStorage>
      </IFSFile>
        ...
    </ProtocolBlock>
</SmartTransducerDescription>
```

FIGURE 16.4 Example STD element.

- *Protocol information*: This block holds protocol-specific information, such as version of the communication protocol, supported baud rates, Universal Asynchronous Receiver/Transmitter (UART) types, and IFS layout.
- *Node service information*: The information in this block specifies the behavior and the capabilities of a node. In the current approach, a service plays a role similar to that of a functional profile (see Section 16.5.1) or function block. Such functional units are especially important for supporting the creation of applications. They conform to the interface model of the CORBA STI standard, since a service consists of a service identifier (e.g., name), input and output parameters, configuration parameters, and management parameters [12]. Parameters are specified by data type and multiple constraints (range, precision, minimum interval time, maximum runtime).

Figure 16.4 shows the description of a file in the IFS, consisting of the name of the file, its length (in records), and the location of the data, i.e., the memory type (RAM, Flash, ROM) where the file should be located (e.g., `data` specifies that a file is mapped into the internal RAM of the microcontroller). The prefix `rodl:` is shorthand for an XML name space. Name spaces allow the reuse of element definitions in multiple places. For example, the elements from the `rodl` (round descriptor list) name space are defined once separately and used in smart transducer descriptions as well as in additional related formats, such as the cluster configuration descriptions (CCDs). While the STD focuses on the nodes, the CCD format deals with system-level aspects.

It is not always possible to store all relevant information outside the node, but by focusing on reducing the amount of required information on the node to the minimum, extensive external meta-information can be used without size constraints. The reference to this external information is the unique combination of series and serial numbers of the node. The series number is identical for all nodes of the same type. The serial number identifies the instance of a node among all nodes of a series.

The advantages of this approach are twofold:

1. The overhead at the node is very low. Current low-cost microcontrollers provide internal RAM and EPROM memory of around 256 bytes. This is not sufficient to store more than the most basic parts of data sheets according to, for example, IEEE 1451.2 without extra hardware like an external memory element. With the proposed description approach, only the memory for storing the series and serial numbers, which is 8 bytes is necessary.
2. Instead of implicitly representing the node information with many predefined data structures mapped to a compact format, it is possible to have an explicit representation of the information in a well-structured and easy-to-understand way. A typical host computer running the configuration and management tools can easily deal with even very extensive generic XML descriptions. Furthermore, XML formats are inherently easy to extend, so the format is open for future extensions of transducer or service types.

```
VARIABLE MeasurmentVar
{ LABEL "Infrared Measurement"
  TYPE FLOAT
  { DISPLAY_FORMAT "3.11";
    MAX_VALUE 140.0;
    MIN_VALUE 0.0;
  }
}
```

FIGURE 16.5 Process variable represented with a device description.

Another interesting description mechanism is the *device description language* (DDL), which has a relatively long history in the fieldbus sector. First drafts emerged around 1990 at Endress+Hauser, where development for a predecessor language, called parameter description language, was already performed in the late 1980s [1]. DDL was first used with the HART fieldbus [5], but later was adopted for the Foundation Fieldbus [13] and most recently for Profibus (where it is called electronic device description). Unfortunately, the different versions are not fully compatible, since they have been extended within the scope of the respective fieldbus protocols.

The syntax of the DDL is similar to the syntax of the C programming language, but conceptually the language strongly relates to specialized markup languages like the hypertext markup language (HTML). In addition to these markup capabilities, DDL also provides enhancements like conditional evaluation and loops. DDL serves several purposes in the description of field devices:

- It describes the information items presented in the memory of the described devices.
- It supports the representation of the described information on different accessing devices (with different displaying capabilities).
- It supports the detailed specification of device properties, such as labels for parameters, engineering units, display precision, help texts, the relationship of parameters, and the layout of calibration and diagnostic menus.

Unlike the other presented approaches, the device descriptions (DDa) based on DDL play a bigger role for system management, since they not only describe the data in the memory of the devices, but also support defining rich meta-information for improving the interaction with devices. Figure 16.5 depicts a process variable defined with DDL. The example DDL fragment defines the representation of a variable on an access device. It specifies a label to represent the variable on the display, the data type, formatting information for the value to be displayed, and constraints on valid inputs for changing the value.

DDs can be stored on devices themselves (using a compact encoding of the information in the DD), as well as externally (e.g., delivered on a disc together with the device or centrally available in a DD repository).

16.6 Application Development

At the center of a fieldbus system is the actual fieldbus application. In the following section, we examine several application development approaches and how they influence system configuration.

A widely used development approach for fieldbus applications is model-based development. The basic idea behind this approach is to create a model of the application that consists of components that are connected via links that represent the communication flow between the components. Different approaches usually differ in what constitutes a component (e.g., function blocks, subsystems, services, functional profiles, physical devices) and the detailed semantics of a link. Many approaches support the recursive definition of components, which allows for grouping multiple lower-level components into one higher-level component. Figure 16.6 depicts a typical small control application consisting of two analog inputs receiving values from two sensors, two production identifiers (PIDs), and one analog output controlling an actuator.

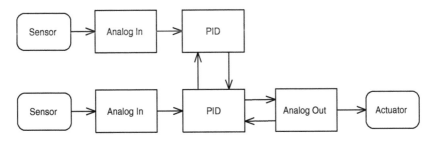

FIGURE 16.6 Example for an application model.

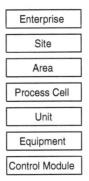

FIGURE 16.7 ANSI/ISA-88.01–1995 hierarchical model.

But the model-based approach is not the only application design approach. Another approach used by multiple fieldbus configuration tools is the ANSI/ISA-88.01–1995 procedural control model [20]. This modeling approach enforces a strictly modular hierarchical organization of the application (Figure 16.7). There should be no or hardly any interaction between multiple process cells. Interaction between components in a process cell is allowed. To make best use of this approach, the structure of the network site and the application should closely correspond to the hierarchy specified by this model.

This modeling approach conceptually follows the typical hierarchy of process control applications with multiple locally centralized programmable logic controllers (PLCs) that drive several associated control devices. This eases transition from predecessor systems and improves overall robustness, since this approach provides fault containment at the process cell level. As a downside, the coupling between the physical properties of the system and the application is rather tight. An example for a fieldbus protocol that supports this modeling approach is the Profibus PA protocol that provides a universal function block parameter for batch identification [4].

Another design approach is two-level design [31], which originated in the domain of safety-critical systems. In this approach, the communication between components must be configured before configuring the devices. While this requires that many design decisions must be made very early in the design process, this approach greatly improves overall composability of the components in the system.

Abstract application models provide several advantages for application development:

- The modular design of applications helps to deal with complexity by applying a divide-and-conquer strategy. Furthermore, it supports reuse of application components and physical separation.
- The separation of application logic from physical dependencies allows hardware-independent design that enables application development before hardware is available, as well as eases migration and possibly allows the reuse (of parts) of applications.

For configuring a physical fieldbus system from such an application model, we must examine (1) how this application model maps to the physical nodes in the network and (2) how information flow is maintained in the network.

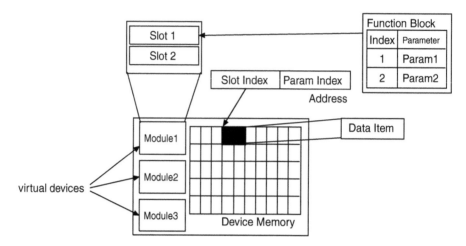

FIGURE 16.8 Mapping of function blocks to a physical device in Profibus DP.

In order to map the application model to actual devices, fieldbuses often provide a model for specifying physical devices as well. For example, in Profibus DP the physical mapping between function blocks and the physical device is implemented as follows (Figure 16.8). A physical device can be subdivided in several modules that take the role as virtual devices. Each device can have one (in case of simple functionality) up to many *slots*. A function block is mapped to a slot, whereas slots may also have associated physical and transducer blocks. Physical and transducer blocks represent physical properties of a fieldbus device. Parameters of a function block are indexed, and the slot number and parameter index cooperatively define the mapping to actual data in the device memory.

In contrast, the *Foundation Fieldbus* (FF) follows an object-oriented design philosophy. Thus, all information items related to configuring a device and the application (control strategy) are represented with objects. This includes function blocks, parameters, and subelements of parameters. These objects are collected in an object dictionary (OD), whereas each object is assigned an index. This OD defines the actual mapping to the physical memory on the respective device.

In order to understand the methods for controlling the communication flow between the application components, we first examine some recurring important communication properties in fieldbus applications:

- The use of state communication as primary communication mechanism for operating a fieldbus [29], i.e., performing the fieldbus application. State communication usually involves cyclically updating the associated application data.
- Support for asynchronous/sporadic communication (event communication) in order to perform management functions and deal with parts of the application that cannot be performed with state communication.

A common method to achieve these properties is scheduling. There are many scheduling approaches with vastly different effects on configuration. Following are some commonly used approaches adopted in fieldbus systems:

- *Multicycle polling*: In this approach, the communication is controlled by a dedicated node that authorizes other nodes to transmit their data [8]. This approach is used, for example, in WorldFIP, FF, and ControlNet. For configuring devices in such a network, the authorization nodes require at least a list of nodes to be polled; i.e., in the case of a master–slave configuration, only one node must be configured with the time information in order to control the whole cluster. For better control on the timely execution of the overall application, a time-division multiplexing scheme is used for bus access.

- *Time triggered*: In a time-triggered communication model, the communication schedule is derived from the progression of physical time. This approach requires a predefined collision-free schedule that defines *a priori* when a device is allowed to broadcast its data and an agreement on the global time, which requires the synchronization of the local clocks of all participating devices [10]. Some examples of protocols that support time-triggered communication are TTP/A [24], TTP/C [35], and the synchronous part of the FlexRay protocol [14]. In order to configure the communication in these systems, the schedules must be downloaded to all the nodes in the network.
- *Event triggered*: Event-triggered communication implements a push model, where the sender decides when to send a message, e.g., when a particular value has changed more than a given *delta*. Collisions on the bus are solved by collision detection/retransmission or collision avoidance, i.e., bitwise arbitration protocols such as Controller Area Network (CAN) [34]. Event-triggered communication does not depend on scheduling, since communication conflicts are resolved either by the protocol at the data link layer (e.g., bitwise arbitration) or by the application.

The scheduling information is usually stored in dedicated data structures that are downloaded to the nodes in the network in order to be available for use by the network management system functions of the node.

The TTP/A protocol deals with both application- and communication-specific configuration information in an integrated way. In this approach, the local communication schedules (called round descriptor lists) as well as the interfaces of application services [12] are mapped onto the same interfacing mechanism, the Interface File System (see Section 16.4.1).

For the representation of the overall system, the cluster configuration description format was developed; it acts as a central and uniform data structure that stores all the information pertinent to the fieldbus system. This information includes:

- *Cluster description meta-information*: This description block holds information on the cluster description itself, such as the maintainer, name of the description file, or version of the CCD format.
- *Communication configuration information*: This information includes round sequence lists as well as round descriptor lists, which represent the detailed specification of the communication behavior of the cluster. Additionally, this part of the CCD also includes (partially physical) properties important for communication, such as the UART specification, line driver, and minimum or maximum signal runtimes.
- *Cluster node information*: This block contains information on the nodes in a cluster, whereas nodes are represented with the smart transducer description format.

16.7 Configuration Interfaces

In the last section we focused on the relation between application and configuration. In the following, we examine aspects of the system configuration that are mostly independent of the application. We will take a brief look at the physical configuration of fieldbus systems, how nodes are recognized by the configuration system, and how the actual application code is downloaded into the fieldbus nodes.

16.7.1 Hardware Configuration

The hardware configuration involves a setup of plugs and cables of the fieldbus system. Several fieldbus systems implement means to avoid mistakes, such as connecting a power cable to a sensitive input, which would cause permanent damage to the fieldbus system or even harm people. Moreover, the hardware configuration interfaces such as plugs and clamps are often subject to failure in harsh environments, e.g., on a machine that induces a lot of vibration.

For hardware configuration the following approaches can be identified:

- The use of special jacks and cables that support a tight mechanical connection and avoid mistakes in orientation and polarity by their geometry. For example, the actuator–sensor interface[*] (AS-i) specifies a mechanically coded flat cable that allows the connection of slaves on any position on the cable by using piercing connectors. AS-i uses cables with two wires transporting data and energy via the same line. The piercing connectors support simple connection, safe contacting, and protection up to class IP 67.
- Baptizing of devices in order to obtain an identifier that allows addressing the newly connected device. This could be done explicitly by assigning an identifier to the device (e.g., by setting dip switches or entering a number over a local interface) or implicitly by the cabling topology (e.g., devices could be daisy chained and obtain their name subsequently according to the chain). Alternatively, it is possible to assign unique identifiers to nodes in advance. This approach is taken, for example, with Ethernet devices where the medium access control (MAC) address is a worldwide unique identifier, or in the TTP/A protocol that also uses unique node IDs. However, such a worldwide unique identifier will have many digits, so that it is usually not feasible to have the number printed somewhere on the device. To overcome this problem, machine-readable identifiers in the form of bar codes or radio frequency (RF) tags are used during hardware configuration.
- Simple configuration procedures, which can be carried out and verified by nonexpert personnel.

16.7.2 Plug-and-Participate

Since the hardware configuration is intended to be simple, a fieldbus system should behave intelligently in order to release human personnel from error-prone tasks.

During the stage of plug-and-participate, the fieldbus system runs an integration task that identifies new nodes, obtains information about these nodes, and changes the network configuration in order to include the new nodes in the communication.

Identification of new nodes can be supported with manual baptizing as described in the previous section. Alternatively, it is also possible to automatically search for new nodes and identify them as described in [11].

If there can be different classes of nodes, it is necessary to obtain information on the type of the newly connected nodes. This information will usually be available in the form of an electronic data sheet that can be obtained from the node or from an adequate repository.

The necessary changes of the network configuration for including the new node greatly depend on the employed communication paradigm. In the case of a polling paradigm, only the list of nodes to be polled has to be extended. In the case of a time-triggered paradigm, the schedule has to be changed and updated in all participating nodes. In the case of an event-triggered paradigm, only the new node has to be authorized to send data. However, it is very difficult to predict how a new sender will affect the timing behavior of an event-triggered system. In all three cases, critical timing might be affected due to a change of the response time, i.e., when the cycle time has to be changed. Thus, in time-critical systems, the extensibility must be taken into account during system design, e.g., by reserving at first unused bandwidth or including spare communication slots.

16.7.3 Application Download

Some frequently reoccurring fieldbus applications, like standard feedback control loops, alert monitoring, and simple control algorithms, can often be put in place like building bricks, since these applications are generically available (e.g., PID controller).

For more complex or unorthodox applications, however, it is necessary to implement user-defined applications. These cases require that code must be downloaded into the target devices.

---

[*]http://www.as-interface.net/.

Ten years ago, the most common method to reprogram a device was to use a socketed EPROM memory chip that was taken out of the circuit, erased under UV radiation, and programmed using a dedicated development system, i.e., a PC with a hardware programming device, and then put back into the system.

Today, most memory devices and microcontrollers provide an interface for in-system serial programming of Flash and EPROM memory. The hardware interface for in-system serial programming usually consists of a connector with four to six pins that is attached to either an external programming device or directly to the development PC. These programming interfaces are often proprietary to particular processor families, but there also exist some standard interfaces that support a larger variety of devices. For example, the Joint Test Action Group (JTAG) debugging interface (IEEE 1149.1) also supports the download of application code.

While the in-system serial programming approach is much more convenient than the socketed EPROM method, both approaches are conceptually quite similar, since it is still necessary to establish a separate hardware connection to the target system.

The most advanced approach for downloading applications is *in-system application download*. In this approach, it is possible to program and configure a device without taking it out of the distributed target system and without using extra cables and hardware interfaces.

In-system configuration is supported by state-of-the-art Flash devices, which can reprogram themselves in part by using a boot loader program. This approach is supported, for example, by state-of-the-art TTP nodes. A cluster consists of a set of at least four TTP/C nodes and a monitoring node that is connected to the development system. Whenever a new application has to be set up, the monitoring node sends a signal that causes the nodes to go into the so-called download mode. In this mode, it is possible to download application code via the fieldbus network. During the download phase, the real-time service is inactive. Misconfigurations that lead to a failure of the download function must be corrected by locally connecting a programming tool.

Alternatively, application code could be downloaded via the fieldbus into the RAM memory at start-up. In this case, only the boot loader resides in the persistent memory of the device, and the user-defined application code has to be downloaded at start-up. This approach has the advantage of being stateless, so that errors in the system are removed at the next start-up. Thus, the engineers could handle many faults by a simple restart of the system. On the other hand, this approach depends on the configuration instance at start-up — the system cannot be started if the configuration instance is down. Moreover, the restart time of a system may be considerably longer.

16.8 Management Interfaces

The ability to perform remote management operations on distributed fieldbus devices is one of the most important advantages of fieldbus systems. Wollschläger [38, p. 89] states that "in automation systems, engineering functions for administration and optimization of devices are gaining importance in comparison with control functions." Typical management operations are monitoring, diagnosis, or node calibration. Unlike the primary fieldbus applications, which often require cyclical, multidrop communication, these management operations usually use a one-to-one (client–server) communication style. For this reason, most fieldbus systems support both communication styles.

A central question is whether and how this management traffic influences the primary application, the so-called *probe effect* [15].

System management operations that influence the timing behavior of network communication are especially critical for typical fieldbus applications (e.g., process control loops) that require exact real-time behavior.

The probe effect can be avoided by reserving a fixed amount of the bandwidth for management operations. For example, in the Foundation Fieldbus and WorldFIP protocols, the application cycle (macrocycle) is chosen to be longer than strictly required by the application, and the remaining bandwidth is free for management traffic.

In order to avoid collisions within this management traffic window, adequate mechanisms for avoiding or resolving such conflicts must be used (e.g., token passing between nodes that want to transmit management information, priority-based arbitration).

In TTP/A, the management communication is implemented by interleaving real-time data broadcasts (implemented by multipartner rounds) with so-called master–slave rounds that open a communication channel to individual devices.

If management traffic is directly mingled with application data, such as in CAN, LonWorks, or Profibus PA, care must be taken that this management traffic does not influence the primary control application. This is typically achieved by analyzing network traffic and leaving enough bandwidth headroom. For complex systems and safety-critical systems that require certain guarantees on system behavior, this analysis can become very difficult.

16.8.1 Monitoring and Diagnosis

In order to perform passive monitoring of the communication of the application, it is usually sufficient to trace the messages transmitted on the bus. However, the monitoring device must have knowledge of the communication scheme used in the network, in order to be able to understand and decode the data traffic. If this scheme is controlled by physical time, as is the case in time-triggered networks, the monitoring node must also synchronize itself to the network.

Some advanced field devices often have built-in self-diagnostic capabilities and can disclose their status to the management system. It depends on the capabilities of the fieldbus system how such information reaches the management framework. Typically, a diagnosis tool or the diagnosis part of the management framework will regularly check the information in the nodes. This method is called *status polling*. In some fieldbus protocols (e.g., FF), devices can also transmit status messages by themselves (*alert reporting*).

In general, the restrictions from the implementation of the management interface of a fieldbus protocol also apply to monitoring, since in most fieldbus systems the monitoring traffic is transmitted using the management interface.

For systems that do not provide this separation of management from application information at the protocol level, other means must be taken to ensure that monitoring does not interfere with the fieldbus application. Since status polling is usually performed periodically, it should be straightforward to reserve adequate communication resources during system design, so that the control application is not disturbed. In the case of alert reporting, the central problem without adequate arbitration and scheduling mechanisms is how to avoid overloading the network in case of "alarm showers," where many devices want to send their messages at once. It can be very difficult to give timeliness guarantees (e.g., the time between when an alarm occurs and the time it is received by the respective target) in such cases. The typical approach to deal with this problem (e.g., as taken in CAN) is to provide much bandwidth headroom.

For in-depth diagnosis of devices, it is sometimes also desirable to monitor operation and internals of individual field devices. This temporarily involves greater data traffic that cannot be easily reserved *a priori*. Therefore, the management interface must provide some flexibility on the diagnosis data in order to dynamically adjust to the proper level of detail using some kind of pan-and-zoom approach [2].

16.8.2 Calibration

The calibration of transducers is an important management function in many fieldbus applications. There is some ambiguity involved concerning the use of this term. Berge [4] strictly distinguishes between *calibration* and *range setting*: "Calibration is the correction of sensor reading and physical outputs so they match a standard" [p. 363]. According to this definition, calibration cannot be performed remotely, since the device must be connected to a standardized reference input.

Range setting is used to move the value range of the device so that the resulting value delivers the correctly scaled percentage value. It does not require applying an input and measuring an output; thus, it can be performed remotely. In the HART bus, this operation is called calibration, whereas calibration is called *trim*.

Fieldbus technology does not influence the way calibration is handled, although information that is required for calibration is stored as part of the properties that describe a device. Such information could be, e.g., the minimum calibration span limit. This is the minimum distance between two calibration points within the supported operation range of a device. Additionally, calibration-related information, i.e., the individual calibration history, can be stored in the devices themselves. This information is then remotely available for management tools in order to check the calibration status of devices. Together with the self-diagnosis capabilities of the field devices, this allows performing a focused and proactive management strategy.

16.9 Maintenance in Fieldbus Systems

Fieldbus maintenance is the activity of keeping the system in good working order. The extensive management functions provided by fieldbus systems, such as diagnosis, and monitoring greatly help in maintaining systems. There are several different maintenance schemes that influence the way these steps are executed in detail. Choice of a particular maintenance scheme is usually motivated by the application requirements [4]:

- *Reactive maintenance* is a scheme in which a device is fixed only after it has been found to be broken. This case should be avoided in environments where downtimes are costly (such as in factory applications). Thus, designers of such applications will usually choose more active maintenance strategies. Nonetheless, fieldbus systems also provide advantages for this scheme, since they support the fast detection of faulty devices.
- *Preventive maintenance* is a scheme in which devices are serviced in regular intervals even if they are working correctly. This strategy prevents unexpected downtime, thus improving availability. Due to the associated costs, this approach will be taken only in safety-related applications such as in aviation, train control, or where unexpected downtimes would lead to very high costs.
- *Predictive maintenance* is similar to preventive maintenance, differing in a dynamic service interval that is optimized by using longtime statistics on devices.
- *Proactive maintenance* focuses on devices that are expected to require maintenance.

Basically, maintenance involves the following steps:

- Recognizing a defective device
- Repairing (replacing) the defective device
- Reintegrating the serviced device

In fieldbus systems, faulty devices will usually be recognized via the network. This is achieved by monitoring the fieldbus nodes and the application or with devices that are capable of sending alerts (refer to Section 16.8).

After the source of a problem has been found, the responsible node must be serviced. This often requires disconnecting the node from the network. Thus, we require strategies of how the system should deal with disconnecting a node, as well as reconnecting and reintegrating the replacement node.

In case the whole system must be powered down for maintenance, a faulty node can be simply replaced and the integration of the new node occurs as a part of the normal initial start-up process. If powering down of the whole system is undesirable or even impossible (in the sense of leading to severe consequences, as in the case of safety-critical applications), this process becomes more complicated. In this case, we have several options:

- *Implementation of redundancy*: This approach must be taken for safety- or mission-critical devices, where operation must be continued after a device becomes defective or during replacement, respectively. A detailed presentation of redundancy and fault-tolerant systems can be found in [30].

• *Shutdown of part of the application*: In the case of factory communication systems that often are organized as multilevel networks or use modular approaches, it might be feasible to shut down a local subnetwork (e.g., a local control loop or a process cell as defined in the ANSI/ISA-88.01–1995 standard).

The replacement node must be configured with individual node data, such as calibration data (these data usually differ between replaced and replacement node), and the *state* of a node. The state information can include:

• Information that is accumulated at runtime (the history state of a system). This information must be transferred from the replaced to the replacement node.
• Timing information, so that the node can synchronize with the network. For example, in networks that use a distributed static schedule (e.g., TTP/A), each node must be configured with its part of the global schedule in order to get a network-wide consistent communication configuration.

One alternative approach for avoiding transferring of system state is to design a stateless system in the first place. Bauer [3] proposes a generic approach for creating stateless systems from systems with state. Another possibility is to provide well-defined reintegration points where this state is minimized. Since fieldbus applications typically use a cyclical communication style, the start of a cycle is a natural reintegration point.

16.10   Conclusion

Configuration and management play an important role in fieldbus systems. The configuration phase can be subdivided into a part that requires local interaction such as connection of hardware and setting dip switches, and a part that can be done remotely via the fieldbus system. An intelligent design requires that the local part is as simple as possible in order to employ nonexpert personal, and that both parts are supported by an adequate architecture and tools that assist the system integrator in tedious and error-prone tasks such as adjusting parameters according to the data sheet of a device. Examples of such an architecture are, among others, IEEE 1451 and the OMG Smart Transducer Standard, which both provide machine-readable electronic data sheets.

Management encompasses functions like monitoring, diagnosis, calibration, and support for maintenance. In contrast to the configuration phase, most management functions are used concurrently with the real-time service during operation. Some management functions, such as monitoring, may even require real-time behavior for themselves. In order to avoid a *probe effect* on the real-time service, the scheduling of a fieldbus system must be designed to integrate management traffic with real-time traffic.

References

1. Borst Automation. Device description language. *The HART Book*, 9, May 1999. Available at http://www.thehartbook.com/
2. Bartram, L., A. Ho, J. Dill, and F. Henigman. The continuous zoom: a constrained fisheye technique for viewing and navigating large information spaces. In *ACM Symposium on User Interface Software and Technology*, 1995, pp. 207–215.
3. Bauer, G. Transparent Fault Tolerance in a Time-Triggered Architecture. Ph.D. thesis, Technische Universität Wien, Institut für Technische Informatik, Vienna, Austria, 2001.
4. Berge, J. *Fieldbuses for Process Control: Engineering, Operation, and Maintenance*. ISA — The Instrumentation, Systems, and Automation Society, Research Triangle Park, NC, 2002.
5. Bowden, R. *HART: A Technical Overview*. Fisher-Rosemount, Chanhassen, MN, 1997.
6. Bühler, D. The CANopen Markup Language: representing fieldbus data with XML. In *Proceedings of the 26th IEEE International Conference of the IEEE Industrial Electronics Society (IECON 2000)*, Nagoya, Japan, October 2000.

7. CAN in Automation e.V. CANopen: Communication Profile for Industrial Systems, 2002. Available at http://www.can-cia.de/downloads/

8. Cavalieri, S., S. Monforte, A. Corsaro, and G. Scapellato. Multicycle polling scheduling algorithms for fieldbus networks. *Real-Time Systems*, 25:157–185, 2003.

9. Eberle, S. XML-basierte Internetanbindung technischer Prozesse. In *Informatik 2000 Neue Horizonte im neuen Jahrhundert.* Springer-Verlag, Heidelberg, 2000, pp. 356–371.

10. Elmenreich, W., G. Bauer, and H. Kopetz. The time-triggered paradigm. In *Proceedings of the Workshop on Time-Triggered and Real-Time Communication,* Manno, Switzerland, December 2003.

11. Elmenreich, W., W. Haidinger, P. Peti, and L. Schneider. New node integration for master-slave fieldbus networks. In *Proceedings of the 20th IASTED International Conference on Applied Informatics (AI 2002),* February 2002, pp. 173–178.

12. Elmenreich, W., S. Pitzek, and M. Schlager. Modeling distributed embedded applications using an interface file system. Accepted for presentation at the *7th IEEE International Symposium on Object-Oriented Real-Time Distributed Computing,* 2004.

13. Fieldbus Technical Overview: Understanding FOUNDATION Fieldbus Technology, 2001. Available at http://www.fieldbus.org

14. Führer, T., F. Hartwich, R. Hugel, and H. Weiler. FlexRay: The Communication System for Future Control Systems in Vehicles. Paper presented at *SAE World Congress 2003,* Detroit, MI, March 2003.

15. Gait, J. A probe effect in concurrent programs. *Software Practice and Experience,* 16:225–233, 1986.

16. Heery, R. and M. Patel. Application Profiles: Mixing and Matching Metadata Schemas. Ariadne, September 25, 2000. Available at http://www.ariadne.ac.uk

17. Institute of Electrical and Electronics Engineers, Inc. IEEE 1451.2–1997, Standard for a Smart Transducer Interface for Sensors and Actuators: Transducer to Micro-Processor Communication Protocols and Transducer Electronic Data Sheet (TEDS) Formats, September 1997.

18. Institute of Electrical and Electronics Engineers, Inc. IEEE 1451.1–1999, Standard for a Smart Transducer Interface for Sensors and Actuators: Network Capable Application Processor (NCAP) Information Model, June 1999.

19. International Electrotechnical Commission (IEC). Digital Data Communications for Measurement and Control: Fieldbus for Use in Industrial Control Systems: Part 1: Overview and Guidance for the IEC 61158 Series, April 2003.

20. ANSI/ISA-88.01, Batch Control Part 1: Models and Terminology, December 1995.

21. Ko, W.H. and C.D. Fung. VLSI and intelligent transducers. *Sensors and Actuators,* 2:239–250, 1982.

22. Kopetz, H., *Real-Time Systems: Design Principles for Distributed Embedded Applications.* Kluwer Academic Publishers, Boston, 1997.

23. Kopetz, H., M. Holzmann, and W. Elmenreich. A universal Smart Transducer Interface: TTP/A. *International Journal of Computer System Science and Engineering,* 16:71–77, 2001.

24. Kopetz, H., et al. Specification of the TTP/A Protocol, Version 2.00. Technical report, Technische Universität Wien, Institut für Technische Informatik, Vienna, Austria, 2002. Available at http://www.ttagroup.org

25. Loy, D., D. Dietrich, and H.-J. Schweinzer (Eds.). *Open Control Networks.* Kluwer Academic Publishing, Boston, 2001.

26. OMG. Smart Transducers Interface, V1.0. Available specification document number formal/2003-01-01, Object Management Group, Needham, MA, January 2003. Available at http://doc.omg.org/formal/2003-01-01

27. Pitzek, S. and W. Elmenreich. Managing fieldbus systems. In *Proceedings of the Work-in-Progress Session of the 14th Euromicro International Conference,* June 2002.

28. Pitzek, S. and W. Elmenreich. Configuration and management of a real-time smart transducer network. In *Proceedings of the 9th IEEE International Conference on Emerging Technologies and Factory Automation,* Volume 1, Lisbon, Portugal, September 2003, pp. 407–414.

29. Pleinevaux, P. and J.-D. Decotignie. Time critical communication networks: field buses. *IEEE Network,* 2:55–63, 1998.

30. Poledna, S. *Fault-Tolerant Real-Time Systems: The Problem of Replica Determinism*. Kluwer Academic Publishers, Boston, 1995.

31. Poledna, S., H. Angelow, M. Glück, M. Pisecky, I. Smaili, G. Stöger, C. Tanzer, and G. Kroiss. TTP Two Level Design Approach: Tool Support for Composable Fault-Tolerant Real-Time Systems. Paper presented at *SAE World Congress 2000*, Detroit, MI, March 2000.

32. Powell, J. The "Profile" Concept in Fieldbus Technology. Technical article, Siemens Milltronics Process Instruments Inc., 2003.

33. Ran, A. and J. Xu. Architecting software with interface objects. In *Proceedings of the 8th Israeli Conference on Computer-Based Systems and Software Engineering*, 1997, pp. 30–37.

34. Bosch, Robert, GmbH, Stuttgart. CAN Specification, Version 2.0, 1991.

35. TTAGroup. Specification of the TTP/C Protocol. TTAGroup, 2003. Available at http://www. tta-group.org

36. Venzke, M. Spezifikation von interoperablen Webservices mit XQuery. Ph.D. thesis, Technische Universität Hamburg-Harburg, Hamburg-Harburg, Germany, 2003.

37. Wollschläger, M. A framework for fieldbus management using XML descriptions. In *Proceedings of the 2000 IEEE International Workshop on Factory Communication Systems (WFCS 2000)*, September 2000, pp. 3–10.

38. Wollschläger, M., C. Diedrich, T. Bangemann, J. Müller, and U. Epple. Integration of fieldbus systems into on-line asset management solutions based on fieldbus profile descriptions. In *Proceedings of the 4th IEEE International Workshop on Factory Communication Systems*, August 2002, pp. 89–96.

39. World Wide Web Consortium (W3C). Extensible Markup Language (XML) 1.0, 2nd ed., October 2000. Available at http://www.w3.org

40. Zadeh, L.A. *The Concept of System, Aggregate, and State in System Theory*, Inter-University Electronics Series, Volume 8. McGraw-Hill, New York, 1969, pp. 3–42.

Section 4.3

Real-Time Ethernet

17

The Quest for Real-Time Behavior in Ethernet*

P. Pedreiras
University of Aveiro, Portugal

Luis Almeida
University of Aveiro, Portugal

Alberto Foneseca
University of Aveiro, Portugal

17.1 Introduction

Nowadays, intelligent nodes, that is, microprocessor-based with communication capabilities, are extensively used in the lower layers of both process control and manufacturing industries [8]. In these environments, applications range from embedded command and control systems, to image processing, monitoring, human–machine interface, etc. Moreover, the communication between the different nodes has specific requirements [9] which are quite different from, and sometimes opposed to, those found in office environments. For instance, predictability is favored against average throughput, and message transmission is typically time and precedence constrained. Furthermore, the nonrespect of such constraints can have a significant negative impact on the quality of the control action in Distributed Computer Control Systems (DCCSs), or on the quality of the observation of the system state in Distributed Monitoring Systems (DMSs). Therefore, to deliver the adequate quality of service, special-purpose networks have been developed, essentially during the last two decades, which are generically called fieldbuses and are particularly adapted to support frequent exchanges of small amounts of data under time, precedence, and dependability constraints [8]. Probably, the most well-known examples existing today are Controller Area Network (CAN), DeviceNet, ProfiBus, WorldFip, P-Net, and Foundation Fieldbus.

In the early days of DCCS, network nodes presented simple interfaces and supported limited sets of actions. However, the quantity, complexity, and functionality of the nodes on a DCCS have been increasing steadily. As a consequence of this evolution, the amount of information that must be exchanged over the network has also increased, either for configuration or for operational purposes.

*This work was partially supported by the Portuguese Government through project CIDER-POSI/1999/CHS/33139.

The increase in the amount of data exchanged between DCCS nodes is reaching the limits that are achievable using traditional fieldbuses due to limited bandwidth, typically between 1 and 5 Mbps [9]. Machine vision, and its growing use, is just one example of a killer application for those systems. Therefore, other alternatives are required to support higher bandwidth demands while retaining the main requirements of a real-time (RT) communication system: predictability, timeliness, bounded delays, and jitter.

Starting in the 1980s, several general-purpose networks, exhibiting bandwidth higher than the traditional fieldbus protocols, have also been proposed for use at the field level. For example, two prominent networks, FDDI and ATM, have been extensively analyzed, for both hard RT and soft RT communication systems. However, due to high complexity, high cost, lack of flexibility, and interconnection capacity, these protocols have not gained general acceptance [10].

Another communication protocol that has been evaluated for use at the field level is Ethernet. The main factors that favor the use of this protocol are [9]: cheap silicon available, easy integration with Internet, clear path for future expandability, and compatibility with networks used at higher layers on the factory structure.

However, the nondeterministic arbitration mechanism used by Ethernet prevents its direct use at field level, at least for hard RT communications. Therefore, in the past, several attempts have been made to allow Ethernet to support time-constrained communications. The methods that have been used to achieve deterministic message transmission over Ethernet range from modifications to the Medium Access Control (MAC) layer (e.g., [11]), to the addition of sublayers over the Ethernet layer to control the instants of message transmission (e.g., [12]) and therefore avoid collisions. More recently, with the advent of switched Ethernet, and therefore the intrinsic absence of collisions, a new set of works with respect to the ability of this topology to carry time-constrained communications has appeared (e.g., [10]).

This chapter presents a brief description of the Ethernet protocol, followed by a discussion of several techniques that have been proposed or used to enforce RT communication capabilities over Ethernet in the last two decades. The techniques referred to include those that support either probabilistic or deterministic analysis of the network access delay. The chapter concludes with a reference to recent work that concerns the use of the Ethernet protocol, highlighting current trends.

17.2   Ethernet Roots

Ethernet originated about 30 years ago, invented by Bob Metcalfe at the Xerox's Palo Alto Research Center. Its initial purpose was to connect two products developed by Xerox: a personal computer and a brand new laser printer. Since then, this protocol has evolved in many ways. For instance, concerning the transmission speed, it has grown from the original 2.94 to 10 Mbps [1–3, 5], then to 100 Mbps [6], and more recently to 1 Gbps [7] and 10 Gbps [26].With respect to physical medium and network topology, Ethernet also has evolved: it started with a bus topology based firstly on thick coaxial cable [2] and afterwards on thin coaxial cable [3]. In the mid-1980s, a more structured and fault-tolerant approach, based on a star topology, was standardized [4], running, however, only at 1 Mbps. In the beginning of the 1990s, an improvement on this latter technology was also standardized [5], running at 10 Mbps over category 5 unshielded twisted pair cable.

Along this way, two fundamental properties have been kept unchanged:

1. Single collision domain, that is, frames are broadcast on the physical medium and all the network interface cards (NICs) connected to it receive them.
2. The arbitration mechanism, which is called Carrier Sense Multiple Access with Collision Detection (CSMA/CD).

According to the CSMA/CD mechanism (Figure 17.1), an NIC with a message to be transmitted must wait for the bus to become idle, and only then does it start transmitting. However, several NICs may have sensed the bus during the current transmission and then attempted to transmit simultaneously

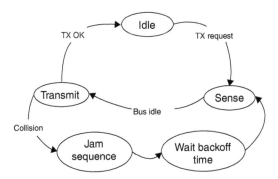

FIGURE 17.1 Ethernet CSMA/CD simplified state diagram.

thereafter, causing a collision. In this case, all the stations abort the transmission of the current message, wait for a random time interval, and try again. The number of retries is limited to 16.

The use of a single broadcast domain and the CSMA/CD arbitration mechanism has created a bottleneck when encountering highly loaded networks: above a certain threshold, as the submitted load increases, the throughput of the bus decreases, a phenomenon referred to as *thrashing*. In the beginning of the 1990s, the use of switches in place of hubs has been proposed as an effective way to deal with thrashing. A switch creates a single collision domain for each of its ports. If a single node is connected to each port, collisions never actually occur unless they are created on purpose, for example, for flow control. Switches also keep track of the addresses of the NICs connected at each port by inspecting the source address in the incoming messages. This allows forwarding incoming messages directly to the respective outgoing ports according to the respective destination address, a mechanism generally known as *forwarding*. When a match between a destination address and a port cannot be established, the switch forwards the respective message to all ports, a process commonly referred to as *flooding*. The former mechanism, forwarding, allows a higher degree of traffic isolation so that each NIC receives the traffic addressed to it only. Moreover, since each forwarding action uses a single output port, several of these actions can be carried out in parallel, resulting in multiple simultaneous transmission paths across the switch and, consequently, in a significant increase in the global throughput.

17.3 Why Use Ethernet at the Fieldbus Level

The first question that should be answered concerns the reasons that make Ethernet appealing for use at the field level, to convey time-constrained traffic. In fact, its designer has not envisaged this kind of application, and therefore some properties of this protocol, such as the nondeterministic arbitration mechanism, pose serious challenges concerning its use for this purpose.

Several works address this subject (e.g., [9, 12, 17]). In [9], Decotignie presents a thorough reasoning on the pros and cons on this issue, highlighting two concise sets of arguments, one in favor and the other against the adoption of Ethernet as a fieldbus. Commonly referred arguments in favor can be summarized as follows:

- It is cheap, due to mass production.
- Integration with Internet is easy (TCP/IP stacks over Ethernet are widely available, allowing the use of application layer protocols such as FTP, HTTP, and so on).
- Steady increases in the transmission speed have occurred in the past, and are expected to occur in the near future.
- Due to its inherent compatibility with the communication protocols used at higher levels, the information exchange with plant level becomes easier.
- The bandwidth made available by existing fieldbuses is insufficient to support some recent developments, like the use of multimedia (e.g., machine vision) at the field level.

- Availability of technicians familiar with this protocol.
- Wide availability of test equipment from different sources.
- Mature technology, well specified, and with equipment available from many sources, without incompatibility issues.

On the other hand, Ethernet does not fulfill some fundamental requirements that are expected from a communication protocol operating at field level. In particular, the destructive and nondeterministic arbitration mechanism has been regarded as the main obstacle faced by Ethernet concerning this application domain. The answer to this concern is the use of switched Ethernet, which allows bypassing the native CSMA/CD arbitration mechanism. In these cases, provided that a single NIC is connected to each port and the operation is full duplex, no collisions occur.

However, just avoiding collisions does not make Ethernet deterministic; for example, if a burst of messages destined to a single port arrive at the switch in a given time interval, they must be serialized and transmitted one after the other. If the arriving rate is greater than the transmission rate, buffers will be exhausted and messages will be lost. Therefore, even with switched Ethernet, full RT behavior still requires some kind of higher-level transmission control. Moreover, bounded transmission delay is not the only requirement of a fieldbus; some other important factors commonly referred to in the literature are temporal consistency indication, precedence constraints, efficient handling of periodic, and sporadic traffic. Clearly, Ethernet, even with switches, does not provide answers to all these demands.

17.4   Making Ethernet RT

In the previous years, a considerably large amount of work has been carried out, leading to many different proposals for adaptations and add-ons to Ethernet in order to support RT communication. In the remainder of this section, some paradigmatic efforts are presented, with a brief discussion of the strong and weak points of each one.

17.4.1   Modification of the Medium Access Control Sublayer

One of the possible approaches consists in modifying the Ethernet MAC sublayer so that a bounded access to the bus may be achieved (e.g., [11, 18]). For instance, the solution presented in [11] (CSMA/DCR) consists of a binary tree search of colliding messages, that is, there is a hierarchy of priorities in the retry.

During normal operation, the CSMA/DCR follows the standard IEEE 802.3 protocol (*Random Access mode*). However, whenever a collision is detected, the protocol switches to the *Epoch mode*. In this mode, lower priority message sources voluntarily cease contending for the bus, and higher priority ones try again. This process is repeated until a successful transmission occurs (Figure 17.2). After all frames involved in the collision are transmitted, the protocol switches back to random access mode.

Figure 17.2, together with Table 17.1, depicts the CSMA/DCR operation in a situation where six messages collide. Considering that lower indexes correspond to higher priorities, after the initial collision the right branch of the tree (messages 12, 14, and 15) ceases contending for the bus. Because there are still three messages on the left branch, a new collision appears, between messages 2, 3, and 5. Thus, the left subbranch is selected again, leaving message 5 out. In the following slot, messages 2 and 3 will collide again. The subbranch selected after this collision has no active message sources, and thus in the following time slot the bus will be idle (step 4). This causes a move to the right subbranch, where messages 3 and 5 reside, resulting in a new collision. Finally, in step 6, the branch containing only the message with index 5 is selected, resulting in a successful transmission. The algorithm continues this way until all messages are successfully transmitted.

Despite assuring a bounded access time to the transmission medium, this approach exhibits two main drawbacks:

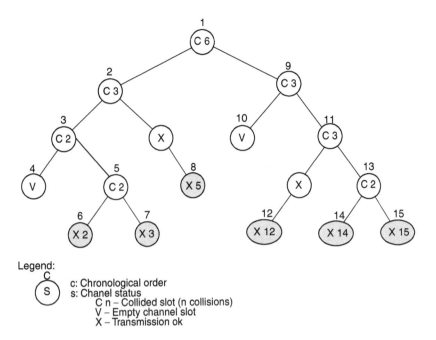

FIGURE 17.2 Example of tree search with CSMA-DCR.

TABLE 17.1 Tree Search Example (Contending Sequence)

Search Order	1	2	3	4	5	6	7	8	9	10	11	12	13	14	15
Channel Status	C	C	C	V	C	X	X	X	C	V	C	X	C	X	X
Source index	2	2	2		2	2	3	5	12		12	12	14	14	15
	3	3	3		3				14		14		15		
	5	5													
	12														
	14														
	15														

1. In some cases (e.g., [11]), the firmware must be modified; therefore, the economy of scale obtained when using standard Ethernet hardware is lost.
2. The worst-case transmission time, which is the main factor considered when designing RT systems, can be orders of magnitude greater than the average transmission time. This forces any kind of analysis to be very pessimistic, and therefore leads to low bandwidth utilization, at least with respect to RT traffic.

17.4.2 Addition of a Transmission Control Layer over Ethernet

Another way to achieve time-constrained communication over Ethernet consists of adding an interface layer above the native medium access control, intended to control the instants of message transmissions, eliminating collisions or, at least, bounding their number. The major advantage of this kind of approach when compared to the modification of the MAC layer is that standard Ethernet hardware can be used.

17.4.2.1 Virtual Time Protocol

One of the approaches present in the literature is known as Virtual Time Protocol and addresses CSMA/CD networks [19, 20]. The protocol allows implementing different scheduling policies (e.g., Minimum-Laxity First [MLF]), and bases its decisions on the assessment of the communication channel status.

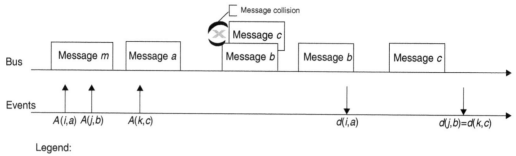

FIGURE 17.3   Example of Virtual-Time protocol (with MLF).

When the bus becomes idle and a node has a message to transmit, it waits for a given amount of time, related to the scheduling policy implemented. For example, if MLF scheduling is used, the waiting time is derived directly from the laxity using a proportional constant. When this amount of time expires, and if the bus is still idle, the node tries to transmit the message. If a collision occurs, then the scheduler outcome resulted in more than one message having permission to be transmitted at the same time (e.g., when two messages have the same laxity in MLF). In this case, the protocol can either recalculate the waiting time using the same rule or use a probabilistic approach according to which the messages involved in a collision are retransmitted with probability p (p-persistent). This last option is important to sort out situations in which the scheduler cannot differentiate messages, for example, messages with the same laxity would always collide.

Figure 17.3 shows the operation of the Virtual-Time protocol, with MLF scheduling. During the transmission of message m, messages a and b become ready. Since the laxity of message a (i.e., deadline minus message transmission time) is shorter than the laxity of message b, message a is transmitted first. During the transmission of message a, message c arrives. Messages b and c have the same deadline and the same laxity. Therefore, an attempt will be made to transmit them at the same time, causing a collision. Then, the algorithm uses the probabilistic approach, with message b having a lower waiting time than message c, and thus being transmitted next. Finally, message c is transmitted on the bus. Since the only global information is the channel status, there is no way to know that there is only a single message pending. For this reason, after the transmission of message b the waiting time corresponding to message c is computed, and only after the expiration of this interval message is c finally transmitted.

Beyond the advantage of using standard Ethernet hardware, this approach also has the advantage of not requiring any other global information but the channel status, which is readily available at all NICs. Thus, a fully distributed and symmetric implementation is possible, which, in this case, also incurs a relatively low computational overhead. Nevertheless, this approach presents some important drawbacks:

- Performance highly dependent on the proportional constant value used to relate the waiting time with the scheduling policy is use, leading to:
  - Collisions if it is too short.
  - Large amount of idle time if it is too long.
- Proportional constant depends on the properties of the message set; therefore, on-line changes to that set can lead to poor performance.
- The waiting times are computed locally using relative parameters only. There is no global time base and, thus, relative phasing is harder to implement.
- Due to possible collisions, worst-case transmission time is much higher than the average transmission time and only probabilistic timeliness guarantees can be given (soft RT systems).

17.4.2.2 Windows Protocols

Another possible approach is to use the *Windows protocols*. These protocols have been proposed both for CSMA/CD and token ring networks [19].With respect to the CSMA/CD implementation, the operation is as follows. The nodes on a network agree on a common time interval (referred to as *window*). All nodes synchronize upon a successful transmission, restarting the respective window. The bus state is used to assess the number of nodes with messages to be transmitted within the window:

- If the bus remains idle, there are no messages to be transmitted in the window.
- If only one message is in the window, it will be transmitted.
- If two or more messages are within the window, a collision occurs.

Depending on the bus state, several actions can be performed:

- If the bus remains idle, the window duration is increased in all nodes.
- In the case of a collision, the time window is shortened in all nodes.
- In case of a successful transmission, the window is restarted and its duration is kept as it is.

In the first two cases, the window duration is changed but the window is not restarted. Moreover, the window duration varies between a maximum (initial) and minimum value. Whenever there is a sufficiently long idle period in the bus, the window will return to its original maximum length. If a new node enters dynamically in the system, it may have an instantaneous window duration different from the remaining nodes. This may cause some perturbation during an initial period, with more collisions than expected. However, as soon as an idle period occurs, all windows will converge to the initial length. A probabilistic retry mechanism may also be necessary when the windows are shrunk to their minimum and collisions still occur (e.g., when two messages have the same transmission time).

Figure 17.4 shows an example of the operation of the windows protocol used to implement MLF message scheduling. The top axis represents the latest send times (*lst*) of messages A, B, and C. The *lst* of a message is the latest time instant by which the message transmission must start so that the respective deadline is met. The first window (Step 1) includes the *lst* of three messages, thus leading to a collision. The intervenient nodes feel the collision, and the window is shrunk (Step 2). However, the *lst* of messages A and B are still inside the window, causing another collision. In response to this event, the window size is shrunk again (Step 3). In this case, only message A has its *lst* within the window, leading to a successful transmission.

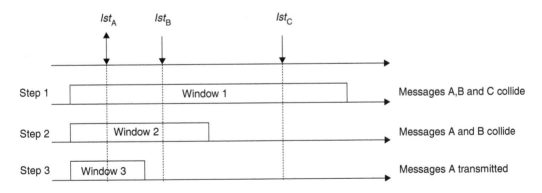

Legend:

lst_X: Latest sending time of message *x*

FIGURE 17.4 Windows protocol example.

This method exhibits properties that are very similar to those of the previous method (virtual time protocol). However, it is somewhat more efficient due to its adaptive behavior. In general, it also aims at soft RT systems and uses a fully distributed symmetrical approach with a relatively low computational overhead. Notice that all message parameters are relative and that there is no global time base again. Moreover, the protocol efficiency is substantially influenced by the magnitude of variations in the window duration, either when increasing or decreasing it.

17.4.2.3 Traffic Shaping

It is a known fact that if the bus utilization is kept low, the probability of collisions is also low (although not zero). Therefore, if the network average load is kept below a given threshold and bursts of traffic are avoided, a given probability of collisions can be obtained, as an estimation of the network-induced delay.

An implementation of this paradigm is presented in [21]. An interface layer, called traffic smoother (Figure 17.5), is placed between the IP network layer and the Ethernet data link layer (for this reason, traffic shaping is also commonly referred to as traffic smoothing). To avoid additional delays, RT traffic is transmitted on demand and it is not handled by the traffic smoother. Notice that real-time (RT) traffic is normally composed of periodic or sporadic message streams with a single packet per message instance, and thus the probability of collisions it induces is low. On the other hand, the Non-Real-Time (NRT) traffic is captured by the traffic smoother, which keeps track of previous message transmissions performed by the node. According to this historical record, it releases NRT messages in a controlled manner to achieve a desired node's traffic generation rate commonly referred to as *station input limit*. Moreover, inside each node, RT messages get higher priority than NRT ones. At the network level, the interference due to NRT traffic is kept under a probabalistic bound.

The station input limit, that is, the parameters of the leaky bucket, can either be static or they can be dynamic, being adjusted to the current network load in an adaptive way. A recent evolution of this latter approach has been presented in [13] in which the network load is estimated using two parameters, the number of collisions and the throughput, both observed in a given interval. These parameters are then used by a fuzzy controller that sets the instantaneous station input limit. The resulting efficiency is substantially higher than with other approaches to traffic shaping, allowing stations (nodes) to take advantage of periods of low throughput to transmit more data over the bus.

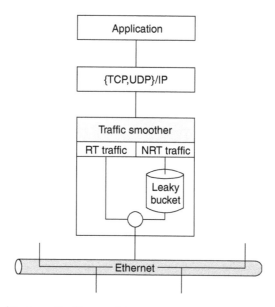

FIGURE 17.5 Software architecture of traffic smoothing.

In any case, traffic shaping supports probabilistic timing guarantees only being suited to soft RT systems. Typical applications can be either multimedia applications or even some robust control applications, which can cope with a few sample losses.

17.4.2.4 Token Passing

Another approach to achieve RT behavior in Ethernet consists in explicitly assigning time slots for exclusive use of a given node. This way, collisions are completely avoided. One possible way of implementing this concept in any shared bus network is by means of a timed-token mechanism [14].

In [12], Venkatramani and Chiueh present an implementation of this concept called RETHER. This protocol operates in normal CSMA/CD mode until the arrival of RT requests, upon which it switches to token-bus mode.

In the token-bus mode, RT data are considered to be periodic and the time is divided into cycles. During the cycle duration, access to the bus is regulated by a token, both for RT and NRT traffic. First, the token visits all nodes that are sources of RT messages. Afterwards, if there is enough time until the end of the cycle, the token visits the sources of NRT data. An on-line admission control policy assures that all accepted RT requests can always be served and that new RT requests cannot jeopardize the guarantees of existing RT messages. Therefore, in each cycle, all RT nodes can send their RT messages. However, with respect to the NRT traffic, no timeliness guarantees are granted.

Figure 17.6 illustrates a possible network configuration with six nodes. Nodes 1 and 4 are sources of RT messages, forming the RT set. The remaining nodes have no such RT requirements and constitute the NRT set. The token first visits all the members of the RT set and after, if possible, the members of the NRT set. A possible token visit sequence could be: cycle I {1 − 4 −1 − 2 − 3 − 4 − 5 − 6}, cycle $i+1$ {1 − 4 − 1 − 2}, cycle $i+2$ {1 − 4 − 1 − 2 − 3 − 4}…. . In the ith cycle, the load is low enough so that the token has time to visit the RT set plus all nodes in the NRT set, too. In the following cycle, besides the RT set, the token only visits nodes 1 and 2 of the NRT set and, in the next cycle, only nodes 1 through 4 of the NRT set are visited.

Due to the complete elimination of collisions, this approach supports deterministic analysis of the worst-case network access delay, particularly for the RT traffic. Furthermore, if the NRT traffic is known *a priori*, it is also possible to bound the respective network access delay, which can be important, for example, for sporadic RT messages. However, since the bandwidth available for NRT messages is distributed according to the nodes order established in the token circulation list, the first nodes always get precedence over the following ones, which end up with very long worst-case network delays. Moreover, this method involves a considerable communication overhead caused by the circulation of the token.

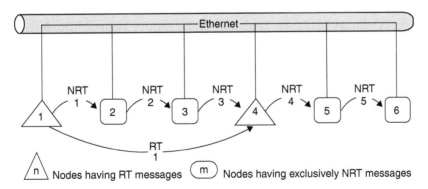

FIGURE 17.6 Sample network configuration for RETHER.

17.4.2.5 Master/Slave Techniques

One of the simplest ways of enforcing RT communication over Ethernet consists in using a master/slave approach in which a special node, the master, controls the access to the medium of all other nodes, the slaves. The traffic timeliness is then reduced to a problem of scheduling that is local to the master. However, this approach typically leads to a considerable under-exploitation of the network bandwidth because every data message must be preceded by a control message issued by the master resulting in a substantial communication overhead. Moreover, there is some extra overhead related to the turnaround time, that is, the time that must elapse between consecutive messages, since every node must fully receive and decode the control message before transmitting the respective data message. Nevertheless, it is a rugged transmission control strategy that has been used in many protocols. It is also the basis for the current Ethernet Powerlink protocol [15], a recently proposed industrial solution for RT communication over Ethernet.

A variant mechanism has also been proposed recently, which allows reducing the referred communication overhead. It is called the master/multi-slave approach [30], according to which the bus time is split into cycles and the master issues one control message for the whole cycle only indicating which data messages must be produced therein. This mechanism has been developed within the FTT framework, Flexible Time-Triggered communication, which has been implemented over different network protocols. One such implementation has been over Ethernet, leading to the FTT-Ethernet protocol [23]. This protocol combines the master/multislave transmission control technique with centralized scheduling, maintaining both the communication requirements and the message scheduling policy localized in one single node, the Master, and facilitating online changes to both, thus supporting a high level of operational flexibility.

The bus time is divided into fixed-duration time slots called elementary cycles (ECs) that are further decomposed into two phases, the synchronous and asynchronous windows (Figure 17.7), which have very different characteristics. The synchronous window carries the periodic time-triggered traffic that is also scheduled by the master node. The expression time-triggered implies that this traffic is synchronized to a common time reference, which in this case is imposed by the master. The asynchronous window carries the sporadic traffic either related to protocol control messages, such as those conveying change requests for the time-triggered traffic, event-triggered messages, such as those related to alarms, and other NRT traffic. There is a strict temporal isolation between both phase so that the sporadic traffic does not interfere with the time-triggered one.

Despite allowing on-line changes to the attributes of the time-triggered traffic, global timeliness is enforced by the FTT–Ethernet protocol by means of on-line admission control. Due to the global knowledge and centralized control of the time-triggered traffic, the protocol may easily support dynamic QoS management as a complement to admission control.

Beyond the flexibility and timeliness properties that this protocol exhibits, there are also some drawbacks that concern the computational overhead required in the master to execute both the message scheduling and the schedulability analysis on-line. This is, however, confined to one node. Nevertheless,

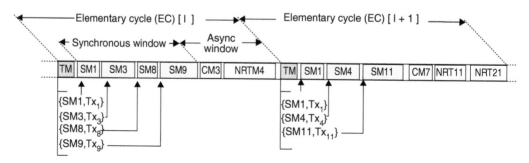

FIGURE 17.7 FTT–Ethernet traffic structure.

slaves also need some computational power to decode the trigger message in time and start the due transmissions at the right moments. Finally, in safety-critical applications, the master must be replicated, requiring further mechanisms to ensure coherency between their internal databases holding the system communication requirements.

17.4.2.6 Switched Ethernet

Since roughly one decade ago, the interest in using Ethernet switches has been growing as a means to improve global throughput, traffic isolation, and reduce the impact of the nondeterministic features of the original CSMA/CD arbitration mechanism. Switches, unlike hubs, provide a private collision domain for each of its ports; that is, there is no "direct" connection between its ports. When a message arrives at a switch port, it is buffered, analyzed with respect to its destination, and moved to the buffer of the destination port (Figure 17.8). The "packet handling" block in the figure includes the switch fabrics, which transfers messages from input to output ports; the switch controller or CPU, which controls the switch fabric and manages the internal traffic; and the switch memory, which is used to hold the port queues.

When the arrival rate of messages at each port, either input or output, is greater than the rate of departure, the messages are queued. Currently, most switches are fast enough in handling message arrivals so that queues do not build up at the input ports (these are commonly referred to as nonblocking switches). However, queues may always build up at the output ports whenever several messages arrive in a short interval and are routed to the same port. In such case, queued messages are transmitted sequentially, normally in FCFS order. This queue handling policy may, however, lead to substantial network-induced delays because higher priority or more important messages may be blocked in the queue while lower priority or less important ones are being transmitted. Therefore, the use of several parallel queues for different priority levels has been proposed (IEEE 802.1p). The number of distinct priority levels is limited to eight, but many current switches that support traffic prioritization offer even a further limited number. The scheduling policy used to handle the messages queued at each port also has a strong impact on the network timing behavior [22].

A common misconception is that the use of switches, due to the elimination of collisions, is enough to enforce RT behavior in Ethernet. However, this is not true in the general case. For instance, if a burst of messages destined to the same port arrives at the switch, output queues can overflow, thus leading to a loss of messages. This situation, despite seeming somewhat unrealistic, can occur with a nonnegligible probability in certain communication protocols based on the producer–consumer model, for example,

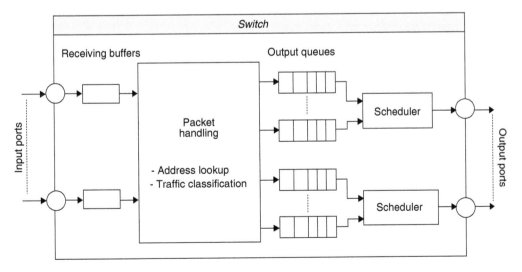

FIGURE 17.8 Switch internal architecture.

Control Information Protocol (CIP) and its lower level protocols such as Ethernet/Industrial Protocol (IP) [24], or based on the publisher–subscriber model such as RTPS [25] used within Interface for Distributed Automation (IDA). In fact, according to these models, each node that produces a given datum (producer or publisher) transmits it to potentially several nodes (consumers or subscribers) that need it. This model is efficiently supported in Ethernet by means of special addresses, called multicast addresses. Each network interface card can define the multicast addresses related to the information that it should receive. However, the switch has no knowledge about such addresses and thus treats all the multicast traffic as broadcasts, that is, messages with multicast destination addresses are transmitted to all ports (flooding).

Therefore, when the predominant type of traffic is multicast/broadcast instead of unicast, one can expect a substantial increase of peak traffic at each output port that increases the probability of queue overflow, causing degradation in network performance. Furthermore, in these circumstances, one of the main benefits of using switched Ethernet, that is, multiple simultaneous transmission paths, can be compromised. A possible way to limit the impact of multicasts is using virtual LANs (VLANs) so that flooding affects only the ports of the respective VLAN [16].

Other problems concerning the use of switched Ethernet are referred in Reference [9], such as the additional latency introduced by the switch in the absence of collisions as well as the low number of available priority levels that hardly supports the implementation of efficient priority-based scheduling.

These problems are, however, essentially technological and are expected to be attenuated in the near future. Moreover, switched Ethernet does alleviate the nondeterminism inherent to CSMA/CD medium access control and paves the way for efficient implementations of RT communication over Ethernet.

17.5 Recent Advances

Most of the recent work performed on RT Ethernet targets switch-based implementations. However, as discussed in section "Switched Ethernet," just replacing a hub by a switch is not enough to make an Ethernet network exhibit RT behavior. One of the issues recently addressed in the literature concerns the way packets are handled by the protocol software (protocol stack) within the system nodes. Most operating systems implement a single queue, usually working according to a First-Come First-Served policy, for both RT and NRT traffic. This approach induces important delays and priority inversions. A methodology that has been proposed to solve this problem is the implementation of multiple transmit/receive queues [27]. With this approach, the RT traffic is intrinsically separated from the NRT traffic. NRT traffic is only sent/processed when the RT queues are empty. It is also possible to build separate queues for each traffic class, providing internal priority-aware scheduling.

Another important issue concerns the degree of freedom in the network topology (e.g., bus, star). The topology impacts the number of switches that messages have to cross before reaching the target, which impacts the temporal properties of the traffic [28, 29]. For instance, the bus (or line) topology, in which each device integrates a simplified switch, eases the cabling but is the most unfavorable topology for RT behavior.

Another aspect concerning switch-based Ethernet networks respects the scheduling policy within the switch itself. Switches support up to eight distinct, statically prioritized traffic classes. Different message scheduling strategies have a strong impact on the RT behavior of the switch [28]. Particularly, strategies oriented toward average performance and fairness, which are relevant for general-purpose networks, may negatively impact the switch RT performance.

Finally, the interest in shared Ethernet is not over yet, either for applications requiring frequent multicasting, in which case the benefits of using switches are substantially reduced, as well as for applications requiring precise control of transmission timing, such as high-speed serving. In fact, switches tend to induce a higher delay and jitter in message forwarding than hubs, caused by internal mechanisms such as MAC address to port translation in forwarding and spanning-tree management protocol. In the previous sections, several examples of this interest were discussed, such as the recent work around adaptive traffic smoothing [13] and master/slave techniques including both the Ethernet Powerlink [15] as well as FTT–Ethernet [23] protocols. This last protocol is also being analyzed for implementation on switched Ethernet, taking advantage of the message queuing in the switch ports and thus simplifying the transmission control. This has the potential to ease the implementation of slave nodes because then it would not be necessary to enforce fine control of the transmission instants of both synchronous and asynchronous messages, strongly reducing the computational overhead. Several existing Ethernet-based industrial protocols, such as Ethernet/IP, are also taking advantage of switches to improve their RT capabilities [16]. Particularly, this protocol is now receiving unprecedented support from major international associations of industrial automation suppliers, such as Open DeviceNet Vendor Association (ODVA), ControlNet International (CNI), Industrial Ethernet Association (IEA), and Industrial Automation Open Networking Alliance (IAONA).

17.6 Conclusion

Due to several reasons, Ethernet became the most popular technology for LANs today. This makes it very attractive even in application domains for which it was not originally designed, in order to benefit from its low cost, high availability, and easy integration with other networks, just to name a few arguments. Some of such application domains, for example, industrial automation, impose RT constraints on the communication services that must be delivered to the applications. This conflicts with the original medium access control technique embedded in the protocol, CSMA/CD, which is nondeterministic and behaves very poorly with medium to high network loads. Therefore, in its nearly 30 years of existence, many adaptations and technologies for Ethernet have been proposed in order to support the desired real-time behavior.

This chapter has presented an overview of those techniques, ranging from changes to the bus arbitration, to the addition of transmission control layers, and also to the use of special networking equipment, such as switches. Such techniques have been described and briefly analyzed in what concerned their pros and cons for different types of application. This chapter concludes with a reference to recent trends, where the growing impact of switches is clear, despite a few specific situations in which shared Ethernet might still be preferable, such as when the traffic is mainly of a multicast nature or a precise transmission timing control is required.

With the current high pressure to bring Ethernet more and more into the world of distributed automation systems, it is likely that such technology will end up taking the place of existing fieldbuses and establishing itself as the *de facto* communication standard for this area. Although its efficiency in terms of bandwidth utilization is still low when considering short messages, particularly lower than with several fieldbuses, its high and still growing bandwidth seems more than enough to supplant such an aspect. Ethernet will then become the long-awaited single networking technology within automation systems, which will support the integration of all levels, from the plant floor to the management, maintenance, supply-chain, etc.

References

1. IEEE, DIX Ethernet V2.0 specification, 1982.
2. IEEE, IEEE 802.3 10BASE5 standard.
3. IEEE, IEEE 802.3 10BASE3 standard.
4. IEEE, IEEE 802.3c 1BASE5 StarLan standard.
5. IEEE, IEEE 802.3i 10BASE-T standard.
6. IEEE, IEEE 802.3c 100BASE-T standard.
7. IEEE, Supplement to IEEE Std 802.3, 1998 Edition: Physical Layer Parameters and Specifications for 1000 Mb/s Operation over 4-Pair of Category 5 Balanced Copper Cabling, Type 1000BASE-T, 1999.
8. Thomesse, J.-P., Fieldbus and interoperability, *Control Engineering Practice*, 7, 81–94, 1999.

9. Decotignie, J.-D., A Perspective on Ethernet as a Fieldbus, in *Proceedings of the 4th IFAC International Conference on Fieldbus Systems and their Applications (FeT'2001)*, IFAC Publications, Elsevier Science, Oxford, England, November 2002, pp. 138–143.

10. Song,Y., Time Constrained Communication over Switched Ethernet, in *Proceedings of the 4th IFAC International Conference on Fieldbus Systems and Their Applications (FeT'2001)*, IFAC Publications, Elsevier Science, Oxford, England, 2002, pp. 152–159.

11. LeLann, G. and N. Rivierre, Real-time communications over broadcast networks: the CSMA-DCR and the DOD-CSMA-CD protocols, Technical Report, INRIA Report RR1863, 1993.

12. Venkatramani, C. and T. Chiueh, Supporting Real-Time Traffic on Ethernet, *in Proceedings of IEEE Real-Time Systems Symposium*, IEEE Press, New York, December 1994, pp. 282–286.

13. Carpenzano, A., R. Caponetto, L.L. Bello, and O. Mirabella, Fuzzy Traffic Smoothing: an Approach for Real-Time Communication over Ethernet Networks, in *Proceedings of the 4th IEEE International Workshop on Factory Communication Systems (WFCS'2002)*, IEEE Press, New York, August 2002, pp. 241–248.

14. Malcolm, N. and W. Zhao, The timed-token protocol for real-time communications, *IEEE Computer,* 27, 35–41, 1994.

15. Ethernet Powerlink Standardization Group, Ethernet powerlink protocol, www.ethernetpowerlink. org

16. Moldovansky, A., Utilization of Modern Switching Technology in Ethernet/IP Networks, in *Proceedings of the 1st International Workshop on Real-Time LANs in the Internet Age*, Edições Politeama, Porto, Portugal, 2002, pp. 25–27.

17. Bello, L.L. and O. Mirabella, Design Issues for Ethernet in Automation, in *Proceedings of 8th IEEE International Conference on Emerging Technologies and Factory Automation (ETFA'2001)*, Vol. 1, IEEE Press, New York, October 2001, pp. 213–221.

18. Court, R., Real-time Ethernet, *Computer Communications,* 15, 198–201, 1992.

19. Malcom, N. and W. Zhao, Hard real-time communication in multiple-access networks, *Real Time Systems,* Vol. 9, Kluwer Academic Publishers, Boston, MA, 1995, pp. 75–107.

20. Molle, M. and L. Kleinrock, Virtual time CSMA: Why two clocks are better than one, *IEEE Transactions on Communications,* 33, 919–933, 1985.

21. Kweon, S.-K. and K. G. Shin, Achieving Real-Time Communication over Ethernet with Adaptive Traffic Smoothing, in *Proceedings of IEEE Real-Time Technology and Applications Symposium, RTAS'00*, IEEE Press, Washington, D.C., June 2000, pp. 90–100.

22. Jasperneit, J. and P. Neumann, Switched Ethernet for Factory Communication, in *Proceedings of 8th IEEE International Conference on Emerging Technologies and Factory Automation (ETFA'2001)*, Vol. 1, IEEE Press, New York, October 2001, pp. 205–212.

23. Pedreiras, P., L. Almeida, and P. Gai, The FTT-Ethernet Protocol: Merging Flexibility, Timeliness and Efficiency, in *Proceedings of 14th IEEE Euromicro Conference on Real-Time Systems*, IEEE Press, New York, June 2002, pp. 134–142.

24. Open DeviceNet Vendor Association, Ethernet/IP (industrial protocol) specification, www.odva.org

25. IDA, Interface for Distributed Automation Group, RTPS, Real-Time Publisher/Subscriber Protocol, www.ida-group.org

26. IEEE, IEEE Standard for Carrier Sense Multiple Access with Collision Detection (CSMA/CD) Access Method and Physical Layer Specifications-Media Access Control (MAC) Parameters, Physical Layer and Management Parameters for 10 Gb/s Operation, 2002.

27. Skeie, T., S. Johannessen, and O. Holmeide, The Road to an End-to-End Deterministic Ethernet, in *Proceedings of the 4th IEEE International Workshop on Factory Communication Systems (WFCS'2002)*, IEEE Press, New York, August 2002, pp. 3–9.

28. Jasperneite, J., P. Neumann, M. Theis, and K. Watson, Deterministic Real-Time Communication with Switched Ethernet, in *Proceedings of the 4th IEEE International Workshop on Factory Communication Systems (WFCS'2002)*, IEEE Press, New York, August 2002, pp. 11–18.

29. Rondeau, E., T. Divoux, and H. Adoud, Study and Method of Ethernet Architecture Segmentation for Industrial Applications, *in Proceedings of the 4th IFAC International Conference on Fieldbus Systems and Their Applications (FeT'2001)*, IFAC Publications, Elsevier Science, Oxford, England, November 2002, pp. 165–172.
30. Almeida, L., P. Pedreiras, and J.A. Fonseca, The ftt-can protocol: why and how, *IEEE Transactions on Industrial Electronics*, 49, 1189–1201, 2002.

18

Principles and Features of PROFInet

Manfred Popp
Siemens AG

Joachim Feld
Siemens AG

Ralph Büsgen
Siemens AG

18.1   Introduction

Automation technology is undergoing continuous change due to the ever shorter innovation cycles for new products. The use of fieldbus technology in recent years has represented a significant innovation. It has enabled the migration of automation systems from centralized to decentralized systems. In this regard, PROFIBUS has set the standard as the market leader for more than 15 years.

Moreover, in today's automation technology, information technology (IT) with established standards, such as Transmission Control Protocol (TCP)/Internet Protocol (IP) and Extensible Markup Language (XML), is increasingly dictating changes. Integration of information technology into automation is opening up significant advances in communication options between automation systems, far-reaching configuration and diagnostic options, and network-wide service functions. These functions have been a fixed component of PROFInet from the start.

PROFInet is the open standard for industrial automation based on the Industrial Ethernet. PROFInet enables problem-free realization of distributed automation, integration of existing field devices, and operation of demanding, time-critical applications (such as motion control).

In addition to utilization of IT technology, protection of investment also plays an important role with PROFInet. PROFInet enables existing fieldbus systems such as PROFIBUS to be integrated without modifications with existing devices. This protects the investments of plant operators, machinery/plant construction firms, and device manufacturers.

Automation technology requirements are thoroughly covered by PROFInet. It was possible to transfer the many years of experience in the PROFIBUS sphere to PROFInet standardization. The use of open standards, simple handling ability, and integration into existing plant units have defined PROFInet from the start. PROFInet is currently integrated in IEC 61158.

A long-term perspective is offered to users through continuous advancements in PROFInet.

Costs incurred by plant or mechanical system engineers for installation, engineering, and start-up are minimized through the use of PROFInet. For plant operators, PROFInet enables plants to be easily expanded and achieve a high level of availability through independently operating plant units.

Establishment of certification by the PROFIBUS User Organization (PNO) guarantees a high standard of quality for PROFInet products.

This chapter describes in detail how experience gained with IT standards in the PROFIBUS sphere has been converted to PROFInet.

18.2 PROFInet at a Glance

The motivation to create PROFInet comes from the user requirements outlined in Section 18.1 and the anticipated cost reduction resulting from manufacturer-independent, plantwide engineering.

With PROFInet, a modern automation concept has emerged that is based on Ethernet and enables simple integration of existing fieldbus systems (in particular, PROFIBUS). This represents an important aspect for satisfying uniformity of requirements from the corporate management level to the field level.

18.2.1 Decentralized Field Devices (PROFInet IO)

Simple field devices are integrated in PROFInet using PROFInet IO and described by the familiar input/output (I/O) view in PROFIBUS DP. Decentralized periphery for PROFInet is also integrated using this approach. The essential feature of this integration is the use of decentralized field devices with their input and output data, which are processed in the programmable logic controller (PLC) user program.

PROFInet IO describes a device model that differentiates slots and channels in a way similar to that of the model for PROFIBUS DP. The device properties are described by an XML-based description file (general station description (GSD)).

PROFInet IO devices are engineered using the same approach that has long been familiar to system integrators of PROFIBUS DP. This includes assignment of the decentralized field devices to a controller during configuration. Productive data are then exchanged between the controller and the assigned field devices (Figure 18.1).

18.2.2 Distributed Automation (Component Model)

Distributed automation systems typically consist of several subunits that act autonomously for the most part and coordinate with each other using signals for synchronization, sequence control, and information exchange.

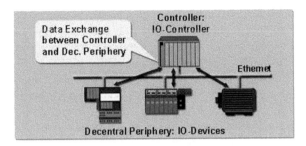

FIGURE 18.1 Architecture of PROFInet IO.

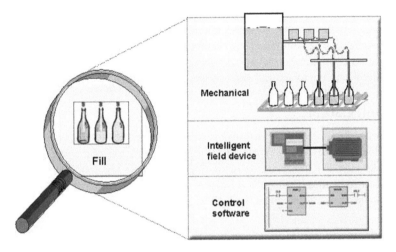

FIGURE 18.2 Mechanical, electrical/electronic, and software aspects are combined in technological modules.

The PROFInet component model refers to these subunits as *technological modules.* The technological modules form an intelligent functional unit. Through the use of the component technology proven in the IT sphere, the overall functionality of a technological module is encapsulated in an associated software component. Such a component is modeled as an object and regarded as a black box. An outside technological component interface is defined in order for the component to communicate with other components within the distributed system.

A distributed automation system designed in this way constitutes the prerequisite for modularization of plants and machinery, and hence for reuse of plant and machine parts. This significantly reduces engineering costs.

A technological module is described in PROFInet within the component model using PROFInet component description (PCD). PCD is XML based and produced by either a component generator of a manufacturer-specific configuration tool or the PROFInet component editor.

A manufacturer-neutral *engineering concept* is available for user-friendly configuration of a PROFInet system.

Engineering of distributed automation systems distinguishes between programming of control logic for individual technological modules (manufacturer-specific configuration tools) and the technological configuration of the entire system (interconnection editor). A systemwide application is formed in three steps: create components, interconnect components, and download interconnection information (Figure 18.2).

18.2.3 Communication

Different performance levels are available for PROFInet communication.

Parameters, configuration data, and interconnection information that are not critical with respect to time are transferred in PROFInet via the standard channel based on TCP/User Datagram Protocol (UDP) and IP. This satisfies the prerequisites for interfacing the automation levels with other networks (manufacturing execution systems (MES), enterprise resource planning (ERP)).

For transfer of time-critical process data within the production plant, the real-time channel known as soft real time (SRT) is available. This channel is implemented as software on the basis of existing controllers.

For isochronous applications, isochrone real-time (IRT) communication is available that enables clock pulse rates of less than 1 ms and a jitter accuracy of 1 μs.

18.2.4   Network Installation

PROFInet network installations are oriented toward specific requirements for Ethernet networks in industrial environments. The "PROFInet Installation Guideline" provides plant construction engineers and plant operators with simple rules for installing Ethernet networks and associated cabling. This guideline provides device manufacturers with clear specifications for device interfaces.

18.2.5   IT Integration

The network management includes functions for administration of PROFInet devices in Ethernet networks. This includes the device configuration, network configuration, and network diagnostics. In the case of Web integration, PROFInet makes use of the Ethernet base technologies and enables access to a PROFInet component by means of standard Internet technologies. In order to preserve an open connection to other system types, PROFInet supports OPC DA (data access) and DX (data exchange).

18.2.6   Fieldbus Integration

An important aspect of PROFInet is the seamless transition from existing fieldbus solutions such as PROFIBUS DP to Ethernet-based PROFInet. This contributes significantly to protection of investments by the device manufacturer, the plant construction/mechanical system engineer, and the end user (Figure 18.3).

PROFInet offers two alternatives for integrating fieldbus systems:

- Integration of fieldbus devices by means of proxies: The proxy is the representative for the lower-level field devices on the Ethernet. Through the proxy principle, PROFInet offers a completely transparent transition from existing to newly installed plant units.
- Integration of whole fieldbus applications: A fieldbus segment represents a self-contained component. The representative for this component is the PROFInet device that operates a fieldbus such as PROFIBUS DP at a lower level. The entire functionality of a lower-level fieldbus is thereby implemented in the form of a component in the proxy, which is available on the Ethernet.

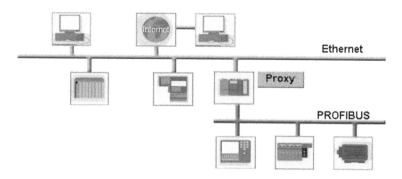

FIGURE 18.3 PROFIBUS systems can be integrated in PROFInet using a proxy.

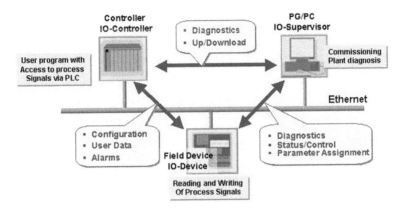

FIGURE 18.4 Device types in PROFInet IO.

18.3 Decentralized Field Devices (PROFInet IO)

Decentralized field devices are integrated directly on the Ethernet using PROFInet IO. To accomplish this, the master–slave system familiar in PROFIBUS DP is transferred over to a provider–consumer model. Though all devices on the Ethernet have equal communication rights, the configuration specifies which field devices are assigned to a centralized controller. In this way, the familiar user view in PROFIBUS is transferred to PROFInet IO. IO signals are read in and processed by the PLC and then re-sent to the outputs.

18.3.1 Functional Scope

PROFInet IO distinguishes between three device types: PN-IO controller, PN-IO device, and PN-IO supervisor (Figure 18.4):

- PN-IO controller: A PLC on which the automation program runs
- PN-IO device: Decentralized field device that is assigned to a PN IO controller (such as remote IO, valve terminals, frequency converters)
- PN-IO supervisor: Programming device or PC with commissioning and diagnostic functions

Data can be transferred between the IO controller and IO device by means of the following channels:

- Cyclic user data via the real-time channel
- Event-triggered interrupts (diagnostics) via real-time channel
- Parameter assignment and configuration as well as reading of diagnostic information via the standard channel based on UDP/IP

At the start, a communication relationship called application relation (IO-AR) is established between the IO controller and the IO device based on the acyclic UDP/IP channel (Figure 18.5). Then the IO

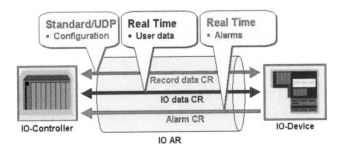

FIGURE 18.5 Communication relationships in PROFInet IO governed by the consumer–provider model.

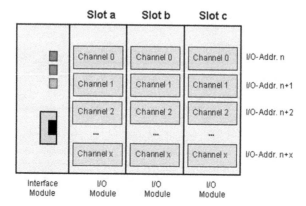

FIGURE 18.6 PROFInet IO device model is similar to that of PROFIBUS DP.

controller transfers the configuration data for the IO device by means of this established channel. Based on the configuration data, (1) the correct operating mode is determined, for example, and the IO device is uniquely identified; and (2) high-speed, cyclic useful data exchange via the real-time channel (IO-CR) is started. If a diagnostic event occurs (such as a wire break), an interrupt is sent to the IO controller via the high-speed, acyclic real-time channel (interrupt CR) for processing in the PLC program located on the IO controller.

18.3.2 Device Model

A uniform device model has been specified for the PROFInet IO device (Figure 18.6). This model enables modular and compact field devices to be modeled. This model is oriented toward the main features in PROFIBUS DP and extends the advantages of these features into the future.

An IO device with a modular configuration consists of slots in which modules are inserted. The modules contain channels over which process signals are read in or read out. The representative of the IO device is the interface module, which receives data from the IO controller and forwards it to the modules via the backplane bus. Conversely, it receives the process and diagnostic information from the modules via the backplane bus and forwards this information to the IO controller. Each IO device receives a global device identification that is uniquely assigned within the framework of PROFInet IO. This 32-bit device ID number is divided into a 16-bit manufacturer identifier and a 16-bit device identifier. This device ID is assigned by the PROFIBUS User Organization (PNO).

18.3.3 Device Description (GSD)

As in PROFIBUS, a device description is used to integrate a PROFInet device into the configuration tool of an IO controller. The properties of an IO device are described in the form of general station description (GSD), which contains all necessary information:

- Properties of the IO device (e.g., communication parameters)
- Insertable modules (number of type)
- Configuration data for individual modules (e.g., 4- to 20-mA analog input)
- Parameters of modules
- Error texts for diagnostics (e.g., wire break, short circuit)

XML is the description basis for the GSD of PROFInet IO devices. Because XML is an open, widespread, and accepted standard for describing data, appropriate tools and derived properties are automatically available, including:

- Creation and validation through a standard tool

- Foreign language integration
- Hierarchical structuring

The GSD structure corresponds to ISO 15745 and consists of a header, the device description in the application layer (e.g., configuration data and module parameters), and the communication properties description in the transport layer.

18.3.4 Configuration and Data Exchange

The description files of the IO devices are imported into the configuration tool. IO addresses are assigned to the individual IO channels of the field devices. The IO input addresses contain the received process values. The user program evaluates and processes these values. The user program forms the IO output values and outputs them to the process via the IO output addresses. In addition, parameters are assigned to the individual IO modules or channels in the configuration tool, e.g., 4- to 20-mA current range for an analog channel.

After conclusion of the configuration, the configuration data are downloaded to the IO controller. The IO devices are assigned and configured automatically by the IO controller and then enter into the cyclic data exchange (Figure 18.7).

18.3.5 Diagnostics

PROFInet IO supports a multilevel diagnostic concept that enables efficient fault localization and correction. When a fault occurs, the faulty IO device generates a diagnostic alarm to the IO controller. This alarm triggers a call in the PLC program to the appropriate program routine in order to be able to respond to the fault. If a device or module defect requires a complete replacement of the device or module, the IO controller automatically performs a parameter assignment and configuration of the new device or module.

The diagnostic information is structured hierarchically:

- Slot number (module)
- Channel number
- Channel type (input/output)
- Coded fault cause (e.g., wire break, short circuit)
- Additional manufacturer-specific information

When an error occurs in a channel, the IO device generates a diagnostic alarm to the IO controller. This alarm triggers a call in the control program to the appropriate fault routine. After processing of the fault routine, the IO controller acknowledges the fault to the IO device. This acknowledgment mechanism ensures that sequential fault processing is possible in the IO controller.

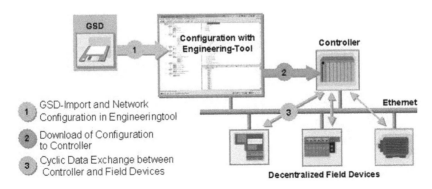

FIGURE 18.7 Configuration path for data exchange.

18.4   Distributed Automation

Automation development has given rise to modular plants and machinery. This structuring has triggered further development in automation to produce distributed automation systems. PROFInet also offers a solution in this area. The PROFInet solution involves separation of plant units into technological modules.

18.4.1   Technological Modules

The function of an automated plant or machine — for a goods manufacturing process — is produced by a defined interaction of mechanical, electrical/electronic, and control logic/software aspects. According to this principle, PROFInet defines the mechanical, electrical/electronic, and control logic/software aspects for a *technological module* (see Figure 18.2). A technological module is in turn modeled by a software component, i.e., the PROFInet component.

18.4.2   PROFInet Components

The representative of a technological module during plant engineering is the so-called PROFInet component. Each PROFInet component has an interface that contains the technological variables to be exchanged with other components.

The PROFInet components are modeled using the standardized component object model (COM) technology. COM is an advanced object orientation that enables applications to be developed on the basis of preassembled components. The components are characterized by the formation of complete units that can be in relationship with other components. Like blocks, the components can be combined flexibly and easily reused, irrespective of how they are implemented internally. Access mechanisms to the component interfaces are uniformly defined in PROFInet.

18.5   Granularity of Technological Modules

When the granularity of modules is being specified, the ability to reuse the modules in different plants must be examined with cost and availability in mind. The objective is to be able to merge individual components into an overall plant as flexibly as possible according to the building block principle. On the one hand, with too fine a granularity the technological view of the plant can become more complex, resulting in higher engineering costs. On the other hand, with too coarse a granularity, the degree of reusability is reduced. This, in turn, results in higher implementation costs.

The machine or system manufacturer creates the software component. The *component design* has a major influence on reduction of engineering and hardware costs and the time response of the automation system (Figure 18.8). When a component is being designed, the granularity (i.e., size of machine/plant or the specified machine parts/plant units) can extend from an individual device to a complete machine containing a number of devices.

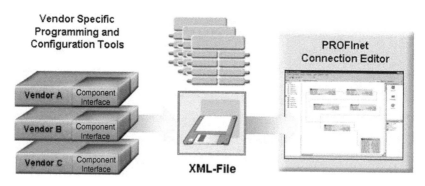

FIGURE 18.8 Component creation is standardized in PROFInet.

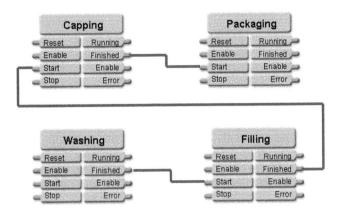

FIGURE 18.9 Interconnected components.

18.5.1 PROFInet Engineering

A manufacturer-neutral engineering concept has been created for user-friendly configuration of a PROFInet system. For one thing, this engineering concept enables development of configuration tools that can be used for components of different manufacturers. For another, it permits manufacturer-specific and application-specific function expansions.

The engineering model distinguishes between programming of control logic for individual technological modules and the technological configuration of the overall plant. A plantwide application is generated in three stages.

18.5.1.1 Component Creation

Components are created as an image of the technological modules by the machinery or plant construction firms. Devices are programmed and configured as before with the respective manufacturer-specific tools. That way, available user programs and the know-how of programmers and service personnel can continue to be used. Then the user software is encapsulated in the form of a PROFInet component. In so doing, a component description (PCD) in the form of an XML file is created. The contents of the component description are specified in PROFInet. These component descriptions are imported into the library of the interconnection editor.

18.5.1.2 Component Interconnection

The created PROFInet components are moved from a library to an application and interconnected at the click of a mouse using the PROFInet *interconnection editor* (Figure 18.9).

Interconnection replaces the previous labor-intensive programming of communication relationships with simple graphical configuration. During programming, detailed knowledge about the integration and sequence of communication functions in the device is required. At the time programming is performed, the following must already have been specified: which devices will communicate with one another, when the communication will occur, and which bus system will be used for the communication. By contrast, knowledge of the communication functions is not required during configuration because these functions run automatically in the devices. The interconnection editor consolidates the individual distributed applications on a plantwide basis. Operation of the interconnection editor is manufacturer neutral, that is, the editor interconnects any manufacturer's PROFInet components.

18.5.1.3 Downloading

Following component interconnection, the interconnection information as well as the code and configuration data for the components are downloaded to the PROFInet devices at the click of a mouse (Figure 18.10). As a result, each device knows all of its communication peers, communication relationships, and information to be exchanged. The distributed application can be executed afterward.

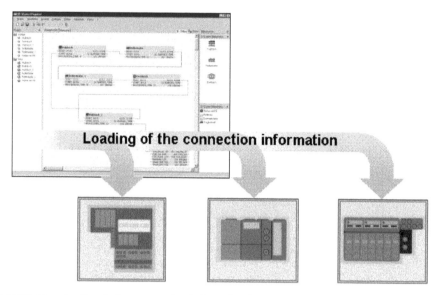

FIGURE 18.10 Downloading the interconnection information for PROFInet devices.

18.5.2 Component Description (PCD)

The PROFInet component description (PCD) is an XML file, which is created by manufacturer-specific tools. This assumes that these tools have a component generator. Alternatively, a manufacturer-neutral PROFInet component editor, available for download on the PROFIBUS Web site, can be used to create the PCD file. The PCD file contains information on the functions and objects of the PROFInet component. Specifically, this information includes:

- Description of components as library elements: component ID, component name
- Hardware description: IP address storage, diagnostic data access, interconnection download
- Description of software functionality: software hardware assignment, component interface, properties of variables such as technological name, data type, and direction (input or output)
- Storage location of component project

Component libraries are generated to support reusability.

18.5.3 Interconnection Editor

In general, an interconnection editor has two views: plant view and network view.

In the plant view, required components are imported from the library and placed on the screen, and the individual interconnections are established (Figure 18.11). This yields a *technological* structure and its local relationships within a plant. By contrast, the *topological* structure of the automation system is created in the network view. Here, the field devices and automation devices are assigned to a communication system or bus system, and the device addresses are specified according to the rules for the underlying bus system (Figure 18.12).

18.5.4 PROFInet Runtime

The PROFInet runtime model defines functions and utilities that require cooperating automation components to accomplish an automation task. This model establishes and monitors the interconnections between PROFInet components that have been configured by engineering. The PROFInet runtime model sets up a provider–consumer model in which the provider creates and sends data and the consumer receives and processes data.

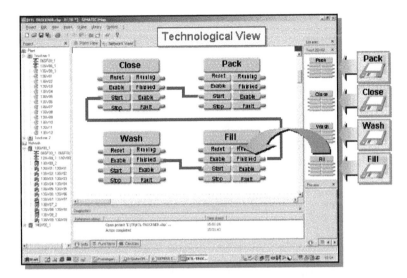

FIGURE 18.11 Plant view in the interconnection editor shows the interconnected components.

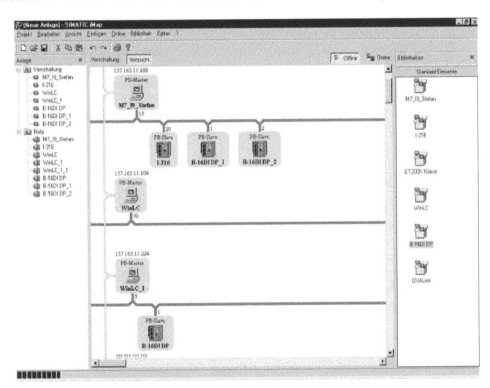

FIGURE 18.12 Network view in the interconnection editor shows the connected field devices.

18.6 PROFInet Communication

Ethernet-based communication is scalable in PROFInet. The following three performance levels are differentiated:

1. TCP/UDP and IP for data that are not time critical, such as parameter assignment and configuration data

2. Soft real time (SRT) for time-critical process data in factory automation
3. Isochrone real time (IRT) for particularly challenging application requirements, such as those for motion control

These three PROFInet communication performance levels cover the spectrum of automation applications in their overall diversity. The PROFInet communication standard is characterized by the following in particular:

- Coexisting utilization of real-time and TCP-based IT communication on one line
- Uniform real-time protocol for all applications both for communication between components in distributed systems and for communication between the controller and the distributed field devices
- Scalable real-time communication from performant to high performant, and isochronous mode

Scalability and a uniform communication basis represent two of the major strengths of PROFInet. They guarantee continuity to the corporate management level and fast response times in the automation process.

18.6.1   Standard Communication with TCP/UDP

PROFInet uses Ethernet and TCP/UDP with IP as the basis for communication. TCP/UDP with IP is a *de facto* standard with respect to communication protocols in the IT landscape. However, for interoperability (i.e., the interaction between applications), establishment of a common communication channel over the field devices, based on TCP/UDP (layer 4), is insufficient. TCP or UDP represents only the foundation on which Ethernet devices can exchange data via a transport channel in local and distributed networks. Therefore, additional specifications and protocols beyond TCP/UDP — the so-called application protocols — are required. Interoperability is only guaranteed if the same application protocol is used on the devices. Typical application protocols are Simple Mail Transfer Protocol (SMTP, for e-mail), File Transfer Protocol (FTP, for file transfer), and Hypertext Transfer Protocol (HTTP, used on the Internet).

18.6.2   Real-Time Communication

Real-time applications in manufacturing automation require update, or response, times in the range of 5 to 10 ms. The update time refers to the time that elapses when a variable is generated by an application in a device, then sent to a peer device via the communication system, and then received updated by the application. For devices, the implementation of a real-time communication causes only a small load on the processor so that execution of the user program continues to take precedence. From experience, in the case of Fast Ethernet (100 Mb/s Ethernet), the transmission rate on the line in proportion to the execution in the devices can be disregarded. Most of the time gets lost in the application. The time it takes to provide data to the application of the provider is not influenced by the communication. This also applies to processing of data received in the consumer. As a result, noteworthy improvements in update rates, and thus the real-time performance, can be obtained primarily through proper optimization of the communication stack in the provider and consumer.

18.6.2.1   Soft Real Time (SRT)

In order to satisfy real-time demands in automation, PROFInet uses an optimized real-time communication channel (Figure 18.13).

This channel is based on the Ethernet (layer 2). The solution minimizes the throughput time in the communication stack considerably and increases performance in terms of the process data update rate. By doing away with several protocol layers, the message frame length is reduced. By not including these layers, the data to be transmitted from the provider can be sent sooner, or are available earlier to the application on the consumer side. At the same time, this significantly reduces the processor power required for communication in the device.

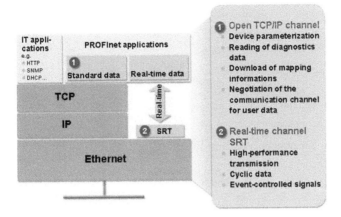

FIGURE 18.13 Communication channels in PROFInet.

18.6.2.1.1 Optimized Data Transmission through Prioritization

In addition to the minimized communication stack in the automation devices, the transmission of data in the network is also optimized in PROFInet. In order to achieve an optimal result, the packets are prioritized in PROFInet according to IEEE 802.1Q. The network components use this priority to control the data flow between the devices. The standard priority for real-time data is based on Prio 6. Thus, priority handling over other applications, such as Internet telephony, is guaranteed because they are using Prio 5.

18.6.2.2 Isochronous Real Time (IRT)

The presented solutions are not sufficient for motion control applications, in particular. These applications require update rates in the range of 1 µs along with a jitter for consecutive update cycles of 1 µs for cases involving up to 100 stations. To satisfy these requirements, PROFInet defines the IRT time slot-controlled transmission process on layer 2 for Fast Ethernet. That means every device knows exactly in which time slot it is allowed to send data over the bus. Through synchronization of the devices involved (network components and PROFInet devices) with the accuracy indicated above, a time slot can be specified during which data critical for the automation task are transferred. The communication cycle is split into a deterministic part and an open part. The cyclic real-time message frames are dispatched in the deterministic channel, while the TCP/IP message frames are transported in the open channel. The process is comparable to the traffic on a highway where the left lane is reserved for time-critical traffic (real-time traffic) and the remaining traffic elements (TCP/IP traffic) are prevented from switching to this lane. Even if there is a traffic jam in the right lane, the time-critical traffic is not impacted.

Isochronous data transmission is realized based on hardware; e.g., it is burned in an ASIC. Such an ASIC covers the cycle synchronization and time slot reservation functionality for the real-time data. Realization in hardware ensures that the accuracy requirements will be achieved within the order of magnitude necessary. Furthermore, the processor in the PROFInet device is relieved from communication tasks. The resulting additional runtime can be made available for automation tasks.

18.6.3 Communication for PROFInet IO

For PROFInet IO, RPC based on UDP/IP is used in the start-up phase for initiation of the data exchange between devices, parameter assignment of distributed field devices, and diagnostics. Through the open and standardized RPC protocol, human machine interface (HMI) stations or engineering systems (IO supervisor) can also access PROFInet IO devices. The PROFInet real-time channel is employed for transmission of user data and alarms.

In a typical IO configuration, there is an IO controller that exchanges the user data cyclically by means of communication relationships that are established during the start-up phase with multiple distributed

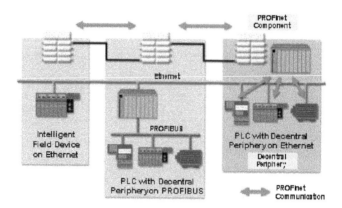

FIGURE 18.14 PROFInet communication between PROFInet components and PROFInet IO devices.

field devices (IO devices). In each cycle, the input data are sent from the assigned field devices to the IO controller, and in return, the output data are sent back to the appropriate field devices. The communication relationship is monitored by keeping track of cyclic messages. If, for example, cyclic messages fail in three cycles, the IO controller recognizes that the corresponding IO device has failed.

The data transmission layer of PROFInet is defined in IEEE 802.3, which describes the protocol design and malfunction monitoring. A user data message frame consists of a minimum of 64 bytes and a maximum of 1500 bytes. The overall protocol overhead for real-time data is 28 bytes.

18.6.4   Communication between Technological Modules

Distributed COM (DCOM) is specified in the PROFInet component view as the common TCP/IP-based application protocol between PROFInet components (Figure 18.14). DCOM represents an enhancement to COM for distribution of objects and their interaction in a network. DCOM is based on the standardized RPC protocol. DCOM is used for loading of interconnections, reading of diagnostic data, device parameter assignment and configuration, establishing interconnections, and, to some extent, exchanging user data between components in PROFInet. However, DCOM does not have to be used for exchange of user data between PROFInet components. Whether user data exchange is to take place via DCOM or the real-time channel is configured by the user in the engineering system, devices can then negotiate the use of a real-time-capable protocol. This is because communication between such plant or machinery modules can require real-time conditions that cannot be satisfied through TCP/IP and UDP.

TCP/IP and DCOM form the common language that can be used to start communication between the devices in all cases. The PROFInet real-time channel is then used for real-time communication between individual stations in time-critical applications. In the configuration tool, the user can select the update time, the so-called quality of service, to determine whether the values are to be transferred between components cyclically during operation or when a change is made. A cyclic transfer is more advantageous for high-frequency update times since checking for changes and acknowledging them imposes less processor load than cyclic sending.

18.7   Installation Technology for PROFInet

The international standard ISO/IEC 11801 and its European equivalent EN 50173 define an application-neutral, information-oriented standard networking for a complex of buildings. The content of the two standards is essentially identical. Both standards assume the buildings are similar to an office environment and assert a claim to being application neutral. The following specific requirements for Ethernet networks in the industrial environment are not taken into consideration in the two standards (Figure 18.15):

- Plant-specific cable routing

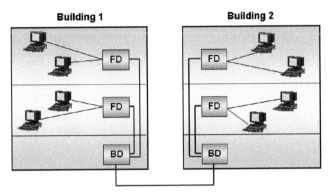

FIGURE 18.15 Structure of Ethernet networks in office systems.

Office Area	Production Area
Fixed basic installation in a building	Largely system-related cabeling
Laid under raised floors	System-related cable routing
Variable device connection at workplace	Connection points are seldom changed
Prefabricated device connection cable	Field-preparable device connections
Tree-shape network structures	Quite often: line-form network structures and (redundant) ring structures
Large data packets (e.g., images)	Small data packets (measured values)
Medium network availability	Very high network availability
Moderate temperatures (from 0 to 50°C)	Extreme temperatures (from –20 to +70°C)
No moisture	Moisture possible (IP65)
Virtually no vibrations	Vibrating machines
Low EMC burden	High EMC burden
Low mechanical danger	Danger of mechanical damage
Virtually no chemical danger	Chemical burden from oily or aggressive atmospheres

FIGURE 18.16 Differences in cable installation in office systems and automation systems.

- Individual degree of networking for each machine/plant
- Linear network structures
- Robust industrial cable and connectors with special requirements for EMC, temperature, moisture, dust, and vibration

For this reason, the "PROFInet Transmission Technology and Cabling" guideline defines an industrial cable installation for the Fast Ethernet application based on the fundamental requirements in IEC 11801 (Figure 18.16).

18.7.1 PROFInet Cable Installation

18.7.1.1 PROFInet Cable Installation with Symmetrical Copper Cable

Signals are transmitted over symmetrical copper cable (twisted pair) in accordance with 100BASE-TX at a transmission rate of 100 Mbits/s (Fast Ethernet). The transmission medium is a two-pair, twisted-conductor, shielded copper cable (twisted pair or star quad) with a characteristic impedance of 100 ohm (Figure 18.17).

Only shielded cable and connection elements are permitted. The individual components must satisfy category 5 requirements according to IEC 11801. The overall transmission line must satisfy the Class D requirements according to IEC 11801. Removable connections are produced using an RJ45 or M12 connector system. Sockets are used as device connections. Connecting cables (device connection cables, marshalling cables) are provided with suitable plugs at both ends. An active network component is used to connect all devices. To ensure that installation is as simple as possible, the transmission cable has been

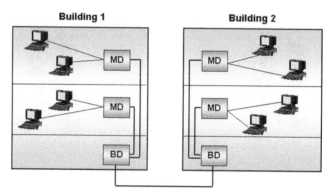

FIGURE 18.17 Ethernet Networks in an industrial environment.

defined to be identical at both ends. This connecting cable satisfies the function of a cable assembly with two identical ends.

The maximum segment length is 100 m.

18.7.1.2 PROFInet Cable Installation with Optical Fibers

PROFInet can be operated with multimode or single-mode fiber-optic cables. Signals are transferred over two-strand optical fibers in accordance with 100BASE-FX at a transmission rate of 100 Mbits/s. The optical interfaces comply with the ISO/IEC 9314-3 (multimode) or ISO/IEC 9314-4 (single-mode) specifications. For applications external to the control cabinet, the outer sheath must satisfy the applicable requirements at the point of use (mechanical, chemical, thermal). The maximum segment length is 2 km for multimode and 14 km for single mode.

18.7.2 Plug Connectors

Plug connectors for M12 and RJ45 are available in PROFInet. They can be easily cut to length and terminated on site (Figure 18.18).

RJ45 type is used in control cabinet settings in PROFInet. It is compatible with an office connector. Plug connectors external to the control cabinet must notably accommodate industrial requirements. The RJ45 types in IP65 or IP67 or the M12 type is utilized in this case (Figure 18.19).

Fiber-optic connections are implemented according to ISO/IEC 11801, preferably with a duplex SC plug connector system. This system is described in IEC 60874-14. Devices are equipped with the socket, and the connecting cable is equipped with the plug. Alternatively, the fiber-optic plug connector BFOC/2,5 in accordance with IEC 60874-10 can be used.

FIGURE 18.18 Example of an RJ45 plug connector in IP20.

FIGURE 18.19 Example of an RJ45 plug connector in IP67.

FIGURE 18.20 Example of a hybrid plug connector with RJ45 in IP67.

The hybrid plug connector is used in situations where decentralized field modules are connected using a combined plug connector containing data and power supply (Figure 18.20). A plug connector that is completely protected against accidental contact enables the use of plug connectors that are identical at both ends because a pin–socket reversal is not required due to the integrated accidental contact protection.

This example involves an RJ45 in IP67 with a two-pair, shielded data cable for communication and a 4 copper conductor for voltage supply.

18.7.3 Switches as Network Components

PROFInet uses general switches as a network component. Switches are devices that are situated in the transmission path between the end devices and that regenerate received signals and forward them selectively. They are used to structure networks. ISO/IEC 15802-3 contains the basic specifications. Switches suitable for PROFInet are sized for Fast Ethernet (100 Mbits/s, IEEE 802.3u) and full-duplex transmission. During full-duplex operation, a switch receives and sends data simultaneously at the same port. No collisions occur when switches are utilized. As a result, bandwidth is not lost due to the Ethernet collision process. Network configuration is simplified significantly because route length checking is not required within a collision domain. To ensure compatibility with old plants or individual, older-end devices or hubs, 10BASE-TX (10 Mbits/s, carrier-sense multiple access with collision detection (CSMA/CD)) is supported.

Moreover, a PROFInet switch supports prioritized message frames in accordance with IEEE 802.1Q, standardized diagnostic paths, and autopolarity exchange, autonegotiation mode, and autocrossover function. Port mirroring for diagnostic purposes is optional.

As a general rule, office-type switches cannot be used, even if the functionality described above is satisfied. Special switches are applied for industrial use. For one thing, these are designed for harsh industrial use based on their mechanical (IP degree of protection, etc.) and electrical (24-V power supply, etc.) properties. For another, they must fulfill the EMC requirements for industrial machine applications to enable safe operation.

18.8 IT Integration

In addition to the automation functionalities described, the use of Ethernet as a communication medium in the PROFInet context enables IT functions to be integrated in PROFInet as well.

As is the case in the fieldbus world, Ethernet places additional network management requirements in connection with TCP/IP. In order to manage all technical aspects of integration of PROFInet devices in such networks, a concept for network management has been specified in PROFInet. The topics of network infrastructure, IP management, network diagnostics, and time-of-day synchronization are the primary elements of the concept. Network management simplifies Ethernet administration and management through the use of standard protocols from the IT world.

The use of Internet technologies in automation systems represents an additional aspect. PROFInet has specified a concept under the scope of Web integration that enables access to a PROFInet component. This access is achieved by using Web utilities based on standard Internet technologies, such as HTTP, XML, and Hypertext Markup Language (HTML).

18.8.1 Network Management

Network management comprises all functions for administering the network, such as configuration (assignment of IP addresses), fault monitoring (diagnostics), and performance optimization.

18.8.1.1 IP Management

The use of TCP/IP in the PROFInet context means that an IP address has to be assigned to the network stations (i.e., PROFInet devices):

- *Address assignment with manufacturer-specific configuration systems*: This alternative is required because a network management system is not always available. In PROFInet, the Discovery and Basic Configuration Protocol (DCP) is specified, enabling IP parameters to be assigned using manufacturing-specific configuration/programming tools or during plantwide engineering (e.g., in the PROFInet interconnection editor). The use of DCP is mandatory for PROFInet devices. In this way, uniform behavior of PROFInet devices is ensured.
- *Automatic address assignment with DHCP*: The Dynamic Host Configuration Protocol (DHCP) has been established for assigning and managing IP addresses in office networks with network management systems. PROFInet provides for the use of this standard and describes how DHCP can be applied in a useful way in the PROFInet environment. Implementation of DHCP in PROFInet devices is optional.

18.8.1.2 Diagnostics Management

The reliability of network operation has a very high priority in network management. The *Simple Network Management Protocol* (SNMP) has been established in existing networks as the *de facto* standard for maintaining and monitoring network components and their functions. In order for PROFInet devices to be monitored with established management systems, it is useful to implement SNMP. SNMP provides for both read access from (monitoring, diagnostics) and write access to (administration) a device.

To begin with, only read access of device parameters was specified in PROFInet. Like the IP management functions, SNMP is optional. When SNMP is implemented in components, only the usual standard information for SNMP is accessed (management information base 2 (MIB-2)).

A specific diagnostic for PROFInet components is possible by means of the mechanisms described in the PROFInet specification. In this context, SNMP will not open an additional diagnostic path. Rather, SNMP enables integration in network management systems that do not normally process PROFInet-specific information.

18.8.2 Web Utilities

In addition to the use of modern Ethernet-based technologies in PROFInet, it is also possible to access a PROFInet component using Web clients based on standard Internet technologies, such as HTTP, XML, HTML, or scripting.

Data are transmitted in standardized format (HTML, XML) and through standardized front ends (browsers such as Netscape, MS Internet Explorer, Opera, etc.). This enables integration of information from PROFInet components into modern, multi-media-supported information systems. The advantages of Web integration in the IT sphere — such as utilization of browsers as uniform user interfaces, access to information on any number of clients from any location, platform independence of clients, and reduced effort for installing and servicing software on the client side — are thus also made available for PROFInet components.

18.8.2.1 Functional Properties

Web integration of PROFInet has been designed with emphasis on commissioning and diagnostics. Web-based concepts can be used very effectively within these application areas.

- No special tools are required to access components. Established standard tools can be used.
- Global accessibility means that user support can be easily obtained when commissioning is performed by the component manufacturer.
- Autodescription of components enables access with a standard tool without configuration information.

Possible scenarios for PROFInet Web integration in the areas of commissioning and maintenance include testing and commissioning, overview of device database, device diagnostics, and plant and device documentation.

The information provided should be represented in both a human-readable format (e.g., via a browser) and a machine-readable format (e.g., via an XML file). Both variants can be provided in integral form using the PROFInet Web integration. For certain information, the PROFInet Web integration also provides standardized XML schemes.

18.8.2.2 Technical Properties

The basic component of Web integration is the Web server. The Web server forms the interface between the PROFInet object model and the basic technologies for Web integration.

The PROFInet Web integration can be scaled according to the performance and properties of the Web server. That means that, in addition to a PROFInet device with an MS Internet Information Server or Apache Web Server, small PROFInet devices equipped with only an embedded Web server can also participate in the Web integration with equal rights.

The Web integration for PROFInet has been created in such a way that it can be made available optionally for each device. Certain functions are optional and can be used for the device depending on its load capacity. This enables implementation of scalable solutions that are adapted to the respective use case to the maximum extent possible. The PROFInet-specific elements can be integrated seamlessly in an existing component Web implementation.

Based on uniform interfaces and access mechanisms, the creator of a technological component can provide his technology-related data via the Web. Through the name space specified in the PROFInet Web integration and the addressing concept, elements of the PROFInet component object model (technological variables) can be referenced by the Web server. In this way, dynamic Web sites configured with current data from the PROFInet component can be created.

18.8.2.3 Scope

The basic architecture model for the Web integration concept is shown in Figure 18.21. Web integration is optional for PROFInet.

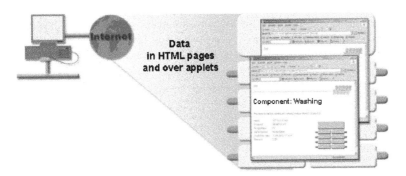

Data
in HTML pages
and over applets

Component: Washing

FIGURE 18.21 Web integration enables Web access to PROFInet components.

Regarding the system architecture of an automation system with PROFInet, all architectural forms are supported by Web integration. In particular, the use of proxies for interfacing to any fieldbus is supported. The specification contains appropriate models that describe the relationships among the PROFInet components, the existing Web components, and the PROFInet Web integration elements.

18.8.2.4 Security

The PROFInet Web integration specification has been created in such a way that access to the PROFInet devices is the same whether it occurs from the Internet or an intranet. As a result, all advantages of Web integration can be reaped even if the device itself is not connected to the Internet. In the case of local access, there is little risk for unauthorized access (comparable to today's HMI systems).

For networking within a larger factory or over the Internet, the PROFInet Web integration is based on a graded security concept. PROFInet Web integration recommends a security concept that has been optimized for the specific use case and includes one or more upstream security zones. No structural restrictions are imposed in the Web integration concept because the security measures are always arranged around PROFInet devices. As a result, PROFInet devices are not burdened, and the security concept can be optimized to the changing security requirements for a continuous automation solution.

The best-practice recommendations for PROFInet Web integration contain scenarios and examples of how requirement-dependent security measures can be implemented around PROFInet devices.

In this way, for example, security mechanisms can be used in the transport protocols (TCP/IP and HTTP). In addition, encoding, authentication, and access management can be scaled in the utilized Web servers. Advanced security elements such as application gateways can be added for Web services, if required.

18.9 OPC

The PROFInet component model and OPC have the same technological basis: DCOM. This yields user-friendly options for data interchange between different plant units. OPC is a widely used interface for data exchange between applications in automation systems. OPC enables flexible selection of stations of different manufacturers and data exchange between the stations without programming. OPC is not object oriented like PROFInet, but rather is tag oriented. That is, the automation objects do not exist as COM objects but as names (tags).

18.9.1 OPC DA (Data Access)

OPC DA (data access) is an industry standard that defines a uniform user interface to access process data. As a result of this standard, the following are harmonized: access to data of process and control devices, the locating of OPC servers, and simple browsing in the name spaces of the OPC servers.

18.9.2 OPC DX (Data Exchange)

OPC DX (data exchange) defines a communication standard for the higher-level exchange of non-time-critical user data on the system level between controllers of different manufacturers and types (e.g., between PROFInet and EtherNet/IP) (Figure 18.22). However, OPC DX does not permit any direct access to the field level of another system. OPC DX represents an expansion of the OPC DA specification and defines interfaces for interoperable data exchange and server-to-server communication in Ethernet networks.

OPC DX is commonly used by the following:

- User and system integrators who integrate devices, controllers, and software of different manufacturers and want to enable access to jointly used data in multivendor systems
- Manufacturers that want to offer products that are based on an open industry standard for interoperability and data exchange

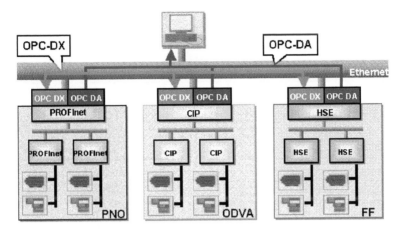

FIGURE 18.22   Cross-system data exchange with OPC DA and OPC DX.

18.9.3   OPC DX and PROFInet

OPC DX was developed with the objective of having a minimum degree of interoperability between the different fieldbus systems and Ethernet-based communication protocols. In order to maintain an open connection to other systems, OPC DX was integrated in PROFInet. Integration is accomplished as follows:

- Each PROFInet node can be referenced as an OPC server because the basic capacity already exists in the form of the PROFInet runtime implementation.
- Each OPC server can be operated as a PROFInet node by means of a standard adapter. This is accomplished by the OPC objectizer, an SW component that implements a PROFInet device on the basis of an OPC server in a PC. This SW component need only be implemented once and can then be used for all OPC servers.

The functionality and performance of PROFInet is significantly greater than OPC. Moreover, PROFInet offers the required real-time capability for automation solutions. On the other hand, OPC exhibits a higher degree of interoperability.

18.10   Integration of Fieldbus Systems

PROFInet offers a model for integrating the existing PROFIBUS and other fieldbus systems. This enables any combination of fieldbus and Ethernet-based subsystems to be established. In this way, a continuous technology transfer from fieldbus-based systems to PROFInet is possible.

18.10.1   Migration Strategies

In view of the large number of existing PROFIBUS systems, it is essential for purposes of protection of investment that these systems be able to be easily integrated into PROFInet (migrated) without any changes (Figure 18.23). The following cases are distinguished:

- The *plant operator* would like to be able to easily integrate his existing installations into a new PROFInet automation concept that is to be installed.
- The *plant construction engineer* would like to be able to use his proven and documented range of devices for PROFInet automation projects without any changes.
- The *device manufacturer* or OEM would like to be able to integrate his existing field devices in PROFInet systems without expending any effort on changes.

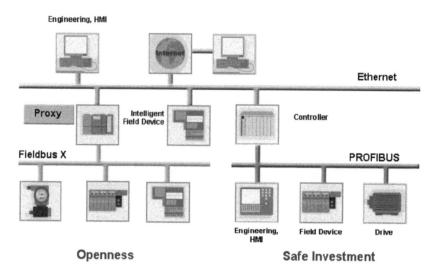

FIGURE 18.23 PROFInet offers openness and protection of investment to device manufacturers and end customers.

There are two ways available in PROFInet for connection of fieldbus systems:

- Integration of fieldbus devices by means of proxies
- Integration of fieldbus applications

18.10.2 Integration by Means of Proxies

The proxy concept in PROFInet enables simple, highly transparent integration of existing fieldbus systems (Figure 18.24). The proxy is the representative on the Ethernet for one or more fieldbus devices (e.g., on the PROFIBUS). This representative ensures transparent communication (no protocol tunneling) between the networks. For example, the representative forwards cyclical data to the fieldbus devices transparently. In the case of PROFIBUS DP, the proxy is, on the one hand, the PROFIBUS master that coordinates data exchange among the PROFIBUS stations and, on the other hand, an Ethernet station with PROFInet communication. A proxy can be implemented, for example, as a PLC or PC-based control or purely as a gateway.

The DP slaves on the PROFIBUS are handled as IO devices in the PROFInet IO context. In the component view, the intelligent DP slaves are used as stand-alone PROFInet components. Within the PROFInet interconnection editor, such PROFIBUS components cannot be distinguished from the com-

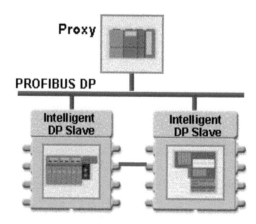

FIGURE 18.24 Principles of integration of individual fieldbus devices using a proxy: PROFIBUS example.

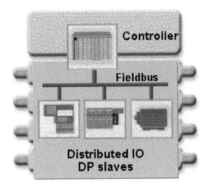

FIGURE 18.25   Principles of integration of fieldbus applications: PROFIBUS example.

ponents on the Ethernet. The use of proxies enables transparent communication between devices on different bus systems.

18.10.3   Integration of Fieldbus Applications

An entire fieldbus application can be copied as a PROFInet component within the framework of the component model. This is always important if an existing operating plant is to be expanded using PROFInet. Which fieldbus was used to automate the plant unit plays no role in this case (Figure 18.25).

To communicate with the existing plant using PROFInet, the fieldbus master in the PROFInet component must be PROFInet capable. Consequently, the existing fieldbus mechanisms (e.g., PROFIBUS DP) are used within the component and the PROFInet mechanisms are used outside the component.

This migration option ensures that the investment of the user (plant operators and plant construction engineers) in existing plants and cabling is protected. In addition, existing know-how in the user programs is protected. This enables a seamless transition to new plant units with PROFInet.

18.10.4   PROFInet and Other Fieldbus Systems

The proxy concept can be used to integrate other fieldbus systems besides PROFIBUS in PROFInet (e.g., Foundation Fieldbus, DeviceNet, CC-Link, etc.). A bus-specific image of the component interfaces must be defined and stored in the proxy for all possible data transmissions on each bus. This enables any fieldbus to be integrated in PROFInet with a manageable amount of effort.

18.11   PNO Offer

Optimal support by the PROFIBUS User Organization (PNO) is important for rapid dissemination of PROFInet in the market (Figure 18.26). In order to guarantee this, a powerful offer of services and products has been established.

18.11.1   Technology Development

18.11.1.1   PROFInet IO

A specification is available for PROFInet IO, which provides a detailed description of the device model and the behavior of a field device in the form of protocols and communication sequences (so-called state machines). This type of description has already been proven with PROFIBUS DP. The level of detail in the PROFInet IO specification permits software creation of a standard stack from different stack suppliers.

It must therefore be anticipated that different implementations by a few firms will be offered. For example, Siemens offers implementation in the form of a development package.

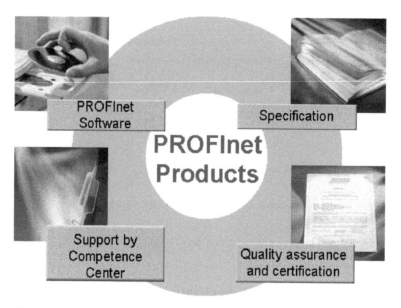

FIGURE 18.26 PROFIBUS User Organization (PNO) offer.

18.11.1.2 Component Model

Like PROFInet IO, the PROFInet component technology is available as a detailed specification. The specification includes all aspects: communication, device model, engineering, network management, Web integration, and fieldbus connection. In addition to the specification, the PROFIBUS User Organization is offering PROFInet software (in the form of source code) for the component technology. The PROFInet software includes the entire runtime communication. The combination of a specification and operating system-independent software (as source code) has created the opportunity for easy, timesaving integration of PROFInet into a wide range of device operating system environments.

The PROFInet runtime software is structured in such a way that it supports a simple integration of existing application software in the runtime object model. Sample portings for Win32, Linux, VxWorks, and WinCE are already available for PROFInet. The PROFInet runtime software has a modular design and consists of various layers that must be adapted to each system environment. These adaptations are confined to the porting interfaces for the different functional parts of the environment, the operating system (e.g., WinCE), and the device application (e.g., PLC). Instructions for porting are available, enabling the device developer to easily understand the individual porting steps.

18.11.2 Quality Measures

From the start, PNO has developed PROFInet with the assurance that the entire life cycle from the PROFInet specification to plant engineering is supported by measures that guarantee a high level of quality in each phase.

18.11.2.1 QA for the Specification and Implementation Processes

The PROFInet specification and software are created in a working group represented by multiple companies (PROFInet Core Team) whose mission is to ensure that the entire development process ranging from the initial formulation of requirements to the release of the PROFInet software is conducted under the auspices of quality management (QM). Quality measures are governed by a quality assurance (QA) manual adapted to the boundary conditions of the multicompany development team. This ensures that the source code corresponds to the quality management rules in effect at that time. The QA manual describes the process model to be applied. It defines the terms, methods, and tools that must be applied in the QA measures. It also specifies the responsibilities in the overall QA process. Defect management represents a major component. Defect management includes a unique error classification and understandable error message communication.

18.11.2.2 Testing and Certification

To ensure the interaction of all PROFInet devices and a high level of product quality from the start, a certification system has been established. Proven models for PROFIBUS products in accordance with the certification system have been established. Certification tests by accredited test laboratories authorized by PNO form the core of the process. The test for obtaining a certificate by these competent test laboratories ensures that the products offered conform to specifications and are free of errors.

18.11.2.3 Defect Database

In order for defects and requests from end customers and device manufacturers to be systematically addressed in the runtime software, a defect database has been set up by PNO. A defect database containing a record of all defects and their status is available. The entries in the database comply with the QA process rules.

18.11.3 Technical Support

An important factor for the success of PROFInet is that a sufficient number of PROFInet products from different manufacturers are available on the market within a short time.

18.11.3.1 Competence Center

The PROFInet Competence Center has been established for support of the product development process. This ensures that porting to the different operation systems and adaptation to product-specific boundary conditions occur in an optimal manner, particularly in the initial phase when companies lack experience. The Competence Center builds up know-how in interested companies so that development departments can develop additional products skillfully without additional support. The services of the PROFInet Competence Center also include a telephone hotline and customized workshops.

18.11.3.2 Tools

A tool for device manufacturers is required for creation of a component description of Ethernet devices in the form of an XML file. The PROFIBUS User Organization offers the PROFInet component editor — similar to the GSD editor in PROFIBUS DP — for download on it Web site (Figure 18.27).

In order to prepare newly developed products for certification, PNO offers a PROFInet testing tool for download on its Web site (Figure 18.28). The PROFInet testing tool enables the device manufacturer to perform static tests prior to certification.

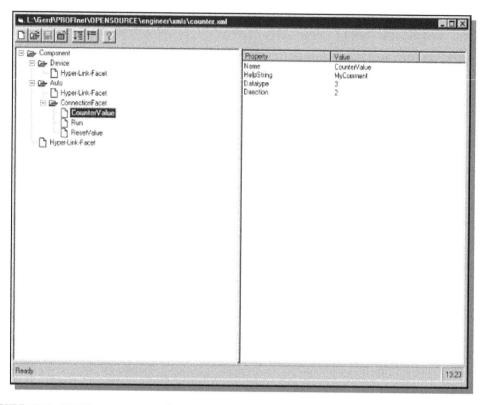

FIGURE 18.27　PROFInet component editor.

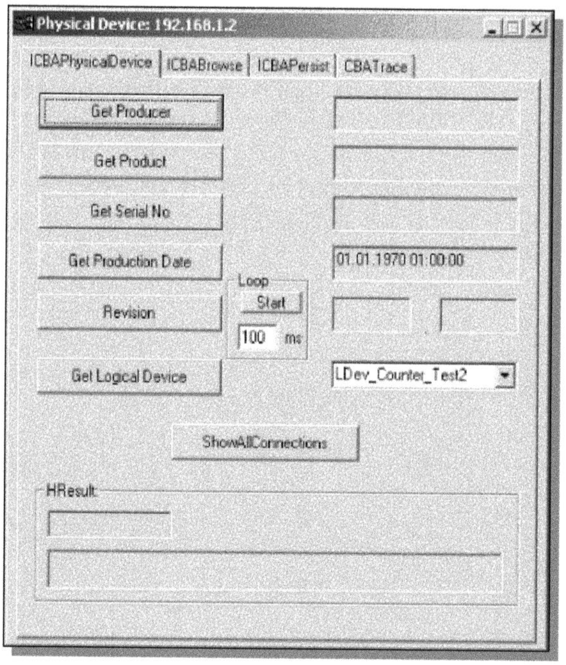

FIGURE 18.28　PROFInet test tool.

Section 4.4

Wireless Technology

19

Wireless Local and Wireless Personal Area Network Technologies for Industrial Deployment

Kirsten Matheus
Carmeq GmbH

19.1 Introduction

The convenience of true mobility offered by wireless connectivity is the main factor behind widespread acceptance of wireless technologies. The global system for mobile communication (GSM), a second-generation cellular system designed mainly for mobile telephony, currently has more than one billion users worldwide. Systems like GSM or the third-generation universal mobile telecommunication system (UMTS) nevertheless require extensive infrastructure. The commercial and industrial deployment of systems that function on a smaller scale and do not require costly frequency licensing or infrastructure has become more appealing; such systems include wireless personal area networks (WPANs) and wireless local area networks (WLANs).

As a consequence, the Bluetooth and IEEE 802.11 technologies and the newly emerging ZigBee have received a significant amount of public and scientific attention. Bluetooth, like ZigBee, is a typical WPAN representative that is inexpensive, consumes little power, is small in size, and supports voice and data services. The different IEEE 802.11 variants are WLAN representatives that provide comparably high user data rates at the cost of a higher battery power consumption. ZigBee is limited to small data rates, but at the same time consumes very little power. With their original purposes fulfilled, new areas of deployment are being developed for these technologies. Bluetooth plays a larger role in markets like warehousing, retailing, and industrial applications [5]. IEEE 802.11 is considered for seamless coverage of complete cities [4, 59].

Depending on the exact application, users of these wireless technologies have certain expectations concerning the quality of the systems. The application requirements have to be considered carefully in order to be able to choose the most suitable technology. The main criteria are generally throughput, delay, and reliability. In addition, cost, power consumption, security, and, last but not least, availability can be important issues. Note, though, that owing to the possibility of interference, adverse radio conditions, or range limits, hard quality-of-service (QoS) guarantees for throughput cannot be provided by wireless systems. In industrial environments, the radio conditions can be especially difficult because metal walls have a significant impact on the transmission. Metal shields radio transmissions while causing respectively more reflections. Systems requiring a certain amount of data rate within a strict time window, for example, because they are security dependent, should not be wireless.

In addition to the parameters discussed above — factors like unit density, traffic demand, mobility, environmental changes during deployment, interference, frequency range, etc., determine how well a technology satisfies the requirements. Thus, both the individual link performance and the overall network capacity should be optimized.

This chapter first describes in Section 19.2 the basic differences between WLANs, WPANs, cellular networks, and *ad hoc* networks. In Sections 19.3, 19.4, and 19.5 the technologies Bluetooth, IEEE 802.11, and ZigBee are described in more detail. Each of those sections provides the technical background on the technology under consideration, as well as investigations on the performance of the systems and their suitability for industrial applications/factory floor environments. Section 19.6 shows how Bluetooth and IEEE 802.11b/g, which are placed in the same frequency band and are possibly used at the same time in the same location — coexist. Section 19.7 provides a summary and the conclusion.

19.2   WLAN, WPAN, Cellular Networks, and *Ad Hoc* Networks

The expressions *wireless local area network* (WLAN), *wireless personal area network* (WPAN), *cellular networks*, and *ad hoc networks* are commonly used, often though without consistency or precision. In the following, a clarification of the terminology is given:

WLAN: A wireless LAN has the same functionality as a wired LAN with the difference that the wires are replaced by air links. This means that within a *restricted area* (home, office, hot spot), *intercommunication* between all devices connected to the network is possible and the focus is on *data communication* as well as *high data rates*. The definition of WLAN says nothing on how the network is organized. Often an infrastructure of mounted access points (APs) enables wireless access to the wired LAN behind the APs, thus representing a cellular network structure. Nevertheless, a wireless LAN can also function on an ad hoc basis.

WPAN: In a wireless PAN, all devices are *interconnectable*. The difference is that all units are somehow *associated* with someone or something alike (either because they are yours or because they are shared or public devices you want to use) and are very *nearby*. A PAN can consist of a variety of devices and can even include different technologies. The applications are therefore not limited to *data transmission*, but *voice communication* can be used in a PAN as well.

While you move *within* the WLAN, you can generally move *with* your WPAN. This means that several independent WPANs can coexist in the same area, each being self-sufficient without any infrastructure. Thus, they generally function on an *ad hoc* basis.

The difference between cellular and ad hoc networks is visualized in Figure 19.1. As can be seen, there are several steps that lead to ad hoc networking: a pure ad hoc network employs neither any infrastructure nor a specific unit (like access point or base station) for the organization of coverage, synchronization, and services. Nevertheless, a network can be ad hoc, when it supports single hops only.

It can be seen that WLAN technologies like IEEE 802.11 in the infrastructure mode or HIPERLAN/2 are in the same classification as typical cellular systems like GSM or UMTS/wireless code-division multiple

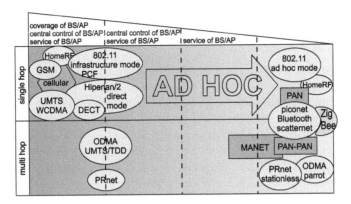

FIGURE 19.1 Classification of wireless technologies between cellular and ad hoc; MANET stands for mobile (or multihop [38]) ad hoc network (designed for Internet applications [31]), PRnet for public radio network [46], and ODMA for opportunity-driven multiple access [1, 54]. A Bluetooth scatternet means that several Bluetooth piconets are interlinked. BS = base station.

access (WCDMA); all are based on infrastructure and use a specific unit for central control. It would thus be correct to call these WLAN systems cellular systems. Despite this, there are distinct differences with regard to coverage. The wide (instead of local) area coverage of cellular systems like GSM has caused cellular systems to be associated with complete coverage and access everywhere, even though this is not correct. The description *wireless wide area network* (WWAN) gives a better idea of the difference from WLAN systems. The fact that the existing WWAN technologies like GSM and UMTS focus on voice communication, while WLAN technologies focus on data transmission, is not such an important difference. More important is that WWAN technologies are designed to support user mobility (and roaming) up to very high velocities, while WLAN systems support stationary or portable access. Because of the licensing regulations and costs for extensive infrastructure, WWAN systems like GSM and UMTS are not of interest for industrial applications. Thus, they are not discussed any further. From the radio network point of view, the most important technical distinction with respect to the discussed terminology — with far-reaching consequences to system design, network optimization, etc. [52] — does not have to be made between WPAN, WLAN, or WWAN technologies. It has to be made between systems organized in cells and ad hoc systems:

- As cellular networks are systematically laid out, the minimum distance to the next co-channel interferer (i.e., the next uncoordinated user transmitting at the same time on the same frequency) is generally controllable, known, and fixed. In contrast, in ad hoc networks, the next co-channel interferer can be very close from one moment to the next, without any possibility of influencing the situation (Figure 19.2).
- The centralized control in cellular networks allows for effective and fair distribution of resources, because of the accurate knowledge of the overall traffic demand and the possibility of a more global resource management. For ad hoc networks or several coexisting WPANs, knowledge of overall traffic demand is generally not available and the systems compete for the resources.

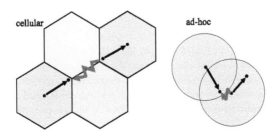

FIGURE 19.2 Next closest co-channel interferer in cellular and ad hoc networks.

19.3   Bluetooth Technology

19.3.1   Technical Background

Bluetooth (BT) is first of all a cable replacement technology aiming at effortless wireless connectivity in an ad hoc fashion. It supports voice as well as data transmission [7, 8, 10, 11, 26]. Its desired key associations are easy to use, low in power consumption, and low in cost, with the aim that the Bluetooth functionality is integrated in as many devices as possible.

To meet these goals, the BT special interest group (SIG) placed the technology in the unlicensed ISM (industrial, scientific, and medical) band at 2.4 GHz. This allows close to worldwide deployment without the need to pay fees for frequency licensing. Nevertheless, it requires complying with the respective sharing rules.* As a consequence, Bluetooth performs a rather fast frequency hopping (FH) over 79 carriers of 1-MHz bandwidth each, such that every Bluetooth packet is transmitted on a newly chosen frequency (which results in a nominal hop rate of 1600 hops/s). To further reduce cost and support the distribution of the Bluetooth technology, the Bluetooth specification is an open standard that can be used without even needing to pay for the use of its key patents, on the term that the strict qualification procedure is passed.** The latter is to ensure the acceptance of the technology. For the same purpose, the specification contains application profiles. The profiles describe in detail the implementation of the foreseen applications, thus enabling units of different manufactures to communicate. The most important characteristics on the physical layer are as follows. The data are Gaussian frequency shift keying (GFSK) modulated at 1 Mbps and organized in packets consisting of access code, header, and payload. The employment of forward error correction (FEC) for the payload is optional. Inter-packet interleaving is not performed. This allows for lower chip prices, because memory can be saved.

Bluetooth uses a master–slave concept in which the unit that initiates a connection is temporarily assigned master status (for as long as the connection is up). The master organizes the traffic of up to seven other active units, called slaves, of this piconet. From the master's device address the identity of each piconet, and with it the frequency-hopping sequence, can be derived. The header of a packet contains the actual addressee, the length of the packet, and other control information. Note that within one piconet, the slave can only communicate with the master (and not directly with the other slaves) and this — in case of data connections — only after having been asked (i.e., polled).***

The channel is organized in a time-division multiple access (TDMA)/time-division duplex (TDD) [24] scheme (Figure 19.3). It is partitioned into 625-μs time slots. Within this slot grid the master can start transmission only in the odd-numbered slots, while the slaves can respond in only even-numbered ones. When a unit is not already in one of the specific power save modes (sniff, hold, park), the slotting is power consumption friendly, because every unit has to listen only during the first 10 μs of its receive slot whether there is a packet arriving (and if not, can close down until the next receive slot****). This means it needs to listen into the channel only 10 μs/(2 · 625 μs) = 0.8% of the time, during an active connection in which no packets are sent. Yet another facet to the long battery life is the low basic transmit power of 0 dBm (resulting in a nominal range of about 10 m). Bluetooth can also be used with up to 20-dBm transmit power. This results in a larger range but requires the implementation of power control to fulfill the Federal Communications Commission (FCC) sharing rules.

---

*For the U.S. [20, Part 15], for Europe [16], for Japan [48].

**The Bluetooth specification has also been adopted by the IEEE. It can be found under IEEE 802.15.1.

***Every BT unit can simultaneously be a member of up to four piconets (though it can be master in only one of them). A formation in which several piconets are interlinked in that manner is called scatternet. Aspects like routing, which are of interest in this constellation, will not be covered in this chapter. The chapter will thus focus on the properties of a single or multiple independent piconets.

****This is quite different from channel access schemes like CSMA, as used in IEEE 802.11 (see Section 19.4). Unless asleep, IEEE 802.11 always has to listen into the channel.

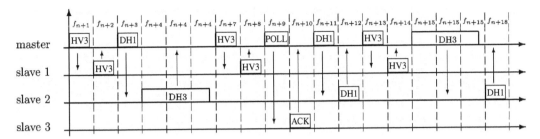

FIGURE 19.3 Example slot occupation within one piconet consisting of a master and three slaves; to slave 1 there is an SCO link, to slave 2 (best-effort) traffic is transmitted in both directions, and slave 3 currently has nothing to send (but has to respond to the POLL packet with an acknowledgment (ACK)). During the transmission of multislot packets, the frequency is not changed.

Bluetooth provides two in principle different types of connections: asynchronous connectionless (ACL) links foreseen for data transmission and synchronous connection-oriented (SCO) links foreseen for speech transmission.

For ACL links there are six packet types defined. The packets occupy either one, three, or five (625-μs) time slots, and their payloads are either uncoded (called DH1, DH3, or DH5, respectively) or protected with a 2/3 rate FEC using a (15, 10) shortened Hamming block code without any interleaving (called DM1, DM3, DM5, respectively). An automatic repeat request (ARQ) scheme initiates the retransmission of a packet in case the evaluation of the cyclic redundancy check (CRC) included in each ACL payload shows inconsistencies. This secures error-free reception of the transmitted information. Table 19.1 gives an overview of the throughput values achievable with ACL connections. The maximum (unidirectional) Bluetooth throughput is 723 kbps.

As speech transmission is delay sensitive, the original SCO links support three different packet types that are transmitted at fixed intervals. These types were designed to transport continuous-variable slope delta (CVSD) encoded speech at 64 kbps. The packet types occupy always just one (625-μs) time slot, but they are differentiated by their payload FEC. The packet payloads are either unprotected (called HV3) or 2/3 rate FEC encoded (HV2) or protected with a 1/3 rate repetition code (HV1). For an HV3 connection, a packet is transmitted every sixth slot (Figure 19.3); for HV2, every fourth slot; and for HV1, every second slot (meaning that with one HV1 connection no other traffic can be transmitted in the piconet). Up to the Bluetooth Specification 1.1 [9] there was no ARQ scheme for SCO links. In case of an erroneous reception of the packet overhead, the SCO packet was replaced by an erasure pattern. In case noncorrectable bit errors occurred in the payload only, these errors were forwarded to the speech decoder. The latest specification, Bluetooth Specification 1.2 [10], includes an enhanced SCO link. This link allows very flexible deployment of the SCO link, providing for a reserved bandwidth for several transmission rates and a limited number of retransmissions.

TABLE 19.1 Throughput Values for ACL Connections

Name	No. of Slots	FEC?	Max. No. of User Bytes	Unidirectional Throughput		Bidirectional Throughput	
				Forward	Reverse	Forward	Reverse
DH1	1	No	27	172.8k	172.8k	172.8k	172.8k
DH3	3	No	183	585.6k	86.4k	390.4k	390.4k
DH5	5	No	339	723.2k	57.6k	433.9k	433.9k
DM1	1	2/3	17	108.8k	108.8k	108.8k	108.8k
DM3	3	2/3	121	387.2k	54.4k	258.1k	258.1k
DM5	5	2/3	224	477.8k	36.3k	286.7k	286.7k

Note: The reverse link in the unidirectional case transmits DH1 or DM1 packets, depending on whether the forward link uses a DH or DM packet type.

To further improve coexistence with other systems in the ISM band, Bluetooth version 1.2 includes the possibility to perform adaptive frequency hopping (AFH), i.e., to exclude carrier frequencies used by other systems from the hop sequence. With AFH the nominal hop rate will be halved, because the specification has been changed such that the slave responds on the same frequency on which it received the packet from the master [10].

Security is supported in Bluetooth by the specification of authentication and encryption.

For the future, a high-rate mode is envisioned that allows direct slave-to-slave communication at an about 10-fold transmission rate. The transmission takes place on 4-MHz channels that are chosen at specifically good locations within the 79-MHz bandwidth.

19.3.2 Performance

On the factory floor Bluetooth can be used as a wireless add-on to wired systems or as a replacement of existing cabling. It can cover machine-to-machine communication, wireless/remote monitoring, or tracking and some type of positioning of moving entities [5, 23]. Considering the comparably short range of Bluetooth and the likely association with a specific unit (represented by a machine, person, or task), it is possible that several independently active Bluetooth piconets coexist and overlap in space. The use of frequency hopping helps to mitigate the effects of interference among these piconets. When assuming more or less time-synchronized piconets, a worst-case approximation of the loss rate can be made with Equation 19.1. It calculates the probability $P(x, n)$ that of x other piconets, n hop onto the same frequency as the considered piconet:

$$P(x,n) = \binom{x}{n}\left(\frac{1}{79}\right)^{n}\left(\frac{78}{79}\right)^{x-n} \tag{19.1}$$

The probability that at least one of the other x piconets transmits on the same frequency is then $P(x) = 1 - P(x, 0)$. The smaller the number of interfering piconets, the better the approximation offered by this approach, because for larger numbers, the distances to some of the interferers are likely to be too large to be harmful.

In [52, 61, 62] a more sophisticated approach has been chosen, and Bluetooth–Bluetooth coexistence results have been obtained with the help of detailed radio network simulations that include traffic, distribution, and fading models, as well as adjacent channel effects. All results have been obtained for an office of 10×20 m², assuming an average master–slave distance of 2 m. Naturally a factory floor is likely to be significantly larger than 10×20 m². Nevertheless, the increased delay spread on the factory floor does not really effect Bluetooth due to its small range (which is different for WLAN technologies; see Section 19.4.2). On the factory floor it is thus possible to place the Bluetooth units with the same density as in the investigated office scenario without loss in performance. Because a factory floor is larger than the investigated office, the overall number of piconets that can be used simultaneously on the factory floor is larger too. Additionally, location and traffic of the factory floor units are likely to be more predictable. Directive antennas also help to improve the performance. The results of the aforementioned publications thus give a good idea of what performance is achievable:

- A 10×20 m² room supports 30 HV3 simultaneous speech connections with an average packet loss rate of 1% (a limit that still allows for acceptable quality).
- HV3 packet types are preferable to HV2 and HV1. The subjective quality will not increase with additional payload coding. Using a coded HV packet just increases (unnecessarily) the interference in the network and the power consumption.
- One hundred simultaneous World Wide Web (WWW) sessions (bursty traffic with an *average* data rate of 33.2 kbps each) in the 10×20 m² size room result in only a 5% degradation of the aggregate throughput.
- The maximum aggregate throughput in the room is 18 Mbps (at 50 fully loaded piconets). These piconets then transmit at a unidirectional data rate of 360 kbps each.

- Long and uncoded packets are preferable to shorter and coded ones. It takes 60 interfering piconets using the same packet type, 10 interfering HV1 connections (worst case), or a link distance of 27 m (which is far beyond the envisioned range of 10 m) before another packet type yields a larger throughput than DH5 [52]. It is advisable not to use the optional FEC (DM packet types). As the coding is not appropriate to handle the almost binary character of the Bluetooth (*ad hoc*) transmission channel,* the additional power that would be needed for the coding can be saved.

Bluetooth is inexpensive and consumes significantly less power than IEEE 802.11 systems. The ACL link is reliable with best-effort traffic (with a maximum throughput of 723 kbps). The SCO link has reserved bandwidth, though the packets might contain residual bit errors (even when using the enhanced SCO link). In principle, Bluetooth is very robust against other Bluetooth interference and good performance can be achieved even in very dense environments. Note that customized implementations, which cannot be based on existing profiles, might be difficult to realize, as the regulations do not allow the implementation of proprietary solutions. The specification of new profiles, though, can be quite time-consuming.

19.4 IEEE 802.11

19.4.1 Technical Background

IEEE 802.11 includes a number of specifications that define the lower layers (mainly the physical (PHY) and medium access control (MAC) layers) for WLANs [32–34, 37, 53]. Being part of the IEEE 802 group means that an interface can be used (IEEE 802.2) to connect to the higher layers, which are then not aware of the — with IEEE 802.11 wireless — network that is actually transporting the data. The key intentions of IEEE 802.11 are thus to provide a high-throughput and continuous-network connection like that available in wired LANs.

To encourage the wide employment of the technology, the use of IEEE 802.11 does not incur frequency licensing fees. IEEE 802.11 either uses infrared (IR) or transmits in the unlicensed ISM band at 2.4 GHz (like Bluetooth) or in 5-GHz bands that are license-exempt in Europe and unlicensed in the U.S. (UNII bands). In contrast to Bluetooth, the companies holding key patents within IEEE 802.11 can charge developers of IEEE 802.11 products for using the patents "on reasonable terms" [36].

In principle, it is possible to have an IEEE 802.11 WLAN consisting of mobile stations (MSs) only. It is more likely, though, that IEEE 802.11 is used as a wireless access technology to a wired LAN to which the connection is made by IEEE 802.11 access points (APs). Should the access point only employ the distributed coordination function,** the MAC layer supports collision avoidance by employing carrier-sense multiple access (CSMA). This means that before transmitting a packet, the respective unit has to listen for the availability of the channel.*** If the channel is sensed free after having been busy, the unit waits a certain period (called DIFS) and then enters a random backoff period**** of

---

*The reasons are manifold. Without interference, the channel varies already due to hopping over 79 relatively narrowband channels. Additionally, with the wavelength used in Bluetooth, even small changes in position can cause large changes in the received signal strength. When there is interference, the effect becomes more pronounced. The existence or nonexistence of a close co-channel interferer can make the channel change from very good to very bad within the fraction of a moment (Figure 19.2).

**Which is likely and assumed in the following of this chapter. In theory, the standard also provides the use of a centralized point coordination function.

***The implementor can choose whether the units react (1) on just other IEEE 802.11 traffic, (2) on just other IEEE 802.11 traffic above a certain receive signal strength, or (3) on any signal above a certain receive signal strength [34, Section 18.4.8.4].

****The random backoff period is entered only when the channel was busy before. Otherwise, the unit will transmit at once after DIFS.

$$\underbrace{random\ (0\ldots\min(2^{n_{PHY}+n_r}-1,1023))}_{CW}\cdot t_{slot} \hspace{2cm} (19.2)$$

with n_{PHY} a parameter depending on the type of physical layer chosen, n_r the index of the retransmission of the packet, t_{slot} the slot duration, and CW the contention window (with $CW_{min} = 2^{n_{PHY}} - 1$). If the channel is available after this period, the unit transmits its packet (consisting of a PHY header, MAC header, and payload). Upon correct reception, the addressee responds with an ACK packet a short period (called SIFS) later (Figure 19.4). The realized ARQ mechanism ensures reliable data.

Obviously, the IEEE 802.11 WLAN MAC concept was designed for best-effort data traffic. Services for which strict delay requirements exist — like speech — are not supported well by the current IEEE 802.11 specifications. To be able to provide QoS in the future, there is an ongoing activity within the IEEE that extends the MAC protocol with the necessary parameters (see Table 19.3). At the moment, QoS is difficult to provide, especially when multiple units coexist in the network. The IEEE 802.11 MAC concept also includes a mechanism to solve the hidden-terminal problem. Whether this ready-to-send/clear-to-send (RTS/CTS) packet exchange saves more bandwidth (due to avoided retransmissions) than it needs depends on the terminal density and payload packet length [6]. As the RTS/CTS mechanism is optional and consumes additional bandwidth, it will be assumed in the following that the RTS/CTS mechanism is not used.

IEEE 802.11 has a significantly larger power consumption than Bluetooth. Note that this is due not only to the higher transmit power (20 dBm in Europe, 30 dBm in the U.S.) but also to the CSMA concept. IEEE 802.11 units not specifically in sleep status have to listen to the channel *all* the time (unlike Bluetooth, which listens only at the beginning of the receive slot). Naturally, the higher transmit power allows for a larger range of about 50 m (with 20 dBm).

There are six different options for the physical layer implementation of IEEE 802.11:

IR: The infrared mode transmits near-visible light at 850- to 950-nm wavelength. The data are pulse position modulated at 1 or 2 Mbps. In principle, the signal needs line of sight (LOS) and cannot go through walls. This and the nonvisibility of IEEE 802.11 IR products are the reasons for not covering the IR mode further in this chapter.

FHSS: The frequency-hopping spread-spectrum mode is placed (like Bluetooth) in the 2.4-GHz ISM band. The data are GFSK modulated using two levels for the 1-Mbps and four for the 2-Mbps modulation rates. The FHSS mode divides the 79 hop frequencies into three distinct sets with 26 different sequences each. The hopping rate can be as slow as 2.5 hops/s. Despite its comparably good interference robustness [56], the popularity of the FHSS mode is limited due to its comparably low transmission rates. Note, though, that its principles have been incorporated in the HomeRF standard [11, 29].

DSSS: The direct-sequence spread-spectrum mode is also used in the 2.4-GHz ISM band. The nominal bandwidth of the main lobe is 22 MHz. The transmit power reduction in the first and residual side lobes is supposed to be 30 and 50 dB, respectively (see Figure 19.5 for a measured spectrum).

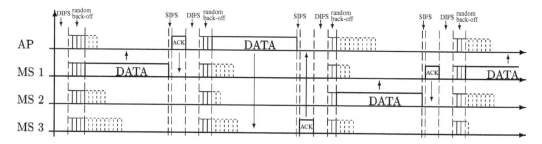

FIGURE 19.4 Principle time behavior of IEEE 802.11 under the distributed coordination function; note that the random backoff timer from nontransmitting units continues after the next DIFS with the remaining number of slots.

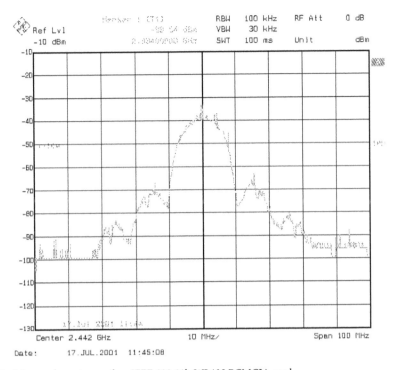

FIGURE 19.5 Measured spectrum of an IEEE 802.11b WLAN PCMCIA card.

In principle, 11/13 (U.S./Europe) center frequencies are available for the DSSS system. Nevertheless, using several systems in parallel requires a spacing of 25/30 MHz (U.S./Europe), which consequently only allows three systems to be used simultaneously.

The DSSS mode includes the original version (specified in IEEE 802.11) and a high-rate extension (specified in IEEE 802.11b). In the original mode the chipping of the baseband signal is performed with 11 Mz, employing an 11-chip pseudorandom code (Barker sequence). For the 1-Mbps modulation rate, a 1-bit DBPSK symbol is spread with the Barker sequence; for the 2-Mbps modulation rate, a 2-bit DQPSK symbol is spread with the same sequence.* The high-rate extension IEEE 802.11b was, at the time of writing, the most popular and widespread WLAN technology. For the PHY header, IEEE 802.11b uses the same 1- and 2-Mbps modulations as the plain DSSS mode. Note, though, that a shortened header of 96 µs can be used. For the IEEE 802.11b PHY payload (consisting of the MAC header and the user data), a 5.5- and 11-Mbps complementary code keying (CCK) modulation is used. The CCK employs a variation of M-ary orthogonal signaling (complex Walsh–Hadamard functions) with eight complex chips in each spreading code word. For the 5.5-Mbps modulation rate, 4 bits are mapped onto 8 chips, and for 11 Mbps, 8 bits are mapped onto 8 chips.

PBCC: The packet binary convolution code (PBCC) physical layer is one of three additional possibilities standardized in IEEE 802.11g as an even higher rate extension to IEEE 802.11b in the 2.4-GHz band. In this optional PHY a single-carrier modulation scheme is used that encodes the payload using a 256-state PBCC. The foreseen modulation rates are 22 and 33 Mbps.

OFDM: The orthogonal frequency-division multiplexing (OFDM) physical layer was originally designed for 5-GHz bands (also referred to as IEEE 802.11a) but has now been adopted for the 2.4-GHz band (as part of IEEE 802.11g). The parameters of the IEEE 802.11 OFDM PHY had at the time

---

*Note that in contrast to a typical CDMA system like UMTS, *all* users use the same spreading code.

of standardization been harmonized with those of HIPERLAN/2.* Seven modi are defined ranging from BPSK with a rate of R = 1/2 FEC (lowest modulation rate with 6 Mbps) to 64-QAM with a rate of R = 3/4 FEC (highest modulation rate with 54 Mbps; Table 19.2). The OFDM technique is based on a 64-point IFFT/FFT, while only using 52 of the subcarriers (48 for user data, 4 for pilot carriers). The subcarrier spacing is $f = 20\ MHz/64 = 0.3125\ MHz$. Note that full OFDM symbols always have to be transmitted. This means that they possibly have to be filled up with dummy bits. To transmit one OFDM symbol $t_{sym} = 1/f + 1/4 \cdot 1/f = 4\ \mu s$ are needed, with the latter part representing the time used for the guard interval to combat multipath propagation. For synchronization, channel estimation, and equalization, a training sequence is transmitted, which consists of often repeated short and two repeated long OFDM symbols [35, 58].

DSSS-OFDM: This optional physical layer format of IEEE 802.11g combines the DSSS PHY with the OFDM PHY such that for the header DSSS is used while the payload employs OFDM (including the OFDM preamble).

Table 19.2 compares the theoretical maximum throughput (TP) values of the different IEEE 802.11 PHY versions after the MAC. The maximum payload length is 4095 bytes (which has to include the 34-byte MAC header). Fifteen hundred bytes is the common length of an Ethernet packet (plus 34 bytes for the MAC header and checksum), 576 is a typical length for a Web-browsing packet, and 60 bytes is the length of a Transmission Control Protocol (TCP) acknowledgment. The throughput TP is calculated as follows:

$$TP = \frac{PayBytes \cdot 8}{\underbrace{DIFS}_{2t_{slot}+SIFS} + \underbrace{\frac{CW_{min}}{2} \cdot t_{slot}}_{average\ back\text{-}off} + t_{data\ packet} + SIFS + t_{ACK}} \tag{19.3}$$

The durations needed to transmit the data packet $t_{data\ packet}$ and acknowledgment t_{ACK} vary depending on the physical layer chosen. For the FHSS and DSSS modes they are calculated as follows:

$$t_{data\ DSSS/FHSS} = t_{PHYh} + \underbrace{\frac{34 \cdot 8}{ModRate}}_{MACheader} + \frac{PayBytes \cdot 8}{ModRate}; \ t_{ACK\ DSSS/FHSS} = t_{PHYh} + \frac{14 \cdot 8}{ModRate} \tag{19.4}$$

For the OFDM physical layer in the 5-GHz bands, Equation 19.5 needs to be calculated. For the OFDM mode in the 2.4-GHz band, an additional 6-μs signal extension has to be added to both. "Ceil" stands for the next larger integer.

$$t_{data\ OFDM\ a} = t_{PHYh} + t_{sym} ceil \left(\frac{16 + (34 + PayBytes) \cdot 8 + 6}{ModRate / 12\ Mbps \cdot 48} \right)$$

$$\tag{19.5}$$

$$t_{ACK\ OFDM\ a} = t_{PHYh} + t_{sym} ceil \left(\frac{16 + 14 \cdot 8 + 6}{ModRate / 12\ Mbps \cdot 48} \right)$$

The packet and acknowledgment durations for the DSSS-OFDM PHY are calculated quite similarly to Equation 19.5:

---

*Originally HIPERLAN/2 was intended to be the WLAN technology for the European market, while IEEE 802.11 was the pendant for North America. Owing to delays in development, HIPERLAN/2 lost the chance to establish itself on the market, despite its better overall network performance (from which the user would have had the advantage of higher user data rates). Publications on HIPERLAN/2 include [17–19, 24, 30, 40, 41, 43, 45, 47, 49, 57].

TABLE 19.2 Comparison of Different Achievable Maximum Throughput Rates (in Mbps) for the Different IEEE 802.11 PHY Modes[a]

Mode	Frq. Band	t_{slot}	SIFS	CW_{min}	t_{PHYh}	Modulation	ModRate (Mbps)	TP (Mbps) for PayBytes			
								60	576	1500	4061
FHSS	2.4 GHz	50 μs	28 μs	15	128 μs	GFSK (2 level)	1	0.29	0.79	0.91	0.96
						GFSK (4 level)	2	0.39	1.40	1.72	1.89
DSSS	2.4 GHz	20 μs	10 μs	31	192 μs	DBPSK	1	0.30	0.80	0.91	0.97
						DQPSK	2	0.40	1.42	1.72	1.89
					96 μs	CCK (QPSK)	5.5	0.67	3.14	4.26	4.97
						CCK (QPSK)	11	0.75	4.54	7.11	9.16
OFDM	5 GHz (2.4 GHz)	9 μs (9/20 μs)	16 μs (10 μs)	15 (31)	20 μs $t_{sym}=4$ μs	BPSK, R1/2	6	1.51 (1.24/0.83)	4.57 (4.29/3.64)	5.37 (5.2/4.8)	5.76 (5.68/5.49)
						BPSK, R3/4	9	1.81 (1.44/0.91)	6.32 (5.81/4.67)	7.79 (7.43/6.64)	8.51 (8.35/7.96)
						QPSK, R1/2	12	2.02 (1.55/0.96)	7.86 (7.05/5.44)	10.0 (9.45/8.21)	11.2 (10.9/10.3)
						QPSK, R3/4	18	2.25 (1.70/1.01)	10.4 (8.97/6.53)	14.1 (13.0/10.8)	16.3 (15.8/14.4)
						16-QAM, R1/2	24	2.38 (1.76/1.03)	12.3 (10.3/7.22)	17.6 (15.9/12.7)	21.2 (20.2/18.1)
						16-QAM, R3/4	36	2.59 (1.86/1.07)	15.2 (12.3/8.14)	23.7 (20.8/15.6)	30.3 (28.4/24.3)
						64-QAM, R1/2	48	2.64 (1.89/1.07)	17.1 (13.7/8.7)	28.5 (24.3/17.5)	38.4 (35.4/29.3)
						64-QAM, R3/4	54	2.70 (1.92/1.08)	17.9 (14.2/8.90)	30.8 (26.0/18.3)	42.2 (38.6/31.4)

[a] Except for the optional DSSS-OFDM and PBCC.

TABLE 19.3 Overview of Activities within IEEE 802.11

Group	Subject	Status
a	PHY in the 5 GHz bands	Completed
b	High rate mode in 2.4 GHz band	Completed
c	Extensions for specific MAC procedures	Completed
d	Supplements for new regulatory regions	Completed
e	Enhancements for QoS	Ongoing
f	To achieve multivendor access point interoperability	Completed
g	Enhancements of 802.11b data rates	Completed
h	Extensions for channel selection for 802.11b	Almost completed
i	Enhancements for security and authentication algorithms	Ongoing
j	Enhancements for the use of 802.11a in Japan	Ongoing
k	Definition of radio resource management measurements	Ongoing, initialized in 2003
l	Nonexistent	
m	Maintenance of 802.11-1999	Ongoing, initialized in 2003

$$t_{\text{data DSSS-OFDM}} = t_{\text{PHYh DSSS}} + t_{\text{Preamble OFDM}} + t_{sym}ceil\left(\frac{16 + (34 + \text{PayBytes}) \cdot 8 + 6}{\text{ModRate}/12 \text{ Mbps} \cdot 48}\right) + 6\mu s$$

$$t_{\text{ACK DSSS-OFDM}} = t_{\text{PHYh DSSS}} + t_{\text{Preamble OFDM}} + t_{sym}ceil\left(\frac{16 + 14 \cdot 8 + 6}{\text{ModRate}/12 \text{ Mbps} \cdot 48}\right) + 6\mu s \tag{19.6}$$

For the PBCC mode at 22 Mbps, the data packet and acknowledgment durations are given in Equation 19.7. In the case of 33 Mbps, a 1-μs clock switch time has to be added to both:

$$t_{\text{data PBCC 22}} = t_{\text{PHYh}} + \frac{(34 + \text{PayBytes} + 1)}{\text{ModRate}}; \quad t_{\text{ACK PBCC 22}} = t_{\text{PHYh}} + \frac{14 \cdot 8}{\text{ModRate}} \tag{19.7}$$

For small payload sizes (60 and 576 bytes), the throughput values are not very good. When considering Ethernet packets, the highest theoretical throughput rates are 7.11 Mbps for IEEE 802.11b and 30.8/26.0 Mbps for the OFDM modes. Naturally, these wireless throughput rates are smaller than the wired ones (where 70 to 80 Mbps is possible); but at least for the higher modulation rates with 1500-byte Ethernet packets, the throughput values are reasonably good. Note that the real-life throughput values for IEEE 802.11b systems are still smaller than the theoretically possible ones: values around 5 Mbps have been measured [50, 51]. This is because for actual implementations used, higher protocol layers like TCP/IP cause additional overhead and delays.

For security, IEEE 802.11 WLANs support several authentication processes, which are listed in the specification (none are mandatory).* Table 19.3 lists the standardization activities involving IEEE 802.11.

19.4.2 Performance

Next to aspects like individual link throughput, network capacity, and interference robustness, the transmission environment has to be taken into consideration when contemplating the use of IEEE 802.11 on the factory floor. Because the scenarios envisioned for IEEE 802.11 were placed primarily in homes and offices, some differences occur when looking at the delay spread. While in homes and offices the delay spread is assumed to be <50 and <100 ns, respectively, it takes on values of 200 to 300 ns on factory floors [53].

---

*Neither Bluetooth nor IEEE 802.11 is renowned for its security concepts, and both have been criticized. To improve the situation, the IEEE 802.11i standardization activities were initiated.

In case of IEEE 802.11b, a conventional RAKE receiver supports (only) about a 60-ns delay spread in the 11-Mbps mode and 200 ns in the 5.5-Mbps version [58]. When employing an IEEE 802.11b system with such a conventional receiver on a factory floor, intersymbol and interchip/codeword interferences (ISI and ICI) are likely to degrade the performance. Nevertheless, with more complex receiver algorithms like those presented in [13, 44], IEEE 802.11b can compensate well for delay spreads of 1 μs and even mobility of the user. When wanting to use IEEE 802.11b on the factory floor, the specific performance concerning the used equipment and environment needs to be measured. Another option, of course, is to use IEEE 802.11a or g. Because of the guard interval inherent in the OFDM technology, delay spreads of several hundred nanoseconds can be easily supported without paying attention to the receiver algorithms implemented [58].

When considering the overall network performance and not just the individual link performance (or interference performance, as discussed in Section 19.6), note first of all that the number of publications presenting well-founded results is very limited. It seems to be general thinking that capacity in an IEEE 802.11-only network is not an issue. The few publications that do exist do not at all support this. In [42] it is shown that co-channel interference with a carrier-to-interference ratio (CIR) of 5 dB still results in a packet loss rate of 10 to 20% (bit error rate [BER] = 10^{-5}), and that with a frequency offset of 5 MHz, still CIR = 3 dB is required to achieve the same result. References [6, 55] present how the aggregate throughput in a single network decreases with the number of users, due to either hidden- or exposed-terminal problems or additional RTS/CTS overhead. With only 10 stations [6] or a hidden node probability of 5% [55], the system throughput is about halved in the case of 1500-byte payloads. Only when there are more than 25 stations is the RTS/CTS implementation justified, while the throughput is still reduced.

Thus, when installing IEEE 802.11 in a cellular fashion, some kind of frequency planning should be performed. For IEEE 802.11 systems a respective mechanism has to be added (e.g., refer to [2] for frequency allocation algorithms); for IEEE 802.11a, a mechanism is currently being standardized in IEEE 802.11h. Note that the most relevant parameter for WLAN frequency planning is the number of mobile terminals that have to be served. From it the optimum number of access points (APs) and distance between APs can be determined. If it is desired that mobile stations can seamlessly change between APs (due to mobile deployment), handover algorithms have to be added.

IEEE 802.11 provides reliable best-effort traffic. The (theoretical) maximum transmission rates for IEEE 802.11a, g, and b are 30.8, 26.0, and 7.11 Mbps, respectively. To achieve network capacity values anywhere near three times these values (three parallel systems are possible for IEEE 802.11b and g; for IEEE 802.11a network capacity is not critical due to the larger-frequency band at 5 GHz), appropriate receivers and some sophisticated frequency planning are needed. The WLAN systems of IEEE 802.11 allow for higher data rates than the WPAN technology Bluetooth. Nevertheless, IEEE 802.11 consumes (significantly) more power than Bluetooth and is not really suitable for speech connections. Currently, IEEE 802.11b chips are also larger in size and more expensive than Bluetooth chips.

19.5 ZigBee

19.5.1 Technical Background

The idea behind ZigBee* was to create a very low cost, very low power, two-way wireless communication solution that meets the unique requirements of sensors and control devices needed in consumer electronics, home and building automation, industrial controls, PC peripherals, medical sensor applications, toys, and games. To allow long battery lives of 2 years and more (and thus minimizing the efforts in maintenance), ZigBee supports low data rates at low-duty cycles. Owing to its nominal range of 10 m, ZigBee is considered a WPAN technology [3, 14, 39].

---

*Note that the technology includes two specifications: the specifications published under IEEE 802.15.4 for the physical and MAC layers as well as the ZigBee specification, which covers the upper layers from network to application layers [12].

TABLE 19.4   IEEE 802.15.4 Parameters for the Different Frequency Bands

Frequency band	868 MHz	915 MHz	2.4 GHz
Location	Europe	North America	Worldwide
Number of center frequencies	1	10	16
Carrier spacing		2 MHz	5 MHz
Gross bit rate	20 kbps	40 kbps	250 kbps
Bit modulation	BPSK	BPSK	16-ary orthogonal
Symbol rate	20 kbps	40 kbps	62.5 kbps
Spreading	15-chips M-sequence		32 chips PN code
Chip modulation	BPSK	O-QPSK	O-QPSK
Chip rate	300 kchips/s	600 kchips/s	2 Mchips/s

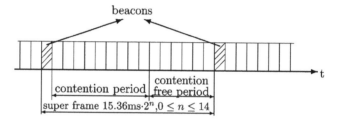

FIGURE 19.6   Optional ZigBee superframe.

To encourage deployment, ZigBee is also placed in unlicensed frequency bands. Like Bluetooth and the 802.11b and g systems, ZigBee can be used almost globally in the 2.4-GHz band. Additionally, ZigBee has been specified for the ISM bands at 868 MHz in Europe and 915 MHz in North America. To comply with the respective sharing rules and to allow simple analogue circuitry, ZigBee uses DSSS. Table 19.4 gives an overview of the respective physical layer parameters. Note that the maximum *user* data rate is about 128 kbps.

A ZigBee network can have a star or a peer-to-peer topology. Each network needs a PAN coordinator unit, which can handle up to 255 devices in the case of a 16-bit address pad, and even more in the case of 64-bit addressing. Of the two types of ZigBee devices, full function device (FFD) and reduced function device (RFD), only FFD can function as a network coordinator. Direct communication between two RFDs is not possible. RFD packets have to be passed to an FFD first.

For medium access, there are in principle three possibilities: CSMA/CA without beacons, CSMA/CA in the contention period of a beacon system, or a guaranteed data rate in the contention-free period of a beacon system. In the case of a beacon system, there are so-called superframes that can be between 15.36 ms and 251.65 s long. If desired, this period is divided between a contention and a contention-free period of limited length (Figure 19.6). The beacons are used for synchronization, for identification, and to describe the structure of the superframe.

For security, ZigBee provides authentication, encryption, and integrity services. The developer can choose between no security, an access control list, and a 32- to 128-bit advanced encryption standard (AES) with authentication.

19.5.2   Performance

Because of its newness, not much information is yet available on the performance of ZigBee. When comparing the system design choices made for ZigBee with those made for Bluetooth, it can be expected that Bluetooth will have superior interference behavior over ZigBee because of Bluetooth's spectral selectivity and purity [27]. Furthermore, the frequency-hopping approach of Bluetooth will be responsible for a different range performance. While for Bluetooth there is — with increasing range — likely to be a smooth degradation in throughput, a loss of connection at a certain distance is more probable for ZigBee. Nevertheless, ZigBee has been developed (and should thus be the best choice) for applications in which the network is static and has many devices of infrequent use that only want to transmit small data packets.

19.6 Coexistence of WPAN and WLAN (Bluetooth and IEEE 802.11b)

It would be most desirable for every mobile unit to connect effortlessly using whatever technology is most suitable at the time. Multiple wireless technologies will in the near future coexist: within an enterprise, for example, WLAN technologies could be used for flexible access to large corporate databases, while WPAN technologies could handle specific tasks (and cellular systems the voice communication). This is generally not a problem unless the technologies used are placed in the same frequency band or are linked by the application with each other (e.g., someone uses a Bluetooth headset with the mobile telephone). The two technologies that are placed in the same frequency band* and that are, due to their popularity, quite likely to be used in such a scenario (and possibly even in the same device) are Bluetooth and IEEE 802.11b/g.**

Numerous publications cover the mutual interference and performance impairments of the systems (e.g., [15, 22, 25, 28, 50, 51, 63]). Depending on the investigated scenarios, the assessment of the situation varies from "good reliability even in fairly dense environments" to "the effects of interference can be quite severe." A relative agreement exists in the causes that determine the systems' performances: link distances (BT–BT, 802.11b–802.11b, BT transmitter–802.11b receiver, BT receiver–802.11b transmitter), traffic load, Bluetooth packet type, density of units, and local propagation conditions. Some of the results are briefly summarized below.

IEEE 802.11b requires a CIR of about 10 dB to cope with a (narrowband) Bluetooth hop in its (wide-) passband. IEEE 802.11b has the disadvantage that its backoff procedure was designed to optimize the IEEE 802.11b WLAN performance but not to handle external interferers: each loss of a packet due to a collision with Bluetooth will increase the backoff window size by a factor of 2 (causing an unnecessary throughput reduction). Furthermore, the protocols overlaying WLAN often incorporate TCP, which includes the risk that packet losses on the air link are mistaken for network congestion, which then might initiate a slow start. In contrast, the main disadvantage of Bluetooth is that its transmit power is 20 dB below that of IEEE 802.11b. Another disadvantage is that the BT reverse link packet, which contains the acknowledgment, is transmitted on a different frequency than the forward link packet. This increases the packet loss probability in case of (frequency static) IEEE 802.11b interference. The packet loss rate for Bluetooth PLR_{BT} then yields to

$$PLR_{BT} = PLR_{forward} + \underbrace{(1 - PLR_{forward}) \cdot PLR_{header}}_{PLR_{reverse}} \qquad (19.8)$$

It would otherwise — were the forward and reverse links to use the same hop frequency — be $PLR_{BT} \approx PLR_{forward}$ (depending on the IEEE 802.11b system load).

There are in principle three different approaches to assist otherwise interfering systems in coexisting: separation in time, separation in frequency, and separation in space:***

- *Separation in the frequency domain*:
  1. IEEE 802.11b can be used with an improved transmit filter that reduces the interference power on the side lobe frequencies and thus enhances the separation on those carriers.

---

*Next to other technologies like HomeRF networks [29], RF-ID systems [21], microwave ovens, etc.

**Investigations on the coexistence between Bluetooth, IEEE 802.11b, IEEE 802.11g, and ZigBee are not yet available.

***The approaches separation through code (keyword CDMA), through the channel (keyword MIMO), or through the modulation (I vs. Q) allow the unlinking of several users of the *same* system. It is not obvious, though, to apply any of these latter methods to improve the coexistence of *different* systems, which is what is needed here.

2. Bluetooth can perform adaptive frequency hopping (AFH), i.e., exclude the most heavily inter-fered frequencies from its hop sequence. Note that in a lot of realistic situations, where the IEEE 802.11b and Bluetooth units are not in the same device, AFH is sufficient to combat the interference effects. AFH has thus become part of Bluetooth Specification 1.2 [10].

- *Separation in the time domain*:

1. The IEEE 802.11b carrier-sensing algorithm could consider Bluetooth signals,* and Blue-tooth could be extended with a carrier-sensing algorithm [60]. The principle problem of carrier sensing, though, is that to be really effective, the transmitter has to sense the situation at the receiver correctly; i.e., the correlation between transmitter and receiver has to be high. In an uncoordinated WLAN–WPAN scenario, the hidden- and exposed-terminal problems are likely to countermeasure any advantage there might be.

2. A joint scheduler can allot alternating transmit time-shares to both systems in a (to be specified) fair way. This, of course, only works (and with AFH is necessary only) when both systems are in one device, or even on one chip.

- *Separation in space*:

1. Should IEEE 802.11b and Bluetooth coexist in the same unit, intelligent antenna design and placement can optimize the isolation between the Bluetooth and IEEE 802.11b antennas. This will not prevent collisions, but will minimize their impact by maximizing the CIRs for the two systems.**

2. Antenna diversity can help each of the two technologies to individually improve its performance.

19.7   Summary and Conclusions

When deciding on a wireless technology to use, you must first clarify the characteristics of the foreseen application. Do you want to move with or within the network? Do you have mobility within a small range or in a larger area? At what speed do the units move? Do you need access to large databases or just locally? Is battery life a critical issue? What maximum distance should the wireless link cover? What distance does it cover on average? As a next step, the existing technologies can be viewed for their applicability.

Bluetooth is a quite power-efficient WPAN technology. Like all wireless systems, it cannot provide hard throughput guarantees. Bluetooth is nevertheless quite robust for best-effort traffic in coexistence environments. One hundred Bluetooth piconets can transmit at an average data rate of $95\% \cdot 33.2$ kbps in an area of 10×20 m². Fifty fully loaded piconets can transmit at an average, unidirectional transmission rate of 360 kbps. Similar results are likely to be achieved on larger factory floors (provided the piconet density is comparable), as the disadvantage of a larger number of units can be outweighed by the more structured and predictable unit location. Of the discussed technologies, Bluetooth is the only one that supports voice in addition to data transmission.

IEEE 802.11 is a WLAN technology that enables higher data rates but is not too power consumption friendly. Additionally, in a relatively dense network scenario, the maximum aggregate throughput of 3 times 7.11/30.8 Mbps is likely to be seriously impaired. To aid the WLAN performance on a factory floor, it is thus advisable to take the following two measures: apply means to combat the increased delay spread (in case of IEEE 802.11b) and (for all IEEE 802.11 systems) carefully plan the frequency layout and access point placement.

---

*Helpful to minimize the interference power is, of course, power control (when applied by the interfering unit). Nevertheless, even though it can be recommended to implement power control (let alone to save power [52]), in real-life situations it cannot be relied upon that the interfering unit can indeed live with less power.

**As has been mentioned before, the IEEE 802.11b specification already provides for the possibility that IEEE 802.11b senses, and not transmits, if systems other than IEEE 802.11b are active [34, Section 18.4.8.4]. Next to the fact that most implementations do not seem to support this option, the principle problem is that IEEE 802.11b will refrain from transmission when it senses that Bluetooth *transmits*. To improve coexistence, though, IEEE 802.11b should refrain from transmission when nearby Bluetooth units *receive*.

The newly emerging ZigBee technology has been designed specifically for sensor data and control information at low data rates. ZigBee supports long battery lives. Nothing can be said yet about the robustness and effectiveness of ZigBee. Yet for low-bit-rate applications in industrial environments, ZigBee seems to be a promising approach.

The most efficient measure to aid the coexistence of Bluetooth and IEEE 802.11b is to use Bluetooth with adaptive frequency hopping. If this is not sufficient, hardware-related improvements and a common scheduler have to be added.

References

1. 3rd Generation Partnership Project, Technical Specification Group Access Network. Opportunity Driven Multiple Access, 3G TR 25.924, version 1.0.0. December 1999.
2. Hamid Aghvami, A. *Resource Allocation in Hierarchical Cellular Systems*. Artech House Publishers, Boston, 2000.
3. Bahl, Venkat, ZigBee and Bluetooth: Competitive or Complementary?, ZigBee Alliance, September 2002.
4. Beaumont, Daniel. Citywide 802.11b networks gain momentum. *Planet Wireless*, July 2002.
5. Beaumont, Daniel. More Bluetooth products but key profiles delayed. *Planet Wireless*, July 2002.
6. Bianchi, Giuseppe. Performance analysis of the IEEE 802.11 distributed coordination function. *IEEE Journal on Selected Areas in Communications*, 18:535–547, 2000.
7. Bisdikian, Chatschik. An overview of the Bluetooth wireless technology. *IEEE Communications Magazine*, 39:86–94, 2001.
8. http://www.bluetooth.com
9. Bluetooth Special Interest Group. Specification of the Bluetooth System, Version 1.1. December 1999.
10. Bluetooth Special Interest Group. Bluetooth 1.2 Core Specification. November 2003.
11. Bray, Jennifer and Charles F. Sturman. *Bluetooth: Connect without Cables*. Prentice Hall, Englewood Cliffs, NJ, 2000.
12. Callaway, Ed. Low Power Consumption Features of the IEEE 802.15.4/ZigBee LR-WPAN Standard. Presentation slides, November 2003.
13. Clark, Martin V., Kin K. Leung, Bruce McNair, and Zoran Kostic. Outdoor IEEE 802.11 cellular networks: radio link performance. In *Proceedings of the IEEE International Conference on Communication (ICC)*, May 2002.
14. Cambridge Consultants. Unleashing Renenue with ZigBee. Presentation slides, 2003.
15. Ennis, Greg. Impact of Bluetooth on 802.11 Direct Sequence, Technical Report IEEE 802.11-98/319. IEEE, September 1998. Available at http://grouper.ieee.org/groups/802/Regulatory/Regulatory%20 Filings/831947a-Impact-of-Bluetooth-on-80211.pdf
16. European Telecommunications Standards Institute (ETSI). ETSI EN 300 328-1, V1.2.2. July 2000.
17. European Telecommunications Standards Institute (ETSI). HIPERLAN Type 2: Data Link Control Specification: Part 1: Basic Data Transport Functions. November 2000.
18. European Telecommunications Standards Institute (ETSI). HIPERLAN Type 2: Data Link Control Specification: Part 2: Radio Link Control (RLC) Sub-Layer. April 2001.
19. European Telecommunications Standards Institute (ETSI). HIPERLAN Type 2: Physical (PHY) Layer. February 2001.
20. Federal Communications Commission (FCC). *Code of Federal Regulations*. 2002.
21. Finkenzeller, Klaus. *RFID-Handbuch*. Hanser, Munich, 2000.
22. Fumolari, David. Link performance of an embedded Bluetooth personal area network. In *Proceedings of the IEEE International Conference on Communication (ICC)*, Helsinki, June 2001.
23. Green, Jeremy, Rob Gear, and Nick Harman. Bluetooth: users, applications and technologies. *Ovum Report*, June 2001.

24. Ajay Chandra V. Gummalla, and John O. Limb. Wireless medium access control protocols. *IEEE Communications Surveys and Tutorials*, 3: 2000. Available at http://www.comsoc.org/pubs/surveys.

25. Golmie, N., R.E. van Dyck, and A. Soltanian. Bluetooth and 802.11b Interference: Simulation Model and System Results, Technical Report IEEE802.15-01/195R0. IEEE, April 2001. Available at http://grouper.ieee.org/groups/802/15/pub/2001/May01/01195r0P802-15_TG2-BT-802-11-Model-Results.pdf

26. Haartsen, Jaap. Bluetooth: the universal radio interface for ad hoc, wireless connectivity. *Ericsson Review*, 3:110–117, 1998.

27. Haartsen, Jaap. ZigBee OR Bluetooth or ZigBee AND Bluetooth. Presentation slides, June 2003.

28. Howitt, Ivan. IEEE 802.11 and Bluetooth Coexistence Analysis Methodology. In *Proceedings of the IEEE Vehicular Technology Conference (VTC)*, Rhodes, May 2001.

29. http://www.homerf.org

30. Huschke, Jörg and Gerd Zimmermann. Impact of decentralized adaptive frequency allocation on the system performance of HIPERLAN/2. In *Proceedings of the IEEE Vehicular Technology Conference (VTC)*, Tokyo, May 2000.

31. http://www.ietf.org/html.charters/manet-charter.html

32. http://grouper.ieee.org/groups/802/11/index.html

33. Institute of Electrical and Electronic Engineering. Part 11: Wireless LAN Medium Access Control (MAC) and Physical Layer (PHY) Specifications, ANSI/IEEE 802.11. September 1999.

34. Institute of Electrical and Electronic Engineering. Part 11: Wireless LAN Medium Access Control (MAC) and Physical Layer (PHY) Specifications: Higher-Speed Physical Layer Extension in the 2.4 GHz Band, IEEE 802.11b-1999. September 1999.

35. Institute of Electrical and Electronic Engineering. Part 11: Wireless LAN Medium Access Control (MAC) and Physical Layer (PHY) Specifications: Higher-Speed Physical Layer in the 5 GHz Band, IEEE 802.11a. September 1999.

36. IEEE-SA Standards Board Bylaws. September 2002. Approved by Standards Association Board of Governors.

37. Institute of Electrical and Electronic Engineering. Part 11: Wireless LAN Medium Access Control (MAC) and Physical Layer (PHY) Specifications, Amendment 4: Further Higher Data Rate Extension in the 2.4 GHz Band, ANSI/IEEE 802.11. June 2003.

38. Ji, L., M. Ishibashi, and M.S. Corson. An approach to mobile ad hoc network protocol kernel design. In *Proceedings of the IEEE Wireless Communications and Networking Conference (WCNC)*, New Orleans, September 1999.

39. Kinney, Patrick. ZigBee technology: wireless control that simply works. Paper presented at the *Communication Design Conference*, October 2003.

40. Khun-Jush, Jamshid, Peter Schramm, Udo Wachsmann, and Fabian Wenger. Structure and performance of the HIPERLAN/2 physical layer. In *Proceedings of the IEEE Vehicular Technology Conference (VTC)*, pp. 2667–2671, Amsterdam, Fall 1999.

41. Kadelka, Arndt, Erkan Yidirim, and Bernhard Wegmann. Serving IP mobility with HIPERLAN/2. In *Proceedings of the European Mobile Communications Conference (EPMCC)*, Vienna, February 2001.

42. Lindgren, Magnus. Physical Layer Simulations of the IEEE 802.11b Wireless LAN-Standard. Master's thesis, Lulea Technical University, Sweden, 2001.

43. Li, Hui, Jan Lindskog, Göran Malmgren, Gyorgy Miklos, Fredrik Nilsson, and Gunnar Rydnell. Automatic repeat request (ARQ) mechanism in HIPERLAN/2. In *Proceedings of the IEEE Vehicular Technology Conference (VTC)*, Tokyo, May 2000.

44. Leung, Kin K., Bruce McNair, Leonard J. Cimini, and Jack H. Winters. Outdoor IEEE 802.11 cellular networks: MAC protocol design and performance. In *Proceedings of the IEEE International Conference on Communication (ICC)*, New York, April–May 2002.

45. Li, Hui, Göran Malmgren, Mathias Pauli, Jürgen Rapp, and Gerd Zimmermann. Performance of the radio link protocol of HIPERLAN/2. In *Proceedings of the IEEE International Symposium on Personal, Indoor and Mobile Radio Communication (PIMRC)*, London, 2000.

46. Maltz, David A. On-Demand Routing in Multi-Hop Wireless Mobile Ad Hoc Networks. Ph.D. thesis, School of Computer Science, Carnegie Mellon University, Pittsburgh, PA, May 2001. Available at http://reports-archive.adm.cs.cmu.edu/anon/2001/CMU-CS-01-130.pdf

47. Matheus, Kirsten. Wireless local area networks and wireless personal area networks (WLANs and WPANs). In Richard Zurawski, editor, *Industrial Information Technology Handbook*. CRC Press, Boca Raton, FL, 2004, chap. 38.

48. Ministry of Telecommunications (MKK), Japan. RCR STD-33A. 2002.

49. Malmgren, Göran, Jamshid Khun-Jush, Peter Schramm, and Johan Torsner. 6:3 HIPERLAN type 2: an emerging world wide WLAN standard. In *Proceedings of the International Symposium on Services and Local Access*, Stockholm, June 2000.

50. Mobilian. Wi-Fi™ (802.11b) and Bluetooth™: An Examination of Coexistence Approaches, 2001. Available at http://www.mobilian.com

51. Matheus, Kirsten and Stefan Zürbes. Co-existence of Bluetooth and IEEE 802.11b WLANs: results from a radio network testbed. In *Proceedings of the IEEE International Symposium on Personal, Indoor and Mobile Radio Communication (PIMRC)*, Lissabon, Portugal, September 2002.

52. Matheus, Kirsten, Stefan Zürbes, Rakesh Taori, and Sverker Magnusson. Fundamental properties of ad-hoc networks like Bluetooth: a radio network perspective. In *Proceedings of the IEEE Vehicular Technology Conference (VTC)*, Orlando, FL, September 2003.

53. O'Hara, Bob and Al Petrick. *IEEE 802.11 Handbook: A Designer's Companion*. Standards Information Network, IEEE Press, New York, 1999.

54. Rouse, T.S., I.W. Band, and S. McLaughlin. Capacity and power analysis of opportunity driven multiple access (ODMA) networks in CDMA systems. In *Proceedings of the IEEE International Conference on Communication (ICC)*, pp. 3202–3206, 2002.

55. Shreyas Sadalgi. A Performance Analysis of the Basic Access IEEE 802.11 Wireless LAN MAC Protocol (CSMA/CA), May 2000. Available at http://paul.rutgers.edu/~sadalgi/network.pdf

56. Stranne, André, Fredrik Florén, Ove Edfors, and Bengt-Arne Molin. Throughput of IEEE 802.11 FHSS networks in the presence of strongly interfering Bluetooth networks. In *Proceedings of the IEEE International Symposium on Personal, Indoor and Mobile Radio Commununication (PIMRC)*, Lissabon, Portugal, September 2002.

57. Torsner, Johan and Göran Malmgren. Radio Network Solutions for HIPERLAN/2. In *Proceedings of the IEEE Vehicular Technology Conference (VTC)*, Houston, Spring 1999, pp. 1217–1221.

58. van Nee, Richard, Geert Awater, Masahiro Morikura, Hitoshi Takanashi, Mark Webster, and Karen W. Halford. New high-rate wireless LAN standards. *IEEE Communications Magazine*, 37:82–88, 1999.

59. U.S. city on the verge of being covered completely with wireless Internet. Available at www.web.de, message of February 12, 2003.

60. Zhen, Bin, Yongsuk Kim, and Kyunghun Jang. The analysis of coexistence mechanisms of Bluetooth. In *Proceedings of IEEE Vehicular Technology Conference (VTC)*, Spring 2000.

61. Zürbes, Stefan, Wolfgang Stahl, Kirsten Matheus, and Jaap Haartsen. Radio network performance of Bluetooth. In *Proceedings of the IEEE International Conference on Communication (ICC)*, New Orleans, June 2000.

62. Zürbes, Stefan. Considerations on link and system throughput of Bluetooth networks. In *Proceedings of the IEEE International Symposium on Personal, Indoor and Mobile Radio Communication (PIMRC)*, London, September 2000, pp. 1315–1319.

63. Zyren, Jim. Extension of Bluetooth and 802.11 Direct Sequence Interference Model, Technical Report IEEE 802.11-98/378. IEEE, November 1998. Available at http://www.ieee802.org/15/pub/SG.html

20

Interconnection of Wireline and Wireless Fieldbuses

Jean-Dominique
Decotignie
CSEM (Centre Suisse d'Electronique
et de Microtechnique)

20.1 Introduction

Networking at the factory floor is now commonplace. Solutions are available at different levels: sensory, cell, plant, etc. In this chapter, we focus on the sensory level where networks usually called fieldbuses (Pleinevaux and Decotignie, 1988) link sensors and actuators to the first level of automation. This kind of network is no longer the exclusivity of factories and finds its way into cars, planes, and buildings.

Since the first solutions were designed in the early 1980s, a number of proposals have flourished and the field is well established in terms of both research and industrial use. Most solutions use wired transmission (twisted pairs, coaxial cables, optical fibers). Very early, the need for mobile nodes and the difficulty to install cables pushed for wireless solutions based on radio or light transmission. Today, most solutions use radio transmission. Due to the special properties of wireless transmission, it is not desirable to have all nodes of the fieldbus be wireless. On the other side, the wired devices need to communicate with the wireless nodes, thus prompting the necessity for interconnection means. However, the differences between the properties of wired and wireless transmissions introduce constraints in this interconnection. This is exacerbated by the special requirements put on fieldbuses.

The chapter is organized as follows. Section 20.2 gives some definitions that will be used in the rest of the chapter. It also explains what are the relevant properties of wireless transmission. Section 20.3 details the different interconnecting mechanisms that can be used. Section 20.4 presents the different architectural options that are used when designing an interconnection architecture. Section 20.5 details how the various options are used in the different proposals and the associated difficulties. Section 20.6 uses the model to derive solutions that were not yet proposed and discusses their applicability. Some conclusions and possible future work are given in the last section.

20.2 Context and Definitions

As mentioned above, the chapter concentrates on field-level devices that communicate through a mix of wireless and wireline transmissions. Although this is a very important topic, in this chapter we do not consider the issue of power consumption in the wireless nodes.

20.2.1 Fieldbus Requirements

Requirements for fieldbuses are numerous and cover all the life cycle activities. Here we restrict ourselves to those properties that are particularly relevant for the interconnection.

- Handle periodic traffic with different period durations. In many fieldbus solutions, this requirement is translated into some cyclic traffic. The real need is to be able to transport the information well before the end of the period at which the data has been sampled.
- Handle sporadic traffic with bounded latency.
- Allow for quasi-simultaneous sampling of a number of inputs on different network nodes.
- Provide indication for temporal consistency. The fact is that control or acquisition systems expect that different sensed values correspond to sampling instants, which should be within a few percent of the sampling period. The network should thus provide ways to know if a set of values exhibits this property, named relative temporal consistency. Sometimes the age (time elapsed since sampling) of data, also called absolute temporal consistency (Kopetz, 1988), is also important to its users.
- For sporadic traffic, provide ways to know the order in which events have occurred. An application will make different decisions depending on the order in which events have occurred. As the events are potentially detected on different nodes of the network, there should be a way to find out the order.
- Transfer data from one node to another or from one node to a number of others.
- Rugged solutions in terms of resistance to interference, vibrations, etc.

It is important to notice the special temporal requirement, which will have a direct effect on the interconnection.

20.2.2 Important Radio Transmission Properties

Wireless transmissions come in two main categories: radio- and light-based systems. They both exhibit characteristics that make them very different from cable-based transmission.

For radio transmission, the main properties are:

- Property 1: Compared to cables, radio transmissions suffer from bit error rates (BERs) that are some orders of magnitude higher. BERs of 10^{-3} to 10^{-4} are usual, whereas in cables one may expect BERs ranging from 10^{-7} to 10^{-9}. Error detection schemes should thus be enhanced accordingly. This is especially a concern for token-based systems because the token recovery takes quite a long time compared to the temporal constraints.
- Property 2: Spatial reuse is low, as spectrum is limited. This means that either coexistence of several systems in the same area should be planned (code or frequency allocation) or the medium access control (MAC) should be designed in a way that takes care of the interference between systems.
- Property 3: Perturbing systems can easily jam radio transmission. This is especially true in the ISM (instrument, scientific, and medical) bands. For instance, in the 2.4-GHz band, high-power medical devices are allowed. They may completely suppress all communications for long periods.

- Property 4: Transmission distances are smaller. Typical range are a few tenths of meters indoors and up to 300 m outdoors. Obstacles may further limit this distance.
- Property 5: Collisions cannot be detected while emitting. The power of remote emitters is much lower than the power of the transmitter emission that masks the others.
- Property 6: A transceiver needs longer time (up to a few milliseconds) to switch from emission to reception and vice versa. This has to be taken into account when designing the protocol. In particular, protocols that require immediate answers to incoming requests cannot be used.
- Property 7: Radio transmissions suffer from frequency-selective multipath fading. Waves may follow different paths that interfere destructively at the receiver site. Communication may be impossible at some points.

Optical waves may be used as an alternative to radio transmission. Here we will restrict the discussion to infrared transmission, as it is the most commonly used. Its special properties are:

- Property 8: Transmission operates only in line of sight. An emitter and a receiver should have direct visibility. This constraint is often relaxed by using satellite-like techniques. A special device that acts as a repeater is located in a place where it "sees" all the other devices. This location is often on the ceiling of the room and thus is named satellite. However, communication may be impossible at some points.
- Property 9: Sources of heat (sun, machines, heaters, etc.) interfere with transmission and induce errors.
- Property 10: Spectrum reuse is limited as all systems share the same wavelength.

As a summary, wireless transmission means exhibiting properties that are significantly different from those of wires. This will have an impact on the solutions that can be used to interconnect wired and wireless nodes.

20.2.3 Definitions

In order to clarify the different architectural options, it is useful to define some terms:

- Data circuit (ISO/IEC 7498:1, 1996) — "A common path in the physical media for OSI among two or more physical-entities together with the facilities necessary in the physical layer for the transmission of bits on it."
- Subnetwork (ISO/IEC 7498:1, 1996) — "An abstraction of a real subnetwork."
- Real subnetwork (ISO/IEC 7498:1, 1996) — "A collection of equipments and physical media which form an autonomous whole and which can be used to interconnect real systems for the purpose of data transfers."
- Data link (ISO/IEC 8802.2, 1998) — "An assembly of two or more terminal installations and the interconnecting communication channel operating according to a particular method that permit information to be exchanged. In this context, the term *terminal installation* does not include the data source and the data sink."
- LAN — A data link using the same physical layer and medium access control protocols.
- Segment — Synonymous with data circuit when the nodes are connected through wires.
- Cell — Synonymous with data circuit but in the case of a wireless medium.

20.3   Interconnection Means

Generally speaking, networks may be interconnected in various ways — repeaters, bridges, routers, and gateways (Perlman, 2000).

20.3.1   Repeaters

Repeaters operate above the physical layer. Conventionally, they work bit by bit receiving the input signal, regenerating the signal, and emitting it on the other side. In the last decade, we have seen the blossoming of repeaters with multiple ports that are called hubs (Ethernet, Universal Serial Bus (USB)). In the context of wireline to wireless (or vice versa) repeaters, this may imply changing the encoding scheme (i.e., NRZI to Manchester). However, theoretically, repeaters operate transparently to the protocols above the physical layer.

As wireless transmission is more error-prone than transmission on cables, a different kind of repeater may be designed (Morel, 1996), namely, a word repeater. Instead of repeating the incoming signal bit by bit (bit repeater), the word repeater waits until a number of bits have been received from the wired side, calculates a forward error correction (FEC) code, and transmits the bits together with the error correction code. A different modulation scheme (for instance, spread spectrum) may be used to improve the spectral efficiency. When receiving from the wireless side, the word repeater uses the FEC code to correct possible errors and retransmits the (possibly) corrected information bits to the wire side. This kind of repeater introduces a longer delay, although shorter than the delay of a bridge, router, or gateway. It reduces the error rate on the wireless side and brings it closer to the wired side bit error rate (Morel, 1996).

Some practical considerations limit the use of repeaters. Repeaters should in principle forward what comes from one port to the other port and vice versa. This would be possible with full-duplex lines with one line for each direction of transfer. However, all popular wireline fieldbuses use a single line for both directions (half duplex). The repeater must thus switch from one direction to the other one according to the flow of data. This may be performed using additional lines but is not practical. In practice, wireline repeaters use some mechanism to sense the direction of travel (Murdock and Goldie, 1989) and autonomously switch accordingly. When an incoming signal is sensed on one side, the repeater is switched so that the signal is regenerated and emitted on the other side. The opposite happens when the signal is sensed on the other side. When signals are sensed on both sides, the repeater is put in isolation mode. This behavior is adequate as long as the medium access control does not rely on collisions. In such a case, when a signal is sensed on both sides, the repeater must emit a jamming signal on both sides so that the collision can be detected on both sides. It must also send this jamming signal on the input port when, after switching to one direction of travel, it senses a collision on the output port. A repeater is thus not completely independent of the medium access control scheme used in the overlying network.

Using radio communications on one side of the repeater does not much change this picture unless the medium access control uses contention. Remember that collisions cannot be detected directly on the wireless side (Property 5). The repeater thus has no means to detect the collision and propagate it on the other side (assumed to be wired). This precludes the use of protocols exclusively based on collisions. Property 6 will also lengthen the delay incurred in the repeater.

It is important to notice that, if the bit rate is different on both sides of a repeater, the repeater must buffer the information. As repeaters are usually bidirectional, the buffering policy will differ from one direction to the other. Let us assume that the bit rate is higher on side A than on side B. As soon as information is received from side A, emission may start on side B. However, bits are emitted more slowly on side B and the repeater must buffer the incoming bits before they can be emitted on side B. Conversely, when something is received from side B, the repeater cannot start relaying the information on side A immediately. It must wait until a sufficient number of bits have been received to emit the complete packet at the side A bit rate. This means that the repeater must know the maximum size of a packet.

20.3.2 Bridges

A bridge is a data link layer relay. A bridge receives a complete MAC or link layer controller (LLC) frame, checks it, and possibly forwards it on the other side. Contrary to repeaters, bridges filter the information that is relayed from one local area network (LAN) to another. The different types of interconnected LANs lead to various categories of bridges (Varghese and Perlman, 1990):

- Pass-through bridges can be used when the LANs on both sides offer identical data link layer functions and addressing. The frames can then be passed unchanged. Ethernet switches are a recent example of this category of bridges.

- Translation bridges are used when both LANs have data link layer function and addressing that are sufficiently similar to allow a direct translation of data link layer protocol data units (PDUs).
- Encapsulation (or tunneling) bridges may be used when the translation is not possible. An incoming frame is encapsulated in the data link layer format on the other link before being forwarded. This is the kind of bridge used when two or more LANs of identical technology need to be interconnected through another one of a different kind.

A bridge participates as a node in each of the LANs it interconnects. It receives a copy of all frames transmitted on each one. Obviously, not all frames need to be relayed. For instance, a frame received on side A, whose destination is also on side A, does not need to be relayed on side B. The bridge has thus to learn which frames should be relayed and which ones should not. In particular, special care should be taken to avoid frame looping when several paths exist between two nodes that communicate (ISO/IEC 7498:1, 1996).

20.3.3 Routers

Routers operate at the network layer level. The main difference between bridges and routers is that the latter are not transparent. Routers modify the packets they forward, in particular their address fields. Routers exchange information between themselves in order to find a route on which the packet is conveyed. They can thus find an optimum path between two nodes, whereas bridges only use a subset of the available topology. As the network layer is most of the time absent in fieldbuses, we will not deal further with this case.

20.3.4 Gateways

Gateways relay the information on top of the application layer. When a gateway node receives some application service indication, it converts it in a service request on the other LAN on which it is connected. When the corresponding confirmation is received, it transforms it in a reply on the other side. Depending on the available services, an invocation may correspond to more than one request. Similarly, the reply may be constructed from a number of confirmations.

In order to overcome the additional delay in the reply to an indication on the LAN on one side, some gateways may make requests on the other side in advance. This is especially true in fieldbuses where the gateway may keep an image of all inputs on one side so that when a read indication is received on the other side, the cached value is returned in the response. We call this kind of gateway a proxy gateway.

20.4 Major Design Alternatives

If the interconnection mean is the prime choice, there are other degrees of freedom in the architecture of a mixed wired–wireless network:

- Single vs. multiple wireline segments. The wireline portion of the system may be composed of a single fieldbus segment on which all the devices are hooked, or it may be made of several segments that need to be interconnected.
- Single vs. multiple wireless cells. As in the previous case, the nodes that are connected through wireless means may be organized in a single cell or multiple cells.
- Separate or integrated wireless and wireline subnetworks. In the first case, the various nodes connected through wireless means form a single subnetwork that is interconnected to the wireline network at some points. This is very similar to the integrating of the wireless cellular network and the telephone system. The other solution is to have one or more subnetworks in which wireless and wired nodes are mixed.
- Single point or multiple points of interconnection. The interconnection between wireless and wireline transmissions can be done at a single point or at a number of points.

- *Ad hoc*, single-base station or multiple-base stations for the wireless subnetwork. In the first case, all nodes cooperate without any one playing a different role than the others. In the second option, a single node acts as coordinator for the traffic between the various wireless nodes. A third option is to have a number of these stations. And yet another option is whether the base station is the node that serves as the interconnection point.

Wireless networks vary along other parameters such as:

- Total or partial interconnection. All wireless nodes can see each other when total interconnection is achieved.
- Multiple hops or single hop. In the case of multiple hops, traffic from a node addressed to a node that cannot be seen (out of transmission range) is routed by other nodes to the destination.
- Satellite. The absence of full visibility can be compensated by a satellite node that retransmits on a different frequency all the emissions from the nodes. This is equivalent to a network of mobile nodes using a satellite to relay the communications.

However, we believe that these additional parameters are not prime parameters in an architectural model. We will not consider them further.

Although a given interconnection solution is a combination of these options, not all combinations lead to a feasible system. For instance, a wireless cell (data circuit) and a wired segment cannot be interconnected through more than a single repeater per cell. In the next paragraph, we will detail and discuss a number of proposals. When available, relevant references will be mentioned and commented upon.

20.5 Solutions for the Interconnection

In this section, we explore the design space and look at the feasibility of various solutions. The practicality of each solution is discussed, and reference to existing work is given when available. Solutions based on routers will not be explored, as most of the fieldbuses do not offer any network layer, thus excluding this solution. The only noticeable exception is LON (EIA-709.1, 1998), which has a seven-layer stack and provides routing between wired segments and wireless nodes.

20.5.1 Repeater-Based Solutions

A solution based on a repeater gives the impression that all nodes share the same medium. The same medium access control scheme must be used on the wireless part and on the wired part. This kind of interconnection has a number of advantages. There is no need for a base station in the wireless cells. It is easier to ensure periodicity because a single MAC scheme is used. The same applies to the guarantees for the latency bounds. However, the latency may increase as an effect of a higher number of nodes sharing the same medium. Furthermore, the higher BER on the wireless part has a direct impact on the frame error rate on the network. This causes higher latency bounds because of the necessary retransmissions. Finally, repeaters introduce some restrictions in the design freedom. Between one segment and another or a wireless cell, only a single repeater can be used. Furthermore, loops are excluded from the topology (for instance, S6 could not be a repeater in Figure 20.2). This precludes using multiple interconnection points.

The properties of wireless communications impact in the same way all the possible solutions. Assuming that the same bit rate is used on both sides of the repeaters, properties 1, 3, and 7 will limit the throughput and increase the latency. Property 5 imposes restrictions on the type of MAC protocol. All protocols that need collision detection cannot be used directly. Finally, Property 6 will increase the latency and response time for request–response protocols.

According to the degrees of freedom explained above, several solutions are possible:

- RPS1: Single wireless cell and single wired segment. This is an interesting solution to interconnect a number of isolated wireless nodes to a single segment. An example of this solution is given in

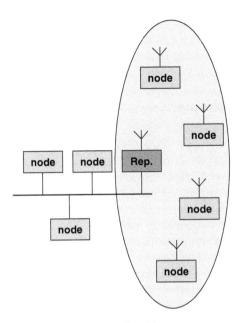

FIGURE 20.1 Repeater-based single-segment multiple-cell architecture.

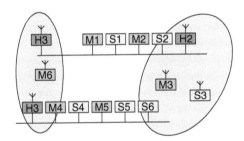

FIGURE 20.2 R-Fieldbus architecture. M = master station; S = slave station; H = repeater.

Morel et al. (1996) (Figure 20.1). The solution can support multiple wireless cells as long as the wireless nodes in two different cells do not receive the radio signals of each other.

- RPS2: Single wireless cell and multiple wired segments. When there is no isolated wireless node, this solution can be used to interconnect two or more wired distant segments together when no cable can be installed in between.
- RPS3: Multiple wireless cells and single wired segment. The wired segment may be considered a backbone for the wireless cells. Alternately, the cells may form a single subnetwork (using wireless-to-wireless repeaters) connected to the wired segment through a single repeater. For this, the wireless cells must overlap, and it is necessary to use a different channel for each cell. When wireless nodes may move and traverse different cells, they have to change the frequency of emission and reception. To trigger this process, special messages are sent regularly by a given node of the network. A mobile station that does not receive such a message starts to listen on other channels until it gets one. More sophisticated procedures are also used, in particular to reduce the time necessary to find the right channel (Rauchhaupt, 2002).
- RPS4: Multiple wireless cells and multiple wired segments. This solution is basically a generalization of the previous one. The integrated wireless and wireline case is the base of the R-Fieldbus approach (Rauchhaupt, 2002) (Figure 20.2). R-Fieldbus uses a Token Bus medium access control and is thus very sensitive to token losses. Because tokens go through the wireless cells, it is necessary to reduce the BER on the wireless part. This has been done using a different encoding,

direct-sequence spread spectrum, and special countermeasures on the receiver side (RAKE receiver). As the bit rate is not the same on the wired part (1.5 Mbit/s) and on the wireless part (2 Mbit/s), buffering is necessary, as explained above. Peter 1999 gives another example of this architecture.

20.5.2 Bridge-Based Solutions

Bridge-based solutions may be used when repeaters cannot be used. For instance, in the presence of a large number of nodes, a repeater-based solution may lead to unacceptable periods or latencies. Using a bridge will partition the nodes into two data links. The number of nodes that will share each link will hence be reduced. The latency will follow. Bridge-based solutions may offer a significant reduction in latency compared to repeater-based architectures. This is particularly the case when most of the traffic remains on each data link (cell or segment) and the traffic that goes through the bridge is minimal. Bridges also remove the need to use similar bit rates on the wired part and on the wireless one.

However, they are not without drawbacks. In a bridge-based solution, a frame emitted on the segment (or cell) on one side is retransmitted on the other using the medium access control protocol of the latter. There is hence an additional waiting time before the frame is actually emitted on the second segment. The maximum delay depends on the medium access control protocol used. With a token-passing protocol, the maximum delay is at minimum the token rotation time. If a master–slave configuration is used, the maximum delay is the maximum time that may elapse between any two successive polls.

In many cases, this delay has to be doubled. Many fieldbuses proceed by transactions in which one station, the initiator, sends a frame and the responder replies with another frame that can be an acknowledge or contain some data. In order to bound the transfer latency, the delay that may elapse between the initiator frame and the response is bounded by an upper limit. This is also used to set a timer that will trigger a retry if it times out. The bridge introduces an additional delay in the initiator frame as well as in the response when the initiator and responder are on different segments or cells. These two delays must be added to the time-out limit, increasing accordingly the maximum latency of the fieldbus. In other words, bridges have a negative impact on the real-time guarantees.

The properties of wireless communication have a lower impact than on repeater-based solutions. Property 1 may be mitigated using forward error correction on the wireless cell. Using error detection and consecutive retries would not be a good solution because this would increase further the maximum additional delay in the bridges. Imagine, for instance, the architecture depicted in Figure 20.3. A station on the wired segment initiates a transaction. The responder is a wireless node. One of the bridges retransmits the request over the corresponding wireless cell after some medium access delay. The

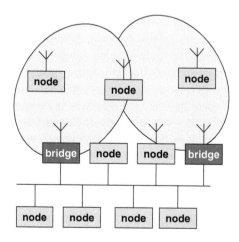

FIGURE 20.3 Isolated node approach.

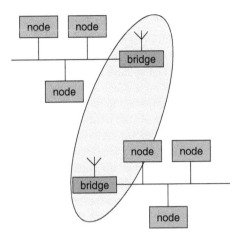

FIGURE 20.4 Interconnection of two wired fieldbuses through a wireless cell using tunneling bridges.

responder replies before the maximum response delay. The response is not received correctly by the bridge, which retries, thus extending the maximum delay before the response is sent to the initiator. The same happens if the reply is lost as a result of Property 7 or 9.

According to the degrees of freedom explained above, several solutions are possible:

- BRS1: Single wireless cell and single wired segment. The bridge may be used as the base station for the wireless cell. In the office world, this is typically what is obtained using an 802.11 cell coordinated by an access point connected to an 802.3 segment. In the fieldbus domain, this architecture would be an alternative to RPS1 that relaxes the constraints of the repeater but is likely to increase the achievable latencies and periods.
- BRS2: Single wireless cell and multiple wired segments. This solution has been suggested to interconnect two fieldbuses that cannot be linked through wires (Cavalieri and Panno, 1998). It is depicted in Figure 20.4. In the proposed solutions both segments use the same protocols and the connection is made through tunneling bridges. The solution offers a better reliability than solution RPS2 because the wireless part may be rugged. Buchholz et al. (1991) report a similar approach in the in-building context. The system is made of a number of Ethernet segments that are interconnected through a number of wireless cells. The options are identical, with two exceptions: there may be more than a single wireless cell, and due to the higher bit error rate, each Ethernet frame is segmented before it is encapsulated in an LLC frame of the wireless protocol.
- BRS3: Multiple wireless cells and single wired segment. This architecture comes in two flavors. Each cell may use a different channel, as presented by Leung (1992). In this approach, a number of wireless nodes are connected to a wired segment through a number of bridges. The set of wired nodes (including the bridges) uses a different protocol than the wireless protocol, but the functions of the data link layer are identical and the addressing is common to all nodes whether wireless or wired. The bridges are thus of the transparent or translating categories. To avoid duplicate transmission, bridges have to learn the topology so as to decide whether to forward a received frame. A second option is to use the same channel for all cells and rely on time or code division to separate the traffic.
- BRS4: Multiple wireless cells and multiple wired segments. Contrary to RPS4, in this solution there may exist more than a single path between any two nodes. Bridge protocols will ensure that no message looping will occur.

Bridges are sometimes interesting solutions either when repeaters cannot be used or to partition the network into smaller domains. Note that in general a bridge requires that the upper layers (network to application) are identical on both sides.

20.5.3   Gateway-Based Solutions

With gateways, operation of each segment or cell may be governed by completely different protocols at all Open System Interconnection (OSI) levels. This is useful when the wireline segment and the wireless cell are built around protocols that are not compatible at the data link layer, precluding the deployment of bridges. For instance, a protocol based on a client–server model such as Profibus (CENELEC EN 50170, 1996a) will not include the same information at the data link layer than a protocol based on producer–consumer on top of broadcast source addressing such as WorldFIP (CENELEC EN 50170, 1996b; Solvie, 1994). The first one will include the address of the source node and the address of the destination node. The second will only include the identification of the data. It is clearly impossible to use a bridge in such a case. It is also necessary when the application layers differ on each side. A wireless WorldFIP (Roberts, 1993) on one side and a wired CANOpen (CENELEC EN 50325-4, 2002) on the other, despite their compatible medium access control frames, cannot be bridged.

As both networks operate asynchronously, in terms of latency, gateway-based solutions suffer from the same drawbacks as the bridge-based solutions. The latency is even a little higher because of the additional overhead introduced by higher protocol layers. Except in special cases (see below), the worst-case latencies (worst-case response time) may be estimated as the sum of the values for the individual networks traversed increased by the penalty introduced by the gateways themselves. This may or may not be worse than using bridges, although as a rule of thumb, worse is more likely. Let us give a case in which the worst-case latency is better. In a bridge-based solution, the time-out length for transactions will be increased to take into account the longer response times of nodes that are on another data link (segment or cell). If we take into account these new values for time-out and the higher probability of loss due to the wireless part, the worst-case latency may be increased significantly. Basing the approach on a gateway means that the original values for time-outs are preserved. The higher probability is accounted on the wireless part (not on all transactions). The result is that the bridge-based solution will give a higher worst-case latency than the solution using a gateway.

An interesting solution to improve the worst-case latency or decrease the sampling period for process values is to have a gateway that acts as a proxy. The gateway responds on one of its sides as if it was a node of the network on its other side. An example of this approach is illustrated in Morel and Croisier (1995) (Figure 20.5). A wireless cell is connected through a single gateway to a WorldFIP (CENELEC EN 50170, 1996b) fieldbus. The gateway acts as a base station for the wireless cell. The gateway approach has been selected because of the strict temporal constraints in WorldFIP. This fieldbus requires that a maximum of 70 bit times separate the last bit of a request and the first bit of the corresponding response. This requirement cannot be fulfilled using bridges. It is very difficult to fill using repeaters. The gateway acts as a proxy representing the wireless nodes to the wireline part. The wireless cell is exploited in a round-robin manner according to a constant cycle. At the beginning of the cycle, the gateway broadcasts to all wireless nodes their update values in a single message. The wireless cells use this message as a sampling order, capture the values, and prepare their responses. Each node sends its response after the previous one. No explicit polling is done because of Property 6, which would increase the response time.

20.6   Amenability to Comply with the Fieldbus Requirements

Table 20.1 gives a summary of the capability to satisfy the fieldbus requirements (see Section 20.2.1) with the different architectures. For events and messages, latency is what matters. This issue has been discussed at length above and the results are just summarized in the table.

The objective to have periodic sampling and exchange of process data can be easily satisfied using some broadcast mechanism to sample and an adequate control of the medium access control (Fonseca, 1999) when repeaters are used. With bridges and gateways, the same level of control is possible in each individual segment or cell, but the combination does not give the required periodicity and simultaneous sampling. The solution lies in the definition of a distributed clock synchronization algorithm.

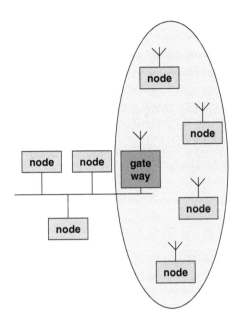

FIGURE 20.5 Interconnection using a proxy gateway.

TABLE 20.1 Repeater Ability to Help in Fulfilling the Fieldbus Requirements

Fieldbus Requirement	Degree of Fulfillment Using		
	Repeaters	Bridges	Gateways
Bounded latency	Easier to prove if MAC is adequate. Longer delay due to higher number of stations in the same link	Possible but more complex to calculate than with repeaters	Possible, sum of figures for each individual network
Periodicity	Easier as the same MAC scheme is used	Difficult to achieve unless synchronized clocks are used	Difficult to achieve unless synchronized clocks are used
Simultaneous sampling	Easily implemented through broadcast	Requires synchronized clocks	Requires synchronized clocks
Consistency indication	Simple to implement (i.e., CENELEC EN 50170, 1996)	May be obtained using time stamps	May be obtained using time stamps
Event ordering	May be based on frame ordering	Requires synchronized clocks	Requires synchronized clocks
Rugged	Higher BER in wireless part will increase overall frame error rate and latency	Better than repeater because wireless part may use a different protocol	Similar to bridge based case

Simultaneous sampling is then based on this clock. Accuracy is highly dependent on the medium access control protocol, the network adapter interface hardware, and the software implementation.

Absolute and relative temporal consistency may be obtained easily using mechanisms like those of WorldFIP (CENELEC EN 50170, 1996b) when repeaters are used. They fail when bridges and routers are used. Here again, synchronized clocks can be used, each information being stamped with its production instant, the stamp being transported with the data. By comparing the stamp values, the two kinds of consistency may be evaluated.

Event ordering may use the distributed clock. When repeaters are used, one can save the burden of implementing such protocols and use the frame sequencing provided the required granularity is not too small. Finally, the repeater-based solution is more sensitive to the higher bit error and frame drop rates (properties 1, 3, and 7 to 9), unless some of the wireless link is made more robust by some form of forward error correction or by improving the receiver design (Rauchhaupt, 2002).

20.7 Conclusion

There are numerous architectures that combine wireless cells and wireline segments. The choice of a solution for a given problem depends on the protocols used, the kind of guarantees offered, and the constraints, in particular with regard to time. Here we have explored the solution space and discussed the applicability of the various options, as well as their pros and cons. This study has also shown that some architectural choices may not be offered for some fieldbus protocols and some combinations of protocols. In general, the lower the interconnection in the OSI model the interconnection, the better the performances are. We have seen that this general rule must be carefully checked in each particular case. This can be verified in a few cited examples.

References

Buchholz, D. et al., Wireless in-building network architecture and protocols, *IEEE Network Magazine*, 5, 1991, 31–38.

Cavalieri, S. and Panno, D., On the integration of fieldbus traffic within IEEE 802.11 wireless LAN, in *Proceedings of the 1997 IEEE International Workshop on Factory Communication Systems*, 1997, pp. 131–138.

Cavalieri, S. and Panno, D., A novel solution to interconnect fieldbus systems using IEEE wireless LAN technology, *Computer Standards and Interfaces*, 20, 1998, 9–23.

CENELEC EN 50170, General Purpose Field Communication System, Vol. 2/3 (Profibus), 1996a.

CENELEC EN 50170, General Purpose Field Communication System, Vol. 3/3 (WorldFIP), 1996b.

CENELEC EN 50325-4, Industrial Communications Subsystem Based on ISO 11898 (CAN) for Controller-Device Interfaces: Part 4: CANopen, 2002.

EIA-709.1, Control Network Specification, March 1998.

Fonseca, J.A. and Almeida, L.M., Using a planning scheduler in the CAN network, in *7th IEEE International Conference on Emerging Technologies and Factory Automation*, October 1999, pp. 815–821.

ISO/IEC 7498:1, Information Processing Systems–Open Systems Interconnection, Basic Reference Model: The Basic Model, 1996.

ISO/IEC 8802.2, Information Technology: Telecommunications and Information Exchange between Systems: Local and Metropolitan Area Networks: Specific Requirements: Part 2: Logical Link Control, 1998.

Kopetz, H., Consistency constraints in distributed real time systems, in *Proceedings of the 8th IFAC Workshop on Distributed Computer Control Systems*, Vitznau, Switzerland, September 1988, pp. 29–34.

Leung, V., Diversity interconnection of wireless terminals to local area networks via radio bridges, *Electronics Letters*, 28, 1992, 489–490.

Morel, Ph., Intégration d'une Liaison Radio dans un Réseau Industriel, Ph.D. thesis 1571, Swiss Federal Institute of Technology (EPFL), Lausanne, 1996.

Morel, Ph. and Croisier, A., A wireless gateway for fieldbus, in *Sixth IEEE International Symposium on Personal, Indoor and Mobile Radio Communications PIMRC '95*, 1995, pp. 105–109.

Morel, Ph., Croisier, A., and Decotignie, J.D., Requirements for wireless extensions of a FIP fieldbus, in *Proceedings of 1996 IEEE Conference on Emerging Technologies and Factory Automation EFTA '96*, 1996, pp. 116–122.

Murdock, G. and Goldie, J., Build a direction-sensing bidirectional repeater, *Electronic Design*, 37, 105–108, 110, 1989.

Perlman, R., *Interconnections: Bridges and Routers*, 2nd edition, Addison-Wesley, Reading, MA, 2000.

Peter, M., The use of radio technology in the fieldbus area: using Interbus as an example, in *Proceedings of FeT '99*, Magdeburg, Germany, September 1999, pp. 55–60.

Pleinevaux, P. and Decotignie, J.D., Time critical communication networks: field busses, *IEEE Network Magazine*, 2, 1988, 55–63.

Rauchhaupt, L., System and device architecture of a radio based fieldbus: the R-Fieldbus system, in *Proceedings of the 2002 IEEE International Workshop on Factory Communication Systems*, 2002, pp. 185–192.

Roberts, D., OLCHFA: a distributed time-critical fieldbus, in *IEE Colloquium on Safety Critical Distributed Systems*, 1993, pp. 6/1–6/3.

Saba, G., Mammeri, Z., and Thomesse, J.P., Some solutions for FIP networks interconnection, in *Proceedings of WFCS '95*, Leysin, Switzerland, October 1995, pp. 13–20.

Solvie, M., Configuration of distributed time-critical fieldbus systems, in *Proceedings of 2nd International Workshop on Configurable Distributed Systems*, 1994, p. 211.

Varghese, G. and Perlman, R., Transparent interconnection of incompatible local area networks using bridges, *IEEE Journal on Selected Areas in Communications*, 8, 42–48, 1990.

Section 4.5

SEMI

21

SEMI Interface and Communication Standards: An Overview and Case Study

A.M. Fong
Singapore Institute of Manufacturing Technology

K.M. Goh
Singapore Institute of Manufacturing Technology

Y.G. Lim
Singapore Institute of Manufacturing Technology

K. Yi
Singapore Institute of Manufacturing Technology

O. Tin
Singapore Institute of Manufacturing Technology

21.1   Introduction to SEMI

Semiconductor Equipment and Materials International (SEMI), founded in the U.S. in 1970, is a global industry association with a worldwide membership of 2500 companies, representing semiconductor-related industries. SEMI has 17 subcommittees that include, among others, the Subcommittees for Automated Test Equipment; Environment, Health and Safety (EHS); and Information and Control.

This chapter will give an overview of the fundamentals of the SEMI Equipment Communication Standard, commonly referred to as SECS, its interpretation, the available software tools, and case study applications.

21.2   The SEMI Equipment Communication Standard

The SEMI Equipment Communication Standard (SECS) defines computer-to-computer communication protocols intended to assist in the automation of electronics manufacturing facilities. SECS enables the automation, communications, and control of semiconductor process equipment, from a variety of vendors, using a set of standards and reliable protocols. These standards have the capacity to resolve/define

TABLE 21.1 SEMI Equipment Communication Standard

Protocol	SECS-I	SECS-II	GEM	HSMS
Date of launch Message definition	1978 → Communication protocol and physical definition	1982 → Message format	1992 → Additional SECS-II sequences Defines document for communication	1994 → Supersede SECS-I for TCP/IP compatibility

the physical layers, signals, block semantics, etc., thus ensuring repeatable, nondeadlocked communications from simple to extended-hierarchy networks.

In its basic form, SECS provides a one-to-one link between a particular unit of equipment and a host computer. The equipment can be in the form of wafer processing equipment, test and metrology equipment, die/package assembly equipment, surface mount equipment, and others.

Similarly, the host computer can range from minicomputers, small dedicated desktop PCs, and laptops, to a network of embedded systems. In many cases, the host computer system manages and coordinates equipment on the factory floor via cell controllers, line controllers, and material controllers. These must be compatible with current *Manufacturing Execution Systems and Manufacturing Legacy Systems* (MLS).[1]

21.2.1   The Four Main SEMI Communications/Control Standards

SEMI Equipment Communications are based on four main standards: SEMI Equipment Communication Standard I (SECS-I), SEMI Equipment Communication Standard II (SECS-II), High-Speed SECS Message Services Standard (HSMS), and Generic Equipment Model Standard (GEM), as shown in Table 21.1. Both SECS-I and HSMS define the message transport layer of the protocol. SECS-II defines the message format, and GEM defines the equipment behavior.

21.2.1.1   The Components for Executing the SECS Protocol

Host systems are linked to different equipment that have to be SECS compliant (i.e., based on SECS). The connectivity to the host can be through serial or Transmission Control Protocol (TCP)/Internet Protocol (IP) ports. GEM is incorporated in some of the newer equipment. Older equipment is generally not GEM compliant, having only serial connectivity. An example of *SECS Implementation Architecture* is shown in Figure 21.1.

21.2.1.1.1   Host System
The main function of the host system is to manage the connectivity to a group of equipment via a configuration file, recipe management, equipment load balancing, equipment monitoring, and connection to enterprise-level applications. The host systems that manage the same type of equipment are referred to as cell controllers. Similarly, the host systems that manage and coordinate different equipment are referred to as line controllers.

21.2.1.1.2   Terminal Server
This is a serial hub (also sometimes called a serial device server) that allows equipment with serial port connections to connect to this hub and provides information sharing among other equipment through Ethernet-ready TCP/IP connections.

21.2.1.1.3   Equipment 1 (Non-GEM Compliant)
This is normally the older semiconductor equipment that is not GEM compliant. Most of the older equipment will only provide serial connections to the host system.

21.2.1.1.4   Equipment 2 (HSMS Compliant)
This is typically the newest semiconductor equipment that has point-to-point communication protocol for TCP/IP Ethernet communications. Some of the equipment is already GEM compliant for host/equipment communications.

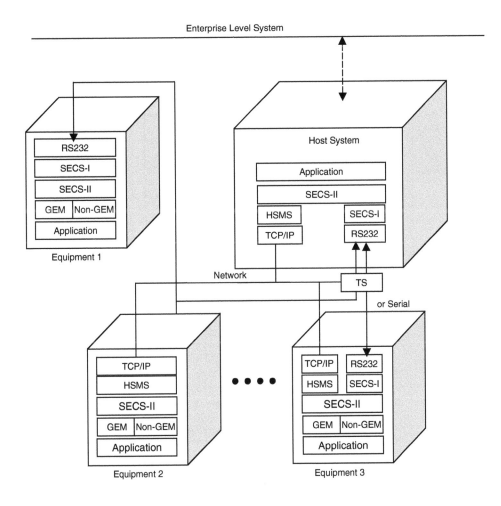

FIGURE 21.1 The components for the execution of SECS protocol.

21.2.1.1.5 Equipment 3 (GEM and HSMS Compliant)

These are typically the up-to-date models of semiconductor equipment that are both GEM and HSMS compliant, having both serial and TCP/IP connections. Nevertheless, some of this equipment is expected to be simultaneously serial port and TCP/IP capable. Usually there is a problem of *connectivity to the Manufacturing Legacy System*, which we cannot easily ignore.

21.2.2 SEMI Equipment Communication Standard I

SECS-I is also called SEMI E4. It was launched in 1978 and released in 1980. It describes a transmission interface for passing messages between the equipment and a host over the RS-232 communication, as shown in Figure 21.2. This standard defines point-to-point communication of data, utilizing a subset of the international standard EIA RS-232-C for the connector and voltage levels. The communication is bidirectional and asynchronous, but is half duplex. The communication speed normally varies from 9,600 to 19,200 baud rate. SECS-I protocol establishes multiblock transfers based on blocks of 256 bytes.

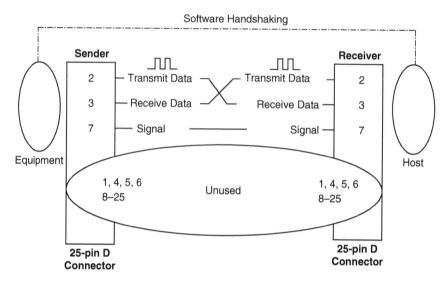

FIGURE 21.2   Three-wire serial data configuration for SECS-I (with software handshake).

TABLE 21.2   Layer Definition of the SECS-I Standard

SECS-I Layers	Definitions for
Transaction layer	Primary message
Message layer	Block header
Block layer	Block format, block protocol, packet format, packet protocol
Physical layer	Physical connector, voltage, bit coding, baud rate

SECS-I includes specifications for the electrical connection between two communicating entities. It defines the message header structure, the block transfer protocol, and the message transfer protocol. The block transfer protocol is the procedure used by the serial line to establish the direction of communication and to provide the environment for passing messages. The message transfer protocol details how the block transfer protocol is used to send and receive messages.

Table 21.2 shows the layer definition of SECS-I. The physical layer defines the physical connectors, electrical characteristics, bit and character coding, and baud rate. The block layer defines the block or packet format and protocols that are used to transfer a single block. The message layer defines the block header. The transaction layer defines how primary messages and their corresponding reply messages are related.

21.2.2.1   SECS-I Message Structure

Figure 21.3 shows the structure of the SECS-I message. This message structure allows multiblock messages with one stream and function. The alphabet in Figure 21.3 can be described as follows:

a. The first byte represents the length of the message.
b. Bytes 0 and 1 of the header represent the device ID. The device ID is to identify the equipment in the factory. There is no device ID for the host system. Bit 8 of the first byte is a reserved bit or R-Bit, which is used for indicating the direction of the message, i.e., host to equipment or equipment to host. If the R-Bit is 0, the message is from the host to the equipment. If it is 1, the message is from the equipment to the host.
c. Byte 2 of the header represents the stream number for the message. The stream represents the category of the messages that are defined in SECS-II.
d. Byte 3 of the header represents the function numbers within the stream.

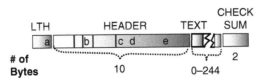

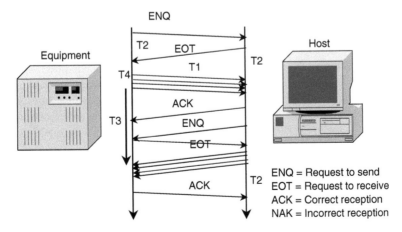

a: Length of message
b: Device ID
c: Stream
d: Function
e: Block number
f: System Bytes

FIGURE 21.3   Structure of SECS-I message.

   e. Bytes 4 and 5 of the header represents the block number of the message. If the message is longer than 244 bytes, the message will be broken into multiple blocks. The system will use these two bytes to identify the sequence of the block number.

   f. Bytes 6 to 9 of the header represent the system bytes. The purpose of system bytes is to avoid duplication of a message.

21.2.2.2 Synchronization Mechanism

SECS-I uses software-based time-outs to handle communication synchronization between host system and equipment. As shown in Figure 21.4, time-outs (T1 to T4) are used in SECS-I for managing the transmission of blocks between the sender and receiver. Each time-out has a special function, as explained below:

   T1 defines intercharacter time-out in the block transfer protocol. It measures the duration of characters transferred within the same block. T1 time-out will be activated if the host system does not receive the block of characters within the defined time.

   T2 defines the protocol time-out. It limits the time required to communicate among the following messages:

    • Request to send (ENQ) and request to receive (EOT)

    • EOT and the receipt of the block of characters

    • Sending block of characters and correct reception (ACK)

   T3 defines the time-out for the reply message. It sets the maximum waiting time for multiple-block message communications. Within a specified time, if the last block is not received, T3 time-out will be triggered.

FIGURE 21.4   Establishing synchronization via time-outs (T1 to T4).

TABLE 21.3 Time-Outs (T1–T4) Definition

Time-Out	Description
T1	Intercharacter time-out
T2	Protocol waiting time time-out
T3	Time-out for reply message
T4	Interblock communication time-out

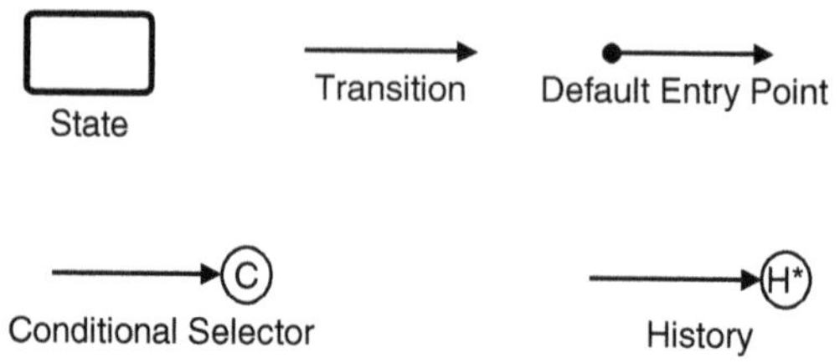

FIGURE 21.5 Harel's notation of state models.

T4 defines the time-out for interblock communications. It sets the time limit to receive the next block of messages within the multiblock communications. If the next block of messages is not received within the specific time, T4 will be activated.

SECS-I provides a mechanism for synchronization and software deadlock resolution via four time-outs (T1 to T4), as shown in Table 21.3.

21.2.3 High-Speed SECS Message Services Standard

With the increasing trend toward network communication for high-performance components and hardware, HSMS was introduced in 1995. HSMS (also referred to as E31) defines a communication interface suitable for the exchange of messages between computers via TCP/IP. HSMS uses TCP/IP streams to provide reliable two-way simultaneous transmission of streams of adjoining bytes. It can be used as a replacement for SECS-I communication as well as other more advanced communication environments.

It was intended as an alternative for applications where higher-speed communication is needed or when a simple point-to-point topology is insufficient. Most of the new SECS/GEM implementations use this standard for the message transport layer. HSMS was designed with the intention that existing SECS-II- and GEM-compliant applications could be easily converted to use HSMS instead of SECS-I. HSMS also provides the means for independent equipment manufacturers to produce their own SECS-II/GEM implementations while still maintaining interequipment compatibility.

21.2.3.1 HSMS State Model

The *state model* is a methodology developed by David Harel and adapted by SEMI to describe the expected functionalities in the standard. As shown in Figure 21.5, a state is a static condition of an operation. A state could have multiple substates. The arrows represent the transitions from one state to another state. In Figure 21.6, the round black dot is the start of the process. When HSMS first starts, it will be in the "TCP/IP not connected" state. If the TCP/IP connection is successful, then the process will move to the "TCP/IP connected" state. In the "TCP/IP connected" state, there are two substates: "not selected" and "selected." The system could be "connected on TCP/IP but HSMS has not been selected." When HSMS is selected, it will go into a substate called "selected" and stays there until a T3 time-out occurs. When "connection terminated or selection fails" occurs, the state will go back to "TCP/IP not connected."

21.2.3.2 HSMS Message Structure

Figure 21.7 shows the structure of the HSMS message. The message structure is similar to SECS-I, but it is for TCP/IP network communication. The alphabet in Figure 21.7 can be described as follows:

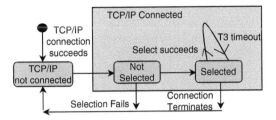

FIGURE 21.6   State model of HSMS.

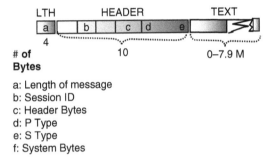

a: Length of message
b: Session ID
c: Header Bytes
d: P Type
e: S Type
f: System Bytes

FIGURE 21.7   Structure of HSMS message.

a. First through fourth bytes represent the length of the message. HSMS can handle about 8 Mb of data. The length defines the total length of the header and the actual message.
b. Bytes 0 and 1 of the headers represent session ID. Session ID is used to identify whether the communication message is a control message or subsequent data information.
c. Bytes 2 and 3 of the header represent the stream and function number or status code of the message.
d. Byte 4 of the header represents the presentation type (P-type) of how the message is encoded. A value 0 means that it will be used in SECS-II. Values 1 to 127 are reserved for subsidiary standards. Values 128 to 255 are reserved but not used.
e. Byte 5 of the header represents the session type (S-type). Value 0 means that the session is sending data information. Other values indicate that the message is control information.
f. Bytes 6 to 9 of the header represent the system bytes. The purpose of system bytes is to avoid duplication of a message.

21.2.3.3   HSMS Subclassification

SEMI has subclassified HSMS into single session (HSMS-SS) and general session (HSMS-GS), as shown in Figure 21.8. HSMS-SS defines a single-session equipment connection so that the implementation does not have to consider other equipment or knowledge of other systems. It will be similar to the SECS-I connection, which is a one-to-one connection. The HSMS generic definition and structure will still apply. HSMS-GS is used for complex systems that have a lot of independent equipment as subsystems. HSMS-GS defines the convention for identifying independent system entities.

21.2.4   SEMI Communication Standard II

SECS-II was released in 1982. SECS-II (also referred to as E5) defines the message content passing between the equipment and the host. SECS-II specifies the format of the text and the minimal unit of semantics, but it does not specify the exact action that a recipient machine should take. With the definition of SECS-II, the host–equipment software may be constructed with minimal knowledge of each other. In addition, SECS-II allows the creation of user-specific messages.

SECS-II defines the method of conveying information between equipment and host in the form of messages. Every SECS-II message is identified by a stream and a function number. Each of the streams

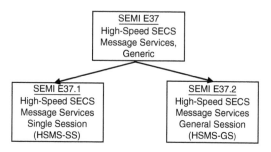

FIGURE 21.8　Subclassification of HSMS.

TABLE 21.4　Streams Defined in SECS-II

Stream	Description
1	Equipment status
2	Equipment control and diagnostics
3	Material status
4	Material control
5	Exception handling
6	Data collection
7	Process program management
8	Control program transfer
9	System errors
10	Terminal services
11	Host file services
12	Wafer mapping
13	Data set transfer
14	Object services
15	Recipe management
16	Processing management
17	Equipment control and diagnostics

represents the category of service available, as shown in Table 21.4. The stream will contain specific messages called functions. It also defines the request for information and the corresponding data transmission. Each combination of stream and function represents a distinct message classification. In every stream, function 0 is reserved for aborting transactions and the others are constructed in pairs. For example, Table 21.5 shows that each pair of odd-numbered function codes is reserved for the primary message and even-numbered function codes for the reply, or the secondary message. If the primary message does not require a reply message, the even-numbered message is discarded.

TABLE 21.5　Odd/Even Message Number of SECS-II Function

Stream	Function	Description
1	0	Abort transaction (S1 F0)
1	1	Are you there request
1	2	Online data
1	3	Select equipment status request
1	4	Select equipment status data
1	5	Formatted status request
1	6	Formatted status data
…	…	…

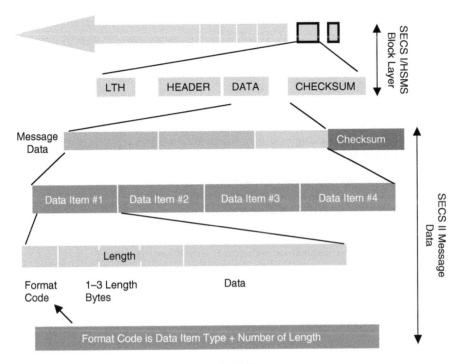

FIGURE 21.9 Data structures of SECS-I, SECS-II, and HSMS.

21.2.4.1 Data Structure

The messages that are sent through the physical layer through serial (RS-232) or TCP/IP (Ethernet network) are broken down by SECS-I or HSMS definitions, yielding the length, header, and data blocks. SECS-II defines the structure of messages into entities called items and lists of items. This structure allows for a self-describing data format to guarantee proper interpretation of the message. Interchange of messages is governed by a set of rules for handling messages called the transaction protocol. SECS-II further defines the data into a multiple-block structure, yielding the format code, length, and data item. All messages are packed into multiple data blocks with headers and checksums, as shown in Figure 21.9.

21.2.4.2 Streams and Functions in SECS-II

Table 21.4 shows the category of various SECS-II messages that are classified under the different streams. As an example, all messages in stream 1 will be part of the equipment status category. This category will help the user and integrator in understanding, designing, and developing their SECS solutions.

Each stream can specify multiple functions, as shown in Table 21.5, where an example for stream 1 is explained. Function 0 is reserved for abort purposes. The remaining functions are operated upon in pairs of requests (odd function or primary message) and reply/acknowledgments (even function or secondary message). The primary message is the request message sent by either the host or equipment. All the odd-numbered functions, e.g., S1 F1, S2 F1, etc., are examples of primary messages. The user can indicate if a reply is needed for the primary message using the wait bit (W-Bit). The W-bit is the first bit from the header byte, which is used to indicate that the sender of a primary message expects a reply. The secondary message is the reply message. All the even-numbered functions, e.g., S1 F2, S2 F2, etc., are examples of secondary messages. For any messages that cannot be processed by the equipment, the appropriate error message on stream 9 will be used. Upon detection of a transaction time-out, the equipment sends S9 F9 to the host. Upon receipt of function 0 as a reply to a primary message, the related transaction is terminated. No error message should be sent to the host by the equipment.

21.2.4.3 Conversation Protocols

A conversation is a series of one or more related SECS-II messages used to complete a specific task. A conversation should include all transactions necessary to accomplish the task and leave both the originator and interpreter free of resource commitments at its conclusion. There are a few types of conversations and they are discussed below.

21.2.4.3.1 Primary Message with No Reply
Once the W-Bit is set to zero, the primary message will not require a reply. It is the simplest conversation. The primary message will send a single block of information without reply. This conversation is used where the originator can do nothing if the message is rejected.

21.2.4.3.2 Primary Message with Data Return (Request/Data Conversation)
The originator will need to get information from the interpreter. The W-Bit will have to be set to 1 and data are returned as a secondary reply message.

21.2.4.3.3 Send/Acknowledge Conversation
The originator sends data in a single block and expects an acknowledgment.

21.2.4.3.4 Inquire, Grant, Send, and Acknowledge Conversation
The originator must ask for permission from the interpreter. The interpreter will grant permission as well as set the time-out for getting the information. If the wait time is longer than the time-out time, S9 F13 will be sent out. Once the information is received, the acknowledgment will be sent. This type of conversation is commonly used in equipment-to-host conversations, and only the equipment should send error messages to the host.

21.2.4.3.5 Unformatted Data Set Conversation
Stream 13 is used for unformatted data conversation. Unformatted data could be in the form of a bitmap file or a binary file.

21.2.4.3.6 Material Handling Conversation
Stream 4 is used for material control and handling of material between equipment.

21.2.4.3.7 Conversation with Delay
The originator may request information from the interpreter, which requires some time to obtain. The interpreter could collect the information immediately and return. The interpreter could also indicate if the information has be obtained or returned in the subsequent transaction.

Messages can also be transferred as single-block messages or multiblock messages. The SECS-I maximum message length for a single block is 244 bytes. The maximum message length for HSMS is 7.9 Mbytes. For longer messages, SECS-II will have to use multiblock messages. For example, if messages that are longer than 244 bytes are needed, they have to be sent by multiple blocks or packets. Each packet will have a block number so that the receiver will be able to join all the blocks together.

21.2.4.4 Sample of SECS-II Message

21.2.4.4.1 Stream 1 Function 1

S1, F1 Are You There Request (R)[a]   S,H <> E, Reply[b]

Description[c]: To establish if the equipment is online. A function 0 response to this message means the communication is inoperative. In the equipment, a function 0 is equivalent to a time-out of the receive timer after issuing S1, F1 to the host.

Structure: Header only[d]

The above example is from the SEMI standard E5.

S1, F2 Online Data (D)   S,H <>E

Description: Data signifying that the equipment is alive.

Structure:[e] L, 2

1. <MDLN>

2. <SOFTREV>

Exception: The host sends a zero-length list to the equipment.

a. Shows the stream (S), function (F), and title of the function (R).
b. Shows the type of message. S represents single-block message. Multiblock message will be represented by M. H <> E means the message could be sent from either equipment (E) or host (H). Other alternatives: E > H, equipment sends to host, or H > E, which is from host to equipment. The message requires "reply." Other alternatives are "optional to reply" or "blank," which means no reply is needed.
c. The description of the message.
d. Shows the structure of the message. In this case, it is just the header of the message and there is no data.
e. Shows the structure of S1, F2. L, 2 means that it has two listed items. The first listed item is <MDLN>, which represents the equipment model type with maximum of six bytes in length. The second item is <SOFTREV>, which represents software revision with a maximum of six bytes as well.

As shown in Table 21.6, SECS-II only uses streams 1 to 14 and functions 1 to 63. In order to provide for expansion of features, streams 15 to 63 are reserved for future SEMI standards. Streams 64 to 127 and functions are available for custom messages.

While SECS-II makes room for vendor-independent compatibility, problems do arise when a vendor applies different semantics for the same message pairs. This creates intervendor incompatibility. GEM was created to define generic equipment features so that intervendor compatibility can be established.

21.2.5   Generic Equipment Model

The Generic Equipment Model Standard (also referred to as E30) was published in 1992. It defines which SECS-II messages should be used, in which situation, and what the resulting activity should be.

GEM is intended to specify the following:

- A behavior model of semiconductor equipment in a SECS-II communication environment
- A description of information and control functions needed in a semiconductor manufacturing environment
- A definition of the basic SECS-II communication capabilities

TABLE 21.6   Status of Streams and Functions (Reserved/Unreserved)

Stream	Status
1–14	Currently defined in the standard
15–63	Reserved for standard messages
64–127	Reserved for user-defined messages

Function	Status
1–63	Reserved for standard messages
64–255	Reserved for user-defined messages

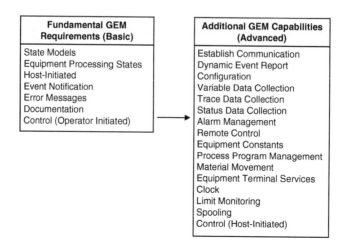

FIGURE 21.10　GEM requirements.

- A single consistent means of accomplishing an action when SECS-II provides multiple possible methods
- Standard message dialogues that are necessary to achieve useful communication

The GEM standard contains two types of requirements:

- Fundamental (or foundation) requirements
- Requirement for additional GEM capabilities

As shown in Figure 21.10, the fundamental GEM requirements form the foundation of the GEM standard, which are basic requirements for all types of semiconductor equipment. The additional GEM capabilities provide the functionality required for some instances of factory automation or functionalities that are applicable to specific types of equipment. Capabilities are operations performed by semiconductor manufacturing equipment. Each of the capabilities consists of a statement of purpose, pertinent definitions, a detailed description, requirements, and various scenarios. A scenario is a group of SECS-II messages arranged in a sequence to perform a function. These scenarios handle communications that involve more than one pair of messages.

21.2.5.1　GEM Fundamental Requirements

21.2.5.1.1　State Models

State models describe the behavior of the equipment from a host perspective. The same model can be used in different equipment for identical functions. The fundamental GEM specifies the state model of process (equipment specific), communication, alarm, control, material movement, and spooling.

Figure 21.11 shows a sample state model for equipment processing. There are also step numbers defined in the diagram. Here is how the state model can be interpreted:

Step 1: The process will be initialized (INIT). After being initialized, the process could be in the "idle" state.

Step 2: The next state is "processing active." It will be in the substate of "process" doing an operation of "setup." The setup operation could be semimanual, as the operator needs to log on and fine-tune the equipment.

Step 3: After setup, the equipment will be in the "ready" state for operation.

Step 4: In normal processes, the equipment should be either in the "executing" state, producing some part or component, or in step 5, which is "idle," waiting for the next part or component.

Step 5: The equipment is idle and waiting for the next part.

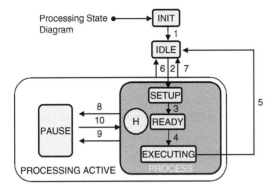

FIGURE 21.11 Equipment processing states.

TABLE 21.7 Example of GEM Transactions for Event Reporting Setup

Step	Host	Direction	Equipment
1	Send report definitions	→	Send acknowledgment
	S2, F33	←	**S2, F34**
2	Send event and report link definitions	→	Send acknowledgment
	S2, F35	←	**S2, F36**
3	Send event-enabling information	→	Send acknowledgment
	S2, F37	←	**S2, F38**

Step 6: The equipment could be in "processing active" and receive a "STOP" command from the operator or host system. This will put the equipment back to the "idle" state.

Step 7: The equipment could be in "processing active" and receive an "ABORT" command from the operator or host system. This will put the equipment back to the "idle" state.

Step 8: The equipment may receive an alarm and it will then be in the "pause" state.

Step 9: The equipment may receive a pause command by the operator or host system.

Step 10: The equipment may receive a resume command by the operator or host system.

An example of the GEM transactions for event-reporting setup is illustrated in Table 21.7.

Step 1: Host defines the report and request for reports.
- The host initiates a report definition by sending S2, F33. Stream 2 is the equipment control and diagnostics stream and function 33 is to "define report." This message allows the host system to define and request the group of reports available from the equipment. A unique ID identifies each report.
- The equipment receives and acknowledges by replying with S2, F34. If the primary message has errors or the report ID is not found, the primary message will be rejected.

Step 2: Host sets an ID for the required reports.
- The host initiates an "event and report" link definition by sending S2, F35. Function 35 will only link the reports that are requested by the host, with an ID. The link will be disabled until the equipment is ready to send.
- The equipment receives and acknowledges by replying S2, F36. Again, if there is any error condition, the whole message will be rejected.

Step 3: The host enables the sending of the reports.
- The host sends the event-enabling definition by sending with S2, F37. At this stage, the equipment knows what reports the host requires and the host will enable the sending process.

- The equipment receives and acknowledges by replying with S2, F38. After the equipment acknowledges the message from the host system, it will then proceed to send the reports to the host system.

21.2.5.1.2 Host Initiates S1, F13; S1, F14

It is a fundamental requirement in GEM that the required equipment be able to accept a connect request from the host. The host can request connections by sending S1, F13 to the equipment. Upon receipt of the message, the equipment will send the S1, F14 "connect acknowledge" reply.

21.2.5.1.3 Event Notification

This capability provides data to the host at specified points during equipment operation. The equipment will send S6, F5 to get permission from the host. The host will reply with S6, F6 to grant permission. Once the permission is granted, the equipment will send the report using S6, F11. The host will acknowledge the report using F6 F12.

21.2.5.1.4 Online Identification

Online identification is the most basic SEMI requirement. The originator (which could be host or equipment) will send S1, F1 to ask if the interpreter (which could be equipment or host) is there. The interpreter will then reply using S1, F2, which has the information of the equipment type and revision.

21.2.5.1.5 Error Messages

Fundamental GEM requires the system to be able to detect communication link errors in SECS-I and to detect SECS-II and GEM format errors. It will also need to support all stream 9 messages and provide appropriate error-handling routines.

21.2.5.1.6 Documentation

Fundamental GEM requires that the equipment should be able to provide a detailed specification document for the SECS–GEM interface on the equipment.

21.2.5.1.7 Operator-Initiated Control

Fundamental GEM requires that the equipment should have operator-initiated control-related capabilities to allow the configuration and manipulation of the control state model. In this way, the host or user may modify the equipment's control-related behavior. The operator at the equipment will send S1, F1 to the host. The host will send S2, F18 to grant an online link to equipment. The equipment will send a control state local event report using S6, F11. The host will acknowledge the report with S2, F12.

21.2.5.2 Additional GEM Requirement

"Additional GEM" requires that the system be able to support communication state model, operator communication state display, operator enable and disable commands, enabled or disabled power-up states, and establish communications timer equipment constants.

21.2.5.2.1 Establish Communications

The "establish communications" capability provides a means of formally establishing communications upon system initialization or upon any loss of communication between communication partners. Notification of the period of noncommunication is then possible. The host system initiates communication by sending S1, F13, and the equipment will reply using S1, F14.

21.2.5.2.2 Event Notification

This capability provides data to the host at specified points during equipment operation. The equipment will ask the host for permission to send by using the S6, F5 message. The host will give permission using the S6, F6 message. Once the permission is given, the equipment will send the event report using S6, F11, and the host will acknowledge the report using S6, F12.

21.2.5.2.3 *Dynamic Event Report Configuration*

This capability provides the data-reporting flexibility required in some manufacturing environments. The host will send S2, F39 to request permission from the equipment and will get the permission from the equipment using S6, F40. Having received the permission, the host will send S2, F33 to define the report requirements. The equipment will acknowledge the report requirements by S6, F34. The host will send S2, F35 to collect the report from the equipment, and the equipment will then send the report using S6, F36.

21.2.5.2.4 *Variable Data Collection*

This capability allows the host to query for the equipment data variables and is useful during initialization and synchronization. The host will send S2, F19 to ask for the variable data, and the equipment will send the data using S6, F20.

21.2.5.2.5 *Trace Data Collection*

The trace data collection provides a method of sampling data on a periodic basis. The host will initiate the trace of certain data using S2, F23. The equipment will initialize the trace data and acknowledge using S2, F24.

21.2.5.2.6 *Status Data Collection*

This capability allows the host to query equipment for selected information and is useful in synchronizing with equipment status. The host will send S1, F11 to request for the variable equipment status, and the equipment will reply the variable status using S1, F12.

21.2.5.2.7 *Alarm Management*

This capability provides for host notification and management of alarm conditions occurring on the equipment. The host will ask for alarm data and text using S5, F5. The equipment will send the alarm data to the host using S5, F5.

21.2.5.2.8 *Remote Control*

This capability provides the host with a level of control over equipment operations. The host will send the remote command using S2, F41. The equipment will acknowledge using S2, F42 and send the data report using F6, F11. The host will acknowledge the report using S6, F12.

21.2.5.2.9 *Equipment Constants*

This capability provides a method for the host to read and change the value of selected equipment constants on the equipment. These equipment constants include nonvolatile storage, validation/verification of equipment constants, host equipment constants name, list request messages, etc. The host will send the new equipment constant using S2, F15, and equipment will send the equipment constant using S2, F16. The host will then send S2, F13 to request for equipment constants, and the equipment will send the equipment constants using S2, F14.

21.2.5.2.10 *Process Program Management*

This capability provides a means to transfer process programs and to share the management of those process programs between the host and equipment. The host will ask for the equipment process program using S7, F19, and the equipment will send the process program data using S7, F20. To upload the process program, the host will send S7, F5 and the equipment will send the process program using S7, F6. To download the process program, the host will send S7, F1 and the equipment will grant permission using S7, F2. Once permission is received, the host will send the process program using S7, F3 and the equipment will acknowledge with S7, F4.

21.2.5.2.11 *Material Movement*

This capability includes the physical transfer of material among equipment, buffers, and storage facilities. The transfer of material can be performed by the operator, AGV (autonomous guided vehicle) robots, tracks, or dedicated fixed material-handling equipment. The equipment will send collection of event using S6, F11, and the host will acknowledge with S6, F12.

21.2.5.2.11.1 Equipment Terminal Services — This capability allows the host to display information on the equipment's display device or the operator of the equipment to send information to the host. The host will send textual information to the equipment using S10, F3. The equipment will display the information and acknowledge with S10, F4.

21.2.5.2.12 Clock

The clock capability enables host management of time-related activities and occurrences associated with the equipment and across multiple pieces of equipment. The equipment will send S2, F17 to request time information from the host, and the host will respond with a time value using S2, F18.

21.2.5.2.13 Limits Monitoring

This capability relates to the monitoring of selected equipment variables. The host will get the equipment's current limit-monitoring status using S2, F47, and the equipment will send the status using S2, F48.

21.2.5.2.14 Spooling

Spooling is a capability whereby the equipment can query a message intended for the host during times of communication failure and subsequently deliver this message when communication is restored. The host will define messages that need spooling in case of communication failure using S2, F43. The equipment will reset the spooling message and acknowledge with S2, F44.

21.2.5.2.15 Host-Initiated Control

The control-related capabilities allow for configuration and manipulation of the control state model. In this way, the host or user may modify the equipment's control-related behavior. The host will request to be offline using S1, F15. The equipment uses S1, F0 to abort the transaction, sends S1, F16 to acknowledge, and uses S6, F11 to send the event report. The host will acknowledge with S6, F12.

21.2.6 Limitations of SECS/GEM

The limitations faced by the currently defined SECS/GEM standards are:

a. *Manual integration cannot be totally avoided.* The factory automation system cannot query SECS/GEM software to determine its full capabilities because different vendors have customized the same message pairs for different applications. Therefore, SECS requires significant manual intervention/customization efforts for each application function.

b. *Weak security mechanism.* SECS is weak in its security mechanism. There is no concept of client authentication or access permission in SECS. Issues of security are becoming important operational considerations because of the need for remote equipment engineering and remote diagnostics.

c. *Single-client architecture.* The implementation of SECS is based on a point-to-point link, and there is only one software process (i.e., one point) on the factory side. This is a critical limitation because the newly developed factory applications such as advanced process control (APC), factory automation, and advanced planning and scheduling systems (APS) require a steady stream of accurate real-time data, even in the form of peer-to-peer communications, to facilitate a decision support process.

21.2.7 Emerging and Evolving New Standards

The industry associations such as SEMI and SEMATECH, as well as universities, have tried to develop a new standard that can handle the various applications needed by today's semiconductor manufacturing environment. SEMI communication and interface standards have adopted some of the popular object-oriented programming, like Extensible Markup Language (XML) and distributed computing techniques, but these are still not very commonly used. Table 21.8 shows the trend in the SEMI communication standard.

TABLE 21.8 Trend of the SEMI Standard

First Generation	Second Generation	Third Generation	Next Generation
Characteristics	Characteristics	Characteristics	Characteristics
Simple text-based reporting	Some recipe management	Full recipe management	Intelligent control
	Some equipment performance matrices	Equipment performance matrices	Remote diagnostics/ monitoring
	Low standardization	SEMI standardization	Mainstream: distributed computing standards
	Standards	Standards	Standards
	SECS-I	SECS-I/SECS-II	Object-based standards
	De facto	GEM	
	Proprietary	HSMS	
	Technology	Technology	Technology
	Serial	Serial	Middleware
	TCP/IP	TCP/IP	
	Issues	Issues	Issues
	Limited solution opportunities	Highest integration cost	Non-semiconductor- specific PC-to-PC
	Highest integration cost	Long lead time	communication
	Longest development	Semiconductor-specific interfaces	Lower integration cost
			Object-oriented approach

21.3 A Survey of SECS/GEM-Compliant Tool Kits

SECS/GEM-compliant software comes in different layers. In this section, a survey of different tool kits that are available in the market is presented together with their different features. The different features that are provided by each vendor are summarized in Table 21.9 to highlight the capabilities among them.

21.3.1 Overview of Equipment Integration Software

The equipment integration software is an important part in semiconductor manufacturing or wafer fabrication. The factory automation software must accommodate various factory applications, such as equipment configuration, online and remote diagnostics, remote control, and equipment simulation. These applications are the essential functionalities in today's semiconductor wafer fabrication. Most of the software vendors would like to provide solutions across the span of process control, system automation, production data acquisition, and communication systems.

The factory automation software can be divided into two levels: the application level and the communication-enabling level. The function of the application level is to carry out the various applications such as remote diagnostics, data collection, tracking of materials, testing and simulation, and more. The function of the communication enabler is to interface the factory networks or host with the equipment, which maybe SECS/GEM compatible or non-SECS/GEM compatible equipment, such as programmable logic controller (PLC) controlled machines. Figure 21.12 gives an overview of the factory automation software.

The software tool kit comes with different features and functionalities. They can be conveniently classified as tool kits for manufacturing execution systems (MES) application, factory automation software/solution, and SECS/GEM drivers.

21.3.2 SEMI Equipment Communication Software

Some of the commercially available factory automation software is described in this section. These products are described from company-provided information. The survey is based on several important

TABLE 21.9 The Survey of the Available Software

Software	Yokogawa — EQBrain[a]	Yokogawa — EQBrain 300[a]	Yokogawa — CellBrain[a]	CIMETRIX — CIM Connect	CIMETRIX — GEM Host Manager	ABAKUS — AHEAD ActiveX GEM interface, SCI Spy	Kinesys — GEMBox	Kinesys — ALPS	ErgoTech — Tran SECS	ErgoTech — JAVA SECS	SDI — sdiRelayer, EDAGateway, sdiStation Host Developer, SMS Developer, Reseller/Integrator of Cimetrix TESTConnect, CIMConnect, CIM300
SECS-I	✓	✓	✓	✓		✓	✓	✓	✓		✓
SECS-II	✓	✓	✓	✓	✓	✓	✓	✓	✓	✓	✓
HSMS	✓	✓		✓	✓	✓	✓	✓	✓	✓	✓
GEM			✓	✓	✓	✓	✓	✓	✓	✓	✓
OBEM				✓				✓			
O/S	Win32	Win32	Win32	Win32	Win32	Win	Win	✓	Win UNIX	Win UNIX	MS Windows, UNIX, Linux
Language	VC++, VB, .NET	VC++, VB	VC++, VB	DCOM	DLL		C++	C++	JAVA	JAVA	C++, C, VB
API	✓	✓	✓	✓	✓		C++	Win	JAVA	JAVA	✓
Recipe management				✓	✓				✓	✓	
ARAM				✓	✓						
Testing tools				✓	✓		✓	✓	✓		✓
Middleware				ActiveX/COM					JAVA Bean	JAVA Bean	CORBA, TIBCO, SOAP, XML, ACE, TAO
Good user interface		✓	✓	✓	✓		✓	✓	✓	✓	✓
GEM to XML								✓	✓	✓	✓

[a] Auto code generator.

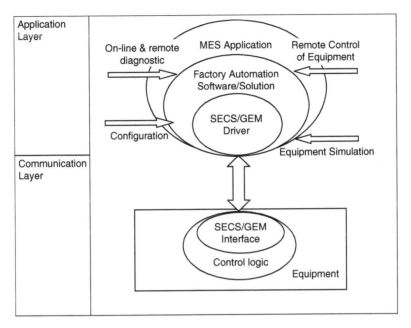

FIGURE 21.12 Application and communication layers of factory automation software.

features listed in Table 21.9. A few software vendors provide the complete solution, but most provide solutions for the application level. They generally use the communication-enabler component from prominent players in the market, such as Cimetrix, in conjunction with their application package.

21.3.3 An Analysis of Current Available SECS/GEM Solutions

An analysis of the SECS/GEM solutions is shown below, based on the features of the software and the intended application.

21.3.3.1 The Analysis Based on the Features of the Software

The analysis on the current available solutions is consolidated from public information provided by the companies. Feedback from vendors is also used to compile this analysis.

21.3.3.2 An Analysis Based on the Intended Application

Due to the fact that SECS/GEM communication software has the propensity to be tailored to a variety of applications, it is necessary to map such applications to varying communication levels, as shown in Figure 21.13 and Table 21.10. For example, it can be used by the equipment manufacturer for developing the host interface, just as it can also be used by the semiconductor manufacturer to communicate with the equipment, to test the equipment for its interface, etc. Figure 21.8 shows one example that highlights the application of the SECS/GEM software. Further, Table 21.10 compares the application intent of the software.

21.3.3.3 Case Study: Design and Implementation of SECS/GEM for a New Semiconductor Back-End Machine (Implementation of SECS/GEM Compliance on a Virgin Machine)

This section presents a case study in the design and implementation of SECS and GEM for a case where a back-end semiconductor machine is to be built with SECS/GEM compliance. We shall see how easy it is to implement SECS-II and GEM for a PC-based equipment controller by leveraging on commercially available software. We have chosen a package from GW, Inc. We will show the steps toward the implementation of a SECS/GEM, in the case where the equipment initiates the communication with the host system.

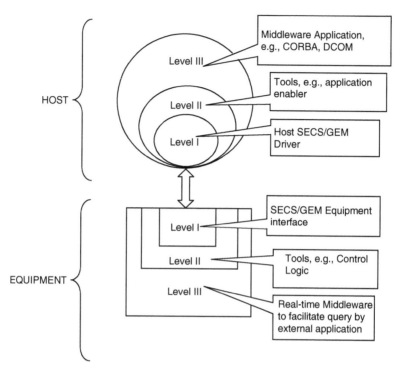

FIGURE 21.13 The functionality of SECS/GEM software.

a. *A Statement of the Specifications and Requirements*

The equipment maker is required to develop GEM-compliant equipment. This equipment is required to be directly integrated into an existing SECS-II-enabled host (cell control) system. The detailed requirements and capabilities must be able to:

1. Network connected to the host system with both SECS-II and HSMS protocols
2. Upload and download process programs for equipment configuration
3. React to the host's requests, based on the process state of equipment
4. Send event or alarm reports to the host system in the event of predefined events or alarms occur
5. Time-synchronize with host system on equipment power-up
6. Implement GEM fundamental requirements

b. *System Architecture and Software Tools*

Figure 21.14 shows the recommended system architecture and communication interfaces. The SECS/GEM software package from GW, Inc., is utilized in this example. It includes:

- GCD file
- GWGEM daemon
- GWGEM extension task
- GWGEM primary message handler task
- GWGEM Application Programming Interface (API) in C++

As we mentioned earlier, in order to run the GEM application that is built on the above five major GWGEM components, the SDR (SECS Driver) must be installed and run as a message driver (HSMS) to communicate with the host system.

c. *Functional Mapping of Modules*

The GWGEM daemon process coordinates all the communications between user applications, GWGEM extension, and GWGEM message handler tasks. The GCD (GEM configuration data) file defines all SECS variables such as status variables, equipment constants, and data variables. The extension task is a process that contains various GWGEM extension routines that are called

TABLE 21.10 A Survey of the Software Based on Communication Levels

Software Vendor	Yokogawa			CIMETRIX		Kinesys	ErgoTech		SDI	ABAKUS
Software	EQBrain	EQBrain 300	CellBrain	CIM Connect	GEM Host Manager	GEMBox	Tran SECS	JAVA SECS	sdiRelayer, EDAGateway, sdiStation Host Developer, SMS Developer, Reseller/Integrator of Cimetrix TESTConnect, CIMConnect, CIM300	AHEAD ActiveX GEM Interface, SCI Spy
Level III (HOST)			✓						✓	
Level II (HOST)			✓		✓			✓		✓
Level I (HOST)			✓		✓			✓		✓
Level I equipment	✓	✓		✓			✓			✓
Level II equipment	✓	✓		✓		✓	✓			✓
Level III equipment	✓	✓				✓				

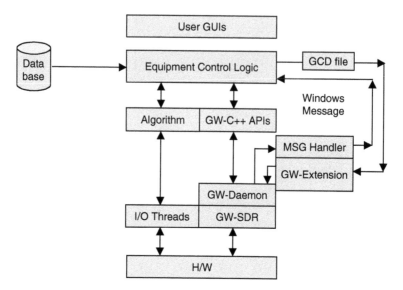

FIGURE 21.14 System architecture for SEMI interface creation.

by the GWGEM daemon process from time to time. Extension routines are user-written application programs that provide application-specific or equipment-specific handling of GEM messages.

d. *Message Partition*

A primary message handler task is used to process a particular type of incoming SECS primary message from the host system. It must provide for the following conditions:

- Messages that are beyond the GEM standard set
- Messages that vary from one equipment type to another
- The ability to override GWGEM built-in handling of a particular incoming primary message
- The ability to scan (peek at) an incoming SECS-II message without actually processing it

e. *Levels of Implementation*

In this project, the following GEM fundamental requirements are implemented:

1. State model
2. Equipment processing status
3. Host-initiated S1F13/F14 scenario
4. Event notification
5. Online identification
6. Error message
7. Documentation
8. Control (operator initiated)

By implementing the GEM fundamental requirements, the equipment is GEM compliant, which means that it is able to communicate with any GEM-compliant equipment no matter which hardware vendor it comes from. In addition to the above, some additional GEM capabilities are also implemented. These include:

1. Initialization and establishment of communications
2. Alarm management
3. Equipment terminal services
4. Process program management
5. Clock

GWGEM C++ encapsulates the GEM requirements into several classes. The developer treats the GEM functionalities as objects. GWGEM C++ for Windows packs as a Windows dynamic link library (DLL). It takes advantage of window programming for WIN32-based applications such as

Microsoft Foundation Class (MFC). GWGEM C++ uses Windows event notification and message service to notify or trigger the WIN32 application to handle incoming SECS-II primary messages.

f. *Integrate GWGEM C++ with an MFC-Based Equipment Controller*

Normally, the development team of GEM-compliant semiconductor equipment is made up of mechanical engineers, control engineers, and SECS/GEM consultants. The first step for the consultant is to work with the mechanical and control engineers to collect equipment, status, event, and alarm data. The consultant needs to translate these data related to the equipment's operation into variables defined in the GCD file (see GCD file sample in Figure 21.15a).

Once the control engineer completes code writing, including Graphical User Interface (GUI) portions and I/O control logic, it is time to integrate GWGEM C++ with the control program. To improve the real-time performance, GWGEM C++ maps events such as equipment state change, process state change, or the arrival of SECS-II primary messages with Window's message handler so that these events are handled concurrently without performance trade-offs. Due to limited space, the code samples in Figure 21.15 only show the necessary steps for declarations, initialization, and deletion of GEGEM objects in an MFC-based application, based on the hierarchy of messages in Figure 21.16.

21.3.3.3.1   Code Samples

By using the GCD compiler, these equipment constants in the GCD file are converted to a standard C++ header file, which is recognized by standard C++ compilers.

Note that these header files are provided with a GWGEM C++ package for the Windows platform.

By using GWGEM C++ for Windows, there is no need for the developer to code the low-level communication details. GWDaemon and its supporting components, such as GWGEM extension tasks, do the processing of incoming SECS-II messages. As a result, it significantly shortens the development cycle and reduces software bugs.

Typically, after 2 months of development and testing, the equipment successfully passes the GEM tests and is declared fit for delivery.

21.4   Conclusion

This chapter has given a brief overview of the standards of SEMI SECS-I, SECS-II, and GEM. While keeping within the special (and at time strange) nomenclature and conventions adopted by SEMI, the authors have provided ample comments and interpretations on the pertinent points to enable the reader to understand the complexity and integrity of SECS equipment. A survey of existing SECS-compliant software is also conducted.

A case study is introduced to illustrate how SECS/GEM compliance can be implemented on virgin machines.

Presently, there are no systematic alternatives to SEMI standards. Although the SEMI communication standard is still evolving, it continues to be critical as a means of sustaining maximum functionality and availability in the wake of new semiconductor processes and new semiconductor manufacturing automation.

Acknowledgments

The authors express their sincere thanks to the following people for various technical discussions and suggestions leading toward the completion of this chapter: Dr. Lim Khiang Wee, Executive Director, SIMTech; Phua Geok Hong; William Tan; and Prof. Sim Siang Kok.

(a)
```
/* GWGEMCPP Base-line GCD Source File */

Constant fixup GemOfflineSubstate = <U1 3>      /* data type is 1 byte unsigned
                                                   integer */

vid = 43                                        /* variable ID is 43 */
name = "OFFLINESUBSTATE"                         /* 1: offline/equipment offline */
units = ""                                       /* no unit */
min = <U1 1>                                      /* minimum value is 1*/
max = <U1 3>                                      /* maximum value is 3 */
default = <U1 3>                                  /* default value is 1 */
event ProcessRunningEvent                         /* event name */
ceid = 122                                        /* constant  event ID is 122 */
event ProcessAbortEvent
ceid = 123
status AramsStateVariable = <A [0..5] "">        /* character array with null value */
vid = 500                                        /* variable ID is 500 */
name = "AramsStateVariable"                       /* status name */
units = ""
```

(b)
```
# include "gwgemcpptemplate.h";   /* for template class */
# include "gemcppbase.h";         /* base class definition */
# include "GemVirtImp.h";         /* virtual classes for application-specific
                                    overwrite*/
# include "GemMessages.h"         /* SECS II message header structures, etc. */
# include "gwcontrol.h            /* equipment control state */
```

(c)
```
/* This is a sample program to be used in conjunction with GWGEM class library
to demonstrate how to create, initialize, access and remove GWGEM object in
MFC code */
CGWGemCPP* pGem = new CGWGemCPP;        /*create a GWGemCPP instance */
pGlobalGem = pGem;                       /* setup global pointer */
```

(d)
```
/*Create a control state message, set initial condition states and call the message handler.*/
pGlobalGem -> GWControl().GetControlState(ControlState); /*get current control state */
pGemConfControlStateChange pControlStateChange = new GemConfControlStateChange;
pControlStateChange->NewControlState = ControlState; /*set state */
pControlStateChange->ControlStateName = getStrControlState(ControlState);
OnGemConfControlStateChange(0, (LPARAM) pControlStateChange); /*notify state
change */
```

FIGURE 21.15 (a) Sample GCD file. (b) C++ sample, header files. (c) C++ sample; create GWGEM object. (d) C++ sample; set up control state. (e) C++ sample; set up communication state. (f) C++ sample; set up spooling state. (g) C++ sample; remove GWGEM object. (h) C++ sample; fire an event. (i) C++ sample; disable the communication link. (j) C++ sample; send S1, F13 to host.

(e)

```
pGlobalGem -> GWLink().GetLinkState(LinkState);  /* get initial communications state */
SetDlgItemText(IDS_LINKSTATE, getStrLinkState(LinkState)); /* display on GUI */
```

(f)

```
pGlobalGem -> GWSpool().GetSpoolState(SpoolState); /* get the spool state */
SetDlgItemText(IDS_SPOOLSTATE, getStrSpoolState(SpoolState)); /*display */
```

(g)

```
delete pGEM; /* remove GWGEM object */
```

(h)

```
int status = pGlobalGem->GWEvent().Send(EventID); /*send event with ID = EventID*/
```

(i)

```
pGlobalGem->GWLink().Disable();
```

(j)

```
/* following code sends S1F13 to the host system */

        PSDRMSG pmsg;                            /* message structure declared
                                                   in gwmessage.h */
        SDRTKT tkx = 0;                          /* set SDR ticket value to 0 */

        unsigned char buffer[512] ;              /* Message text buffer  */

        unsigned char ModelNum[7] = "SDR";       /* set model number */

        unsigned char SoftRev[7] = "Rev10";      /* software version */

        pmsg->stream = 1;                        /* set stream to 1 */

        pmsg->function = 13;                     /* set function to 13 */

        pmsg->wbit = 1;                          /* request reply */

        pmsg->buffer = buffer;                   /* pointer to message buffer */

        pmsg->length = sizeof(buffer);

pGlobalGem->GWSdr().SdrItemInitO( pmsg);     /* fill up SECS II message */

pGlobalGem->GWSdr().SdrItemOutput( pmsg, GWS2_L, NULL,
(SDRLENGTH)2);

pGlobalGem->GWSdr().SdrItemOutput(pmsg,GWS2_STRING,ModelNum,
(SDRLENGTH)6);

pGlobalGem->GWSdr().SdrItemOutput( pmsg,GWS2_STRING, SoftRev,
(SDRLENGTH)6);

int status = pGlobalGem->GWSdr().SdrRequest(0, pmsg, &tkx);
/* send S1F13 out */
```

FIGURE 21.15 Continued.

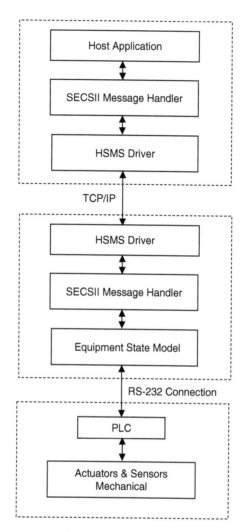

FIGURE 21.16   Design of intercommunication process.

References

1. Tin, O., Competitive Analysis and Conceptual Design of SEMI Equipment Communication Standards and Middleware Technology, Master of Science (Computer Integrated Manufacturing) dissertation, Nanyang Technological University, 2003.
2. SEMATECH, *Generic Equipment Model (GEM) Specification Manual: The GEM Specification as Viewed from Host*, Technology Transfer 97093366A-XFR, 2000, pp. 4–39.
3. SEMATECH, *High Speed Message Services (HSMS): Technical Education Report*, Technology Transfer 95092974A-TR, 1999, pp. 11–34.
4. GW Associates, Inc., Solutions for SECS Communications, Product Training (PowerPoint slides), 1999.
5. *SEMI International Standards*, CD-ROM, SEMI, 2003.
6. *Semiconductor Equipment and Materials International Equipment Automation/Software*, Volumes 1 and 2, SEMI, 1995.
7. SEMATECH, *CIM Framework Architecture Guide 1.0*, 97103379A-ENG, 1997, pp. 1–31.
8. SEMATECH, *CIM Framework Architecture Guide 2.0*, 1998, pp. 1–24.

9. SEMI, Standard for the Object-Based Equipment Model, SEMI Draft Document 2748, 1998, pp. 1–52.

10. SEMI E98, Provisional Standard for the Object-Based Equipment Model.

11. Weiss, M., *Increasing Productivity in Existing Fabs by Simplified Tool Interconnection*, 12th edition, Semiconductor FABTECH, 2001, pp. 21–24.

12. Yang, H.-C., Cheng, F.-T., and Huang, D., Development of a Generic Equipment Manager for Semiconductor Manufacturing, paper presented at 7th IEEE International Conference on Emerging and Factory Automation, Barcelona, October 1996, pp. 727–732.

13. Feng, C., Cheng, F.-T., and Kuo, T.-L., *Modeling and Analysis for an Equipment Manager of the Manufacturing Execution System in Semiconductor Packaging Factories*, 1998, pp. 469–474.

14. ControlPRo™, *Developer Guide*, Realtime Performance, Inc., 1996.

15. Kaufmann, T., *The Paradigm Shift for Manufacturing Execution Systems in European Projects and SEMI Activities*, 8th edition, Semiconductor FABTECH, 2002, pp. 17–25.

16. GW Associates, Inc., *SECSIMPro GEM Compliance Scripts User's Guide*, 2001.

17. GW Associates, Inc., *SECSIMPro, SSL Reference Guide*, 2001.

18. GW Associates, Inc., *SECSIMPro, User's Guide*, 2001.

19. SEMATECH, *SEMASPEC GEM Purchasing Guidelines 2.0*, Technology Transfer 93031573B-STD, 1994, pp. 10–30.

Web References

1. www.cimetrix.com   Home page of Cimetrix Software
2. www.abakus-soft.de   Home page of Abakus Software
3. www.kinesysinc.com   Home page of Kinesys Software — The GEM Box
4. www.secsandgem.com   Home page of Ergo Tech Software
5. www.sdiusa.com   Home page of SDI Software
6. www.yokogawa.com.sg   Home page of Yokogawa Software
7. www.asyst.com   Home page of Asyst Software
8. www.siautomation.com   Home page of SI Automation Software
9. www.ais-dresden.de   Home page of VECC Product
10. www.agilent.com   Home page of Agilent Software

Part 5

Agent-Based Technologies in Industrial Automation

22

From Holonic Control to Virtual Enterprises: The Multi-Agent Approach

Pavel Vrba
Rockwell Automation

Vladimir Marik
Czech Technical University

22.1 Introduction

Both the complexity of manufacturing environments as well as the complexity of tasks to be solved are growing continuously. In many manufacturing scenarios, traditional centralized and hierarchical approaches applied to production control, planning and scheduling, supply chain management, and manufacturing and business solutions in general are not adequate and can fail because of the insufficient means to cope with the high degree of complexity and practical requirements for *generality* and *reconfigurability*.

These issues naturally lead to a development of new manufacturing architectures and solutions based on the consideration of highly distributed, autonomous, and efficiently cooperating units integrated by the plug-and-play approach. This trend of application of multi-agent systems (MAS) techniques is clearly visible at all levels of manufacturing and business. On the lowest, real-time level, where these units are tightly linked with the physical manufacturing hardware, we refer to them as *holons* or *holonic agents* [1].

Intelligent agents are also used in solving production planning and scheduling tasks, both on the workshop and factory levels. More generic visions of intensive cooperation among enterprises connected via communication networks have led to the ideas of *virtual enterprises.*

A virtual enterprise is a temporary alliance of enterprises that come together to share skills or core competencies and resources in order to better respond to business opportunities, and whose cooperation is supported by computer networks [2].

The philosophical background of all these highly distributed solutions is the same: the community of autonomous, intelligent, and goal-oriented units efficiently cooperating and coordinating their behavior in order to reach the global level goals. The decision-making knowledge stored and exploited locally in the agents/holons invokes the global behavior of the system that is not deterministic, but rather emergent — such a behavior cannot be precisely predicted at the design time of the community. The experimental testing of the global behavior with the physical manufacturing/control environment being involved is not only extremely expensive, but nonrealistic as well. The only possible solution is the simulation, both of the controlled process of the manufacturing facility as well as the simulation of the inter-agent interactions.

22.2 Technology Overview

The architecture of an agent usually consists of the *agent's body* and the *agent's wrapper.* We can also say that the body, the functional core of an agent, is encapsulated by the wrapper to create an agent [3]. The wrapper accounts for the inter-agent communication and real-time reactivity. The body is an agent's reasoning component, responsible for carrying out the main functionality of the agent. It is usually not aware of the other members of the community, their capabilities, duties, etc. This is the wrapper, which is responsible for communicating with the other agents, for collecting information about the intents, goals, capabilities, load, reliability, etc., of the other units in the agents' community.

From the implementation point of view, there are two types of agents: (i) *custom-tailored agents*, which are implemented in order to provide a specific service to the community (e.g., service brokering) and (ii) *integrated agents*, which encapsulate a preexisting, "inherited," or "legacy" piece of software/hardware by the agent's wrapper into the appropriate agent structure. In this case, the wrapper provides a standardized communication interface enabling one to plug the legacy system into the corresponding agent community. From the outside, such a wrapped software system cannot be distinguished from the *custom-tailored agents* as it communicates in a standard way with the others and understands the predefined language used for the inter-agent communication (e.g., Agent Communication Language — ACL defined by the Foundation for Intelligent Physical Agents [FIPA]). These *agentification processes* provide an elegant mechanism for *system integration* — a technique supporting the *technology migration* from the centralized systems toward the distributed agent-based architectures.

We distinguish three different concepts of agency that we need to explain in more detail.

22.2.1 Mobile Agents

Mobile agents are generally pieces of code traveling freely inside a certain communication network (usually inside the Internet). Such agents, being usually developed and studied in the area of computer science, are stand-alone, executable software modules, which are being sent to different host computers/ servers to carry out specific computational tasks (usually upon the locally stored data). They are expected to report back the results of the computation process. The mobile agents can travel across the network, can be cloned, or can be destroyed by their own decision when fulfilling their specific task, etc.

22.2.2 MAS

MAS can be described as goal-oriented communities of cooperative/self-interested agents in a certain interaction environment. They have been developed within the domain of the distributed artificial intelligence and they explore the principles of artificial intelligence for reasoning, communication, and cooperation. The important attribute of each agent is its *autonomy* — the agent resides on a computer platform where it autonomously carries out a particular task/functionality. The agent owns only a part of the global information about the goals of the community that is sufficient for its local decision-making and behavior. However, in some situations, for example, when the agent is not capable of fulfilling the requested operation alone, the agent is capable of *cooperation* with other agents (asking them for help) usually via message sending. The agents asked for cooperation are still autonomous in their decisions, that is, can either agree on cooperation or can even refuse to cooperate (e.g., due to lack of their own resources). Another important attribute of MAS is the *plug-and-play* approach — the agents, which can be grouped into different types of communities (such as teams, coalitions, platforms, etc.; see Section 22.3), can freely join and leave the communities. Usually, the community offers a yellow-pages-like mechanism that is used by agents to offer their services to the other agents as well as to find out suitable agents for possible cooperation. This allows one to dynamically change the overall capabilities and goals of the MAS (according to user requests) by adding new agents with desired functionalities.

The cooperation among the agents supported by their *social behavior* is the dominant feature of the activities of the agents in the community. The term "social behavior" means that the agents are able to communicate; to understand the goals, states, capabilities, etc., of the others; and to respect the general rules and constraints of behavior valid for each of the community members. Communication (exchange of information usually in the form of messages) plays the crucial and decisive role in the agents' communities with social behavior of either a cooperative or competitive nature.

22.2.3 Holons

Holons are agents dedicated to real-time manufacturing tasks [4]. These are tightly physically coupled, really "hard-wired" with the manufacturing hardware (devices, machines, workshop cells), thus with a low degree of freedom in their mutual communication. The holons operate in the "hard" real time, and their patterns of behavior are strictly preprogrammed. This means that their reactions in certain situations are predictable and emergent behavior is not highly appreciated (usually not allowed). The Holonic Manufacturing Systems (HMS) are designed mainly to enable a fast and efficient system *reconfiguration* in the case of any machine failure. One of the pioneering features of holonic systems is a complete separation of the data flow from the control instructions. The communication principles, supporting the holons' interoperability, are already standardized (the IEC 61499 standard; see Section 22.6).

22.3 Cooperation and Coordination Models

As we have already mentioned, communication among the agents is an important enabler of their social behavior. The agents usually explore specific communication language with standardized types of messages. The set of messages is chosen so that it represents the most typical communicative acts, often called *performatives* (according to the speech act theory of Searle [5]), used by agents in a particular domain. Examples of such performatives can be request, used by an agent to request the other one to perform a particular operation, agree, used to confirm the willingness to cooperate, or, for example refuse to deny to perform requested action. Along with the performative, the message bears the information about the sender and receiver, the content of the message, and the identification of the language used for the message's content. Additional attributes can be included in the message, for example, the reference to the appropriate knowledge ontology describing semantics of the message (see Section 22.5) or the brief description of the negotiation strategy used (the structure of replies expected).

The communication among the agents is usually not just a random exchange of messages, but the message flow is managed by a set of standard *communication protocols*. These protocols range from a simple "question–reply," via "subscribe–inform," through to more complex negotiation protocols like "Contract-Net-Protocol" (CNP), different versions of auctions (Dutch, sealed, Vickery, etc.). The communication protocols and communication traffic in general can be represented graphically by means of

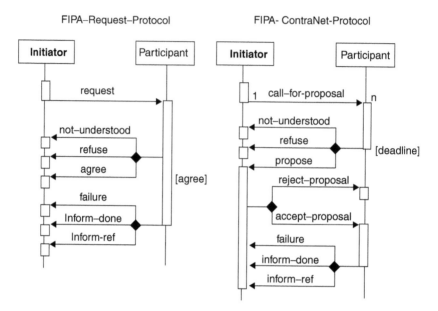

FIGURE 22.1 FIPA's *Request* and *Contract-net* communication protocols.

interaction diagrams. Figure 22.1 shows the *Request* and *Contract-net* protocols from FIPA specifications captured in AUML — Agent-based extensions to the standard UML.

The more knowledge is available locally, which means it is "owned" by individual agents, the smaller the communication traffic needed to achieve cooperative social behavior. One of the crucial questions is how to store and use the knowledge locally. For this purpose, different *acquaintance models* are located in the wrappers of individual agents. These are used to organize, maintain, and explore knowledge about the other agents (about their addresses, capabilities, load, reliability, etc.). This kind of knowledge, which strongly supports collaboration activities among the agents, is called *social knowledge* [3]. These models can be used to organize both the long-term as well as temporary or semipermanent knowledge/data concerning cooperation partners. To keep the temporary and semipermanent knowledge fresh, several *knowledge maintenance techniques* have been developed, namely (i) *periodic knowledge revisions* — the knowledge is updated periodically by regular "question–answer" processes, and (ii) *subscription-based update* — the knowledge update is preordered by a specific subscription mechanism. The field is strongly influenced and motivated by Rao and Georgeff's BDI (Beliefs-Desires-Intentions) model [11] used to express and model agents' beliefs, desires and intentions, and other generic aspects of MAS.

The MAS research community provides various techniques and components for creating the architecture of an agent community. The crucial categories of agents, according to their intra- and inter-community functionalities (e.g., resource agents, order or customer agents, and information agents) have been identified. Services have been defined for specific categories of agents (e.g., white pages, yellow pages, brokerage, etc.). In some of the architectures, the agents do not communicate directly among themselves. They send messages via facilitators, which play the role of communication interfaces among collaborating agents [6]. Other architectures are based on utilization of matchmakers, which proactively try to find the best possible collaborator, brokers that act on behalf of the agent [7], or mediators that coordinate the agents by suggesting and promoting new cooperation patterns among them [8]. Increasing attention has recently been paid to the concept of the meta-agent, which independently observes the interagent communication and suggests possible operational improvements [9].

The techniques for organizing long-term *alliances* and short-term *coalitions,* as well as techniques for planning of their activities (*team action planning*), have been developed recently [10]. These algorithms can help to solve certain types of tasks more efficiently and to allocate the load among the agents in an optimal way. They can be explored with an advantage for automated creation and dissolving of virtual organizations as well as for an optimal load distribution among the individual bodies in a virtual organization. Alliance and coalition formation techniques can be linked with methodologies and techniques for administration and maintenance of *private, semiprivate,* and *public knowledge.* This seems to be an important issue to tackle, especially in virtual organizations of temporary nature where units competing in one business project are also contracted to cooperate in the other one. It is also possible to classify and measure the necessary leakage of private/semiprivate knowledge to reflect this fact in the future contracts.

22.4 Agents Interoperability Standardization — FIPA

The FIPA is a nonprofit association registered in Geneva, Switzerland, founded in December 1995. The main goal of FIPA is to maximize interoperability across agent-based applications, services, and equipment. This is done through FIPA specifications.

FIPA provides specifications of basic agent technologies that can be integrated by agent systems developers to make complex systems with a high degree of interoperability. FIPA specifies the set of interfaces, which the agent uses for interaction with various components in the agent's environment, that is, humans, other agents, nonagent software, and the physical world. It focuses on specifying *external communication among agents* rather than the internal processing of the communication at the receiver.

The FIPA Abstract Architecture defines a high-level organizational model for agent communication and core support for it. It is neutral with respect to any particular network protocol for message transport or any service implementation. This abstract architecture cannot be directly implemented; it should be viewed as a basis or specification framework for the development of particular architectural specifications.

The FIPA Abstract Architecture contains agent system specifications in the form of both descriptive and formal models. It covers three important areas: (i) *agent communication,* (ii) *agent management,* and (iii) *agent message transport.*

22.4.1 Agent Communication and Agent Communication Language

FIPA provides standards for agent communication languages. The messages exchanged among the agents must comply with a FIPA-ACL specification. FIPA-ACL is based on speech act theory [5] and resembles Knowledge Query Manipulation Language (KQML). Each message is labeled by a *performative,* denoting a corresponding *communicative act* (see previous section), such as inform or request. Along with the performative, each message contains information about its sender and receiver, the content of the message, a content language specification, and an ontology identifier. Other important attributes of a message are the information about the conversation protocol that applies to the current message (e.g., FIPA-Request) and the *conversation ID* that uniquely identifies to which "conversation thread" this message belongs (since there can be more conversations between two agents following the same protocol at the same time).

The core of the FIPA ACL message is its content, which is encoded in language denoted in the *language* slot of the message. FIPA offers a Semantic Language (*FIPA-SL*) as a general-purpose knowledge representation formalism for different agent domains. This formalism maps each agent message type (performative) to an SL formula that defines constraints (called feasibility conditions) that the sender has to satisfy, and to another formula that defines the rational effect of the corresponding action. Nevertheless, any other existing or user-defined language can be used as a content language of FIPA messages. One of the most popular languages used today to express the message syntax is the XML.

Bellow is an example of a FIPA message, where the agent "the-sender" requests the agent "the-receiver" to deliver a specific box to a specific location, using the XML content language and the FIPA-Request conversation protocol.

```
(request
:sender (agent-identifier :name the-sender)
:receiver (set (agent-identifier :name the-receiver))
:content
      <deliver item="box017" locX="12" locY="19" />
:protocol fipa-request
:language XML
:conversation-id order567
)
```

22.4.2   Agent Management

The FIPA Agent Management Specification provides a specification of how the members of the multi-agent community shall be registered, organized, and managed. According to the FIPA philosophy, the agents are grouped into high-level organizational structures called *agent platforms* (APs). Members of each platform are usually geographically "close" — for example, they may be run on one computer or be located in a local area network. Each platform must provide its agents with the following two mandatory services: the Agent Management System (AMS) and the Directory Facilitator (DF).

The AMS administers a list of agents registered with the platform. This component implements the creation and deletion of the running agents and provides the agents with a "white-page-list" type of service, for example, a list of all agents accommodated on the given agent platform and their addresses. Unlike the AMS, the DF supplies the community with a "yellow-page-list" type of service. Agents register their services with the DF and can query the DF to find out what services are offered by other agents, or to find all agents that provide a particular service. Thus, agents can find the addresses of others that can assist in accomplishing the desired goals.

22.4.3   Message Transport Service

The Message Transport Service (MTS) is a third mandatory component (besides AMS and DF) that the agent platform has to offer to the agents. The MTS is provided by so-called Agent Communication Channel (ACC), which is responsible for the physical transportation of messages among agents local to a single AP as well as among agents hosted by different APs. For the former case, FIPA does not mandate to use a specific communication protocol or interface — different protocols are being used today in agent platform implementations, such as the TCP/IP sockets, the UDP protocol, or, particularly in JAVA implementations, the JAVA Remote Method Invocation (RMI). In the latter case, the FIPA defines a *message transport protocol* (MTP) that ensures the interoperability between agents from different agent platforms. For this purpose, the ACC must implement the MTP for at least one of the following communication protocols specified by FIPA: the IIOP (Internet Inter-Orb Protocol), WAP, or HTTP.

22.5   Ontologies

Ontologies play a significant role not only in the interagent communication, where the content of messages exchanged among agents must conform to some ontology in order to be understood, but also in knowledge capturing, sharing, and reuse. One of the main reasons why ontologies are being used is the *semantic interoperability* enabling, among others:

- to share knowledge — by sharing the understanding of the structure of information exchanged among software agents and people
- to reuse knowledge — ontology can be reused for other systems operating on a similar domain
- to make assumptions about a domain explicit — for example, for easier communication.

Basically, ontology can be referred to as a vocabulary providing the agents with the semantics of symbols, terms, or keywords used in messages [12]. Thus, if an agent sends a message to another agent using particular ontology, it can be sure that the other agent (of course, if it shares the same ontology) will understand the message.

22.5.1 Ontologies for MAS

Two examples can be selected to illustrate multi-agent-oriented ontological efforts in the area of manufacturing. The first one, FIPA Ontology Service Recommendation, is a part of a set of practical recommendations on how to implement agents in a standardized way. The other one, Process Specification Language (PSL) project [13], tries to develop general ontology for representing manufacturing processes. Its aim is to serve as interlingua for translating between process ontologies. The transformation between ontologies for translation using PSL is expected to be defined by humans.

The former approach seems to be much more general and applicable. FIPA uses Open Knowledge Base Connectivity, (OKBC) [14], as a base for expressing ontologies. OKBC is an API for accessing and modifying multiple, heterogeneous knowledge bases. Its knowledge model defines a meta-ontology for expressing ontologies in an object-oriented, frame-based manner. OKBC can be mapped to the object-oriented languages, so that classes in programming languages can be built on the underlying ontology and be used for exchanging information. Semantics of OKBC constructs is defined in KIF as a description of what the constructs intuitively mean. However, no reasoning engine that would enable to use this information for, for example, ontology integration, is provided. Moreover, it could be difficult to provide a reasoning support for some of the constructs.

Ontologies in FIPA proposals and related ontologies for practical applications [15] are motivated mainly by the need to have something that would work immediately, because currently more attention is paid to the functional behavior of agents. There is nothing wrong with this approach, if we want to have a working solution in a short time where we do not care about the possibility of reasoning about the ontologies and further interoperability. However, the need for reasoning about ontologies can easily arise, for example, when requiring interoperability in open multi-agent systems, that is, systems where new agents with possibly other ontologies can join the community.

22.6 HMS

Over the past 10 years, research has attempted to apply the agent technology to various manufacturing areas such as supply chain management, manufacturing planning, scheduling, and execution control. This effort resulted in the development of a new concept, the *HMS*, based on the ideas of *holons* presented by Koestler [17] and strongly influenced by the requirements of industrial control.

Holons are autonomous, cooperative units that can be considered as elementary building blocks of manufacturing systems with decentralized control [16]. They can be organized in hierarchical or heterarchical structures. Holons, especially those for real-time control, are usually directly linked to the physical hardware of the manufacturing facility, and are able to physically influence the real world (e.g., they may be linked to a device, tool, or other manufacturing unit, or to a transportation belt or a storage manipulator).

Holons for real-time control are expected to provide reactive behavior rather than being capable of deliberative behavior based on complex "mental states" and strongly proactive strategies. They are expected mainly to react to changes in the manufacturing environment (e.g., when a device failure or a change in the global plan occurs). Under "stable circumstances," during routine operation, they are not required to change the environment proactively. The reason for the prevalence of reactive behavior of real-time control holons is that each of them is linked to a physical manufacturing facility/environment, changes to which are not very simple, cheap, or desirable in a comparatively "stable" manufacturing facility. The physical linkage to physical equipment seems to be a strong limiting factor of the holons' freedom in decision-making.

The more generic holonic ideas and considerations have led to the vision of a *holonic factory* [18]. Here, all the operations (starting from product ordering, planning, scheduling, and manufacturing, to invoicing the customer) are based entirely on holonic principles. A holonic factory contains a group of principal system components (holons) that represent physical manufacturing entities such as machines or products as well as virtual entities like orders or invoices. The holons work autonomously and cooperate together in order to achieve the global goals of the factory. Thus, the factory can be managed toward global goals by the activities of individual autonomous holons operating locally. The community of researchers trying to implement the vision of the holonic factory is well organized around the international HMS consortium.

The vision of the holonic factory covers several levels of information processing for manufacturing. We can distinguish at least three separate levels, namely,

1. *real-time control*, which is tightly linked with the physical manufacturing equipment
2. *production planning and scheduling*, both on the workshop and on the factory level
3. *supply chain management*, integrating a particular plant with external entities (suppliers, customers, cooperators, sales network, etc.).

At the lowest RT-control level, the main characteristic of holons is their linkage to the physical manufacturing devices — these holons read data from sensors and send control signals to actuators. Within the HMS activities, the standard IEC-61499 known as *function blocks* has been developed for these RT-control purposes. It is based on function blocks part of the well-known IEC-1131-3 standard for languages in programmable logical controllers (PLCs). The major advantage is the separation between the data flow and the event flow among various function blocks. Multiple function blocks can be logically grouped together, across multiple devices into an application, to perform some process control.

Since the IEC-61499 fits well these RT-control purposes, it does not address the higher level aspects of holons acting as cooperative entities capable of communication, negotiation, and high-level decision making. It is obvious that this is the field where the techniques of multi-agent systems have to be applied. Thus, a general architecture combining function-blocks with agents was presented in Reference [1]. As shown in Figure 22.2, a software agent and function block control application (connected to the physical layer) are encapsulated into a single structure.

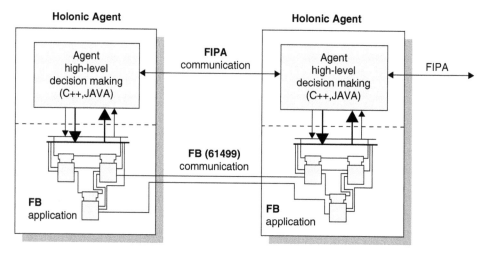

FIGURE 22.2 Holonic agent: combination of function block application and software agent.

In such a holon equipped with a higher-level software component, three communication channels should be considered:

1. *Intraholon* communication between the function block part and the software agent component.
2. *Interholon* communication that is aimed at communication among the agent-based parts of multiple holons — FIPA standards are used more and more often for this purpose.
3. A *direct* communication channel between function block parts of neighboring holons. If we are prepared to break the autonomy of an independent holon, then this communication is standardized by IEC 61499 already; otherwise, a new type of real-time coordination technology is needed to ensure real-time coordination.

As a matter of fact, the holons defined in this way behave — on the level of interholon communication — like standard software agents. They can communicate widely among themselves, carry out complex negotiations, cooperate, develop manufacturing scenarios, etc. We can call them holonic agents (or agentified holons), as they consist of both a holonic part connected with the physical layer of the manufacturing system (operating in hard real time) and a software agent for higher-level, soft real-time, or non-real-time intelligent decision making. It has already been mentioned that the interholon communication is usually standardized by the FIPA approach, and direct communication could be achieved by IEC standards (not necessarily IEC 61499). Let us stress that the FIPA standards are not applicable for the low-level real-time control purposes as they do not take account of the real-time control aspects.

The attention of system developers is currently directed mainly at the intraholon communication, which is usually both application- and company-specific and is usually connected with the solution of the "migration problem" (the problem of exploring the classical real-time control hardware for holonic control). McFarlane et al. [19] introduced a blackboard system for accomplishing the intra-holon communication, while others [20] proposed using a special management service interface function block.

It is expected that the communication among holonic agents will be standardized for many reasons. One reason is that these holonic agents should be involved in global communities of company agents, where they can directly participate in supply chain management negotiations or contribute to virtual enterprise simulation games, etc. The FIPA communication standards are considered preferable for implementing the inter-holon communication. To develop these standards, the HMS community must declare messages, define their semantics, and develop the appropriate *knowledge ontologies* (see Section 22.5). This seems to be quite a demanding task, as manufacturing, material-handling, production planning, and supply chain management requirements differ significantly between different industries and between different types of production.

From a wider perspective, for the FIPA standards to be applicable to holonic manufacturing, they should take account of the preexistence and coexistence of other standards. In manufacturing industry, there is STEP (Standard for the Exchange of Product Model Data), which is a comprehensive ISO standard (ISO 10303) that is used for representation and exchange of engineering product data and specifies the EXPRESS language for product data representation in any kind of industry. Integration of these widely accepted concepts with the HMS and FIPA effort seems to be of high importance. On the level of physical interoperability, there are various standards, such as the TCP/IP and UDP protocols, Common Request Broker Architecture (CORBA), Distributed Common Object Model (DCOM), and others. Similarly, the use of higher-level cooperation standards in the area of multi-agent systems such as KQML, FIPA, and JINI is inevitable for dynamic, flexible, and reconfigurable manufacturing enterprises.

22.7 Agent Platforms

A complex nature of the agents (high-level decision-making units capable of mutual collaboration) requires using a high-level programming language such as C++ or JAVA for their implementation. As mentioned in the previous section, for manufacturing purposes, the software agents' parts of holonic agents have to be able to interact with the low-level control layer. In the majority of current holonic testbed implementations, the low-level holonic control (connected to the physical layer) is usually carried out by IEC 1131-3 (mainly ladder logic) or IEC 61499 function block programs that run on industrial PLC-based automation controllers. However, the software agent parts, implemented in C++ or JAVA, are running separately on a standard PC and communicate, for example, via a *blackboard system* (part of the data storage area in a controller allocated for each holonic agent and shared by the agent- and holonic-subsystem of the holonic agent).

It is obvious that for real industrial deployment, particularly where a high degree of robustness is required, the use of PC(s) for running agent-components of holonic agents is not safe and is also not possibly feasible for certain types of control systems. The only acceptable solution is to *run holonic agents as wholes directly within PLC-based controllers*. One controller can host one or more holonic agents, but not all of them — they have to be distributed in reasonable groups over several controllers and allowed to communicate with each other either within a single controller or among different controllers.

The major issue of such a solution is to extend the current architecture of a PLC in such a way that it is able to run software agents written in a high-level programming language in parallel with the low-level control code and also provide the interface for interactions between these two layers. The programming language in which the software agents should be implemented can either be C++ or JAVA. However, there are many reasons to prefer the JAVA language to be the target one. One of its advantages is the portability of JAVA programs, which the user develops independent of hardware platforms or operating systems — the same application can run either on a PC with Microsoft Windows or Unix/Linux or on a small device like Personal Digital Assistant (PDA) or a mobile phone with Windows CE, Symbian, or other operating systems with JAVA support. Another reason to choose JAVA is that currently there are a large number of JAVA-based agent development tools available, either as commercial products or open-source projects, that simplify the development of agent systems. Moreover, some of them are fully compliant with the FIPA specifications, which insures the desired interoperability.

22.7.1 Agent Development Tools Characteristics

Basically, the agent development tool, often called an *agent platform*, provides the user with a set of JAVA libraries for specification of user agent classes with specific attributes and behaviors. A kind of a *runtime environment* that is provided by the agent platform is then used to actually run the agent application. This runtime environment, implemented in JAVA as well, particularly ensures transport of messages among agents, registration, and deregistration of agents in the community (white pages services) and also registration and lookup for services provided by agents themselves (yellow pages services). Some other optional tools can also be a part of the agent platform runtime, for instance, a graphical viewer of messages sent among agents, etc.

The implementation of JAVA-based agents in the automation controllers obviously requires such an agent platform runtime to be embedded into the controller architecture. Since it is used as a background for the *real-time* holonic agents, there are specific requirements on the properties of the agent platform, such as speed, memory footprint, reliability, etc. The evaluation of available JAVA agent platforms, presented in the following paragraphs, has been conducted [21] in order to find out to what extent they fulfill these criteria and therefore which ones are best suited for the purposes of manufacturing control.

22.7.2 FIPA Compliancy

Compliance with the FIPA standards has been recognized as a crucial property ensuring the interoperability of holonic agents not only at the lowest real-time control level (allowing, e.g., communication of different kinds of holonic agents hosted by PLC controllers from different vendors), but also the interoperability between holonic agents and other agents at higher levels of information processing within the company, for example, data-mining agents, ERP agents, supply chain management agents, and so on.

The FIPA specification of the message transport protocol (see the Section 22.4.3) defines how the messages should be delivered among agents within the same agent community and particularly between different communities. For the latter case, the protocol based on IIOP or HTTP ensures the full interoperability between different agent platform implementations. It means that the agent running, for example, on the JADE agent platform can easily communicate with the agent hosted by the FIPA-OS platform, etc.

22.7.3   Costs and Maintainability of the Source Code

From the cost point of view, the agent platforms that are currently available can basically be divided into two categories: free and commercial ones. Majority of the free agent platforms are distributed under a kind of an open source license (e.g., GNU Lesser General Public License), which means that you are provided with the source codes and allowed to modify them. This is an important characteristic since the integration of the agent platform into the PLC-based controllers certainly requires some modifications to be made, for example, due to different versions of JAVA virtual machine supported by the controller, the specifics of the TCP/IP communication support, or other possible issues and limitations.

On the other hand, in the case of commercial products, the cost in order of thousands USD per each installation, for example, can considerably increase the total cost of the agent-based control solution where a large number of PLC controllers, PCs, and possibly other devices running agents are expected to be deployed. Moreover, the source codes are not available, so that all modifications of the platform that need to be made in order to port it to another device has to be committed to the company developing the agent platform.

22.7.4   Memory Requirements

An issue that has to be taken into account is usually a limited memory available for user applications on the controller. Within the RAM memory of the controller, which can, for example, be about 4 to 8 MB, the agent platform run-time environment, the agents themselves, and also the low-level control code (ladder logic or function blocks) have to fit inside. There are also smaller PLC-like devices that can have only 256 KB of memory available, which would be a strong limitation factor for integrating the run-time part of the agent platform. Fortunately, the agent platform developers, especially in the telecommunication area, are seriously interested in deploying agents on small devices like mobile phones or PDAs, that is, on devices with similar memory limitations. Due to this fact, for some of the agent platforms, their lightweight versions have been developed, usually implemented in Java2 Micro Edition (CLDC/MIDP) [22]. It has been documented [23] that the memory footprint of such an agent platform runtime can be less than 100 KB, that is, small enough to fit well within the memory capacity limits of majority of small mobile devices and thus the PLC-based automation controllers as well.

22.7.5   Message Sending Speed

The last factor considered in this evaluation is the speed of the message sending between the agents. It has already been argued that the holonic agents are expected to be used for real-time control applications where a fast reaction can be a vital characteristic. A direct communication channel between RT control subsystems of neighboring holons is conceded, but it obviously breaks the autonomy of holonic agents. If we are not willing to accept such a violation, communication at the agent level is the only allowable way of interaction among holonic agents. Thus, the agent platform runtime, carrying out such interactions, should be fast enough to ensure reasonable message delivery times (i.e., in the order of milliseconds or tens of milliseconds).

We have conducted a series of tests to compare the message-sending speed of different agent platforms. Detailed information about the benchmarking testbed configuration and the speed measuring results can be found in the subsequent section.

TABLE 22.1 Agent Platforms Overview

JAVA-Based Agent Development Toolkits/Platforms — Overview

Agent Platform	Developer	FIPA Compatibility			Open-Source	J2ME Version (Lightweight)	Security (authent., SSL, …)
		Agent Communication	Agent Management	Inter-Platform Messaging (MTP)			
JADE (Java Agent Development Framework)	CSELT http://jade.cselt.it	✓	✓	✓	✓	✓	✓
FIPA-OS	Emorphia http://fipa-os.sourceforge.net	✓	✓	✓	✓	✓	✓
ZEUS	British Telecom www.labs.bt.com/projects/agents/zeus	✓	✓	✓	✓	×	✓
JACK (Jack Intelligent Agents)	Agent Oriented Software http://www.agent.software.com	✓	✓	✓	×	×	✓
GRASSHOPPER 2	IKV ++ Technologies AG http://www.grasshopper.de	✓	✓	✓	×	✓	✓
ADK (Agent Development Kit)	Tryllian http://www.tryllian.com	✓	×	×	×	?	?
JAS (Java Agent Services API)	Fujitsu, HP, IBM, SUN, … http://java-agent.org	✓	✓	✓	✓	×	?
AgentBuilder	IntelliOne Technologies http://www.agerbuilder.com	×	×	×	×	×	?
MadKit (Multi-Agent Development Kit)	MadKit Team http://www.madkit.org	×	×	×	✓	×	?
Comtec Agent Platform	Communication Technologies http://cas.comtec.co.jp/ap	✓	✓	✓	✓	×	×
Bee-agent	Toshiba http://www2.toshiba.co.jp/beeagent/index.htm	×	×	×	×	×	✓
Aglets	IBM Japan http://lri.ibm.com/aglets	×	×	×	✓	×	✓

22.7.6 Agent Platforms Overview

Table 22.1 gives an overview of majority of currently available agent development tools with respect to the properties discussed in previous paragraphs. A security attribute has been added as a property of the agent platform ensuring secure communication (usually via SSL), authorization, authentication, permissions, etc. The ✓ sign indicates that an agent platform has a particular property; meanwhile, the ✗ sign indicates that such a property is missing. If a ? sign is used, there is no reference to such a property in available sources and it can be assumed that the platform does not have it.

The Java Agent Services (JAS) have also been included in Table 22.1. However, this project is aimed at the development of the standard JAVA APIs (under the `javax.agent` namespace), that is, a set of classes and interfaces for the development of your own FIPA-compliant agent-based systems. From this perspective, JAS cannot be considered as a classical agent platform, since it does not provide any runtime environment that could be used to run your agents (either on a PC or possibly on an automation controller).

The JINI technology [24] has not been considered in this evaluation either. Similar to JAS, JINI is a set of APIs and network protocols (based on JAVA Remote Method Invocation) that can help you to build and deploy distributed systems. It is based on the idea of services providing useful functions on the network and the lookup service that helps clients to locate these services. Although JINI provides a solid framework for various agent implementations (see, e.g., Reference [25]), it cannot itself be regarded as an agent platform.

22.7.7 Message-Sending Speed Benchmarks

It has been discussed earlier that a speed at which the messages are exchanged among agents can be a crucial factor in agent-based real-time manufacturing applications. Thus, we have put selected agent platforms through a series of tests where the message delivery times have been observed under different conditions.

In each test, the so-called average round-trip time (avgRTT) is measured. This is the time period needed for a pair of agents (let say A and B) to send a message (from A to B) and receive a reply (from B to A). We use a JAVA `System.currentTimeMillis()` method, which returns the current time as the number of milliseconds since midnight, January 1, 1970. The round-trip time is computed by the A-agent when a reply from B is received as a difference between the receive time and the send time. An issue is that a millisecond precision cannot be mostly reached; the time grain is mostly 10 or 15 msec (depending on the hardware configuration and the operating system). However, it can easily be solved by repeating a message exchange several times (1000 times in our testing) and computing the average from all the trials.

As can be seen in Table 22.2, three different numbers of agent pairs have been considered: 1 agent pair (A–B) with 1000 messages exchanged, 10 agent pairs (A1–B1, A2–B2, …, A10–B10) with 100 messages exchanged within each pair, and finally 100 agent pairs (A1–B1, A2–B2, …, A100–B100) with 10 messages per pair. Moreover, for each of these configurations, two different ways of executing the tests are applied. In the serial test, the A agent from each pair sends one message to its B counterpart and when a reply is received, the round-trip time for this trial is computed. It is repeated in the same manner N-times (N is 1000/100/10 according to number of agents) and after the Nth round-trip is finished, the average response time is computed from all the trials. The parallel test differs in such a way that the A agent from each pair sends all N messages to B at once and then waits until all N replies from B are received. In both the cases, when all the agent pairs are finished, from their results the total average round-trip time is computed.

As the agent-based systems are distributed in their nature, all the agent platforms provide the possibility to distribute agents on several computers (hosts) as well as run agents on several agent platforms (or parts of the same platform) within one computer. Thus, for each platform, three different configurations have been considered: (i) all agents running on one host within one agent platform, (ii) agents running on one host but within two agent platforms (i.e., within two Java Virtual Machines — JVMs), and (iii)

TABLE 22.2   Message Delivery Time Results for Selected Agent Platforms

JAVA-Based Agent Development Toolkits/Platforms — Benchmark Results

	Message sending - average roundtrip time (RTT)					
Agent Platform	agents : 1 pair messages: 1,000 × ↔		agents: 10 pairs messages: 100 × ↔		agents: 100 pairs messages: 10× ↔	
	serial [ms]	parallel [s]	serial [s]	parallel [s]	serial [s]	parallel [s]
JADE v2.5	0.4	0.36	4.4	0.22	57.8	0.21
JADE v2.5 1 host, 2 JVM, RMI	8.8	4.30	85.7	4.34	1426.5	4.82
JADE v2.5 2 hosts, RMI	6.1	3.16	56.2	3.60	939.7	3.93
FIPA-OSv2.1.0	28.6	14.30	607.1	30.52	2533.9	19.50
FIPA-OS v2.1.0 1host, 2 JVM, RMI	20.3	39.51	205.2	12.50	×	×
FIPA-OS v2.1.0 2 hosts, RMI	12.2	5.14	96.2	5.36	×	×
ZEUS v1.04	101.0	50.67	224.8	13.28	×	×
ZEUS v1.04 1 host, 2 JVM, ?	101.7	51.80	227.9	×	×	×
ZEUS v1.04 2 hosts, TCP/IP	101.1	50.35	107.6	8.75	×	×
JACK v3.51	2.1	1.33	21.7	1.60	22.9	1.60
JACK v3.51 1 host, 2 JVM, UDP	3.7	2.64	31.4	3.65	185.2	2.24
JACK v3.51 2 hosts, UDP	2.5	1.46	17.6	1.28	165.0	1.28
NONAME	141.3	1.98	2209.3	0.47	×	×
NONAME 1 host, 2 JVM, ?	N/A	N/A	N/A	N/A	N/A	N/A
NONAME 2 hosts, IIOP	158.9	×	×	×	×	×

agents distributed on two hosts. The distribution in the last two cases was obviously done by separation of the A–B agent pairs.

The overall benchmark results are presented in Table 22.2. Recall that the results for serial tests are in milliseconds (msec) while for parallel testing, seconds (sec) have been used. Different protocols used by agent platforms for the interplatform communication are also mentioned: Java RMI for JADE and FIPA-OS, TCP/IP for ZEUS, and UDP for JACK. To give some technical details, two Pentium II processor-based computers running on 600 MHz (256 MB memory) with Windows 2000 and Java2 SDK v1.4.1_01 were used. Some of the tests, especially in the case of 100 agents, were not successfully completed mainly because of communication errors or errors connected with the creation of agents. These cases (particularly for FIPA-OS and ZEUS platforms) are marked by a × symbol.

22.7.8   Platforms — Conclusion

On the basis of the results of this study, the *JADE* agent platform seems to be the most suitable open-source candidate for the development tool and the run-time environment for agent-based manufacturing solutions. In comparison with its main competitor, *FIPA-OS*, the JADE platform offers approximately twice the speed in message sending and, above all, a much more stable environment, especially in the case of larger numbers of agents deployed.

Among the commercial agent platforms, to date, only *JACK* can offer full FIPA compliancy and also cross-platform interoperability through a special plug-in — JACK agents can send messages, for example, to JADE or FIPA-OS agents (and vice versa) via the FIPA message transport protocol based on HTTP. In the intraplatform communication, based on the UDP protocol, JACK (unlike other platforms) keeps pace with JADE in case of one host and even surpasses it in other cases, being approximately 2–3 times faster. Considering the full implementation of the Belief–Desire–Intention (BDI) model, JACK can be regarded as a good alternative to both the open-source JADE and FIPA-OS platforms.

22.8 Role of Agent-Based Simulation

The process of developing and implementing a holonic system relies on several phases and widely explores the simulation principles. The simulation process and its fast development using efficient simulation tools represent the key tasks for any implementation of a real-life holonic/agent-based system. The following stages of the design process based on simulation can be gathered like this:

1. *Identification of holons/agents:* The design of each holonic system starts from a thorough analysis of:
   A. The system to be controlled or manufacturing facility to be deployed.
   B. The control/manufacturing requirements, constraints, and hardware/software available.
   The result of this analysis is the first specification of *holon/agent classes* (types) to be introduced. This specification is based on the application and its ontology knowledge. The obvious design principle is that each device, or each segment of the transportation path, or each workcell is represented by a holon.
2. *Implementation/Instantiation of holon classes from the holon/agent type-library.* The holon/agent-type library is either developed (step 1) or reused (if already available). Particular holons/agents are created as instances of the holonic definitions in the holon/agent-type library. Furthermore, the implementation of communication links among these holon/agent instances is established within the framework of initialization from these generic holon/agent classes (for instance, holons are given the names of their partners for cooperation).
3. *Simulation*: The behavior of a holonic system is not deterministic, but rather emergent — the decision-making knowledge stored locally in the agents/holons invokes the global behavior of the system in a way that cannot be precisely predicted. Yet, the direct experimental testing of the global behavior with the physical manufacturing/control environment being involved is not only extremely expensive, but nonrealistic as well. Simulation is the only way out. For this purpose, it is necessary to have:
   A. A suitable tool used to model and simulate the physical processes in the manufacturing facility. Standard simulation tools like, for example, Arena, Grasp, Silk, or MATLAB can be used for these purposes.
   B. A suitable agent runtime environment for modeling the interactions of holonic agent parts. On the basis of the results of agent platforms comparison (Section. 22.7), the JADE platform as open source or JACK as a commercial tool can be recommended.
   C. A good simulation environment to model the real-time parts of multiple holons. There are function block emulation tools from Rockwell Automation (Holobloc [26]) and the modified 4-control platform from Softing [27]. Yet, these tools do not adequately generate and handle real-time control problems. A more sophisticated real-time solution would be to use embedded firmware systems like JBED, with its time-based scheduler, to run JAVA objects (which simulate function blocks) and manage events in a realistic manner.
   D. Human–Machine Interfaces (HMIs) for all the phases of the system design and simulation.
4. *Implementation of the target control/manufacturing system*: In this stage, the target holonic control or manufacturing system is reimplemented into the (real-time) running code. This implementation usually relies on ladder logic, structured text, or function blocks at the lowest level of control. However, some parts of the targeted manufacturing systems (such as resource or operation planning subsystems) are often reused as in Phase 3. For example, in the eXPlanTech production planning MAS [28], there was 70% of the real code reused from the simulation prototype. Therefore, the choice of the multi-agent platform in Phases 3 and 4 is critical (it has been advised to operate with one platform only).

22.9   Conclusions

Why does the agent technology seem to be so important for the area of manufacturing? What are the reasons and advantages of applying them? What do they really bring? Let us try to summarize the current experience shared by both the holonic and multi-agent communities. The main advantages can be summed up as follows:

1. *Robustness and flexibility of the control/diagnostic systems*:
   A. Robustness is achieved mainly due to the fact that there is no central element, no centralized decision making. Any loss of any subsystem cannot cause a fatal failure of any other subsystem.
   B. The agent technology enables one to handle the problems of production technology failures in a very efficient way. Optimal reconfiguration of the available equipment (which remains in operation) can be carried out in a very fast way. Thus, sustainable continuation of the production task or operation or safety stopping of the manufacturing process can be achieved. (Similarly, accomplishment of the life-critical part of a mission — after an important part of the equipment has been destroyed — can be achieved in the military environment).
   C. Changes in the production facility (adding a machine, deleting a transportation path, etc.) can be handled on the fly, without any need to reprogram the software system as a whole. Just a couple of messages are exchanged, and the agents are aware of the change and behave accordingly.
   D. Changes in production plan or schedule can be handled easily, without the need for stopping the process or bringing it back to some of the initial states. The changes in the production plan or schedule can be handled in parallel to solving the tasks connected with changes in the facility equipment and/or failures.
2. The *plug-and-play approach* is strongly supported. This enables to change/add/delete the hardware equipment as well as software modules on the fly. The migration process from the old to the new technology can be carried out smoothly, on a permanent basis, without any need to stop the operation. This also makes the system maintenance costs significantly cheaper.
3. Control and diagnostics are carried out as *near to the physical processes* as possible; control and diagnostic subsystems can cooperate on the lowest level (and in a much faster way). Control and diagnostics can be really fully integrated. This fact improves the behavior of control/diagnostic systems in the hard real-time control/diagnostic tasks. Moreover, it is possible to change the principles of behavior centrally, just by changes in the rules or policies known to each of the agents.
4. *The same agent-based philosophy can be used on different levels*, in different subsystems of the manufacturing facility and company. The same agent-oriented principles and techniques can be, for example, applied on the hard real-time level (holonic control), soft real-time control, strategic decision making for control tasks, integrated diagnostics or diagnostics running as a separate process aside of control, production planning and scheduling, higher-level decision making on the company level, supply-chain management, and for the purposes of virtual enterprises (viewed as coalitions of cooperating companies). Despite the same communication standards and negotiation scenarios being used across all the tasks mentioned above, a very high efficiency resulting from automatic communication and negotiation between units on different levels and located in different subsystems can be obtained.

Besides the advantages of the agent-based solutions, several disadvantages can also be easily identified:

1. The *investments* needed to implement the agents-based manufacturing system *are higher*. Unfortunately, the available flexibility, which is the payoff for these expenses, is usually so enormous that the manufacturing process can leverage just a very small portion of it.
2. As there is no central control element present (in an ideal agent-based factory), in the society of mutually communicating agents, unpredictable, *emergent behavior* can be expected. This causes several obstacles for the agent-based solutions to be easily accepted by the company management.

The only way out seems to be a very thorough *simulation of the agent-community behavior.* From the authors' own experience, the simulation detects just a limited number of patterns of emergent behavior. The "dangerous" patterns can be avoided by introducing appropriate policies across the system. Thus, the system simulation helps to understand the patterns of emergent behavior and their nature and to find protective measures, if necessary.

3. The current control systems offered by all the important vendors support the centralized control solutions only. The *migration toward autonomous, independent controllers* communicating asynchronously (when needed) among themselves in the peer-to-peer way, seems to be the necessary technology enabler for a wider application of the agent-based solutions. Rockwell Automation, as a pioneering company in solving the migration process, is currently extending the classical PLC controller architecture to enable to run JAVA agents as well as a JAVA-based agent run-time environment directly within existing PLCs in parallel with the classical real-time scan-based control. Thus, the concept of holonic agents presented in Section 22.6 shifts from mainly academic considerations to the actual implementation.

4. Nearly the entire community *of control engineers has been educated to design*, run, and maintain strictly *centralized solutions.* This is quite a serious obstacle, as the engineers with the "classical" centralization-oriented approach (stressed in the last three decades under the CIM label) are really not ready and able to support the agent-based solutions. Much more educational efforts will be needed to overcome this serious hurdle.

5. Not all the tasks can be solved by the agent-based approach (the estimates talk about 30% of the control tasks and 60% of the diagnostic tasks to be suitable for application of the agent-based techniques). But certain areas with a higher degree of applicability of the agent-based technology have been identified already (see below). In general, *applying the agent-based technology in inappropriate tasks can lead to frustration.*

The areas suitable for application of the agent-based techniques are as follows:

1. *Transportation of material/material handling.* The transportation paths (conveyors, pipelines, AGVs, etc.) and their sensing and switching elements (diverters, crossings, storages, valves, tag readers, pressure sensors, etc.) can be easily represented by agents; their mutual communication can be defined and organized in a quite natural way. Interesting pioneering testbeds have been built. They document the viability and efficiency of this approach in the given category of tasks.

2. *Intelligent control of highly distributed systems,* namely in the *chemical industry* and in the area of *utility distribution* control (electrical energy, gas, waste water treatment, etc.). Many decisions can be made locally, in a very fast way; the communication among the autonomous unit is carried out only if really needed.

3. *Flexible manufacturing in automotive industry.* For this industry (aimed at mass production of individually customized products), very variable customization requirements, changes in the plans and schedules, changes in technology, as well as equipment failures seem to be quite obvious features of everyday operation. All these requirements and emergency situations can be easily handled by the agent technology.

4. *Complex military systems* (like aircraft and their groups, ships, army troops in the battlefield) can be *modeled and managed as groups of agents.* For instance, very high flexibility of the technical equipment on board a ship enables to accomplish at least a part of its mission if certain subsystems are destroyed or permanently out of operation.

The research in the field of agent-based control and diagnostic systems for manufacturing has been concentrated namely around the HMS consortium within the frame of the international initiative Intelligent Manufacturing Systems. Currently, there can be recognized several leading academic and industrial centers active in this field and bringing important results.

Let us mention the following academic sites: University of Cambridge, Center for Distributed Automation and Control (CDAC), Cambridge, U.K.; University of Calgary, Department of Mechanical

Engineering, Canada; Katholieke University of Leuwen, Department of PMA, Belgium; Vienna University of Technology, INFA Institute, Vienna, Austria; University of Hannover, IPA, Hannover, Germany; Czech Technical University, Gerstner Lab, Prague, Czech Republic.

Among the industrial leaders in agent-based control and diagnostics, the following companies should be mentioned: Rockwell Automation, Milwaukee, WI; Rockwell Scientific Company, Thousand Oaks, CA; Daimler-Chrysler, Central Research Institute, Stuttgart, Germany; Toshiba + Fanuc, Japan; ProFactor, Steyr, Austria; SoftIng, Munich, Germany; CertiCon, a.s., Prague, Czech Republic; CSIRO, Melbourne, Australia.

The agent-based technology for manufacturing is developing in a very fast way. This development trend strictly follows the current trend in MAS research in the field of Artificial Intelligence as well as all the recommendations of the FIPA standardization consortium. But a long way remains in front of us: it is necessary, for example, (i) to change the way of thinking of industrial designers and engineers of control systems; (ii) to document the reliability and manageability of the emergent behavior of the agent-based systems (for this purpose, much more robust simulation tools should be developed); (iii) to support the migration processes from the centralized to agent-based control, which concern both the hardware and software; (iv) to solve the technical problems of interoperability, communication, and negotiation among the agents; and (v) to work toward widely acceptable ontology structures and languages.

References

1. Marik, V., M. Pechoucek, P. Vrba, and V. Hrdonka, FIPA standards and holonic manufacturing, in *Agent Based Manufacturing: Advances in the Holonic Approach*, Deen, S.M., Ed., Springer-Verlag, Berlin, 2003, pp. 89–121.
2. Camarinha-Matos, L.M. and H. Afsarmanesh, Eds., Interim Green Report on New Collaborative Forms and Their Needs, September 2002.
3. Marik, V., M. Pechoucek, and O. Stepankova, Social knowledge in multi-agent systems, in *Multi-Agent Systems and Applications*, Lecture Notes in Artificial Intelligence, Michael Luck, Vladimir Marik, Olga Stepankova, Robert Trappl, Eds., Vol. 2086, Springer-Verlag, Heidelberg, 2001, pp. 211–245.
4. Deen, S.M., Ed., *Agent Based Manufacturing: Advances in the Holonic Approach*, Springer-Verlag, Berlin, 2003.
5. Searle, J.R., *Speech Acts*, Cambridge University Press, Cambridge, 1969.
6. McGuire, J., D. Kuokka, J. Weber, J. Tenebaum, T. Gruber, and G. Olsen, SHADE: technology for knowledge-based collaborative engineering, *Concurrent Engineering: Research and Applications*, 1, 137–146, 1993.
7. Decker, K., K. Sycara, and M. Williamson, Middle Agents for Internet, in *Proceedings of the International Joint Conference on Artificial Intelligence 97*, Nagoya, Vol. 1, 1997, pp. 578–583.
8. Shen, W., D.H. Norrie, and J.A. Barthes, *Multi-Agent Systems for Concurrent Intelligent Design and Manufacturing*, Taylor & Francis, London, 2001.
9. Marik, V., M. Pechoucek, O. Stepankova, and J. Lazansky, ProPlanT: multi-agent system for production planning, *Applied Artificial Intelligence Journal*, 14, 727–762, 2000.
10. Pechoucek, M., V. Marik, and J. Barta, A knowledge-based approach to coalition formation, *IEEE Intelligent Systems*, 17, 17–25, 2002.
11. Rao, A.S. and M.P. Georgeff, An Abstract Architecture for Rational Agents, in *Proceedings of Knowledge Representation and Reasoning KR&R-92*, 1992, pp. 439–449.
12. Obitko, M. and V. Marik, Mapping between ontologies in agent communication, in *Multi-agent Systems and Applications III*, Lecture Notes in Artificial Intelligence, Vol. 2619, Springer-Verlag, Heidelberg, 2003, pp. 177–188.
13. Knutilla, A., C. Schlenoff, and R. Ivester, A Robust Ontology for Manufacturing Systems Integration, *in Proceedings of the 2nd International Conference on Engineering Design and Automation*, 1998.

14. Chaudhri, A.F., R. Fikes, P. Karp, and J. Rice, OKBC: A Programmatic Foundation for Knowledge Base Interoperability, in *Proceedings of AAAI-98*, 1998.

15. Vrba, P. and V. Hrdonka, Material Handling Problem: FIPA compliant agent implementation, in *Multi-Agent Systems and Applications II*, Lecture Notes in Artificial Intelligence, Vladimir Marik, Olga Stepankova, Hana Krautwurmova, Michael Luck, Eds., Vol. 2322, Springer-Verlag, Berlin, 2002.

16. Van Leeuwen, E.H. and D. Norrie, Intelligent manufacturing: holons and holarchies, *Manufacturing Engineer*, 76, 86–88, 1997.

17. Koestler, A., *The Ghost in the Machine*, Arkana Books, London, 1967.

18. Chirn, J.L. and D.C. McFarlane, Building holonic systems in today's factories: a migration strategy, *Journal of Applied Systems Studies*, 2, 82–105, 2001.

19. McFarlane, D.C., M. Kollingbaum, J. Matson, and P. Valckenaers, Development of Algorithms for Agent-based Control of Manufacturing Flow Shops, *in Proceedings of the IEEE International Conference on Systems, Man and Cybernetics*, Tucson, 2001.

20. Fletcher, M. and R.W. Brennan, Designing a Holonic Control System with IEC 61499 Function Blocks, *in Proceedings of the International Conference on Intelligent Modeling and Control*, Las Vegas, 2001.

21. Vrba, P., Agent platforms evaluation, in *Holonic and Multi-Agent Systems for Manufacturing*, Lecture Notes in Artificial Intelligence, Vladimir Marik, Duncan McFarlane, Paul Valckenaers, Eds., Vol. 2744, Springer-Verlag, Heidelberg, 2003, pp. 47–58.

22. Java2 Platform, Micro Edition: website http://java.sun.com/j2me

23. Berger, M., S. Rusitschka, D. Toropov, M. Watzke, and M. Schlichte, Porting Distributed Agent-Middleware to Small Mobile Devices, in *Proceedings of the Workshop on Ubiquitous Agents on Embedded, Wearable, and Mobile Devices*, Bologna, Italy, 2002.

24. Jini technology: website http://www.jini.org

25. Ashri, R. and M. Luck, Paradigma: Agent Implementation through Jini, *in Proceedings of the Eleventh International Workshop on Database and Expert Systems Applications*, Tjoa, A.M., R.R. Wagner, and A. Al-Zobaidie, Eds., IEEE Computer Society, Silver Spring, MD, 2000, pp. 53–457.

26. Holobloc, Rockwell Automation, http://www.holobloc.com

27. 4-Control, Softing AG, http://www.softing.com

28. Riha, A., M. Pechoucek, H. Krautwurmova, P. Charvat, and A. Koumpis, Adoption of an Agent-Based Production Planning Technology in the Manufacturing Industry, in *Proceedings of the 12th International Workshop on Database and Expert Systems Applications*, Munich, Germany, 3–7 Sept. 2001, pp. 640–646.

Part 6

Security in Industrial Automation

23

IT Security for Automation Systems

Martin Naedele
ABB Corporate Research,
Switzerland

23.1 Introduction

These days, more and more automation systems, both systems for automating manufacturing processes and for controlling critical infrastructure installations, for example, in power and water utilities, are directly or indirectly connected to public communication networks like the Internet. While this leads to productivity improvements and faster reaction on market demand, it also creates the risk of attacks via the communication network. This chapter surveys how network-connected plants and automation systems can be secured against information system and network-based attacks by state-of-the-art defensive means, and it will provide an outlook on future research.

23.2 Motivation

The influence of automation systems pervades many aspects of everyday life in most parts of the world. In the form of factory and process control systems, they enable high productivity in industrial production, and in the form of electric power, gas, and water utility systems, they provide the backbone of technical civilization.

Up to now, most of these systems were isolated, but for the last couple of years, due to market pressures and novel technology capabilities, a new trend has emerged to interconnect automation systems to achieve faster reaction times, to optimize decisions, and to collaborate between plants, enterprises, and industry sectors. Initially, such interconnections were based on obscure, specialized, and proprietary communication means and protocols. Now, more and more open and standardized Internet technologies are used for that purpose.

In security terminology, a risk exists if there is a vulnerability, that is, an opportunity to cause damage, together with a threat, the possibility that someone will try to find and exploit a vulnerability in order to inflict damage. The importance of automation systems for the functioning of modern society, together with market pressure and competition on the one hand and geopolitical tensions on the other, make the existence of security threats from terrorism, business competitor sabotage, and other criminal activity appear likely. The pervasiveness of automation systems that are nowadays accessible from anywhere in the world via communications and information technologies — for which there are thousands of experts worldwide and that have a large number of well-known security issues — creates many IT security vulnerabilities. As a consequence, there are good reasons to investigate and invest regarding how to reduce the IT security vulnerabilities of automation systems, and thus the resulting risks of large financial damage, deteriorated quality of life, and potentially physical harm to humans. This chapter presents an overview of state-of-the-art best practices in that respect and an outlook on future opportunities.

23.3 Scope

The scope of automation systems considered in this chapter ranges from embedded devices, potentially in isolated locations, via plant control systems to plant- and enterprise-level supervisory control and coordination system, both in the distributed control system (DCS) flavor, more common in factory automation, and the supervisory control and data acquisition (SCADA) flavor, widespread in utility systems [5].

In the associated types of applications, in contrast to commercial and administrative data processing, often not typical data security issues (e.g., confidentiality, integrity) as such are the most important goal; but IT security is one component of the safety and fault-tolerance strategy and architecture for the plant.

23.4 Security Objectives

IT security has a number of different facets that are, to some extent, independent of each other. When defining the security requirements for a system, these facets, on which risk analysis and in turn design of countermeasures are based, can be expressed in terms of the eight security objectives explained in the following subsections.

23.4.1 Confidentiality

The confidentiality objective refers to preventing disclosure of information to unauthorized persons or systems. For automation systems this is relevant both with respect to domain-specific information, such as product recipes or plant performance and planning data, and to the secrets specific to the security mechanisms themselves, such as passwords and encryption keys.

23.4.2 Integrity

The integrity objective refers to preventing modification of information by unauthorized persons or systems. For automation systems, this applies to information coming from and going to the plant, such as product recipes, sensor values, or control commands, and information exchanged inside the plant control network. This objective includes defense against information modification via message injection, message replay, and message delay on the network. Violation of integrity may lead to safety issues, that is, equipment or people may be harmed.

23.4.3 Availability

Availability refers to ensuring that unauthorized persons or systems cannot deny access/use to authorized users. For automation systems, this refers to all the IT elements of the plant, such as control systems, safety systems, operator workstations, engineering workstations, manufacturing execution systems, as well as the communications systems between these elements and to the outside world. Violation of availability may lead to safety issues, as operators may lose the ability to monitor and control the process.

23.4.4 Authorization

The authorization objective, also known as access control, is concerned with preventing access to or use of the system or parts by persons or systems without permission to do so. In the wider sense, authorization refers to the mechanism that distinguishes between legitimate and illegitimate users for all other security objectives, for example, confidentiality, integrity, etc. In the narrower sense of access control, it refers to restricting the ability to issue commands to the plant control system. Violation of authorization may lead to safety issues.

23.4.5 Authentication

Authentication is concerned with determination of the true identity of a system user (e.g., by means of user-supplied credentials such as username/password combination) and mapping of this identity to a system-internal principle (e.g., a valid user account) under which this user is known to the system. Authentication is the process of determining who the person trying to interact with the system is, and whether he really is who he claims to be. Most other security objectives, most notably authorization, distinguish between authorized and unauthorized users. The base for making this distinction is to associate the interacting user by means of authentication with an internal representation of his permissions used for access control.

23.4.6 Nonrepudiability

The nonrepudiability objective refers to being able to provide irrefutable proof to a third party of who initiated a certain action in the system. This security objective is mostly relevant to establish accountability and liability with respect to fulfillment of contractual obligations or compensation for damages caused. In the context of automation systems, this is most important with regard to regulatory requirements, for example, FDA approval. Violation of this security objective has typically legal/commercial consequences, but no safety implications.

23.4.7 Auditability

Auditability is concerned with being able to reconstruct the complete behavioral history of the system from historical records of all (relevant) actions executed on it. While in this case it might very well be of interest to also record who initiated an action, the difference between the auditability security objective and nonrepudiability is the ability of proving the actor identity to a third party, even if the actor concerned is not cooperating. This security objective is mostly relevant to discover and find reasons for malfunctions in the system after the fact, and to establish the scope of the malfunction or the consequences of a security incident. In the context of automation systems, this is most important in the context of regulatory requirements, for example, FDA approval. Note that auditability without authentication may serve diagnostic purposes but does not provide accountability.

23.4.8 Third-Party Protection

The third-party protection objective refers to averting damage done to third parties directly via the IT system, that is, damage that does not involve safety hazards of the controlled plant. The risk to third parties through possible safety-relevant failures of the plant arising out of attacks against the plant automation system is covered by other security objectives, most notably the authorization/access control objective. However, there is a different kind of damage only involving IT systems: the successfully attacked and subverted automation system could be used for various attacks on the IT systems or data or users of external third parties, for example, via distributed-denial-of-service (DDOS) or worm attacks. Consequences could range from damaged reputation of the automation system owner up to legal liability for the damages of the third party. There is also a certain probability that the attacked third party may retaliate against the subverted automation system causing access control and availability issues. This type of counter attack may even be legal in certain jurisdictions.

23.5 Differences to Conventional IT Security

As the security objectives in the previous section are generally valid, many security issues are the same for automation systems and conventional, office-type IT systems, and many tools can be used successfully in both domains. However, there are also major differences between these two domains with respect to requirements and operating environment, characteristics, and constraints, which make some security issues easier and others more difficult to address. In the following, some of these differences are explained.

23.5.1 Requirements

While office IT security requirements center around confidentiality and privacy issues, for any automation and process control system the foremost operational requirement is safety, the avoidance of injury to humans. Second after that is availability: the plant and the automation system have to be up and running continuously over extended periods of time, with hard real-time response requirements in the millisecond range. In many cases, this precludes standard IT system administration practices of system rebooting for fixing problems, and makes the installation of up-to-date software (SW) patches, for example, addressing security problems in the running application or the underlying operating system, difficult if not impossible.

On the other hand, in contrast to e-commerce applications, connectivity to outside networks, including the company intranet, is normally not mandatory for the automation system, and although extended periods of disconnection are inconvenient, they will not have severe consequences — after all, many automation systems nowadays still run completely isolated.

23.5.2 Operational Environment

The configuration, both of HW and SW, of the automation system part, which contains the safety critical automation and control devices, is comparatively static. Therefore, all involved devices and their normal, legitimate communication patterns (regarding communication partners, frequency, message size, message interaction patterns, etc.) are known at configuration time, so that protection and detection mechanisms can be tailored to the system. Modifications of the system are rare enough to tolerate a certain additional engineering effort for reconfiguring the security settings and thus being able to trade in the convenience of dynamic, administration-free protocols like DHCP against higher determinism and, as a consequence, security of, for example, statically setting up tables with communication partners/addresses in all devices. Static structure and behavioral patterns also make the process of anomaly discovery for intrusion detection easier.

The hosts and devices in the automation system zone are not used for general-purpose computing, preventing the risks created by mainstream applications like e-mail, instant messaging, office application macroviruses, etc. Often, they are even specialized embedded devices dedicated to the automation functionality, such as power line protection in substation automation. All appropriate technical and administrative means are taken to ensure that only authorized and trustworthy personnel have physical access to the automation equipment. Automation system personnel are accustomed to a higher level of care and inconvenience when operating computer systems than office staff. This increases the acceptance and likelihood of correct execution of security-relevant operating procedures even if they are not absolutely straightforward and convenient. In many plants, additional nonnetworked (out-of-band) safety and fault-tolerance mechanisms are available to mitigate the consequences of failure of one or multiple components of the automation system.

One problem with security mechanisms is that they cannot prevent all attacks directly themselves, but produce output, like alerts and log entries, which humans need to review to decide on the criticality of an event and to initiate appropriate responses. This is often neglected as the expert IT staff is not available around the clock to monitor the system. Automated plants, in contrast, are usually continuously monitored by dedicated staff. A defense architecture could make use of this fact, even though these plant operators do not have IT or even IT security expertise.

23.5.3 Challenges

On the other hand, the characteristics of automation systems and devices create some additional security challenges.

Automation devices often have a lower processing performance compared to desktop computers, which limits the applicability of mainstream cryptographic protocols.

The operating systems of such devices in many cases do not provide authentication, access control, fine-granular file system protection, and memory isolation between processes — or these features are optional and are not used due to the abovementioned limited processing power.

Especially in telemonitoring applications (e.g., SCADA), communication channels with small to very small bandwidth like telephone, mobile phone, or even satellite phone lines are used, which makes it imperative to reduce communication overhead and thus collides with certain security protocols.

Automation systems tend to have very long lifetimes. This has consequences both for the currently operative systems and for newly implemented systems: those currently operative "legacy" automation systems, as far as IT security was given a thought at all, were designed based on a philosophy of "security by obscurity," assuming that the system would be isolated and only operatable by a small, very trustworthy group of people. This kind of thinking persists even today, as can be seen from the 2002 IEEE 1588 standard on precision time synchronization for automation systems for which it is explicitly stated as a design rationale that security functionality is neglected as all relevant systems can be assumed to be secure. Another consequence of longevity is that automation system installations tend to be very heterogeneous with respect to both subsystem vendors and subsystem technology generations.

For newly built automation systems, the long expected lifetime means that the data communication and authentication/access control functionality must be designed so that it will be able to interoperate with reasonable effort with systems and protocols to appear on the market 10 or 20 years later.

Last but not the least, automation systems are operated by plant technicians and process engineers. Due to their training background, they have a very different attitude toward IT system operation and security than corporate IT staff, and frequently a mutual lack of trust has to be overcome to implement an effective security architecture.

23.6 Building Secure Automation Systems

Building secure systems is difficult, as it is necessary to spread effort and budget so that a wide variety of attacks are efficiently and effectively prevented. For automation systems, the challenges mentioned in the section Challenges create additional difficulties. In the following, two common approaches to secure systems are explained and their effectivity is assessed.

23.6.1 Hard Perimeter

A popular doctrine for defense, be it of cities or IT systems, is the notion of the *hard perimeter*. The idea is to have one impenetrable wall around the system and to neglect all security issues within. In general, however, this approach does not work, for a variety of reasons. The hard perimeter approach does not make use of reaction capabilities: at the time of detection of a successful attack, the attacker has already broken through the single wall and the whole system is open to him. As a consequence, this means that the wall would need to be infinitely strong because it needs to resist infinitely long [15]. Also, monoculture is dangerous: the wall is based on one principle or product. If that principle or product fails for some reason to resist the attack, the whole defense is ineffective. The wall must have doors to be usable, which opens it to both technical and nontechnical (social engineering) security risks. Once the attacker has managed to sneak inside, the system is without defense — the risk of the proverbial Trojan horse. A hard perimeter is also, by definition, ineffective against insider attacks. Progressing technology gives the attacker continuously better wall-penetration capabilities. Last but not the least, humans make mistakes: it is illusory to assume that we can design a wall that is without weak spots either in design, implementation, or operation — various border walls in history serve as example.

23.6.2 Defense-in-Depth

The alternative approach is defense-in-depth. Here, several zones/shells are placed around the object that is to be protected. Different types of mechanisms are used concurrently around and inside each zone to defend it. The outer zones contain less valuable targets; the most precious goods, in this case the (safety critical) automation system, are in the innermost zone. In addition to defense mechanisms, there are also detection mechanisms, which allow the automation system operators to detect attacks, and reactive mechanisms and processes to actively defend against them. Each zone also buys time to detect and fend off the attacker. In the spirit of Schwartau's time-based security [15], this allows to live with the fact of imperfect protection mechanisms, as only a security architecture strength of $P \geq D + R$ has to be achieved, where P is the time during which the protection offered by the security system resists the attacker, D is the delay until the ongoing attack is detected, and R is the time until a defensive reaction on the attack has been completed.

Conclusion: There are two basic approaches for securing systems commonly used today, but only one, defense-in-depth, will result in a secure system, provided it is properly implemented.

23.7 Elements of a Security Architecture

In this section, the most important technical elements of a security architecture for automation systems are surveyed. Note, however, that a system cannot be secured purely using technical means. Appropriate user behavior is essential to ensure the effectiveness of any technical means. Acceptable and required user behavior should be clearly documented in a set of policies that are strictly and visibly enforced by plant management. Such policies should address, among other things, user account provisioning, password selection, virus checking, private use, logging and auditing, etc.

The topic of policies and user behavior will not be further discussed here. Example documents are available from various government agencies, IT security organizations, as well as in a number of books.

According to their physical and logical location in the architecture, in the spirit of a defense-in-depth, security mechanisms can be classified as belonging to one or multiple of the following categories. These categories are orthogonal to the security objectives of Section 23.3.

1. *Deterrence.* Means of pointing out to the potential attacker that his personal pain in case of getting caught does not make the attack worthwhile. However, in most threat scenarios for safety-critical and infrastructure systems, the deterrence component, especially the threat of legal action, is ineffective.

2. *Connection authorization.* Means to decide whether the host trying to initiate a communication is at all permitted to talk to the protected system, and to prevent such connections in case of a negative decision.

3. *User authorization.* Means to decide whether and with which level of privileges a user or application is permitted to interact with the protected system and to prevent such interaction in case of a negative decision.

4. *Action authorization.* Means to decide whether a user or application is permitted to initiate specific actions and action sequences on the protected system or application, and to prevent such inter-action in case of a negative decision. Action authorization is an additional barrier assuming a preceding positive user authorization decision.

5. *Intrusion detection.* Means to detect whether an attacker has managed to get past the authorization mechanisms. Most intrusion detection systems are based on monitoring and detecting whether anything "unusual" is going on in the system.

6. *Response.* Means to remove an attacker and the damage done by him from the system. Means to lessen the negative impact of the attack on the system and its environment. Means to prevent a future recurrence of the same type of attack.

7. *Mechanism protection.* Means to protect the mechanisms for the above-listed categories against subversion. This refers, for example, to not sending passwords in clear text over public networks and hardening operating systems and applications by fixing well-known bugs and vulnerabilities, etc.

23.7.1 Deterrence

The technical mechanisms for deterrence are warning banners at all locations, for example, log in screen, where an attacker — not necessarily an outsider — could access the system. These warning banners should state clearly that unauthorized access is prohibited and legal action may be the consequence. This type of statement could be required to be able to prosecute intruders in certain legislation.

23.7.2 Connection Authorization

The following sections list and explain connection authorization mechanisms for dial-in remote access, WAN, and LAN situations.

23.7.2.1 Firewall

A firewall or filtering router passes or blocks network connection requests based on parameters such as source/destination IP addresses, source/destination, TCP/UDP ports (services), protocol flags, etc. Depending on the criteria, the specific firewall product uses and how clearly the permitted traffic on the network is defined, a firewall (or, more often, dual-firewall architecture) can be anything from a highly effective filter to a fig leaf.

23.7.2.2 Intelligent Connection Switch/Monitor

As the information and message flows between the individual devices and applications in an automation system are often deterministic and well defined at system configuration time, this information can be used by a special, intelligent connection monitoring device to determine the legitimacy of a certain message. This goes beyond the parameters that conventional firewalls use, as it also takes into account criteria like message size, frequency, and correctness of complex interaction sequences. Nonconforming messages can be suppressed in collaboration with a switch. In this case, the device is acting as a connection authorization device. On the other hand, an alert can be raised, in which case the device is acting as an automation system-level intrusion detection system.

23.7.2.3 Personal Firewall

A personal firewall is an application running on a host as an additional protection for the main functionality of this host, for example, as a firewall, proxy server, or workstation. It monitors and controls which applications on the host are allowed to initiate and accept connection requests from the network. In contrast to a (network) firewall, a personal firewall does not control a network segment, but only its own host.

23.7.2.4 Switched Ethernet

With standard Ethernet, all hosts on a network segment can see all traffic, even that not addressed to them, and multiple hosts sending at the same time can cause nondeterministic delays. The first issue is directly security relevant, and the second one is indirectly security relevant (denial-of-service attacks). Using switches to give each host its own network segment and to directly forward each message only to the intended receiver remedies both issues. Note that there exist attacks on switches to force them to operate in hub mode (broadcast).

23.7.2.5 Dial-Back

After authentication, a dial-back modem interrupts the telephone connection and dials back to the telephone preconfigured for the authenticated user. This prevents an attacker from masquerading as an authorized user from anywhere in the world, even if he or she managed to obtain this user's credentials, because the attacker would also need to obtain physical access the authorized user's phone line. Note that in a number of countries, the technical behavior of the public telephone network is such that dial-back modems can be defeated by an attacker.

23.7.2.6 Access Time Windows

For automation systems, it is often true that the remote connection is not necessary for operation, but only to upload configuration changes and download measurement values, which are not urgent and are irregular in timing. Therefore, if a continuous remote connection is not required, the remote access can be restricted to certain time windows only known to authorized users in order to reduce exposure to attacks. Outside these time windows access is disabled, for example, by electrically switching off the modem or router device.

23.7.2.7 Mutual Device Authentication

As all communication partners in the automation system (devices and applications) and message flows are known at configuration time, it is possible to require mutual authentication of each communication relationship at run-time, provided the available computing power of the automation devices tolerates the execution of the necessary protocols. This renders the installation of rogue devices ineffective.

23.7.3 User Authorization

23.7.3.1 Log-in Mechanisms

Most access control schemes rely on passwords ("something you know") as the underlying authentication principle to establish that the user is who he or she claims to be. This makes the selection of suitable passwords, as well as their management and storage one of the most important aspects, and potential weaknesses, of each system. Alternatively, instead of a fixed password, a one-time password, generated just-in-time by individualized devices (tokens, smartcards) that are given to all authorized users, is used. It replaces or augments the "something you know" principle by "something you have." The third option for proving identity are biometrics, for example, fingerprints. This is the most direct mechanism, but expensive, inconvenient, only applicable to humans (not other applications), and there have been incidents where biometrics products have been fooled by an attacker. In any case, if authentication fails, the user has to initiate a new session to continue communication with the protected system, for example, for another login attempt.

23.7.4 Action Authorization

23.7.4.1 Role-Based Access Control in Applications

In many situations, not all authorized users are in the same way permitted to execute all types of activities on the system; therefore, the system log-on alone does not offer a sufficient granularity of access control. With role-based access control, each authorized user has one or multiple roles, which correspond to sets of actions he can execute in the application. Role-based access control needs to be designed into each application, as it cannot be normally provided by external add-on devices or applications. Defining the rights in the application in terms of roles instead of actual users improves scalability and makes maintenance of the security configuration easier.

23.7.4.2 System Architecture for Data Exchange

If the remote access functionality is not necessary for interactive operation and examination of the automation system, but only for upload and download of preconfigured parameters and data, the system can be architected so that direct remote access to the automation device, which is the source or target of the data, is not required. Instead, the remote access occurs only to a less valuable FTP or web server, which acts as a cache and communicates over a series of time-shifted, content-screening data-forwarding operations with the actual data source or target. This strongly restricts the type of interactions that the outside system can have with the automation device.

23.7.4.3 Managed Application Installation

Through centralized administration and monitoring, it is enforced that only the authorized and necessary applications and services are running on the automation system workstations and servers, and that configurations are not changed by the users. The hosts do not carry user-specific data or configurations and thus can be reinstalled from a known-good source at regular, frequent intervals to remove any unauthorized modification in the system, even if the modification was never detected.

23.7.4.4 Application-Level Gateway

The process-level network to which the automation equipment is connected interfaces with the higher level LAN, for example, the one to which the operator workstations are connected, only via a small number of gateway servers for application-level communications. Direct interactive access (log-in) to the automation device is disabled in this scenario. As the purpose, the software applications, and the topology/configuration of the automation system are well known, these gateways can be customized for the authorized applications and communications to screen messages passing the interface for validity based on domain-specific criteria, such as predefined interaction sequences. Illegitimate or invalid messages can be suppressed and alerts can be raised.

23.7.4.5 Dual Authorization

Certain commands issued by a user in an automation system can have drastic consequences. If there are no situations in which these commands need to be entered extremely quickly in an emergency, they can be protected by requiring confirmation from a second authorized user. This mechanism secures the system against intruders, malicious insiders, and serves as a protection against unintentional operator errors.

23.7.4.6 Code Access Security

Code access security is concerned with restricting what an application is allowed to do on a host (e.g., reading or writing files, creating new user accounts, initiating network connections, etc.), even if it executes in the context of an authorized, highly privileged user. This type of "sandboxing" is available for several computing platforms and can serve as additional protection against applications that are not completely trusted, for example, because they have been developed externally.

23.7.5 Intrusion Detection

23.7.5.1 Network-Based Intrusion Detection System

A network-based intrusion detection system (NIDS) tries to discover attacks based on known attack profiles and/or unusual system behavior from communication traffic seen on a network segment (type, content, frequency, path of the transmitted messages).

23.7.5.2 Host-Based Intrusion Detection System

A host-based IDS (HIDS) tries to discover attacks based on known attack profiles and/or unusual system behavior from information seen locally on the host on which it is running. A host-based IDS obtains its information, for example, from file system integrity checkers, which monitor whether important system files change without operational reason, from personal firewall logs, or from application logs, for example, for application-level role-based access control.

23.7.5.3 Honeypot

A honeypot (single host) or honeynet (subnet) [16] is a subsystem that appears particularly attractive to an attacker, for example, from the naming of the host or files on it, or by simulating certain weaknesses in the installation. It is, however, a dedicated and isolated system without importance for the functioning of the automation system, which is especially instrumented with intrusion detection systems. The idea is that an attacker who breaches the first line of defense successfully will be attracted to the honeypot host first, and thus is at the same time delayed and kept away from the really sensitive areas, as well as detected by the intrusion detection systems.

23.7.5.4 Authentication/Authorization Failure Alerts

Alerts are sent to the operators or IT staff whenever one of the authentication mechanisms fails, which could indicate an unsuccessful attempt to attack the system. While this would probably not be useful in an office system, due to the high false alarm rate of legitimate users mistyping their passwords, it may be a feasible security mechanism in an automation system with a very small number of concurrent legitimate users and few login actions per time period.

23.7.5.5 Log Analysis

All actions of the authentications devices are logged, and these logs are manually or automatically screened for unusual occurrences or patterns, for example, that an authorized user is suddenly accessing the system outside his or her normal work hours.

23.7.5.6 Malicious Activity Detection/Suppression Protocol

As is shown in Reference [9], network-based electronic attacks originating from malicious devices in an automation system, for example, in a power substation, can be categorized as either message injection, message modification, or message suppression. Using a suitable communication protocol for detection of invalid messages, one can reduce these three categories to message suppression, which can in many cases be regarded as a system failure that conventional fault-tolerance and fault-response mechanisms such as redundant devices and emergency shutdown sequences can handle.

23.7.6 Response

23.7.6.1 System Isolation

The connections between the compromised subsystem, for example, the outer part of the security zone at the interface between automation system and other networks, and other, more important parts of the automation system, are closed to avoid further spreading of the attack. Depending on system/remote access functionality and importance, and on whether delaying the attacker and collection of further evidence, or quick restoration of operation is of higher importance, it is an option to shut down all remote connections, both for the affected and the not-yet affected systems, until the effects of the attack are removed. Like electric power grids, the automation system should already be architected and designed such that it can be partitioned into zones, which can be isolated with minimum disturbance of the whole system.

23.7.6.2 Collect and Secure Evidence

All logs and media (e.g., hard disks) that contain evidence of the malicious activity are gathered, copied, and stored at a secure location. This evidence can be used to identify the attack and thus the system weaknesses in detail, to locate and assess the amount of damage done by the attack, and also to support legal prosecution of the attacker. In this case, the legally correct handling of evidence is especially important.

23.7.6.3 Trace Back

This refers to activities that aim at discovering the source of the attack, both as part of the evidence collection process (see above) and to enable stronger defense mechanisms for identified sources of attacks (e.g., blocking). Due to various possibilities of faking packet data, such as the originating address, this is technically not easy. Lee and Shields [6] discuss the various technical options for trace back and their obstacles.

23.7.6.4 Active Counter-Attack

An active counter-attack has an aim to selectively disable the attacker's computer to prevent further attacks and to "punish" the attacker. Due to the fact that an unambiguous trace back is difficult (see above), that often "innocent" systems are used as intermediary stages for staging an attack, and that the legality of a counterattack is dubious in most situations and localities, a counterattack response is normally not recommended.

23.7.6.5 Information Sharing

Early sharing of information about ongoing attacks, especially novel types of attacks, with the IT community represents good "Internet citizenship" because it gives more defenders a chance to increase alertness and to remove weaknesses. Many types of attacks, such as distributed denial of service and viruses, rely on the fact that the same weakness can be exploited on a large number of systems, which are then used to launch further attacks. Therefore, reducing the number of systems vulnerable to an attack is in every defender's interest. On the other hand, many companies might be concerned about the effects of making the facts and circumstances of attacks known to competitors and the public. For this purpose, several institutions exist, which receive and distribute information about attacks without disclosing the sources of the information. Examples are the SEI CERT (http://www.cert.org) and various industry branch-specific Information Sharing and Analysis Centers (ISACs).

23.7.6.6 Selective Blocking

If the origin of the attack and the location of entrance into the system can be identified, blocking rules of firewalls, routers, and access servers, perhaps already at the Internet Service Provider, can be temporarily or permanently modified to close the in-roads of the attack.

23.7.6.7 Switching to Backup IT Infrastructure

If a system is compromised and its unavailability during evidence collection and restauration is unacceptable, a backup automation system, perhaps with minimum functionality, should be available for immediate switch-over. This backup system, consisting of automation system workstations and servers, access servers, firewalls, IDS host, etc., should be preconfigured with different passwords, different network addresses, and, even better, also use software different from the primary system, to avoid being immediately subverted by the attacker with the knowledge gained from the primary system.

23.7.6.8 Automated, Periodic Reinstallation of Applications, Operational Data, and Configurations

The SW applications of the automation system, as well as static operational data and configurations, especially of the security components, are periodically reinstalled from a known-good read-only storage device (e.g., CD). This removes applications with attacker-installed backdoors (Trojans) or modified system files, for example, password files, even if the attack was never detected. Of course, this provides only a weak defense if the security vulnerabilities that allowed the system compromise in the first place are allowed to persist, but it may prevent follow-up exploits by other attackers and frustrate the original attacker sufficiently to turn to easier targets.

23.7.6.9 New Passwords

When automatically or manually reinstalling system configurations in case of an attack, the passwords should also be changed, in particular those for the security components, to prevent the attacker from immediately reentering the system with a previously compromised set of credentials.

23.7.6.10 Activation of Safety Mechanisms

If the attack on the automation system endangers the safety of the plant, standard safety mechanisms such as reverting to manual operation or emergency shut-down are a last resort to protect plant equipment and human life. As a consequence, such safety mechanisms should be decoupled from the networked automation system.

23.7.7 Mechanism Protection

23.7.7.1 Dedicated Lines

Instead of connecting the dial-in modems to telephone lines, which are accessible to anybody from anywhere in the world, dedicated telephone cables are used, which connect only secure, authorized systems. Among other things, this protects the information between the protected system and the remote user, in particular, his credentials such as passwords during the authentication process, from eavesdropping. However, this scheme offers no protection against the telecommunication company and (government) organizations that can force access to dedicated lines. Also, there have been incidents where attackers were able to subvert the telephone switches of telecommunication providers to access dedicated lines.

23.7.7.2 Virtual Private Network

A virtual private network (VPN) uses encryption and digital signatures to achieve the effect of a physical dedicated line over a shared medium such as a normal telephone connection or the Internet. A VPN is both cheaper and more secure than a line leased from a telecommunication provider, but the computational overhead for strong cryptography may be unacceptable for certain communicating automation devices.

23.7.7.3 Disable Remote Reprogramming of Dial-In/Dial-Back Modems

The dial-back mechanism described in Section 23.7.2.5 is only effective if the remote attacker cannot redirect the dial-back call to his own telephone.

23.7.7.4 Network Address Translation

With network address translation (NAT), the system uses IP addresses internally different from those shown in the externally visible messages. A border device, for example, the firewall, is responsible for on-the-fly translation of addresses in both directions. NAT makes remote probing of the internal network topology for interesting or vulnerable targets much more difficult and prevents certain attacks that bypass the firewall.

23.7.7.5 Diversity

System diversity, for example, by using different operating systems like Windows and Unix/Linux for the production systems and the intrusion detection systems or other security mechanisms, or by selecting different brands of firewalls for different zones and subzones increases security, as an attacker cannot rely on a single vulnerability in one product to break through all defenses. Thus, it also offers a bit more resilience in the time between publication of an exploit for a vulnerability and the design and installation of corresponding patches in the system.

23.7.7.6 Role-Based Access Control

Role-based access control is also important for the security functionality itself, to ensure that only security administrators and not all inside users can change security settings and read/edit logs. This is the basis of all security precautions against insider attacks and also creates an additional hurdle for attackers that have broken into the account of one authorized user against taking complete control over the system.

23.7.7.7 Hardened Host

Mechanisms like role-based access control, logging, and intrusion detection rely on the basic functions of the operating system on the host not having been corrupted by the attacker. Today's applications, operating systems, and system configurations are often so complex and complicated that they have many security vulnerabilities caused by misconfiguration or by applications with security relevant bugs, which are installed as part of the operating system, but that are not really necessary for the automation system functionality. Hardening a system means to remove all unnecessary applications and services, to fix known bugs, to replace critical applications with more trustworthy ones of the same functionality, and to set all system configuration parameters to secure values. Guidelines for the hardening of various common operating systems and applications are available from multiple sources.

23.8 Further Reading

A large number of technical and research publications exist on the issue of IT security for home and office information systems. Schneier [14] gives a good general introduction to the topic, and [13] is the reference on cryptographic algorithms and protocols. Reference [10] is a comprehensive resource on the practical issues of securing a computer network, while References [1] and [17] address the issue of engineering secure (software) systems from a larger perspective.

On the other hand, apart from some vendor white papers, there is almost no literature on the specific security needs and capabilities of industrial automation systems. Palensky [12] investigates remote access to automation systems, specifically home automation systems, with potentially malicious devices, and proposes the use of smartcards as trusted processors to achieve end-to-end security between each device and its legitimate communication partners. In Reference [8], IT security mechanisms applicable for automation systems are presented, according to which conceptual zone they defend — remote access, operator workstations, or automation devices. Byres [3] investigates networking and network-level security issues for Ethernet networks on the plant floor. Falco [5] motivates and describes efforts to create a protection profile for process control systems according to the Common Criteria security evaluation standard. Moore [7] reports on a survey about IT security conducted among automation system users. Dafelmair [4] promotes the use of a Public key Infrastructure (PKI) for process control systems, and Beaver et al. [2] suggest a lightweight PKI for power-utility SCADA systems. Oman [11] applies standard IT security mechanisms to power utility control systems, with special emphasis on password management.

23.9 Research Issues

As can be seen from the scarcity of published work, automation system IT security is a comparatively new field. Much more research will be necessary until both security requirements and opportunities specific for automation systems have been explored to the current level of home and business system IT security.

The following are just some of the topics to be addressed in the future:

- Considering that plant control systems have a lifetime of 20 to 30 years, how can effective defense-in-depth security mechanisms cost-efficiently be retrofitted onto them?
- What security mechanisms can and should be required for automation systems? This topic is being currently addressed by various industry standard organizations, such as the ISA SP99 working group.
- What criteria should be used for assessing and auditing the security mechanisms in automation devices?
- How to cope with the vulnerabilities inherent in networking protocols like SNTP for time synchronization or XML web services/OPC-XML by modifying the protocols or adapting the system architecture?
- How to ensure appropriate levels of access control, data integrity, and data confidentiality for automation devices with low computing power [18]?
- How can the auditability of every interaction and individual accountability of every member of the plant staff who interacts with the automation system be ensured according to regulatory requirements, while at the same time achieving both ease of use in normal operation and fast response times for emergency interventions?
- How can a plant operator, who is not an IT security expert, effectively contribute to plant security?

23.10   Summary

Nowadays, realistic scenarios for network-based attacks on infrastructure/utility automation systems and manufacturing/plant automation systems with respect to both motivation as well as technical feasibility exist. In contrast to business systems, which need to be available for business and protect their confidential data, for automation system IT security, the most important security objective is the integrity of the control system to prevent physical damage and human injury.

This chapter has presented arguments regarding why a defense-in-depths approach with layered mechanisms is a better strategy for securing systems than the often-used approach of placing a security "wall" around the system and leaving the inside unchanged. It has also given an overview of available security mechanisms, pointing out which are specifically applicable to automation systems due to the specific operational characteristics of automation systems.

It remains to recall that, as the saying goes, security is not a destination, but a journey — both the notifications generated every day by an appropriate security system and the whole architecture of the security system need to be reviewed regularly to detect and adapt to new vulnerabilities and threats.

References

1. Anderson, Ross, *Security Engineering*, Wiley, New York, 2001.
2. Beaver, Cheryl, Donald Gallup, William Neumann, and Mark Torgerson, Key Management for SCADA, Technical report SAND2001-3252, Cryptography and Information Systems Security Department, Sandia National Laboratories, March 2002.
3. Byres, Eric, Designing secure networks for process control, *IEEE Industry Applications Magazine,* 6: 33–39, 2000.
4. Dafelmair, Ferdinand J., Improvements in Process Control Dependability through Internet Security Technology, in *Proceedings Safecomp 2000, Lecture Notes in Computer Science,* Vol. 1943, Springer, Berlin, 2000, pp. 321–332.

5. Falco, Joe, Keith Stouffer, Albert Wavering, and Frederick Proctor, IT Security for Industrial Control Systems, Technical Report, Intelligent Systems Division, (U.S.) National Institute of Standards and Technology (NIST), 2002.

6. Lee, Susan C. and Clay Shields, Technical, legal and societal challenges to automated attack traceback, *IEEE IT Professional*, Vol. 4, No. 3, May/June: 12–18, 2002.

7. Moore, Bill, Dick Slansky, and Dick Hill, Security Strategies for Plant Automation Networks, Technical Report, ARC Advisory Group, July 2002.

8. Naedele, Martin, IT Security for Automation Systems — Motivations and Mechanisms, *atp-Automatisierungstechnische Praxis*, 45: 2003.

9. Naedele, Martin, Dacfey Dzung, and Michael Stanimirov, Network security for substation automation systems, in *Computer Safety, Reliability and Security (Proceedings Safecomp 2001)*, Voges, Udo, Ed., Lecture Notes in Computer Science, Vol. 2187, Springer, Berlin, 2001.

10. Northcutt, Stephen, Lenny Zeltser, Scott Winters, Karen Fredrick, and Ronald W. Ritchey, *Inside Network Perimeter Security: The Definitive Guide to Firewalls, Virtual Private Networks (VPNs), Routers, and Intrusion Detection Systems*, New Riders, Indianapolis, 2003.

11. Oman, Paul, Edmund Schweitzer, and Deborah Frincke, Concerns about Intrusions into Remotely Accessible Substation Controllers and SCADA Systems, Technical Report, Schweitzer Engineering Laboratories, 2000.

12. Palensky, Peter and Thilo Sauter, Security Considerations for FAN-Internet Connections, in *Proceedings 2000 IEEE International Workshop on Factory Communication Systems*, 2000.

13. Schneier, Bruce, *Applied Cryptography*, 2nd ed., Wiley, New York, 1996.

14. Schneier, Bruce, *Secrets and Lies — Digital Security in a Networked World*, Wiley, New York, 2000.

15. Schwartau, Winn, *Time Based Security*, Interpact Press, 1999.

16. [pitzner, Lance, The Honeynet project: trapping the hackers, *IEEE Security and Privacy*, 1:15–23, 2003.

17. Viega, John and Gary McGraw, *Building Secure Software*, Addison-Wesley, Reading, MA, 2001.

18. on Hoff, Thomas P. and Mario Crevatin, HTTP Digest Authentication in Embedded Automation Systems, *9th IEEE International Conference on Emerging Technologies and Factory Automation*, Lisbon, Portugal, 2003.

Author Index

Author	Affiliation	Subsection	Chapter	Title
Almeida, Luis	University of Aveiro, Portugal	4.3	17	The Quest for Real-Time Behavior in Ethernet
Bandyopadhyay, Pulak	GM R&D Center, Warren	2	2	Introduction to e-Manufacturing
Büsgen, Ralph	Siemens AG, Germany	4.3	18	Principles and Features of PROFInet
Decotignie, Jean-Dominique	Centre Suisse d'Electronique et de Microtechnique, Switzerland	4.4	20	Interconnection of Wireline and Wireless Fieldbuses
Diedrich, Christian	Institut für Automation und Kommunikation eV – IFAK, Germany	4.1	11	Integration Technologies of Field Devices in Distributed Control and Engineering Systems
Elmenreich, Wilfried	Vienna University of Technology, Austria	4.2	16	Configuration and Management of Fieldbus Systems
Emerson, David	Yokogawa America	3.1	3	Enterprise–Manufacturing Data Exchange Using XML
Feld, Joachim	Siemens AG, Germany	4.3	18	Principles and Features of PROFInet
Fong, A.M.	Singapore Institute of Manufacturing Technology, Singapore	4.5	21	SEMI Interface and Communication Standards: An Overview and Case Study A.M. Fong, K.M. Goh, Y.G. Lim, K. Yi, and O. Tin (Singapore Institute of Manufacturing Technology, Singapore)
Foneseca, Alberto	University of Aveiro, Portugal	4.3	17	The Quest for Real-Time Behavior in Ethernet
Goh, K.M.	Singapore Institute of Manufacturing Technology, Singapore	4.5	21	SEMI Interface and Communication Standards: An Overview and Case Study
Hanisch, Hans-Michael	University of Halle-Wittenberg, Germany	3.6	8	Achieviing Reconfigurability of Automation Systems Using the New International Standard IEC 61499: A Developer's View
Hu, Zaijun	ABB Corporate Research Center, Germany	3.2	4	Web Services for Integrated Automation Systems – Challenges, Solutions, and Future
Iwanitz, Frank	Softing AG, Germany	3.3	5	OPC — Openness, Productivity, and Connectivity
Jecht, Ulrich	UJ Process Analytics, Germany	4.2	14	PROFIBUS: Open Solutions for the World of Automation
Koç, Muammer	University of Michigan – Ann Arbor	2	2	Introduction to e-Manufacturing
Kruse, Eckard	ABB Corporate Research Center, Germany	3.2	4	Web Services for Integrated Automation Systems – Challenges, Solutions, and Future
Lange, Juergen	Softing AG, Germany	3.3	5	OPC — Openness, Productivity, and Connectivity
Lee, Jay	University of Cincinnati	2	2	Introduction to e-Manufacturing

Author	Affiliation	Subsection	Chapter	Title
Lee, Kang,	National Institute of Standards and Technology	4.1	10	A Smart Transducer Interface Standard for Sensors and Actuators
Lim, Y.G.	Singapore Institute of Manufacturing Technology, Singapore	4.5	21	SEMI Interface and Communication Standards: An Overview and Case Study
Luder, Arndt	University of Magdeburg, Germany	3.5	7	Java Technology and Industrial Applications
Marik, Vladimir	Czech Technical University, Prague, Czech Republic	5	22	From Holonic Control to Virtual Enterprises: The Multi-Agent Approach
Matheus, Kirsten	Carmeq, Germany	4.4	19	Wireless Local and Wireless Personal Area Network Technologies for Industrial Deployment
Meo, Fabrizio	FIDIA, Italy	4.1	12	Open Controller Enabled by an Advanced Real-Time Network (OCEAN)
Naedele, Martin	ABB Research Center, Switzerland	6	23	IT Security for Automation Systems
Ni, Jun	University of Michigan – Ann Arbor	2	2	Introduction to e-Manufacturing
Pedreiras, P.	University of Aveiro, Portugal	4.3	17	The Quest for Real-Time Behavior in Ethernet
Peschke, Jörn	University of Magdeburg, Germany	3.5	7	Java Technology and Industrial Applications
Pitzek, Stefan	Vienna University of Technology, Austria	4.2	16	Configuration and Management of Fieldbus Systems
Popp, Manfred	Siemens AG, Germany	4.3	18	Principles and Features of PROFInet
Sauter, Thilo	Austrian Academy of Sciences, Austria	4.2	13	Fieldbus Systems: History and Evolution
Schiffer, Victor	Rockwell Automation, Germany	4.2	15	The CIP Family of Fieldbus Protocols
Schwarz, Karlheinz	Schwarz Consulting Company, Germany	3.4	6	The Standard Message Specification for Industrial Automation Systems: ISO 9506 (MMS)
Stripf, Wolfgang	Siemens AG, Germany	4.2	14	PROFIBUS: Open Solutions for the World of Automation
Tin, O.	Singapore Institute of Manufacturing Technology, Singapore	4.5	21	SEMI Interface and Communication Standards: An Overview and Case Study
Vetter, Claus	ABB Corporate Research Center, Switzerland	3.7	9	Integration between Production and Business Systems
Vrba, Pavel	Rockwell Automation, Czech Republic	5	22	From Holonic Control to Virtual Enterprises: The Multi-Agent Approach
Vyatkin, Valeriy	University of Auckland, New Zealand	3.6	8	Achieving Reconfigurability of Automation Systems Using the New International Standard IEC 61499: A Developer's View
Wenzel, Peter	PROFIBUS International, Germany	4.2	14	PROFIBUS: Open Solutions for the World of Automation
Werner, Thomas	ABB Corporate Research Center, Switzerland	3.7	9	Integration between Production and Business Systems

Author	Affiliation	Subsection	Chapter	Title
Yi, K.	Singapore Institute of Manufacturing Technology, Singapore	4.5	21	SEMI Interface and Communication Standards: An Overview and Case Study
Zurawski, Richard	ISA Group	1	1	Integration Technologies for Industrial Automated Systems: Challenges and Trends

Index

Milton Keynes UK
Ingram Content Group UK Ltd.
UKHW052028071024
449327UK00027B/2472